As- G · II - 1 - 94

9780070966345

Geology of India

Geology of India

D. N. WADIA
M.A., D.SC., F.R.S., F.G.S., F.N.I.

FOURTH EDITION

Tata McGraw-Hill Publishing Co
NEW DELHI

© Meher D. N. Wadia

First Edition 1919
Revised 1926
Second Edition 1939
Reprinted 1944, 1949
Third Edition 1953
Revised 1957, 1961 *and* 1966
Fourth Edition 1975

This edition can be exported from India only by the Publishers, Tata McGraw-Hill Publishing Company Ltd

THIS BOOK HAS BEEN SUBSIDIZED BY THE GOVERNMENT OF INDIA THROUGH NATIONAL BOOK TRUST, INDIA, FOR THE BENEFIT OF STUDENTS

Rs. 18.00

PUBLISHED BY TATA MCGRAW-HILL PUBLISHING COMPANY LIMITED, C-98A, SOUTH EXTENSION PART II, DELHI, AND PRINTED BY K. K. GHOSH AT THE INDIAN PRESS PRIVATE LIMITED, ALLAHABAD

To A. AND F.

IN HEAVEN,

THESE PAGES ARE INSCRIBED

Preface

As a lecturer in Geology to students preparing for the Punjab University Examinations I have constantly experienced great difficulty in the teaching of the Geology of India, because of the absence of any adequate modern book on the subject. The only work that exists is the one published by the Geological Survey of India in 1887, by H. B. Medlicott and W. T. Blanford, revised and largely rewritten by R. D. Oldham in 1893—a quarter of a century ago. Although an excellent official record of the progress of the Survey up to that time, this publication has naturally become largely out of date (now also out of print) and is, besides, in its voluminous size and method of treatment, not altogether suitable as a manual for students preparing for the University Examinations. Students, as well as all other inquirers, have, therefore, been forced to search for and collect information, piecemeal, from the multitudinous *Records* and *Memoirs* of the Geological Survey of India. These, however, are too numerous for the diligence of the average student—often, also, they are inaccessible to him—and thus much valuable scientific information contained in these admirable publications was, for the most part, unassimilated by the student class and remained locked up in the shelves of a few Libraries in the country. It would not be too much to say that this lack of a handy volume is in the main responsible for the almost total neglect of the Geology of India as a subject of study in the colleges of India and as one of independent scientific inquiry.

The object of the present volume is to remedy this deficiency by providing a manual in the form of a modern text-book, which summarises all the main facts of the subject within a moderate compass. It is principally a compilation, for the use of the students of Indian Geology, of all that has been published on the subject, especially incorporating the later researches and conclusions of the Geological Survey of India since Oldham's excellent edition of 1893.

In a subject of such proportions as the Geology of India, and one round which such voluminous literature exists, and is yearly growing, it is not possible, in a compendium of this nature, to aim at perfection of detail. Nor is it easy, again, to do justice to the devoted labours of the small body of original workers who, since the '50's of the last century, have made Indian Geology what it is to-day.

By giving, however, in bold outlines, the main results achieved up to date and by strictly adhering to a text-book method of treatment, I have striven to fulfil the somewhat restricted object at which I have aimed.

In the publication of this book I have received valuable help from various quarters. My most sincere thanks are due to Sir T. H. Holland, F.R.S., D.Sc., for his warm sympathy and encouragement. To the Director of the Geological Survey of India, I offer my grateful acknowledgments for the loan of blocks and plates from negatives for the illustrations in the book and for permission to publish this volume. My indebtedness to Mr. C. S. Middlemiss, C.I.E., F.R.S., retired Superintendent of the Geological Survey of India, the *doyen* of Indian Geologists, I can never sufficiently acknowledge. His guidance and advice in all matters connected with illustrations, correction of manuscript and text, checking of proofs, etc., have been of inestimable value. Indeed, but for his help several imperfections and inaccuracies would have crept into the book.

In the end, I tender my grateful acknowledgments to Messrs. Macmillan for their uniform courtesy.

D. N. WADIA

JAMMU, KASHMIR,
DECEMBER, 1916.

Postscript

THE revision of the last edition of this book was completed in 1937. Since then the progress of geological investigation in India during the War and succeeding years, especially in the field of economic geology and mineralogy, makes a fresh revision necessary. Some sections of the book have been re-written to bring them up to date.

The recent political division of India and the territorial regrouping of Provinces and States does not affect the terminology of Indian geology, nor the descriptive treatment of their stratigraphy, structure or palaeontology. The sub-continent of India is a well-marked geological as well as geographical entity, a discrete segment of the earth's crust, segregated from Eurasia by Tertiary earth-movements. Its main stratigraphic and structure lines criss-cross over the political boundary-lines of Pakistan, Burma, Ceylon and even Tibet. In a treatise on the geology of India, therefore, these countries, forming natural physical extensions of India proper, have to be included as integral parts of the Indian region.

The chapter on the Geology of Kashmir, which appeared as an appendix in previous editions, is eliminated and its substance incorporated in relevant chapters in the body of the book. The redistribution has not been so thorough as it should have been and a few repetitions and overlaps have been allowed to remain.

It is my pleasant duty to tender my grateful acknowledgments to Mr. Percy Evans and Mr. Wilfred Crompton, both of the Burmah Oil Company, and to Mrs. Evans, all of whom read through the proofs of this edition. Although it was not possible to adopt all their suggestions, their assistance has permitted the bringing up to date of the Tertiary chapters and the correction of oversights which might otherwise have passed unnoticed. But for Mr. Evans's helpful suggestions and criticisms some chapters of the book would have remained imperfectly revised. I am also indebted to my wife Meher Wadia for much help in revising the proofs.

D. N. WADIA

NEW DELHI,
1952

Note on the Fourth Edition

The revision of the 1966 edition of this book was completed by the author a few months before his demise in June, 1969. The present work has since been brought to date for publishing in India for the first time. I am thankful to Mr. K. K. Dar, Director, Atomic Minerals Division, Department of Atomic Energy, Government of India, for his help and for supplying some of the information required for this edition.

Meher D. N. Wadia (Mrs.)

New Delhi,
July, 1973

Contents

CHAPTER I

Physical Features 1

Geological divisions of India; their characters and peculiarities; types of the earth's crust exemplified by these divisions. Physical characters of the plains of India. Rajputana a debatable area. Mountains of India; the Himalayan mountains; physical features of the Himalayas; High Asia; meteorological influence of the Himalayas. Limits of the Himalayas. The syntaxial bends at the N.W. and S.E. Classification of the Himalayan ranges, (1) Geographical, (2) Geological. Orography of Kashmir Himalayas; the Outer Ranges; the "Duns"; the Middle Ranges; the Panjal Range; the Inner Ranges; the transverse gorges of the rivers. Other ranges of extra-Peninsular India. Mountain ranges of the Peninsula; Vindhya mountains; the Satpura range; the Western Ghats; the Eastern Ghats. Glaciers: glaciers of the Himalayas; their size; limit of Himalayan glaciers; peculiarities of Himalayan glaciers; records of past glaciation in the Himalayas. The drainage system: the easterly drainage of the Peninsula; the Himalayan system of drainage not a *consequent* drainage; the Himalayan watershed; the transverse gorges of the Himalayas; river-capture or "piracy"; the hanging valleys of Sikkim. Lakes; the lakes of Tibet, Kashmir and Kumaon; salinity of the Tibetan lakes; their desiccation; the Sambhar lake; the Lonar lake. The Coasts of India; submerged mountain chain and valleys of the Arabian Sea. Volcanoes: Barren Island; Narcondam; Popa; Koh-i-Sultan. Mud-volcanoes: *sub-Recent* volcanic phenomena. Earthquakes: the earthquake zone of India; the Assam earthquake; the Kangra earthquake; Bihar earthquake; Quetta earthquake; the Mekran earthquake of 1945; Assam earthquake, 1950. Local alterations of level; recent elevation of the Peninsular tableland; other local alterations; submerged forest of Bombay; alterations of level in Kutch; the Himalayas yet in a state of tension. Isostasy. Denudation; the monsoonic alternations; the lateritic regolith; general character of denudation in India sub-tropical, desert-erosion in Rajasthan. Peculiarity of river-erosion in India: the river-floods. Late changes in the drainage of Northern India; the Siwalik river, its dismemberment into the Indus and Ganges; reversal of the northwesterly flow of the Ganges. References.

CONTENTS

CHAPTER II

Stratigraphy of India—Introductory 55

Difficulty of correlation of the Indian formations to those of the world; principles involved. The different "facies" of the Indian formations. Provincial faunas. Radio-active minerals as an aid to stratigraphy. The chief geological provinces of India: the Salt-Range; the N. W. Himalayas; the Central Himalayas; Sind; Rajasthan; Burma and Baluchistan; the Coastal tracts. Method of study of the geology of India. Table of Standard Stages of the Geological Record. Table of the geological formations of India. Table of geological formations in the N.W. Himalayas. References.

CHAPTER III

The Archaean System—Gneisses and Schists 73

General. Distribution of the Archaean of India; petrology of the Archaean system; the chief petrological types: gneisses; granites; syenites; Charnockite, Khondalite, Gondite, Kodurite, calc-gneisses and calciphyres, etc. Classification of the Archaean system. Bengal gneiss; types of Bengal gneiss. Bundelkhand gneiss. The Charnockite series; petrological characters of the Charnockite series. Archaean of the Himalayas. The crystalline complex of N.W. Himalayas. References.

CHAPTER IV

Archaean System (continued)—The Dharwar System 89

General. Outcrops of the Dharwar rocks; the lithology of the Dharwars; plutonic intrusions in the Dharwars; crystalline limestones originating by the metasomatism of the gneisses. Distribution of the Dharwar system. Type-area Dharwar-Mysore State; Rajasthan: the Aravalli mountains; the Aravalli series; the Raialo series; the Shillong series; the Dharwar rocks of Madhya Pradesh; the manganiferous deposits of the Dharwar system—the Gondite and the Kodurite series: Bihar and Orissa—the Iron-ore series. Manganese ores of the Dharwar system. The Dharwar system of the Himalayas. The Vaikrita series; Salkhala series; Jutogh series and Daling series. The sedimentary pre-Cambrian system of N.W. Himalayas. Homotaxis of the Dharwar system. The Archaean-Dharwar controversy. Table of correlation of Dharwar formations. Economics. References.

CONTENTS

CHAPTER V

The *Cuddapah* *System* 113

General. The Cuddapah system; lithology of the Cuddapahs; absence of fossils in the Cuddapahs; classification of the system. Distribution. The lower Cuddapah; the Delhi system; Bijawar series; the Cheyair and Gwalior series. The Upper Cuddapahs; the Nallamalai, Kaladgi, Kistna, etc., series. Economics. Stratigraphic position of the Cuddapahs. References.

CHAPTER VI

The *Vindhyan* *System* 121

Extent and thickness; rocks: structural features. Life during the Vindhyan period. Classification. Distribution of the lower Vindhyan; Semri series; the Kurnool series, Malani series, etc. Meaning of "Lower" and "Upper" Vindhyan. The distribution of the Upper Vindhyan. Vindhyan sandstones. Economics of the system. The Himalayan Vindhyans. The relation of the Himalayan unfossiliferous systems to the Peninsular Puranas. Homotaxis of the Vindhyan system. References.

CHAPTER VII

The *Cambrian* *System* 132

The Cambrian of India. (i) The Salt-Range. The principal geological features of the range. The Cambrian of the Salt-Range; the salt-marl and salt, the purple sandstone; the *Neobolus* beds; magnesian sandstone. Salt-pseudomorph shales. (ii) The Spiti area—the Spiti geosyncline. The Cambrian of Spiti. Haimanta system; Cambrian fossils. Autoclastic conglomerates. The Cambrian of Kashmir. Stratigraphy of Kashmir and Simla Himalayas. References.

CHAPTER VIII

The *Ordovician,* *Silurian,* *Devonian* *and* *Lower* *and* *Middle* *Carboniferous* *Systems* 147

General. (i) The Spiti area; Ordovician and Silurian; the Devonian; the Carboniferous—Lipak and Po series; the Upper Carboniferous unconformity. Table of Palaeozoic systems in Spiti. (ii) Kashmir area. (iii) Chitral. (iv) Burma—the Northern Shan States; Ordovician; Silurian—Namshim series, Zebingyi series; Silurian fauna of Burma; Devonian; the Devonian fauna; the Wetwin slates. Carboniferous of Burma; the Plateau limestone; Fusulina limestone. Table of the Palaeozoic formations of Burma. Physical changes at the end of the Dravidian era. References.

CHAPTER IX

The Gondwana System 164

General. The ancient Gondwanaland; Leumuria; the Gondwana system of India; the geotectonic relations of the Gondwana rocks; their fluviatile nature; evidences of changes of climate; organic remains in the Gondwana rocks; successive *oras*; land-bridge between Gondwanaland and Angaraland; distribution of the Gondwana rocks; classification of the system. The Lower Gondwana: Talchir series; Talchir fossils; the Damuda series; igneous rocks of the Damuda coal-measures; effects of contact-metamorphism; the Damuda flora; Damuda series of other areas. Homotaxis of the Damuda and Talchir series. Economics. Classification. Lower Gondwanas of the Himalayas.

CHAPTER X

The Gondwana System (continued) 181

The Middle Gondwanas: rocks; the Panchet series; the Pachmarhi or Mahadev series; Maleri series; Parsora series; Triassic age of the Middle Gondwanas. The Upper Gondwanas: distribution; lithology; the Rajmahal series; the Rajmahal flora; Satpura and Madhya Pradesh; Jabalpur stage; Godavari area; Kota stage; Gondwanas of the East Coast; Rajahmundry, Ongole, Madras and Cuttack; Gondwanas of Sri Lanka; Gondwanas of the West Coast—the Upper Gondwanas of Kutch. Umia series. Economics. References.

CHAPTER XI

Upper Carboniferous and Permian Systems 193

The commencement of the Aryan era; the Himalayan geosyncline; the nature of geosynclines. Upper Carboniferous and Permian of India. (i) Upper Carboniferous and Permian of the Salt-Range. Boulder-beds; Speckled sandstones; Productus limestone; Productus fauna. The Anthracolithic systems. (ii) Upper Carboniferous and Permian of the Himalayas. The Permo-Carboniferous of Spiti.

CONTENTS xvii

Productus shales. The Middle Carboniferous of Kashmir—Fenestella shales; the mid-Palaeozoic unconformity; Panjal Volcanic series; Gangamopteris beds. The Permian and Zewan series; the Permian of Jammu ; Krol series of Simla; Karakoram and Chitral; Hazara; Burma. Marine beds of Umaria. References.

CHAPTER XII

The Triassic System 221

General. The principles of classification of the geological record; the view of Professors Chamberlin and Salisbury. (i) The Trias of Spiti. The zonal classification of the system; Triassic fauna. "Exotic" Trias of Malla Johar. (ii) Hazara. (iii) The Trias of the Salt-Range—the Ceratite beds. (iv) Baluchistan. (v) Kashmir. (vi) Burma; Napeng Series. References.

CHAPTER XIII

The Jurassic System 237

Instances of Jurassic development in India. Life during the Jurassic period. (i) Jurassic of the Central Himalayas; the Kioto limestone; Spiti shales; the fauna of the Spiti shales. Mt. Everest region. The Tal series of the Outer Himalayas. (ii) Hazara; the Spiti shales of Hazara. (iii) Baluchistan. (iv) Jurassic of the Salt-Range. Marine Transgressions during the Jurassic period, the nature of marine transgressions. (v) The Jurassics of Kutch—Patcham, Chari, Katrol; the marine Jurassic of Coast. (vi) Rajasthan—Jaisalmer limetstone. (vii) Burma—Namyau beds. References.

CHAPTER XIV

The Cretaceous System 254

Varied facies of the Cretaceous of India, the geography of India during the Cretaceous period. (i) Cretaceous of Spiti; Giumal sandstone; Chikkim series; Flysch. (ii) Chitral. Plutonic and volcanic action during the Cretaceous. Exotic blocks of Johar. (iii) Cretaceous Volcanic series of Kashmir. Ladakh. (iv) Hazara. (v) Cretaceous of Sind and Baluchistan; Hippurite limestone; Parh limestone; Pab sandstone. *Cardita beaumonti* beds. (vi) Salt-Range. (vii) Assam. (viii) Burma. References.

CONTENTS

CHAPTER XV

The Cretaceous System (continued)—*Peninsula* 266

(i) Upper Cretaceous of the Coromandel coast; geological interest of the S. E. Cretaceous; the Utatur stage; Trichinopoly stage; Ariyalur stage; Niniyur stage; Eocene of Pondicherry; Fauna of the S.E. Cretaceous; Utatur, Trichinopoly, and Ariyalur faunas. (ii) The Narmada valley Cretaceous; Bagh beds; conclusions from the fauna of the Bagh beds. (iii) The Lameta series or infra-Trappean beds; metasomatic limestones. Age of the Lameta series. Cretaceous Dinosaurs of India. Western India. References.

CHAPTER XVI

Deccan Trap 275

The great volcanic formation of India. Area of the plateau basalts; their thickness; the horizontality of the lava sheets; petrology; instances of magmatic differentiation; microscopic characters of the Deccan basalts. Stratigraphy of the Deccan Trap. Inter-Trappean beds; a type-section. The mode of eruption of the Deccan Traps—fissure-eruption. Fissure dykes in the Traps. Age of the Deccan Traps. Economics. References.

CHAPTER XVII

The Tertiary Systems 287

General. Physical changes at the commencement of the Tertiary era. The elevation of the Himalayas; three phases of upheaval of the Himalayas. Distribution of the Tertiary systems in India: Peninsula; extra-Peninsula. Dual facies of Tertiary deposits. Geography of India during early Tertiary. (i) Tertiaries of Surat and Broach. (ii) Saurashtra. Dwarka beds, Perim Island Tertiary. (iii) Tertiaries of Kutch. (iv) Rajasthan. (v) The Coromandel Coast —Cuddalore series. (vi) The Malabar Coast—Warkalli and Quilon beds. Tertiary systems of the extra-Peninsular India. (i) Sind. Table of formations. (ii) Salt-Range. Table of formations. (iii) Himalayas—Kashmir Himalayas and Punjab and Kumaon Himalayas. Tertiaries of Inner Himalayas. (iv) Assam. (v) Burma. The Tertiary gulf of Burma. Table of correlation of Tertiary formations. References.

CONTENTS

CHAPTER XVIII

The Eocene System 307

Ranikot series; Fossils of the Ranikot series. Laki series. Kirthar series; Nummulitic limestone; Fossils of the Kirthar series. (i) Sind and Baluchistan. (ii) Salt-Range. (iii) Kohat. (iv) Potwar. (v) Hazara. (vi) Kashmir; the Indus Valley Tertiaries; the Subathu series of Jammu and the Pir Panjal. (vii) The Outer Himalayas; Subathu series. (viii) Assam; economic utility of the Assam Eocene rocks. (ix) Burma; Eocene mammals. References.

CHAPTER XIX

The Oligocene and Lower Miocene Systems 322

Oligocene; restricted occurrence. (i) Baluchistan. (ii) Sind; Nari series. (iii) Assam. (iv) Burma; Pegu series: Petroleum, origin, mode of occurrence, gas, migration; petroleum areas in India. Lower Miocene. (i) Sind; Gaj series; Bugti beds. (ii) Salt-Range, Potwar, Murree series. (iii) Outer Himalayas; Murree series, Kashmir Sub-Himalayas. (iv) Assam; Surma series. (v) Burma; Upper Pegu series. Igneous action. Change in conditions. References.

CHAPTER XX

The Siwalik System—Middle Miocene to Lower Pleistocene 335

General. The nature of the Siwalik deposits. Geotectonic relations of the Siwaliks. The "Main Boundary" Fault. The real nature of the "Main Boundary" Fault; recent views. The Palaeontological interest of the Siwalik system. Evolution of the Siwalik fauna; migrations into India. Lithology; mode of formation of the Siwaliks in the Gangetic trough. Classification. The Bain boulder-bed. The Siwalik zone of Kashmir sub-Himalayas. Siwalik fauna; fossil anthropoid apes. Age of Siwalik system. Parallel series of deposits. References.

CHAPTER XXI

The Pleistocene System—The Ice Age 353

The Pleistocene or Glacial Age of Europe and America. A modified Pleistocene Glacial Age in India. The nature of the evidence for an Ice Age in India; Dr. Blanford's views. Ice Age in the Himalayas; Physical records. The extinction of the Siwalik mammals. Interglacial periods. Table of correlation of Glacial stages with the Upper Siwaliks of N. India. Pleistocene Ice Age deposits of Kashmir; their relation to Karewa series. Man in Ice Age. References.

CHAPTER XXII

The Pleistocene System (continued)—*The Indo-Gangetic Alluvium* 364

The plains of India. Nature of the Indo-Gangetic depression. Extent and thickness of the alluvial deposits. Changes in Rivers. Lithology. Classification; Bhangar; Khadar. The Ganges delta, the Indus delta. Economics. The Rajasthan desert; composition of the desert sand; the origin of the Rajasthan desert. The Rann of Kutch. References.

CHAPTER XXIII

The Pleistocene System (continued)—*Laterite* 376

Laterite a regolith peculiar to India. Composition of laterite; its distribution. High-level laterite and low-level laterite. Theories of the origin of laterite, recent views; secondary changes in laterite; resilicification. The age of laterite. Economics. References.

CHAPTER XXIV

Pleistocene and Recent 381

Examples of Pleistocene and Recent deposits. Ossiferous alluvium of the Upper Sutlej; of the Tapti and Narmada. The Karewas of Kashmir. Coastal alluvial deposits. Sand-dunes; *Teris.* Loess. The Potwar fluvio-glacial deposits. Cave-deposits. *Regur* or black cotton-soil; the origin of Regur. The "Daman" slopes. The Human epoch. References.

CHAPTER XXV

Physiography 390

Principles of physiography illustrated by the Indian region. Mountains; the structure of the Himalayas; recent ideas; the tectonic zones; cause of the syntaxial bends of the Himalayas; Geotectonic features of the N.W. Himalayas; Simla and Garhwal Himalayas; Orographic trend lines of North India; the structure of the Peninsula; the mountains of the Peninsula. Plateaus and plains: plateau of volcanic accumulation; plateau of erosion. Valleys: the valley of

CONTENTS

Kashmir a tectonic valley; erosion-valleys; valleys of the Himalayas; the transverse gorges; configuration of the Himalayan valleys; valleys of the Peninsula. Basins or lakes: functions of lakes; types of lakes; Indian examples. The coast-lines of India. References.

CHAPTER XXVI

Economic Geology 414

General. Water; wells, springs, artesian wells; thermal and mineral springs. Clays; china-clay, terra-cotta, fire-clay, fuller's earth. Sands; glass-sand. Lime; cements; mortar; composition of cements; production. Building-stones; granites; limestones; marbles, serpentine; sandstones, Vindhyan sandstones, Gondwana sandstones; laterite, slates, traps. Coal; production in India; Gondwana coal; Tertiary coal. Estimated reserves. Peat. Petroleum; Burma, Assam, Gujarat, Pakistan; natural gas. Metals and ores. Aluminium; bauxite in laterite; economic value of Indian bauxite; uses. Antimony; stibnite. Arsenic; sulphides of arsenic, orpiment and realgar. Beryllium. Chromium; occurrence; uses. Cobalt and Nickel. Copper; the copper ores of Bihar. Gold; its occurrence; production of vein-gold; alluvial gold. Iron; its occurrence; economic value; production; its distribution. Lead and silver; lead-ores of Bawdwin; production. Magnesium; occurrence of magnesite. Manganese; distribution of manganese in the geological formations of India; production; uses. Strontium. Thorium. Tin; the tin-ore of Mergui and Tavoy. Titanium. Tungsten; wolfram of Tavoy; uses of tungsten. Uranium; ore-occurrences. Vanadium. Zinc. Precious and semi-precious stones. Diamonds; Panna and Golconda diamonds. Rubies and sapphires; Burma rubies; gem-gravels of Sri Lanka; sapphires of Kashmir. Spinel. Jadeite; occurrence; formation. Emeralds and aquamarines; beryl. Chrysoberyl. Garnets. Zircons. Tourmalines. Other gem-stones of India. Agates, rock-crystal, amethyst. Amber. Economic mineral products. Alkaline salts. Alum Asbestos. Barytes. Bauxite. Borax. Corundum; occurrence; distribution; uses; other abrasives; millstones, grindstones. Fluorspar. Graphite; its occurrence; uses. Gypsum. Kyanite and Sillimanite. Limestone and dolomite. Mica; uses of mica; mica-deposits of Nellore and Hazaribagh and Rajasthan. Mineral paints. Monazite; its occurrence; uses. Phosphatic deposits. Pyrite. Rare minerals. *Reh* or *Kalar*. The origin of *reh* efflorescence. Salt; sources of Indian salt; rock-salt mines; other salts. Saltpetre or nitre. Mode of occurrence of nitre; its production; uses. Steatite; mode of origin of steatite. Sulphur; uses of sulphur. Soils; soil-formation; the soils of India; main soil-groups of India.

Index 486

List of Illustrations

PLATES

facing page

I. Alukthang Glacier	22
II. Snout of Sona Glacier from Sona	24
III. Mud Volcano—one of the largest—Minbu, Burma	36
IV. Bellary Granite, Gneiss Country, Hampi	82
V. Banded Porphyritic Gneiss (Younger Archaean), Nakta Nala, Chhindwara District	84
VI. "Marble Rocks" (Dolomite Marble), Jabalpur	100
VII. Upper Rewah Sandstone, Rahutgarh, Sangor District	122
VIII. Overfolding of the Palaeozoic Rocks, Upper Lidar Valley, Central Himalayas	150
IX. Reversed Fault in Carboniferous Rocks, Lebung Pass, Central Himalayas	156
X. Barrier of Coal across Kararia Nala	174
XI. Contorted Carboniferous Limestone, Nankshang Pass, Central Himalayas	204
XII. Folded Trias Beds, Dhauli Ganga Valley, Central Himalayas	222
XIIA. Cretaceous-Eocene Flysch near Kargil, Ladakh	265
XIII. Geological Map of Lidar Valley. Silurian-Trias Sequence in Kashmir	*At end of book*
XIV. Plan of Vihi District, Kashmir	,, ,,
XV. Geological Map of the Pir Panjal	,, ,,
XVI. Sketch Map of the Himalayan Geosyncline and its Relation to adjacent Mountain-systems of Central Asia	,, ,,
XVII. Tectonic Sketch Map of the Garhwal Himalayas	,, ,,
XVIII. Geological Sketch Map of the Syntaxial Bend of the North-West Himalayas	,, ,,
XIX. Geological Map of Hazara	,, ,,

DIAGRAMS AND SKETCHES

1. Diagrammatic Section through the Himalayas to show their relations to the Tibetan Plateau and the Plains of India	5
2. View of the great Báltoro glacier	22
3. The Volcano of Barren Island in the Bay of Bengal	35
4. Diagram showing Contortion in the Archaean Gneiss of Bangalore	74

LIST OF ILLUSTRATIONS

5. Section across the Aravalli Range to the Vindhyan Plateau showing the peneplaned synclinorium of the most ancient mountain range of India ... 98
6. Section across the Singhbhum Anticlinorium, Chota Nagpur ... 102
7. Diagram showing the Relation of Dharwar Schists with the Gneisses ... 107
8. Sketch Section illustrating the Relation of Cuddapah and Kurnool Rocks ... 114
9. Section showing Relation between Gwalior Series and Rocks of the Vindhyan System ... 127
10. Section illustrating the General Structure of the Salt-Range (Block-faults) Section over Chambal Hill (East) ... 133
11. Section across the Dandot scarp from Khewra to Gandhala, Salt-Range ... 134
12. General Section, Naubug Valley, Margan Pass and Wardwan, to show the disposition of the Palaeozoic rocks of Kashmir ... 141
13. Section across Lidar Valley Anticline ... 141
14. Section along the Parahio River, Spiti ... 149
15. Section of Palaeozoic Systems of N. Shan States (Burma), Section across the Nam-tu Valley at Lilu ... 157
16. Sketch Map of typical Gondwana Outcrop ... 168
17. Tectonic Relations of the Gondwana Rocks ... 170
17a. A reconstruction of Gondwaa Continent as it existed in early Mesozoic ... 180
18. Sketch Map of the Gondwana Rocks of the Satpura Area ... 181
19. Generalised Section through the Gondwana Basin of the Satpura Region ... 183
20. Section from the Dhodha Wahan across the western part of Sakesar ridge ... 194
21. Section across the Salt-Range, taken N.E.—S.W., from the exit of the Khanzaman *nala* ... 196
22. Section of the Carboniferous to Trias Sequence in the Tibetan Zone of the Himalayas (Spiti) ... 203
23. Section of the Zewan Series, Guryul Ravine ... 214
24. Palaeozoic Rocks of the N. Shan States ... 219
25. Section of the Trias of Spiti ... 226
26. Diagrammatic Section of Mt. Sirban, Hazara ... 228
27. Continuation of preceding Section further South-East to the Taumi Peak ... 229
28. Section through the Bakh Ravin from Musa Khel to Nammal ... 230
29. Section of the Triassic System of Kashmir ... 232
30. Section of the Jurassic and Cretaceous Rocks of Hundes ... 238
31. Sketch Section in the Chichali Pass ... 246
31a. Sketch section across Mt. Everest region ... 253
32. View of Deccan Trap Country ... 276
33. Sketch and section to show the nummulitic (Laki) limestone scarp in the Salt-Range ... 310

34. Section showing the Relation of Permo Carboniferous and Eocene, Jammu Hills ... 316
35. Section across the Potwar Geosyncline ... 328
36. Diagrams to illustrate the formation of Reversed Faults in the Siwalik Zone of the Outer Himalayas ... 336
37. Section to illustrate the Relations of the Outer Himalayas to the Older Rocks of the Mid-Himalayas (Kumaon Himalayas) ... 338
38. Section across the Sub-Himalayan Zone east of the Ganges River ... 339
38a. Diagrammatic Section across the Brahmputra Valley, Upper Assam ... 350
38b. Section showing trespass of Main Boundary thrust over the Siwaliks ... 352
38c. Siwalik Elephants ... 352
39. Section of the Pir Panjal across the N.E. Slope from Kilnag—Tatakuti ... 359
40. Section across the Outermost Hills of the Sub-Himalaya at Jammu ... 361
41. Diagrammatic Section across the Indo-Gangetic Synclinorium ... 366
42. Diagrammatic Section across the Kashmir Himalaya, showing the broad Tectonic Features ... 396
43. Section through the Simla Himalaya ... 397
44. Diagrammatic representation of the nappe structure of Garhwal Himalaya ... 400
45. Section across Western Rajasthan. To illustrate the peneplanation of an ancient mountain-range ... 405
46. Section through Digboi Oil-field ... 434
47. Diagrammatic Section through the Nahorkatiya Oil-field ... 435
48. Diagrammatic Section through the Badarpur Anticline ... 435

CHAPTER I

Physical Features

BEFORE commencing the study of the stratigraphical, *i.e.* historical, geology of India, it is necessary to acquire some knowledge of the principal physical features. The student should make himself familiar with the main aspects of its geography, the broad facts regarding its external relief or contours, its mountain-systems plateaus and plains, its drainage-courses, its glaciers, volcanoes, etc. This study, with the help of physical or geographical maps, is indispensable. Such a foundation-knowledge of the physical facts of the country will not only be of much interest in itself, but the student will soon find that the physiography of India is in many respects correlated to, and is, indeed, an expression of, its geological structure and history.

Geological divisions of India—The most salient fact with regard to both the physical geography and geology of the Indian region is that it is composed of three distinct units or earth-features, which are as unlike in their physical as in their geological characters. The first two of these three divisions of India have a fundamental basis, and the distinctive characters of each, as we shall see in the following pages, were impressed upon it from a very early period of its geological history, since which date each area has pursued its own career independently. These three divisions are:

1. The triangular plateau of the *Peninsula* (*i.e.* the Deccan, south of the Vindhyas), with the island of Sri Lanka.

2. The mountainous region of the Himalayas which borders India to the west, north, and east, including the countries of Afghanistan, Baluchistan, and the hill-tracts of Burma, known as the *extra-Peninsula*.

3. The great *Indo-Gangetic Plain* of the Punjab and Bengal, separating the two former areas, and extending from the valley of the Indus in Sind to that of the Brahmaputra in Assam.

Their characters and peculiarities—As mentioned above, the Peninsula, as an earth-feature, is entirely unlike the extra-Peninsula. The following differences summarise the main points of divergence between these two regions. The first is *stratigraphic*, or that connected with the geological history of the areas. Ever since the dawn of geological history (Cambrian period), the Peninsula has been a land area, a continental fragment of the earth's surface, which since that epoch in earth-history has never been submerged beneath the sea, except temporarily and locally. No considerable marine sediment of later age than Cambrian was ever deposited in the interior of this land-mass. The extra-Peninsula, on the other hand, has been a region which has lain under the sea for the greater part of its history, and has been covered by successive marine deposits characteristic of all the great geological periods, commencing with the Cambrian.

The second difference is *geotectonic*, or pertaining to the geological structure of the two regions. The Peninsula of India reveals quite a different type of architecture of the earth's crust from that shown by the extra-Peninsula. Peninsular India is a segment of the earth's outer shell that is composed in great part of the most ancient complex of rock-beds that stand upon a firm and immovable foundation and that have, for an immense number of ages, remained so—impassive amid all the revolutions that have again and again changed the face of the earth. Lateral thrusts and mountain-building forces have had but little effect in folding or displacing its original basement. The Deccan is, however, subject to one kind of structural disturbance, *viz.*, fracturing of the crust in blocks, and their radial or vertical movement due to tension or compression. The extra-Peninsula, on the contrary, is a comparatively weak and flexible portion of the earth's circumference that has undergone a great deal of crumpling and deformation. Rock-folds, faults, thrust-planes, and other evidences of movement within the earth are observed in this region on an extensive scale, and they point to its being a portion of the earth that has undergone, at a late geological epoch, an enormous amount of compression and upheaval. The strata everywhere show high angles of dip, a closely packed system of folds, and other violent departures from their original primitive structure.

The third difference is the diversity in the *physiography* of the two areas. The difference in the external or surface relief of Peninsular and extra-Peninsular India arises out of the two above-mentioned differences as a direct consequence. In the Peninsula, the mountains are mostly of the "relict" type, *i.e.* they are not mountains in the true sense of the term, but are mere outstanding portions of the old plateau of the Peninsula that have escaped, for one reason or another, the weathering of ages that has cut out all the surrounding parts of the land; they are, so to say, huge "tors" or blocks of the old plateau. Its rivers have flat, shallow valleys, with low imperceptible gradients, because of their channels having approached to the base-level of erosion. Contrasted with these, the mountains of the other area are all true mountains, being what are called "tectonic" mountains, *i.e.* those which owe their origin

PHYSICAL FEATURES 3

to a distinct uplift in the earth's crust and, as a consequence, have their strike, or line of extension, more or less conformable to the axis of that uplift. The rivers of this area are rapid torrential streams, which are still in a very youthful or immature stage of river development, and are continuously at work in cutting down the inequalities in their courses and degrading or lowering their channels. Their eroding powers are always active, and they have cut deep gorges and precipitous cañons, several thousands of metres in depth, through the mountains in the mountainous part of their track.

Types of the earth's crust exemplified by these divisions—The type of crust segments of which the Peninsula is an example is known as a *Horst*—a solid crust-block or *shield* which has remained a stable land-mass of great rigidity, and has been unaffected by any folding movement generated within the earth during the later geological periods. The only structural disturbances to which these parts have been susceptible are of the nature of vertical, downward or upward, movements of large segments within it, between vertical (radial) fissures or faults. The Peninsula has often experienced this "block-movement" at various periods of its history, most notably during the Gondwana period.

The earth-movements characteristic of the flexible, more yielding type of the crust, of which the extra-Peninsula is an example, are of the nature of lateral (*i.e. tangential*) thrusts which result in the wrinkling and folding of more or less linear zones of the earth's surface into a mountain-chain (orogenic movements). These movements, though they may affect a large surface area, are solely confined to the more superficial parts of the crust and are not so deepseated as the former class of movements characteristic of horsts.

Physical characters of the plains of India—The third division of India, the great alluvial plains of the Indus and the Ganges, though, humanly speaking, of the greatest interest and importance, as being the principal theatre of Indian history, is, geologically speaking, the least interesting part of India. In the geological history of India they are only the annals of yester-year, being the alluvial deposits of the rivers of the Indo-Ganges systems, borne down from the Himalayas and deposited at their foot. They have covered up, underneath a deep mantle of river-clays and silts, valuable records of past ages, which might have thrown much light on the physical history of the Peninsular and the Himalayan areas and revealed their former connection with each other. These plains were originally a deep depression or furrow lying between the Peninsula and the mountain-region. With regard to the origin of this great depression there is some difference of opinion. The eminent geologist Eduard Suess thought it was a "Fore-deep" fronting the Himalayan earth-waves, a "sagging" or subsidence of the northern part of the Peninsula as it arrested the southward advance of the mountain-waves. Colonel Sir S. Burrard, from some anomalies in the observations of the deflections of the plumb-line and other geodetic considerations, has suggested quite a different view.[1] He thinks that the Indo-Gangetic alluvium conceals a great deep rift, or fracture, in the earth's sub-crust, several thousand metres deep, the hollow being subsequently filled up by detrital deposits. He ascribes to such sub-crustal cracks or rifts a fundamental importance in geotectonics, and attributes the elevation of the Himalayan chain to an incidental bending or curling movement of the northern wall of the fissure. Such sunken tracts between parallel, vertical dislocations are called "Rift-Valleys" in geology. The geologists of the Indian Geological Survey have not accepted this view of the origin of the Indo-Gangetic depression.[2]

Rajasthan area—The large tract of low country, forming Rajasthan west of the Aravallis, possesses a mingling of the distinctive characters of the Peninsula with those of the extra-Peninsula, and hence cannot with certainty be referred to either. Rajasthan can be regarded as a part of the Peninsula inasmuch as in geotectonics it shows no post-Cambrian folding, while in its containing marine, fossiliferous deposits of Mesozoic and Cainozoic ages it shows greater resemblance to the extra-Peninsular area. It is really a part of the Deccan block that has time and again been invaded by marine transgressions from the southern sea during the Mesozoic and Cainozoic. In this country, long-continued aridity has resulted in the establishment of a desert topography, buried under a thick mantle of sands disintegrated from the subjacent rocks as well as blown in from the western seacoast and from the Indus basin. The area is cut off from the water-circulation of the rest of the Indian continent, except for occasional storms of rain, by the absence of any high range to intercept the moisture-bearing south-west monsoons which pass directly over its expanse. The desert conditions are hence accentuated with time, the water-action of the internal drainage of the country being too feeble to transport to the sea the growing mass of sands.

There is a tradition, supported by some physical evidence, that the basin of the Indus was not always separated from the Peninsula by the long stretch of sandy waste as at present. "Over a vast space of the now desert country, east of the Indus, traces of an-

---

[1] *The Origin of the Himalayas*, 1912 (Survey of India Publication). *Presidential Address*, the Indian Science Congress, Lucknow, 1916.

[2] See Dr. Hayden, Relationship of the Himalaya to the Indo-Gangetic Plain and the Indian Peninsula, *Rec. G.S.I.* vol. xliii. pt. 2, 1913, and R. D. Oldham, *Mem. G.S.I.* vol. xlii. pt. 2, 1917.

cient river-beds testify to the gradual desiccation of a once fertile region; and throughout the deltaic flats of the Indus may still be seen old channels which once conducted its waters to the Rann of Kutch, giving life and prosperity to the past cities of the delta, which have left no living records of the countless generations that once inhabited them."[1]

MOUNTAINS

The Himalayan mountains—The mountain-ranges of the extra-Peninsula have had their origin in a series of earth-movements which proceeded from outside India. The great horst of the Peninsula, composed of old crystalline rocks, has played a large part in the history of mountain-building movements in Northern India. It has limited the extent and to some degree controlled the form of the chief ranges. Broadly speaking, the origin of the Himalayan chain, the most dominant of them all, is to be referred to powerful lateral thrusts acting from the north or Tibetan direction towards the Peninsula of India. These thrusting movements resulted in the production of arcuate folds of the earth's crust, pressing against the Peninsula. The curved form of the Himalayas[2] is due to this resistance offered by the Peninsular "foreland" to the southward advance of these crust-waves, aided in some measure by two other minor obstacles—an old peneplained mountain-chain like the Aravalli mountains to the north-west and the Assam ranges to the north-east.[3] The general configuration

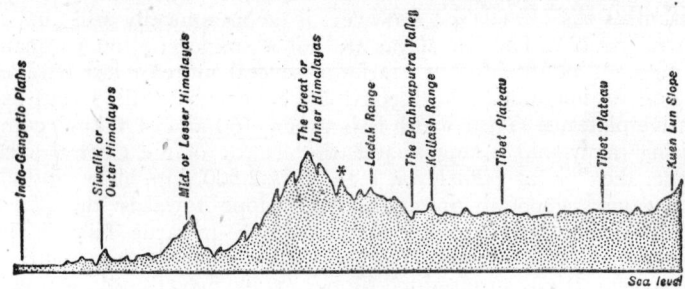

FIG. 1—Diagrammatic section through the Himalayas to show their relations to the Tibetan Plateau and the plains of India.

*Watershed of the Himalayas. (Vertical scale greatly exaggerated).

---

[1] Sir T. H. Holdich, *Imperial Gazetteer*, vol. i.

[2] From Sanskrit, *Him Alaya*, meaning the abode of snow.

[3] Another view is that the curvature is the result of the interference of similar folding movements proceeding from the Iranian or the Hindu Kush system of mountains.

---

to them it ignores the essential physical unity of the hill-ranges beyond the Indus and the Brahmaputra with the Himalayas. They would extend the term Himalaya to all those ranges to the east and west (*i.e.* the Hazara and Baluchistan and the Arakan ranges of Burma) which originated in the same great system of Pliocene orogenic upheavals.

The Syntaxial Bends of the Himalayas—The trend-lines of the Himalayan chain and its east and west terminations possess much interest from a structural point of view and need further remarks. For 2,500 kilometres from Assam to Kashmir, the chain follows one persistent S.E.-N.W. direction and then appears to terminate suddenly at one of the greatest eminences on its axis, Nanga Parbat (8,119 metres), just where the Indus has cut an immensely deep gorge right across the chain. Geological studies have shown that just at this point the strike of the mountains bends sharply to the south and then to the south-west, passing through Chilas and Hazara, instead of pursuing its north-westerly course through Chitral. All the geological formations here take a sharp hair-pin bend as if they were bent round a pivotal point obstructing them. This extraordinary inflexion affects the whole breadth of the mountains from the foot-hills of Jhelum to the Pamirs. On the west of this *syntaxis* (as this acutely reflexed bundle of mountain-folds is termed) the Himalayan strike swings from the prevalent N.E. to a N.-to-S. direction in Hazara and continues so to Gilgit; then it turns E.-to-W., the Pamirs showing a distinct equatorial disposition of their geological formations. To the south-east of this, the main tectonic strike quickly takes on a N.W.-S.E. orientation through Astor and Deosai—a direction which persists with but minor departures to western Assam.

The eastern limit of the Himalayas beyond Assam is not yet quite certain, but from the few geographical and geological observations that have been made in this region it appears that the tectonic strike here also undergoes a deep knee-bend from an easterly to a southerly trend. In the Arakan Yomas the geological axis of the mountains for several hundred kilometres is meridional, bending acutely to the N.E. near Fort Hertz. Beyond this point there is an abrupt swing to the N.W., then to E.N.E.-W.S.W. and finally to E.-W. through Assam and Sikkim.

The cause of these remarkable bends of the mountain-axis is discussed on pp. 395-7.[1]

Classification of the Himalayan Range

(I) **Geographical**—For geographical purposes Burrard has di-

---

[1] D. N. Wadia; The Syntaxis of the N.W. Himalayas: its Rocks, Tectonics and Orogeny. *Rec. G.S.I.*, vol. lxv. pt. 2, 1931.

of the Himalayan chain, its north-west–south-east arcuate trend, the abrupt steep border which it presents to the plains of India with the much more gentle slope towards the opposite or Tibetan side, are all features which are best explained, on the above view, as having been due to the resistances the mountain-making forces had to contend against in (1) the inflexible block of the Deccan and (2) the two older mountain-masses which acted as mighty obstacles in the path of the southwardly advancing mountain-folds. The convex side of a mountain range is, in general, in the opposite direction to the side from which the thrusts are directed, and is the one which shows the greatest amount of plication, fracture, and overthrust. This is actually the case with the outer or convex side of the Himalayan arc, in which the most characteristic structural feature is the existence of a number of parallel, reversed faults, or thrust-planes. The most prominent of this system of thrusts, the outermost, can be traced from the Punjab Himalayas all through the entire length of the mountains to their extremity in eastern Assam. This great fault or fracture is known as the Main Boundary Fault.

Physical features of the Himalayas—The geography of a large part of the Himalayas is not known, because areas within it have not been explored in detail by scientists; much therefore remains for future observation to add to (or alter in) our existing knowledge. Lately, however, the Mt. Everest and other expeditions to Tibet and the Karakoram have made additions to our knowledge of large tracts of the Himalayas. The east (Assam) section of the Himalayas, however, is geographically still almost a *terra incognita*. The Himalayas are not a single continuous chain or range of mountains, but a series of several more or less parallel or converging ranges, intersected by enormous valleys and extensive plateaus. Their width is between 160 and 400 km. comprising many minor ranges, and the length of the Central axial range, the "Great Himalayan range", is 2,500 km. The individual ranges generally present a steep slope towards the plains of India and a more gently inclined slope towards Tibet. The northern slopes are, again, clothed with a thick dense growth of forest vegetation, surmounted higher up by never-ending snows, while the southern slopes are too precipitous and bare either to accumulate the snows or support, except in the valley basins, any but a thin sparse jungle. The connecting link between the Himalayas and the other high ranges of Central Asia—the Hindu Kush, the Karakoram, the Kuen Lun, the Tien Shan and the Trans-Alai ranges—is the great mass of the Pamir, "the roof of the world". The Pamirs (Persian *Pa-i-mir* = foot of the eminences) are a series of broad, alluvium-filled valleys, over 3,600 m. high, separated by linear mountain-masses, rising to 5,200 m. From the Pamirs, the Himalayas extend to the south-east as an unbroken wall of snow-covered mountains, pierced by passes, few of are less than 5,200 metres in elevation. The Eastern Himala Nepal and Sikkim rise very abruptly from the plains of and Oudh, and suddenly attain their great elevation abo snow-line within strikingly short distances from the foot mountains. Thus, the peaks of Kanchenjunga and Evere only a few kilometres from the plains and are visible to their i tants. But the Western Himalayas of the Punjab and Kuma gradually from the plains by the intervention of many ran lesser altitudes; their peaks of everlasting snows are mor 150 kilometres distant, hidden from view by the mid-Hima ranges to the inhabitants of the plains.

To the north of the Himalayas is the block of High As biggest and most elevated land-mass on the earth's surfac rectly to the north is the high plateau of Tibet of 5,000 metres altitude, traversed by the "Trans-Himalaya" and the Aling ranges; farther north are the Kuen Lun and Altyn Tagh and, separated by the great desert basin of Tarim, the Tien range. A peculiarity of these ranges is that they decrease i vexity of arc as we go north till in the Tien Shan the tren becomes nearly straight. The easterly extension of the Karal range into Tibet beyond 80° long. is not known.

Meteorological influence of the Himalayas—This range of mountains exercises as dominating an influence ov meteorological conditions of India as over its physical geogr vitally affecting both its air and water circulation. Its high s ranges have a moderating influence on the temperature and midity of Northern India. By reason of its altitude and its tion directly in the path of the monsoons, it is most favou conditioned for the precipitation of much of their cont moisture, either as rain or snow. Glaciers of enormous magn are nourished on the higher ranges by this precipitation, w together with the abundant rainfall of the lower ranges, fe number of rivers, which course down to the plains in hundre fertilising streams. In this manner the Himalayas protect from the gradual desiccation which is overspreading the C Asian continent, from Tibet northwards, and the desert cond that inevitably follow continental desiccation.

Limits of the Himalayas—Geographically, the Himalay generally considered to terminate, to the north-west, at the bend of the Indus, where it cuts through the Kashmir Hima while the south-eastern extremity is defined by the similar be the Brahmaputra in upper Assam. At these points also the a well-marked bending of the strike of the mountains from general north-west—south-east to an approximately north south direction. Some geographers have refused to accept limitation of the Himalayan mountain system, because acco cient river-beds testify to the gradual desiccation of a once fertile region; and throughout the deltaic flats of the Indus may still be seen old channels which once conducted its waters to the Rann of Kutch, giving life and prosperity to the past cities of the delta, which have left no living records of the countless generations that once inhabited them."[1]

MOUNTAINS

The Himalayan mountains—The mountain-ranges of the extra-Peninsula have had their origin in a series of earth-movements which proceeded from outside India. The great horst of the Peninsula, composed of old crystalline rocks, has played a large part in the history of mountain-building movements in Northern India. It has limited the extent and to some degree controlled the form of the chief ranges. Broadly speaking, the origin of the Himalayan chain, the most dominant of them all, is to be referred to powerful lateral thrusts acting from the north or Tibetan direction towards the Peninsula of India. These thrusting movements resulted in the production of arcuate folds of the earth's crust, pressing against the Peninsula. The curved form of the Himalayas[2] is due to this resistance offered by the Peninsular "foreland" to the southward advance of these crust-waves, aided in some measure by two other minor obstacles—an old peneplained mountain-chain like the Aravalli mountains to the north-west and the Assam ranges to the north-east.[3] The general configuration

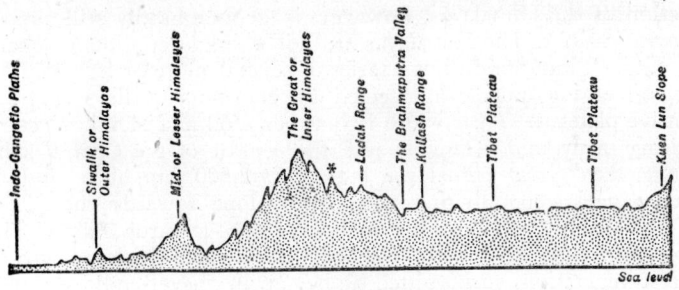

FIG. 1—Diagrammatic section through the Himalayas to show their relations to the Tibetan Plateau and the plains of India.

*Watershed of the Himalayas. (Vertical scale greatly exaggerated).

---

[1] Sir T. H. Holdich, *Imperial Gazetteer*, vol. i.
[2] From Sanskrit, *Him Alaya*, meaning the abode of snow.
[3] Another view is that the curvature is the result of the interference of similar folding movements proceeding from the Iranian or the Hindu Kush system of mountains.

of the Himalayan chain, its north-west–south-east arcuate trend, the abrupt steep border which it presents to the plains of India with the much more gentle slope towards the opposite or Tibetan side, are all features which are best explained, on the above view, as having been due to the resistances the mountain-making forces had to contend against in (1) the inflexible block of the Deccan and (2) the two older mountain-masses which acted as mighty obstacles in the path of the southwardly advancing mountain-folds. The convex side of a mountain range is, in general, in the opposite direction to the side from which the thrusts are directed, and is the one which shows the greatest amount of plication, fracture, and overthrust. This is actually the case with the outer or convex side of the Himalayan arc, in which the most character-istic structural feature is the existence of a number of parallel, reversed faults, or thrust-planes. The most prominent of this system of thrusts, the outermost, can be traced from the Punjab Himalayas all through the entire length of the mountains to their extremity in eastern Assam. This great fault or fracture is known as the Main Boundary Fault.

Physical features of the Himalayas—The geography of a large part of the Himalayas is not known, because areas within it have not been explored in detail by scientists; much therefore remains for future observation to add to (or alter in) our existing knowledge. Lately, however, the Mt. Everest and other expedi-tions to Tibet and the Karakoram have made additions to our knowledge of large tracts of the Himalayas. The east (Assam) section of the Himalayas, however, is geographically still almost a *terra incognita*. The Himalayas are not a single continuous chain or range of mountains, but a series of several more or less parallel or converging ranges, intersected by enormous valleys and ex-tensive plateaus. Their width is between 160 and 400 km. com-prising many minor ranges, and the length of the Central axial range, the "Great Himalayan range", is 2,500 km. The indivi-dual ranges generally present a steep slope towards the plains of India and a more gently inclined slope towards Tibet. The northern slopes are, again, clothed with a thick dense growth of forest vegetation, surmounted higher up by never-ending snows, while the southern slopes are too precipitous and bare either to accumulate the snows or support, except in the valley basins, any but a thin sparse jungle. The connecting link between the Hima-layas and the other high ranges of Central Asia—the Hindu Kush, the Karakoram, the Kuen Lun, the Tien Shan and the Trans-Alai ranges—is the great mass of the Pamir, "the roof of the world". The Pamirs (Persian *Pa-i-mir* = foot of the eminences) are a series of broad, alluvium-filled valleys, over 3,600 m. high, separated by linear mountain-masses, rising to 5,200 m. From the Pa-mirs, the Himalayas extend to the south-east as an unbroken wall of snow-covered mountains, pierced by passes, few of which are less than 5,200 metres in elevation. The Eastern Himalayas of Nepal and Sikkim rise very abruptly from the plains of Bengal and Oudh, and suddenly attain their great elevation above the snow-line within strikingly short distances from the foot of the mountains. Thus, the peaks of Kanchenjunga and Everest are only a few kilometres from the plains and are visible to their inhabitants. But the Western Himalayas of the Punjab and Kumaon rise gradually from the plains by the intervention of many ranges of lesser altitudes; their peaks of everlasting snows are more than 150 kilometres distant, hidden from view by the mid-Himalayan ranges to the inhabitants of the plains.

To the north of the Himalayas is the block of High Asia, the biggest and most elevated land-mass on the earth's surface. Directly to the north is the high plateau of Tibet of 5,000 metres mean altitude, traversed by the "Trans-Himalaya" and the Aling Kangri ranges; farther north are the Kuen Lun and Altyn Tagh ranges and, separated by the great desert basin of Tarim, the Tien Shan range. A peculiarity of these ranges is that they decrease in convexity of arc as we go north till in the Tien Shan the trend-line becomes nearly straight. The easterly extension of the Karakoram range into Tibet beyond 80° long. is not known.

Meteorological influence of the Himalayas—This mighty range of mountains exercises as dominating an influence over the meteorological conditions of India as over its physical geography, vitally affecting both its air and water circulation. Its high snowy ranges have a moderating influence on the temperature and humidity of Northern India. By reason of its altitude and its situation directly in the path of the monsoons, it is most favourably conditioned for the precipitation of much of their contained moisture, either as rain or snow. Glaciers of enormous magnitude are nourished on the higher ranges by this precipitation, which, together with the abundant rainfall of the lower ranges, feeds a number of rivers, which course down to the plains in hundreds of fertilising streams. In this manner the Himalayas protect India from the gradual desiccation which is overspreading the Central Asian continent, from Tibet northwards, and the desert conditions that inevitably follow continental desiccation.

Limits of the Himalayas—Geographically, the Himalayas are generally considered to terminate, to the north-west, at the great bend of the Indus, where it cuts through the Kashmir Himalayas, while the south-eastern extremity is defined by the similar bend of the Brahmaputra in upper Assam. At these points also there is a well-marked bending of the strike of the mountains from the general north-west—south-east to an approximately north and south direction. Some geographers have refused to accept this limitation of the Himalayan mountain system, because according to them it ignores the essential physical unity of the hill-ranges beyond the Indus and the Brahmaputra with the Himalayas. They would extend the term Himalaya to all those ranges to the east and west (*i.e.* the Hazara and Baluchistan and the Arakan ranges of Burma) which originated in the same great system of Pliocene orogenic upheavals.

The Syntaxial Bends of the Himalayas—The trend-lines of the Himalayan chain and its east and west terminations possess much interest from a structural point of view and need further remarks. For 2,500 kilometres from Assam to Kashmir, the chain follows one persistent S.E.-N.W. direction and then appears to terminate suddenly at one of the greatest eminences on its axis, Nanga Parbat (8,119 metres), just where the Indus has cut an immensely deep gorge right across the chain. Geological studies have shown that just at this point the strike of the mountains bends sharply to the south and then to the south-west, passing through Chilas and Hazara, instead of pursuing its north-westerly course through Chitral. All the geological formations here take a sharp hair-pin bend as if they were bent round a pivotal point obstructing them. This extraordinary inflexion affects the whole breadth of the mountains from the foot-hills of Jhelum to the Pamirs. On the west of this *syntaxis* (as this acutely reflexed bundle of mountain-folds is termed) the Himalayan strike swings from the prevalent N.E. to a N.-to-S. direction in Hazara and continues so to Gilgit; then it turns E.-to-W., the Pamirs showing a distinct equatorial disposition of their geological formations. To the south-east of this, the main tectonic strike quickly takes on a N.W.-S.E. orientation through Astor and Deosai—a direction which persists with but minor departures to western Assam.

The eastern limit of the Himalayas beyond Assam is not yet quite certain, but from the few geographical and geological observations that have been made in this region it appears that the tectonic strike here also undergoes a deep knee-bend from an easterly to a southerly trend. In the Arakan Yomas the geological axis of the mountains for several hundred kilometres is meridional, bending acutely to the N.E. near Fort Hertz. Beyond this point there is an abrupt swing to the N.W., then to E.N.E.-W.S.W. and finally to E.-W. through Assam and Sikkim.

The cause of these remarkable bends of the mountain-axis is discussed on pp. 395-7.[1]

Classification of the Himalayan Range

(I) **Geographical**—For geographical purposes Burrard has di-

---

[1] D. N. Wadia; The Syntaxis of the N.W. Himalayas: its Rocks, Tectonics and Orogeny. *Rec. G.S.I.*, vol. lxv. pt. 2, 1931.

wall of snow-covered mountains, pierced by passes, few of which are less than 5,200 metres in elevation. The Eastern Himalayas of Nepal and Sikkim rise very abruptly from the plains of Bengal and Oudh, and suddenly attain their great elevation above the snow-line within strikingly short distances from the foot of the mountains. Thus, the peaks of Kanchenjunga and Everest are only a few kilometres from the plains and are visible to their inhabitants. But the Western Himalayas of the Punjab and Kumaon rise gradually from the plains by the intervention of many ranges of lesser altitudes; their peaks of everlasting snows are more than 150 kilometres distant, hidden from view by the mid-Himalayan ranges to the inhabitants of the plains.

To the north of the Himalayas is the block of High Asia, the biggest and most elevated land-mass on the earth's surface. Directly to the north is the high plateau of Tibet of 5,000 metres mean altitude, traversed by the "Trans-Himalaya" and the Aling Kangri ranges; farther north are the Kuen Lun and Altyn Tagh ranges and, separated by the great desert basin of Tarim, the Tien Shan range. A peculiarity of these ranges is that they decrease in convexity of arc as we go north till in the Tien Shan the trend-line becomes nearly straight. The easterly extension of the Karakoram range into Tibet beyond 80° long. is not known.

Meteorological influence of the Himalayas—This mighty range of mountains exercises as dominating an influence over the meteorological conditions of India as over its physical geography, vitally affecting both its air and water circulation. Its high snowy ranges have a moderating influence on the temperature and humidity of Northern India. By reason of its altitude and its situation directly in the path of the monsoons, it is most favourably conditioned for the precipitation of much of their contained moisture, either as rain or snow. Glaciers of enormous magnitude are nourished on the higher ranges by this precipitation, which, together with the abundant rainfall of the lower ranges, feeds a number of rivers, which course down to the plains in hundreds of fertilising streams. In this manner the Himalayas protect India from the gradual desiccation which is overspreading the Central Asian continent, from Tibet northwards, and the desert conditions that inevitably follow continental desiccation.

Limits of the Himalayas—Geographically, the Himalayas are generally considered to terminate, to the north-west, at the great bend of the Indus, where it cuts through the Kashmir Himalayas, while the south-eastern extremity is defined by the similar bend of the Brahmaputra in upper Assam. At these points also there is a well-marked bending of the strike of the mountains from the general north-west—south-east to an approximately north and south direction. Some geographers have refused to accept this limitation of the Himalayan mountain system, because according to them it ignores the essential physical unity of the hill-ranges beyond the Indus and the Brahmaputra with the Himalayas. They would extend the term Himalaya to all those ranges to the east and west (*i.e.* the Hazara and Baluchistan and the Arakan ranges of Burma) which originated in the same great system of Pliocene orogenic upheavals.

The Syntaxial Bends of the Himalayas—The trend-lines of the Himalayan chain and its east and west terminations possess much interest from a structural point of view and need further remarks. For 2,500 kilometres from Assam to Kashmir, the chain follows one persistent S.E.-N.W. direction and then appears to terminate suddenly at one of the greatest eminences on its axis, Nanga Parbat (8,119 metres), just where the Indus has cut an immensely deep gorge right across the chain. Geological studies have shown that just at this point the strike of the mountains bends sharply to the south and then to the south-west, passing through Chilas and Hazara, instead of pursuing its north-westerly course through Chitral. All the geological formations here take a sharp hair-pin bend as if they were bent round a pivotal point obstructing them. This extraordinary inflexion affects the whole breadth of the mountains from the foot-hills of Jhelum to the Pamirs. On the west of this *syntaxis* (as this acutely reflexed bundle of mountain-folds is termed) the Himalayan strike swings from the prevalent N.E. to a N.-to-S. direction in Hazara and continues so to Gilgit; then it turns E.-to-W., the Pamirs showing a distinct equatorial disposition of their geological formations. To the south-east of this, the main tectonic strike quickly takes on a N.W.-S.E. orientation through Astor and Deosai—a direction which persists with but minor departures to western Assam.

The eastern limit of the Himalayas beyond Assam is not yet quite certain, but from the few geographical and geological observations that have been made in this region it appears that the tectonic strike here also undergoes a deep knee-bend from an easterly to a southerly trend. In the Arakan Yomas the geological axis of the mountains for several hundred kilometres is meridional, bending acutely to the N.E. near Fort Hertz. Beyond this point there is an abrupt swing to the N.W., then to E.N.E.-W.S.W. and finally to E.-W. through Assam and Sikkim.

The cause of these remarkable bends of the mountain-axis is discussed on pp. 395-7.[1]

Classification of the Himalayan Range

(I) **Geographical**—For geographical purposes Burrard has di-

---

[1] D. N. Wadia; The Syntaxis of the N.W. Himalayas: its Rocks, Tectonics and Orogeny. *Rec. G.S.I.*, vol. lxv. pt. 2, 1931.

vided the long alignment of the Himalayan system into four sections : the *Punjab Himalayas,* from the Indus to the Sutlej, 560 km. long; *Kumaon Himalayas,* from the Sutlej to the Kali, 320 km. long; *Nepal Himalayas* from the Kali to the Tista, 800 km. long; and *Assam Himalayas,* from the Tista to the Brahmaputra 725 km. long. Also the Himalayan system is classified into three parallel or longitudinal zones, differing from one another in wellmarked orographical features :

(1) The *Great Himalaya* : the innermost line of high ranges, rising above the limit of perpetual snow. Their average height extends to 6,100 m. On it are situated the peaks, like Mount Everest, Kanchenjunga, Dhaulagiri, Nanga Parbat, Gasherbrum, Gosainthan, Nanda Devi, etc.[1]

(2) The *Lesser Himalayas,* or the middle ranges: a series of ranges closely related to the former but of lower elevation, seldom rising much above 3,600-4,600 km. The Lesser Himalayas form an intricate system of ranges; their average width is eighty km.

(3) The *Outer Himalayas,* or the *Siwalik ranges,* which intervene between the Lesser Himalayas and the plains. Their width varies from eight to fifty km. They form a system of low foot-hills with an average height of 900-1,500 m.

(II) **Geological**—As regards geological structure and age the Himalayas fall into three broad stratigraphical belts or zones. These zones do not correspond to the geographical zones as a rule.

(1) The Northern or *Tibetan Zone* ; lying behind the line of highest elevation (*i.e.* the central axis corresponding to the Great Himalaya). This zone is composed of a continuous series of highly fossiliferous marine sedimentary rocks, ranging in age from the earliest Palaeozoic to the Eocene age. Except near the northwestern extremity (in Hazara and Kashmir) rocks belonging to this zone but rarely occur south of the line of snow peaks.

(2) The Central or *Himalayan Zone,* comprising most of the Lesser or Middle Himalayas together with the Great Himalaya.

[1]
Mount Everest	Nepal Himalaya	-	- 8,870 m.
K^2 - -	- Karakoram	-	- 8,615 ,,
Kanchenjunga	- Nepal Himalaya	-	- 8,585 ,,
Dhaulagiri -	- ,,	-	- 8,172 ,,
Nanga Parbat	- Kashmir Himalaya	-	- 8,119 ,,
Gasherbrum	- Karakoram	-	- 8,073 ,,
Gosainthan -	- Nepal Himalaya	-	- 8,019 ,,
Nanda Devi	- Kumaon Himalaya	-	- 7,822 ,,
Rakaposhi -	- Kailas range	-	- 7,793 ,,
Namcha Barwa	- Assam Himalaya	-	- 7,761 ,,
Badri Nath -	- Kumaon Himalaya	-	- 7,073 ,,
Gangotri -	- ,,	-	- 6,594 ,,

It is mostly composed of crystalline and metamorphic rocks—gneisses, and schists, with unfossiliferous sedimentary Purana and Mesozoic deposits.

(3) The Outer or *Sub-Himalayan Zone*, corresponding to the Siwalik ranges, and composed almost entirely of Tertiary, and principally of Upper Tertiary, sedimentary river-deposits.

The above is a very brief account of a most important subject in the geography of India, and the student must refer to the works mentioned at the end of the chapter for further information, especially to that by Sir Sidney Burrard and Sir Henry Hayden, Second Edition, 1932, revised by Burrard and Dr. A. M. Heron, which contains the most luminous account of the geography and the geology of the Himalayas.

Physical Features of Kashmir Himalayas

Large parts of the Himalayas are yet unexplored; not only the geology, but even the main features of the orography and geography are not well known over vast areas. The only parts that are surveyed with some degree of exactness are the Punjab Himalayas of Kashmir and Simla-Chakrata, a few of the great valleys of the Central Himalayas, the tracks of exploring expeditions and of the Tibetan travellers and traders. Even within these there are large districts which are geologically unknown, *viz*. the terrain between the Ravi and Sutlej, which is mostly unexplored ground, while districts such as Baltistan, Zanskar and Ladakh are imperfectly known. To give an idea of the main features of a section of the N. W. Himalayas, its orography and physical features, the following account of the geography of the Kashmir mountains, which may broadly serve as a type, will be summarised below.

The orographic features—Punjab Himalayas—There is a close uniformity in physical features and geological constitution of the Outer-Middle Himalayan tract from Rawalpindi to Dehra Dun. An admirable account of the geography of the Kashmir-Himalayan region is given by Frederick Drew, who spent many years in this region, in his well-known book, *Jammu and Kashmir Territories* (E. Stanford, London, 1875). What follows in this section is an abridgement of this author's description, modified, to some extent, by incorporating the investigations of later observers. The central Himalayan axis, after its bifurcation near Kulu, has one branch to the north-west, known as the Zanskar Range, terminating in the high twin-peaks of Nun Kun (7110 metres) ("the Great Himalaya Range" of Burrard); the other branch runs due west, a little to the south of it, as the Dhauladhar Range, extending farther to the north-west as the high picturesque range of the Pir Panjal, so conspicuous from all parts of the Punjab. Between these two branches of the crystalline axis of the Himalayas lies a longitudinal valley with a south-east to north-west trend, some 135 km. long and 40 km. broad in its middle, the broadest part. The long diameter of the oval is parallel to the general strike of the ranges in this part of the Himalayas. The total area of this Kashmir valley is 4,921 sq. km, its mean level about 1,585 m. above the sea. The ranges of mountains which surround it at every part, except the narrow gorge of the Jhelum at Baramula, attain, to the north-east and north-west, a high general altitude, some peaks rising above 5,500 m. On the south-western border, the bordering ridge, the Pir Panjal, is of comparatively lower altitude, its mean elevation being 4270 m. The best known passes of the Pir Panjal range, the great highways of the past, are the Panjal Pass, 3,480 m; the Budil, 4,270 m; Golabghar Pass, 3,812 m; Banihal Pass, 2,835 m. Tata Kuti and Brahma Sakal are the highest peaks, above 4,727 m. in elevation.

The Outer Ranges (the Sub-Himalaya or Siwalik Ranges)

The simple geological structure of the outer ranges. The "duns"—The outermost ranges of the Kashmir Himalaya rise from the plains of the Punjab, commencing with a gentle slope from Jammu, attain about 600 m. in altitude, and then end abruptly in steep, almost perpendicular, escarpments inwards. Then follows a succession of narrow parallel ridges with their strike persistent in a N.W.-S.E. direction, separated by more or less broad longitudinal or strike-valleys (the basins of subsequent streams). These wide longitudinal or strike-valleys inside the hills are of more frequent occurrence in the central parts of the Himalayas, and attain a greater prominence there, being known there as "duns" (*e.g.* Dehra Dun, Kothri Dun, Patli Dun, etc.) In the Jammu hills the extensive, picturesque duns of Udhampur and Kotli are quite typical. The Kashmir valley itself may be taken as an exaggerated instance of a dun in the middle Himalaya. These outer hills, formed entirely of the younger Tertiary rocks, rarely attain to greater altitude than 1,220 m. or thereabouts. The outer ranges of the sub-Himalayan zone, bounded by the Ravi and the Jhelum, the two east and west boundaries of the Kashmir State, are known as the Jammu hills. Structurally, as well as lithologically, they partake of the same characters as are seen in the hills to the east and west, which have received a greater share of attention by the Indian geologists. Ranges situated more inwards, and formed of older Tertiary rocks (of the Murree series), reach a higher altitude, about 1,800 to 2,450 m. At the exit of the great rivers, the Chenab and the Jhelum, there is an indentation or a deep flexure inwards into this region corresponding to an abrupt change in the direction of the strike of the hills. In the case of the Jhelum at Muzaffarabad this flexure is far more conspicuous and significant, the result of the syntaxial bend of the whole mountain-system, the strike of the whole Himalayan range there changing from the usual south-east–north-west to north and south and thence undergoing another deflection to north-east–south-west. (See Pl. XVIII.)

The Middle Ranges (Lesser or Middle Himalayas—The Panjal and Dhauladhar Ranges)[1]

The Panjal Range. "Orthoclinal" structure of the Middle ranges—This region consists of higher mountains (3600–4,600 m.) cut into by deep ravines and precipitous defiles. The form of these ranges bears a great contrast to the outer hills described above, in being ridges of irregular direction that branch again and again, and in exhibiting much less correspondence between the lineation of the hills and the strike of the beds constituting them. In the Pir Panjal, a singularly well-defined range of mountains extending from the Kaghan valley to beyond the Ravi valley, which may be taken as a type of the mountains of the Middle Himalaya, these ridges present generally a steep escarpment towards the plains and a long gentle slope towards Kashmir. Such mountains are spoken of as having an "orthoclinal" structure, with a "writing-desk" shape (see Fig. 39, p. 359). To this cause (among several others) is due the presence of dense forest vegetation, the glory of the Middle Himalaya, clothing the northern and north-eastern slopes, succeeded higher up by a capping of snows, while the opposite, southern slopes are, except in protected valley-slopes, barren and devoid of snow, being too steep to maintain a soil-cap for the growth of forests or allow the winter-snows to accumulate. South-east of the Ravi, the Pir Panjal is continued by the Dhauladhar range, passing through Dalhousie, Dharamsala and Simla. Geologically the middle Himalaya of this part are different from the foothills, being composed of a zone of highly compressed and altered rocks of various ages, from the Purana and Carboniferous to Eocene. The axial zone of the Panjal range is composed of the Permo-Carboniferous. For map of the Pir Panjal, see Pl. XV.

Inner Himalayas

The zone of highest elevation. Physical aspects of the inner Himalayas—To the north of the Pir Panjal and Dhauladhar ranges are the more lofty mountain-ranges of the innermost zone of the

---

[1] For a connected account of the geology of Pir Panjal, see Middlemiss, *Rec. G.S.I.* vol. xli. pt. 2, 1911, and Wadia, "Geology of Poonch and Adjoining Area" *Mem. G.S.I.* vol. li. pt. 2, 1928.

PHYSICAL FEATURES

Himalayas, rising above the snow-line into peaks of perpetual snow. The valley of Kashmir is the synclinal basin enclosed between the Pir Panjal range to the south and an offshoot of the Central axial range to the north. In the North Kashmir range, an offshoot of the Zanskar range, which forms the north-eastern border of the valley, there are peaks of from 4,500 to 6,100 m. in height. Beyond this range the country, with the exception of the deep gorges of the Middle Indus, is a high-level plateau-desert, utterly devoid of all kinds of vegetation. Here there are elevated plateaus and high mountain-ranges separated from one another by great depressions, with majestic peaks towering to 7,300 m. The altitude steadily increases farther north, till the peak K^2, on the mighty Karakoram or Mustagh range, attains the culminating height of 8,621 m.—the second highest mountain in the world. The Karakoram chain is the watershed between India and Turkestan. The valleys of these regions show varying characters. In the south-east is the Changchenmo whose width is from eight to ten km. with an average height of 4,270 m. above sea-level. From that to the north-west the height of the valley-beds descends, till in Gilgit on the very flanks of the gigantic peak of Nanga Parbat, Diyamir (8,120 m.), the rivers have cut so deeply through the bare, bleak mountains that the streams flow at an elevation of only 1,525 and, in one case, 1,067 m. above the level of the sea. At places, in north and north-east Kashmir, there are extensive flat, wide plains or depressed tracts among the mountains, too wide to be called valleys, of which the most conspicuous are the plateaus of Deosai, 3,965 m. high, Lingzhitang, 4,880 m, and Dipsang of about the same height. The physical features of this extremely rugged, wind-swept and frost-bitten region vary much in character. They present an aspect of desolate, ice-bound altitudes and long dreary wastes of valleys and depressed lands totally different from the soft harmony of the Kashmir mountains, green with the abundance of forest and cultivation. The rainfall steadily diminishes from the fairly abundant precipitation in the outer and middle ranges to an almost total absence of any rainfall in the districts of Ladakh and Gilgit, which in their bleakness and barrenness partake of the character of Tibet. Ladakh is one of the loftiest inhabited regions of the world, 3,600–4,600 m. Its short but warm summers enable a few grain and fruit crops to ripen. Owing to the great aridity of the atmosphere, the climate is one of fierce extremes, from the burning heat of some of the desert tracts of the Punjab plains in the day to several degrees below freezing-point at night. Baltistan, lying directly to the north of Kashmir, and receiving some share of the atmospheric moisture, has a climate intermediate between the latter and that of Ladakh. In consequence of the great insolation and the absence of any water-action, there has accumulated an abundance of detrital products on the dry uplands and valleys forming a peculiar kind of mantle-rock or regolith of fresh, undecomposed rock-fragments. The bare mountains which rise from them exhibit the exquisite desert coloration of the rocks due to the peculiar solar weathering. Between Ladakh and the Dhauladhar range are the districts of Zanskar, Lahoul and Rupshu, consisting of intricately ramifying glaciated ranges of crystalline rocks, intersected by lofty valleys having but a restricted drainage into a few saline lakes and marshes. This rugged country is crossed by a few trade-routes from Simla and Kulu to Tibet, through high passes, 4,880 to 5,500 m. With the exception of a part of Ladakh, which consists of Tertiary rocks and a basin of Mesozoic sedimentary rocks on the northern flank of the Zanskar mountains, by far the larger part of the inner mountains is composed of igneous and metamorphic rocks—granites, gneisses and schists.

There is no counterpart of the Kashmir basin north of the Dhauladhar in the Simla mountains. East of the Sutlej the Dhauladhar range approaches and closes in with the Great Himalaya Range. The important Spiti basin of Palaeo-Mesozoic sediments lies to the north of the crystalline gneissic axis of the latter.

Valleys of Kashmir

The transverse valleys. The configuration of the valleys— In conformity with the peculiarities of the other Himalayan rivers, briefly referred to above, the great rivers of this area—the Indus, Jhelum, Chenab, Ravi, and Sutlej—after running for variable distances along the strike of the mountains, suddenly make an acute bend to the south and flow directly across the mountains. The Sutlej, like the Indus, takes its origin in Tibet, much to the north of the Indo-Tibet watershed. Just at the point of the bend, a large tributary joins the main stream and forms, as it were, its upward continuation. The Gilgit thus joins the Indus at its great bend to the south; the Wardwan joins the Chenab at its first curve in Kishtwar, and the Ans at its second curve plainwards, above Riasi. The Kishenganga and the Kunhar meet the Jhelum at Domel, where the latter takes its acutest curve southwards before emerging into the Punjab. Similarly the Spiti river joins the Sutlej where the latter takes its final southward turn. These transverse, inconsequent valleys of the Himalayas, as we shall see later, are of great importance in proving the antiquity of the Himalayan rivers, an antiquity which dates before the elevation of the mountain-system (see page 27). The configuration of the valleys in the inner Himalaya of the Kashmir regions is very peculiar, most of the valleys showing an abrupt alternation of deep U-shaped or I-shaped gorges, with broad shelving valleys of an open V-shape. This is due to the scanty rainfall, which is powerless in eroding the slopes of the valley where they are formed of hard crystalline rocks and where the downward corrasion of the large volume of streams produced by the melted snows is the sole agent of valley-formation. The broad valleys which are always found above the gorge-like portions are carved out of soft detrital rocks which, having no cover of vegetation or forest growth to protect them, yield too rapidly to mechanical disintegration. Many of the valleys are very deep. This is particularly seen in Drava, Karnah and Gilgit. By far the deepest of all is the Indus valley in Gilgit, which at places is bordered by stupendous precipices 5,200 metres in height above the level of the water at its bed. That this enormous chasm has been excavated by the river by the ordinary process of river-erosion would be hard to believe were not the fact conclusively proved by the presence of small terraces of river gravels at numerous levels above the present surface of its waters.

At Shipki the Sutlej receives its principal tributary, the Spiti river, which has drained the wide synclinal basin of marine Palaeozoic and Mesozoic sediments Up to this point the Sutlej is a strike-valley, flowing along the whole length of the alluvial plateau of Hundes in a profound 900 metre cañon, excavated through horizontally bedded ossiferous Pleistocene boulder gravel and clay, deposited by itself at a former stage of its history. Below Shipki the river turns south and traverses a variety of geological formations of the Zanskar and the Great Himalaya ranges, in narrow gorges that are 3,050 m. deep at places, with perpendicular rock-cliffs of 1,800 to 2,100 m. sheer fall. Its passage through the sub-Himalayan Tertiary zone below Simla shows that the river at various stages must have been impeded and deflected in its course again and again by its own deposits.

From the presence of numerous terraces of lacustrine silt along the channel, the former presence of a chain of lakes all along the course of the Sutlej through the high mountains is indicated. This feature it shares with the Jhelum, Chenab and the Kunhar.

Lakes of Kashmir

There are very few lakes in the N.W. Himalayas, contrary to what one would expect in a region of its description. The few noteworthy lakes are the Wular, in the valley; the salt-lakes of Ladakh, bearing evidence of a progressive desiccation of the country, *viz*., the Tsomoriri in Rupshu, which is 25 km. long and 3 to 8 km. wide and about 4,600 m. high; the Pangkong in Ladakh, which is 64 km. long, 3 to 7 km. wide and 4,270 m. in elevation. The origin of the two last-named lakes is ascribed by Drew to the damming of old river courses by the growth of alluvial fans or dry deltas of their tributary streams across them. These lakes have got several high-level beaches of shingle and gravel resting on wave-cut terraces, marking their successive former levels at considerable heights above the present level of the water. The wide, level valley-plains of the Changchenmo, Dipsang and Lingzhitang, at an elevation of from 4,800–5,200 m. may be regarded as of lacustrine origin, produced by the desiccation and silting up of saline lake-basins without any outlet. There are a number of smaller lakes or tarns, both in the valley of Kashmir proper and in the bordering mountains, most of which are of recent glacial origin, a few of which may be true rock-basins.

The source of the Sutlej is now known to be the two sacred ice-bound lakes of Manasarowar and Rakas Tal, situated behind the Himalayan water-shed at an altitude of 4,880 m. to the south of the peak of Kailas. Sven Hedin has found that the Sutlej flows from the Rakas Tal, which derives its water by subterranean drainage from the adjacent Manasarowar and not usually through any visible channel.

Glaciers of Kashmir

Transverse and longitudinal glaciers—In Drew's work, already mentioned, there is a snow-map of Kashmir which admirably shows the present distribution of glaciers and snow-fields in the more elevated parts of these mountains beyond 1680 m. elevation. With the exception of a few small glaciers in the Chamba mountains, there are no glaciers in the middle and outer Himalayas at present. In the Zanskar range glaciers are numerous though small in size; only at one centre, on the north-west slopes of the towering Nanga Parbat (8,119 m.), do they appear in great numbers and of large dimensions. One of these (the Diyamir) descends to a level of 2,870 m. above the sea, near the village of Tarshing. North and north-east of these no glaciers of any magnitude occur till the Hunza valley on the south of the Mustagh, or Karakoram, range is reached, whose enormous snow-fields are drained by a number of large glaciers which are among the largest glaciers of the world.[1] The southern side of this stupendous mountain-chain nourishes a number of gigantic glaciers some of which, the Biafo, the Báltoro, the Siachen, the Remo, and the Braldu glaciers, are only exceeded in size by the great Humboldt of Greenland. There are two classes of these glaciers: those which descend transversely to the strike of the mountains and those which descend in longitudinal valleys parallel to the trend of the mountains. The latter are of large dimensions and are more stable in their movements, but terminate at higher elevations (about 3,050 m.)

---

[1] For results of exploration of Karakoram and Baltistan glaciers, papers by Dainelli and Mason may be consulted.

than the former, which, in consequence of their steeper grade, descend to as much as 2,450 to 2,130 m. The Biafo glacier of the Shigar valley reaches nearly 64 km. in length and the Hispar 61 km. The lowest level to which glaciers descend in the Kashmir Himalayas is 2,450 or even 2,130 m. reaching down to cultivated grounds and fields fully 1,220 m. lower than the lowermost limit of the glaciers in the eastern Himalayas of Nepal and Sikkim. Many of these glaciers show secular variations indicative of increase or diminution of their volumes, but no definite statement of general application can be made about these changes (pp. 22-23). The majority of the glaciers, like the Tapsa, are receding, and leaving their terminal moraines in front of them, which have become covered by grass and in some cases even by trees; but others, like the Palma glacier, are steadily advancing over their own terminal moraines.[1]

Proof of Pleistocene Ice Age—There are abundant evidences, here as everywhere in the Himalayas, of the former greater development of glaciers, although there are no indubitable proofs of their ever having descended to the plains of the Punjab, or even to the lower hills of the outer Himalayas. Large transported blocks are frequently met with at various localities, at situations, in one case, but little above 1,220 m. The Jhelum valley between Uri and Baramula contains a number of large boulders of granitoid gneiss brought from the summit of the Kaj Nag range (to the N. W.), some of which are as large as cottages. These are common phenomena in all the other valleys; rock-polishing and grooving are well seen on the cliff-faces of the Lidar, Sind, and their tributaries, while typical *roches moutonnées* are not rare on the hard, resistant rock-surfaces in the beds or sides of these valleys. In the Sind valley, near the village of Hari (1,980 m.) on the road to Sona Marg, Drew has seen a well-grooved *roche moutonnée*. A little higher up, at Sona Marg itself (2,740 m.), are seen undulating valleys made up entirely of moraines. In the valley of Kashmir proper some of the fine impalpable buff-coloured sands and laminated clays, interstratified among the Karewa deposits, are of glacial origin ("rock meal"), formed during melting of the ice in the interglacial periods. The whole north-east side of the Panjal range and to a less extent elevations above 1,980 m. on the south-west are covered thickly under an extensive accumulation of old moraine materials, which have buried all the solid geology (see Fig. 39). In northern Baltistan, where the existing glaciers attain their maximum development, there are other characteristic proofs of old glaciation at far lower levels than the lowest of modern glaciers; polished rock surfaces, rock-groovings, perched blocks, etc. occur abundantly in the Braldu valley of this

---

[1] For glaciers of the Hunza valley, see *Rec. G.S.I.* vol. xxv. pts. 3 and 4, 1907.

district. Many of the valleys of this region in their configuration are of a U-shape, which later denuding agencies are trying to change to the normal V shape.

Other Mountain Ranges of India

Other ranges of the extra-Peninsula—Running transversely to the strike of the Himalayas at either of its extremities, and believed to belong to the same system of upheaval, are the other minor mountain-ranges of extra-Peninsular India. Those to the west are the flanking ranges which form mountain-arcs on the Indo-Afghan and Indo-Baluchistan frontier. Those to the east are the mountain-ranges of Burma. Many of these ranges have an approximate north-to-south but pronouncedly arcuate trend. The names of these important ranges are :

West	East
The Salt-Range	The Assam ranges
The Sulaiman range	The Manipur ranges
The Bugti range	The Arakan Yoma
The Kirthar range	The Tenasserim range
The Mekran range	

With the exception of the Salt-Range and the Assam ranges, the other mountains are all of a very simple type of mountain-structure, and do not show the complex inversions and thrust-planes met with in the Himalayas. They are again principally formed of Tertiary rocks. The Salt-Range and the Assam ranges, however, are quite different and possess several unique features which we shall discuss later on. Their rocks have undergone a greater amount of fracture and dislocation, and they are not composed so largely of Tertiary rocks.

Mountain ranges of the Peninsula—The important mountain ranges of the Peninsula are the Aravalli mountains, the Vindhyas, Satpuras, the Western Ghats (or, as they are known in Sanskrit, the Sahyadris), and the irregular broken and discontinuous chain of elevations known as the Eastern Ghats. Of these, the Aravallis are the only instance of a true tectonic mountain-chain, all the others (with the exception to be mentioned below) are merely mountains of circumdenudation, *i.e.* they are the outstanding remnants, or outliers, of the old plateau of the Peninsula that have escaped the denudation of ages with no axis of upheaval that is coincident with their present strike. The Aravallis were a prominent feature in the old Palaeozoic and Mesozoic geography of India, and extended as a continuous chain of lofty mountains from the Deccan to possibly beyond Garhwal. What we at present see of them are the eroded remnants of these mountains—their mere stumps laid bare by repeated cycles of erosion.

Vindhya mountains. Satpura range—The rocky country which rises gradually from the south of the Gangetic plains culminates in the highlands of Central India, Indore, Bhopal, Bundelkhand, etc. The southern edge of this country is a steep line of prominent escarpments which constitute the Vindhyan mountains and their easterly continuation, the Kaimur range, between 760 and 1,220 m. above sea-level. The Vindhyas are for their greater part composed of horizontally bedded sedimentary rocks of ancient age. South of the Vindhyas, and roughly parallel with their direction, are the Satpura mountains. The chain of ridges commences from Gaya and Rewah, runs south of the Narmada valley and north of the Tapti valley, and stretches westwards through the Rajpipla hills to the Western Ghats. The Vindhya and the Satpura chains form together the backbone of middle India. Very large parts of the Satpuras, both in the west and the east, are formed of bedded basalts; the central part has a core of granitoid and metamorphic rocks, overlain by Gondwana sandstones. Parts of the Satpuras give proof of having been folded and upheaved. It is probable, therefore, they are a weather-worn remnant of an old tectonic chain.

The Western Ghats—The greater part of the Peninsula is constituted by the Deccan plateau. It is a central tableland extending from 12° to 21° North latitude, rising about 600 m. mean elevation above the sea, and enclosed on all sides by hill-ranges. To its west are the Sahyadris, or Western Ghats, which extend unbroken to the extreme south of Malabar, where they merge into the uplands of the Nilgiris, some of whose peaks rise to the altitude of 2,652 m. (the Dodabetta peak). From the Nilgiris the Western Ghats extend (after the solitary opening, Palghat Gap), through the Anaimalai hills, to the extreme south of the Peninsula. The Western Ghats, as the name Ghat denotes, are, down to Malabar, steep-sided, terraced, flat-topped hills or cliffs facing the Arabian sea-coast and running with a general parallelism to it. Their mean elevation is some 900 m. The horizontally bedded lavas of which they are wholly composed have, on weathering, given to them a characteristic "landing-stair" aspect. The physical aspect of the Western Ghats south of Malabar—that is, the portion comprising the Nilgiris, Anaimalai, etc.—is quite different from these square-cut, steep-sided hills of the Deccan proper. The difference in scenery arises from difference in geological structure and composition of the two portions of the Western Ghats. Beyond Malabar they are composed of the most ancient massive crystalline rocks, and not of horizontal layers of lava-flows.

The Eastern Ghats—The broken and discontinuous line of mountainous country facing the Bay of Bengal, and known as the Eastern Ghats, has neither the unity of structure nor of outline characteristic of a mountain-chain. They form a discontinuous line from North Orissa to Madras and thence through the hill-masses of Nallamalais and Shevaroys, they fuse with the Western Ghats in the Nilgiris. Their average altitude is barely 610 m. The component parts belong to no one geological formation, but vary with the country through which the hills pass, and the high ground is made up of several units, which are formed of the steep scarps of several of the South Indian formations. Some of these scarps are the surviving relics of ancient mountain-chains elevated contemporaneously with the Aravallis.

The remaining, less important, hill ranges of the Peninsula are the trap-built Rajmahal hill of western Bengal; the Nallamalai hills near Cuddapah built of gneissose granite, and the Shevaroys and Pachaimalai, south-west of Madras, are built of charnockite gneiss and khondalites.

Though, in the main, immune from fold tectonics, the Deccan Peninsular block is fractured by numerous normal faults which have, in some cases, given rise to block-uplift mountains through epeirogenic forces. The most prominent example of this is the Nilgiri-Palni block of hills and their S. n. continuation, the Cardamom hills, which, for km. along their W. n. and SE. n. flanks, are bounded by gigantic escarpments, 1,070 to 1,830 m. rising above the Archaean peneplain at their foot. The chains of fault-basins filled with Gondwana sediments traversing some anciet valleys of the Deccan, and the Narmada and Tapti rift-valleys are other examples of this fault structure. The Indian peninsula, thus, though a rigid shield, is not an unbroken unit. Perhaps it is due to this multiple block and basin-faulting that it has attained isostatic adjustment and its immunity from major seismic disturbances. ("The Making of India," *Proc. Ind. Sc. Cong.*, 1942).

GLACIERS

The snow-line, *i.e.* the lowest limit of perpetual snow, on the side of the Himalayas facing the plains of India, varies in altitude from about 4,300 m. on the eastern part of the chain to 5,800 m. on the western. On the opposite, Tibetan, side it is about 900 m. higher, owing to the great desiccation of that region and the absence of moisture in the monsoon winds that have traversed the Himalayas. In Ladakh, with a scanty snow-fall, it is 5,500 m. In the Hindu Kush the average snow-line is 5,200 m. high. Owing to the height of the snow-line, the mountains of the Lesser Himalayas, whose general elevation is considerably below 4,600 m, do not reach it, and therefore do not support glaciers at the present day. But in some of the ranges *e.g.* the Pir Panjal there is clear evidence, in the thick masses of moraines covering their summits and upper slopes, in the striated and polished rock-surfaces, in the presence of numerous erratics, and other evidences of mountain-sculpture by glacier-ice, such as *cirques* and numerous small lake-basins, that these ranges were extensively glaciated at a late geological period, corresponding with the Pleistocene Glacial age of Europe and America.

Glaciers of the Himalayas—The Great Himalaya, or the innermost line of ranges of high altitudes reaching beyond 6,000 m. are the enormous gathering grounds of snow which feed a multitude of glaciers, some of which are among the largest in the world outside the Polar circles. Much attention has been expended on the scientific study and observation of the Himalayan snow-fields and glaciers, both by the Indian Geological Survey, meteorologists and by scientific explorers of other countries, *e.g.*, de Filippi, Bullock-Workman, and Dainelli.

Their size—In *size* the glaciers vary between wide limits, from those that hardly move beyond the high recesses in which they are formed, to enormous ice-flows rivalling those of the Arctic circle. The majority of the Himalayan glaciers are from three to five km. in length, but there are some giant streams of forty km. and upwards, such as the Milam and Gangotri glaciers of Kumaon and the Zemu glacier, draining the Kanchenjunga group of peaks in Sikkim. The largest glaciers of the Indian region are those of the Karakoram, discharging into the Indus; these are the Hispar and the Batura of the Hunza valley, 58 to 61 km. long, while the Biafo and the Báltoro glaciers of the Shigar tributary of the Indus are about 60 km. in length; the thickness of the ice-stream in these glaciers greatly varies, from 120 to 305 m. Still more mighty examples are the Siachen glacier, falling into the Nubra affluent of the Indus, some 72 km. long, and the Fedchenko of the Pamir region of about the same dimensions. Some measurements taken at the end of the Báltoro glacier gave a depth of 120 m. of solid ice; the thickness in the middle of the body would be considerably greater. The thickness of ice in the Zemu stream is nearly 200 m. while the Fedchenko has a depth of nearly 550 m. of ice.

These giant ice-streams of the Karakoram are doubtless survivors of the last Ice Age of the Himalayas, as the present-day precipitation of snow in this region is not sufficient to feed these great rivers of ice. Like the dwindling glaciers of the Kuen Lun, these streams also will gradually diminish in size and retreat from continuous defect of "alimentation".[1] The majority of the glaciers are of the type of valley glaciers, but what are known as hanging glaciers are by no means uncommon. As a rule the glaciers descending transversely to the strike of the mountain are shorter, more fluctuating in their lower limits, and, since the grade is steeper, they descend to such low levels as 2,100-2,500 m. in some

---

[1] Prof. Kenneth Mason, *Rec. G.S.I.*, vol. lxiii. pt. 2, 1930.

parts of the Kashmir Himalayas. Those, on the other hand, that move in longitudinal valleys, parallel to the strike of the moun-

FIG. 2—View of the great Báltoro Glacier. (From a drawing by Col. Godwin-Austen.)

tains, are of a larger volume, less sensitive to alternating temperatures and seasonal variations, and, their gradients being low, they rarely descend to lower levels than 3,050 m.

Limit of Himalayan glaciers—The *lowest limit* of descent of the glaciers is not uniform in all parts of the Himalayas. While the glaciers of Kanchenjunga in the Sikkim portion hardly move below the level of 3,965 m. altitude, and those of Kumaon and Lahoul of 3,660 m. the glaciers of the Kashmir Himalayas descend to much lower limits, 2,500 m. not far above villages and fields. In several places recent terminal moraines are observed at so low a level as 2,100 m. A very simple cause of this variation has been suggested by T. D. La Touche. In part it is due to the decrease in latitude, from 36° in the Karakoram to 28° in the Kanchenjunga, and in part to the greater fall of the atmospheric moisture as rain and not as snow in the eastern Himalayas, which rise abruptly from the plains without the intervention of high ranges, than in the western Himalayas where, though the total precipitation is much less, it all takes place in the form of snow.

Peculiarities of Himalayan glaciers—One notable peculiarity of the Himalayan glaciers, which may be considered as distinctive, is the presence of extensive superficial moraine matter, rock-waste, which almost completely covers the upper surface to such an extent that the ice is not visible for long stretches. On many of the Kashmir glaciers it is a usual thing for the shepherds to encamp in summer, with their flocks, on the moraines overlying the glacier

PLATE I. ALUKTHANG GLACIER.

Photo. T. D. La Touche. (Geol. Survey of India, Records, vol. xl.)

ice. The englacial and sub-glacial moraine stuff is also present in such quantity as sometimes to choke the ice. The *diurnal motion* of the glaciers, deduced from various observations, is between 7 and 13 cm. at the sides, and from 20 cm. to about 30 cm. in the middle. Observations on the movement of the great Báltoro glacier by the Italian Expedition of 1909 gave as the velocity of ice at the snout the comparatively much higher figure of 1.75 m. in 24 hours. The diurnal motion of the Fedchenko is about 45 cm. while that of the Zemu is 22.5 cm. In many parts of the Himalayas there are local traditions, supported in many cases by physical evidence, that there is a slow, general retreat of the glacier-ends; at the lower ends of most of the Himalayan glaciers there are enormous heaps of terminal moraines left behind by the retreating ends of the glaciers. The rate of diminution is variable in the different cases, and no general rule applies to all. In some cases, again, there is an undoubted advance of the glacier-ends on their own terminal moraines. Professor Mason's recent study of the Himalayan and Karakoram glaciers has given some valuable results: the velocity of glaciers and their advance and retreat depend on topographical factors and not on climatic factors; the velocity has been found to vary in different glaciers from one inch to many feet per day; variations in glacier activity, as indicated by movements of the snout, may be due to causes which are, in distinct cases, secular, periodic, seasonal, or accidental. Mason observes that the Karakoram and Himalayan glaciers show no evidence whatever of any regular periodic variation corresponding with any supposed weather-cycles.

In the summer months there is a good deal of melting of the ice on the surface. The water, descending by the crevasses, gives rise to a considerable amount of englacial and sub-glacial drainage. The accumulated drainage forms an englacial river, flowing through a large tunnel, the opening of which at the snout appears as an ice-cave.

Records of past glaciation in the Himalayas—Large and numerous as are the glaciers and the snow-fields of the Himalayas of the present day, they are but the withered remnants of an older and much more extensive system of ice-flows and snow-fields which once covered Tibet and the Himalayas. As mentioned already, many parts of the Himalayas bear the records of an "Ice Age" in comparatively recent times. Accumulations of moraine débris are seen on the tops and sides of many of the ranges of the middle Himalayas, which do not support any glaciers at the present time. Terminal moraines, often covered by grass, are to be seen in the Pir Panjal at heights above 1,980 m. while the shapes of the ice-planed mountains and the U-shaped valleys, at times terminating at the heads in amphitheatre-like hollows (*cirques*), are very characteristic features of this range. Ancient moraines are seen before the snouts of existing glaciers, extending to such low elevations as 1,800 m. or even 1,500 m. Sometimes there are grassy meadows, pointing to the remains of old silted-up glacial lakes. These facts, together with the more doubtful occurrences of what may be termed fluvio-glacial drift at much lower levels in the hills of the Punjab, lead to the inference that this part of India at least, if not the Peninsular highlands, experienced a Glacial Age in the Pleistocene period.[1]

RIVERS AND RIVER-VALLEYS

Rivers, with their tributary-systems, are the main channels of drainage of the land-surface; they are at the same time also the chief agents of land-erosion and sculpture and the main lines for the transport of the products of the waste of the land to the sea. The drainage-systems of the two regions, Peninsular and extra-Peninsular India, having had to accommodate themselves to two very widely divergent types of topography, are necessarily very different in their character. In the Peninsula the river-systems, as is obvious, are all of great antiquity, and consequently, by the ceaseless degradation of ages, their channels have approached the last stage of river-development, *viz*. the *base-levelling* of a continent. The valleys are broad and shallow, characteristic of the regions where vertical erosion has almost ceased, and the lateral erosion of the banks, by winds, rain, and streams, is of greater moment. In consequence of their low gradients the water has but little momentum, except in flood-time, and therefore a low carrying capacity. In normal seasons they are only depositing agents, precipitating their silt in parts of their basins, alluvial banks, estuarine flats, etc., while the streams flow in easy, shallow,

---

[1] The principal glaciers of the Himalays :

Sikkim—		*Kumaon*—	
Zemu	26 km.	Milam	19 km.
Kanchenjunga	16 km.	Kedar Nath	14.5 km.
		Gangotri	26 km.
		Kosa	11 km.
Punjab (Kashmir)—			
Rupal	16 km.	*Karakoram*—	
Diyamir	11 km.	Biafo	62.7 km.
Sonapani	11 km.	Hispar	61 km.
Rundun	19 km.	Báltoro	58 km.
Punmah	27 km.	Gasherbrum	39 km.
Rimo	40 km.	Chogo Lungma	39 km.
Chong Kumdan	19 km.	Siachen	72 km.
Niuapin	unknown	Batura	58 km.

(Col. K. Mason)

PLATE II. SNOUT OF SONA GLACIER FROM SONA.

Photo. J. L. Grinlinton. (Geol. Survey of India, Records, vol. xliv.)

meandering valleys. In other words, the rivers of the Peninsula have almost base-levelled their courses, and are now in a mature or adult stage of their life-history. Their "curve of erosion" is free from irregularities of most kinds except those caused by late earth-movements, and is more or less uniform from their sources to their mouths.[1]

Easterly drainage of the Peninsula—One very notable peculiarity in the drainage-system of the Peninsula is the pronouncedly easterly trend of its main channels, the Western Ghats, situated so close to the west border of the Peninsula, being the water-shed. The rivers that discharge into the Bay of Bengal thus have their sources, and derive their head waters, almost within sight of the Arabian Sea. This feature in a land area of such antiquity as the Peninsula, where a complete hydrographic system has been in existence for a vast length of geologic time, is quite anomalous, and several hypotheses have been put forward to account for it. One supposition regards this fact as an indication that the present Peninsula is the remaining half of a land mass, which had the Ghats very near its centre as its primeval water-shed. This watershed has persisted, while a great extension of the country west of it has been submerged underneath the Arabian Sea. Another view, equally probable, is suggested by the exceptional behaviour of the Narmada and the Tapti. These rivers discharge their drainage to the west, while the other chief rivers of the area, from Cape Comorin through the Western Ghats and the Aravallis to the Siwalik hills near Hardwar (a long watershed of 2,736 km.), all run to the east. This exceptional circumstance is explained by the supposition that the Narmada and Tapti do not flow in valleys of their own eroding, but have usurped for their channels two fault-planes, or deep alluvium-filled rifts in the rocks, running parallel with the Vindhyas. These faults are said to have origi-

---

[1] It cannot be said, however, that the channels are wholly free from *all* irregularities, for some of them do show very abrupt irregularities of the nature of *Falls*. Among the best known waterfalls of South India are: the Sivasamudram falls of the Kaveri in Karnataka, which have a height of about 90 m; the Gokak falls of the river of that name in the Belgaum district, which are 55 m. in height; the "Dhurandhar" or the falls of the Narmada at Jabalpur, in which, though the fall is only 9 m., the volume of water is large. The most impressive and best-known of the waterfalls of India are the Jog falls of the river Sharavati in North Kanara, where the river is precipitated over a ledge of the Western Ghats to a depth of 260 m. in one single fall. The Yenna falls of the Mahableshwarhills descend 183 m. below in one leap, while the falls of the Paikara in the Nilgiri hills descend less steeply in a series of five cataracts over the gneissic precipice. Indeed it may be said that such falls are more characteristic of Peninsular than of extra-Peninsular India and bear evidence to some minor disturbances in a late geological age.

nated with the bending or "sagging" of the northern part of the Peninsula at the time of the upheaval of the Himalayas as described before. As an accompaniment of the same disturbance, the Peninsular block, south of the cracks, tilted slightly eastwards, causing the eastern drainage of the area.

This peculiarity of the hydrography of the Peninsula is illustrated in the distribution and extent of the alluvial margin on the two coasts. There is but a scanty margin of alluvial deposit on the western coast, except in Gujarat, whereas there is a wide belt of river-borne alluvium on the east coast, in addition to the great deltaic deposits at the mouths of the Mahanadi, Godavari, Krishna, Kaveri, etc.

Another peculiarity of the west coast is the absence of deltaic deposits at the mouths of the rivers, even of the large rivers Narmada and Tapti. This peculiarity arises from the fact that the force of the currents generated by the monsoon gales and the tides is too great to allow alluvial spits or bars—the skeleton of the deltas—to accumulate. On the other hand, the debouchures of these streams are broad deep estuaries daily swept by the recurring tides.

As a contrast to the drainage of Peninsular India, it should be noted that the island of Sri Lanka has a "radial" drainage, *i.e.* the rivers of the island flow outwards in all directions from its central highlands, as is well seen in any map of Sri Lanka.

The Drainage of the Extra-Peninsular Area

The Himalayan system of drainage not a *consequent* **drainage**—In the extra-Peninsula the drainage system, owing to the mountain-building movement of the late Tertiary age, is of much more recent development, and differs radically in its main features and functions from that of the Peninsula. The rivers here are not only eroding and transporting agents but are also depositing agents during their journey across the plains to the sea. Thus they have built the vast plains of North India out of a part of the silt they have removed from the mountains. The most important fact to be realised regarding the drainage is that it is not in a large measure a *consequent* drainage, *i.e.* its formation was not consequent upon the physical features, or the relief, of the country as we now see them; but there are clear evidences to show that the principal rivers of the area were of an age anterior to them. In other words, many of the great Himalayan rivers are older than the mountains they traverse. During the slow process of mountain-formation by the folding, contortion, and upheaval of the rock-beds, the old rivers kept very much to their own channels, although certainly working at an accelerated rate, by reason of the great stimulus imparted to them by the uplift of the region near their source. The great momentum acquired by this upheaval was expended in eroding their channels at a faster rate. Thus the elevation of the mountains and the erosion of the valleys proceeded, *pari passu*, and the two processes keeping pace with one another to the end a mountain-chain emerged, with a completely devloped valley-system intersecting it in deep transverse gorges or cañons. These long, deep precipitous gorges of the Himalayas, cutting right through the line of the highest elevations, are the most characteristic features of the geography, and are at once the best-marked results, as they are the clearest proofs, of the inconsequent drainage of this region. From the above peculiarities the Himalayan drainage is spoken of as an *antecedent* drainage, meaning thereby a system of drainage in which the main channels of flow were in existence before the present features of the region were impressed on it.

The Himalayan watershed—This circumstance of the antecedent drainage also gives an explanation of the much-noted peculiarity of several of the great Himalayan rivers, *e.g.* the Indus, Sutlej, Bhagirathi, Alaknanda, Kali, Karnali, Gandak, Kosi and the Brahmaputra, that they drain not only the southern slopes of those mountains, but, to a large extent, the northern Tibetan slopes as well, the watershed of the chain being not along its line of highest peaks, but a great distance to the north of it. This, of course, follows from what we have said in the last paragraph. The drainage of the northern slopes flows for a time in longitudinal valleys, in structural troughs parallel to the mountains, but sooner or later the rivers invariably take an acute bend and descend to the plains of India by cutting across the mountains in the manner already described.

The transverse gorges of the Himalayas—These transverse gorges of the Himalayas are sometimes thousands of m. in depth from the crest of their bordering precipices to the level of the water at their bottom. The most remarkable example is the Indus valley in Gilgit, where at one place the river flows through a narrow defile, between enormous precipices nearly 6,000 m. in altitude, while the bed of the valley is only 900 m. above its level at Hyderabad (the head of its delta). This gives to the gorge the stupendous depth of nearly 5,200 m., yet the fact that every centimetre of this chasm is carved by the river is clear from the fact that small patches or "terraces" of river gravel and sand-beds are observed at various elevations above the present bed of the Indus, marking the successive levels of its bed. Other examples of similar gorges are numerous, *e.g.* those of the Sutlej, Gandak, Kosi, Alaknanda, etc. are deep defiles of from 1,800 to 3,700 m. depth and only from 9 to 29 km. width between the summits of the mountains on the sides.

[Although there is not much doubt now regarding the true origin of the transverse gorges of the Himalayas by the process described above, these valleys have given rise to much discussion in the past, it being not admitted by some observers that those deep defiles could have been entirely due to the erosive powers of the streams that now occupy them. It was thought by many that originally they were a series of transverse fissures or faults in the mountains which have been subsequently widened by water-action. Another view was that the elevation of the Himalayas dammed back the old rivers and converted them into lakes for the time being. The waters of these lakes on overflowing have cut the gorges across the mountains, in the manner of retreating waterfalls. The absence of lacustrine deposits at the head of the principal rivers does not lend support to this view, though it is probable that this factor may have operated in a secondary way in some cases. The defile of the Alaknanda, again, is known to have carved a part of its valley along a line of fault.

There is no doubt, however, that some of these transverse valleys, namely those of the minor rivers, have been produced in a great measure by the process of *head-erosion*, from the combined action of the stream or the glacier at the head of the river cutting back into the mountains, whereby the water-shed receded farther and farther northwards. It is necessary to suppose this because the volume of drainage from the northern slopes, in the early stages of valley growth, could not have been large enough to give it sufficient erosive energy to keep its valleys open during the successive uplifts of the mountains.]

River-capture or piracy—Many of the Himalayan rivers, in their higher courses, illustrate the phenomena of river-capture or "piracy". This has happened oftentimes through the rapid head-erosion of their main transverse streams, capturing or "beheading" successively the secondary laterals belonging to the Tibetan drainage-system on the northern slopes of the Himalayas. The best examples of river-capture are furnished by the Bhagirathi and other tributaries of the Ganges, the Arun in the Everest area, the Tista of Sikkim, and the Sind[1] river in Kashmir.

"Hanging valleys" of Sikkim—Some of the valleys of the Sikkim and Kashmir Himalayas furnish instructive examples of "hanging valleys", that is, side-valleys or tributaries whose level is some hundreds or thousands of metres higher than the level of the main stream into which they discharge. These hanging valleys have in the majority of instances originated by the above process of rapid head-erosion and capture of the lateral streams on the opposite slope. A well-known example is that of a former tri-

---

[1] Oldham, *Rec. G.S.I.* vol. xxxi. pt. 3, 1904.

butary of the Tista river of Sikkim, discharging its waters by precipitous cascades into the Rathong Chu, which is flowing nearly 600 m. below its bed. Prof. Garwood, in describing this phenomenon, suggests that the difference in level between the hanging side-valley and the main river is due not wholly to the more active erosion of the latter, but also to the recent occupation of the hanging valley by glaciers, which have protected it from the effects of river-erosion.

LAKES

Lakes play very little part in the drainage system of India. Even in the mountainous regions of the extra-Peninsula, particularly in the Himalayas, where one might expect them to be of frequent occurrence, lakes of any notable size are very few.

Lakes of Tibet, Kumaon and Kashmir—The principal lakes of the extra-Peninsula are those of Tibet, including the sacred Manasarowar and Rakas Tal, the reputed source of the Indus, Sutlej, and Ganges of Hindu traditions (but which have now been proved to be the source of the Sutlej only). Koko Nor is the largest (4,144 sq. km.) amid hundreds of lake-basins in Tibet. The Manasarowar, 518 sq. km. in area, and Rakas Tal, 363 sq. km. are fresh-water lakes, while Gunchu Tso, 48 km. to the east, is a saline lake, 24 km. long, it being a closed basin without any outlet. Other examples are : the lakes of Sikkim, Yamdok Cho, 72 km. in circumference; Chamtodong, 87 km.; the group of small Kumaon lakes (the Nainital, Bhim Tal, etc.); and the few lakes of Kashmir, of which the Pangkong, Tsomoriri, the Salt Lake, the Wular and Dal are the best-known surviving instances. There is some controversy with regard to the origin of the numerous lakes of Tibet, which occupy thousands of square km. of its surface and are the recipients of its inland surface drainage. Many are regarded as due to the damming up of the main river-valleys by the alluvial fans of tributary side-valleys (F. Drew);[1] some are regarded as due to an elevation of a portion of the river-bed at a rate faster than the erosion of the stream (Oldham);[2] while some are regarded as true erosion-hollows, scooped out by glaciers—rock-basins.[3] The origin of the Kumaon lakes is yet uncertain; while a few may be due to differential earth-movements like faulting, others may have been produced by landslips, glaciers, etc. The small fresh-water lakes of Kashmir are ascribed a very simple origin by Dr. Oldham. They are regarded by him as mere inundated hollows in the alluvium of the Jhelum, like the *jhils* of the

---

[1] *Jammu and Kashmir Territories* (London), 1875.

[2] *Rec. G.S.I.* vol. xxi. pt. 3, 1888.

[3] Huntington, *Journal of Geology*, vol. xiv., 1906, p. 599.

Ganges delta. The Manchar lake of Sind, a shallow depression only 2.4–3 m. deep, but attaining an area of 518 sq. km. in the monsoon, is in all probability of like character and origin, forming a part of the drainage system of the Indus in Sind.

Salinity of the Tibetan lakes—The lakes of Tibet exhibit two interesting peculiarities, *viz*. the growing salinity of their waters and their pronounced diminution of volume since late geological times. The former circumstance is explained by the fact that the whole lake-area of Tibet possesses no outlet for drainage. The interrupted and restricted inland drainage, therefore, accumulates in these basins and depressions of the surface where solar evaporation is very active, concentrating the chemically dissolved substances in the waters. All degrees of salinity are met with, from the drinkable waters of some lakes to those of others saturated with common salt, sodium carbonate, and borax.

Their desiccation—The desiccation of the Tibetan lakes is a phenomenon clearly observed by all travellers in that region. Old high-level terraces and sand and gravel beaches, 60 to 90 m. above the present level of their waters, are seen surrounding almost all the basins, and point to a period comparatively recent in geological history when the water stood at these high levels. This diminution of the volume of the water, in some cases amounting to a total extinction of the lakes, is one of the signs of the increasing dryness or desiccation of the region north of the Himalayas following a great change in its climate. This is attributed in some measure to the disappearance of the glaciers of the Ice Age, and to the uplift of the Himalayas to their present great elevation, which has cut off Tibet from the monsoonic currents from the sea.

The well-marked desiccation of the lakes of Skardu, Rupshu and other districts of the north and north-east of Kashmir is a very noteworthy phenomenon and has an important bearing on this question. The former high levels of their waters point to a greater rainfall and humidity connected with the greater cold of a glacial period. The Tsomoriri has a terrace or beach-mark at a height of 12 m. above the present level of its waters. The Pangkong lake has similar beaches at various levels, the highest being 36–37 m. above the surface of the present lake.

Lakes of the Peninsula—Besides the few small fresh-water lakes of the Peninsula, two or three structural hollows in the Salt-Range, filled with saline water—the Son Sakesar, Kallar Kahar, etc.—and the numerous small saline or alkaline basins (*dhands*) of Sind lying amid sand-hills, there are two occurrences of importance because of some exceptional circumstances connected with their origin and their present peculiarities. The one is the group of salt-lakes of Rajasthan, the other is the volcanic hollow or crater-lake of Lonar in the Deccan.

The Sambhar salt-lake—Of the four or five salt-lakes of Rajasthan, the Sambhar lake is the most important. It has an area of 233 square km. when full during the monsoon, at which period the depth of the water is about 1.2 m. For the rest of the year it is dry, the surface being encrusted by a white saliferous silt. The cause of the salinity of the lake was ascribed to various circumstances, to former connection with the Gulf of Cambay, to brine springs, to chemical dissolution from the surrounding country, etc. But Sir T. H. Holland and Dr. Christie[1] have suggested that the salt of the Sambhar and of the other salt-takes of Rajasthan (Didwana, Phalodi and Pachbadra, etc.) is wind-borne; it is derived partly from the evaporation of the sea-spray from the coasts and partly from the desiccated surface of the Rann of Kutch, from which sources the dried salt-particles are carried inland by the prevalent winds. The persistent south-west monsoons which blow through Rajasthan for half the year carry a large quantity of saline mud and salt-particles from the above sites, which is dropped when the velocity of the winds decreases. The occasional rainfall of these parts gathers in this salt and accumulates it in the lake-hollows which receive the drainage of the small streams. It is calculated by these authors, after a series of experiments, that some 132,080 tonnes of saline matter are annually borne by the winds in this manner to Rajasthan during the hot-weather months. From a recent examination of the data regarding direction and strength of surface and upper winds in this region it appears unlikely that a great proportion of the salt could have been wind-borne from the Kutch littoral.

The Lonar Lake—The Lonar lake is a deep circular crater-like hollow in the basalt-plateau of the Deccan, in the district of Buldana. The depression is about 168 m. in depth and over 1.5 km. in diameter. It is surrounded on the south side by a 30 m. high rim formed of blocks of basalts. The depression contains at the bottom a shallow lake of saline water. The chief constituent of the salt water is sodium carbonate, together with a small quantity of sodium chloride. These salts are thought to have been derived from the surrounding trap country by the chemical solution of the disintegrated products of the traps and subsequent concentration.

The origin of the Lonar lake hollow has been ascribed to a crypto-volcanic explosion unaccompanied by any lava eruption. This is one of the rare instances of volcanic phenomena in India within recent times. On this view the lake-hollow is an explosion-crater or a *caldera*. Another explanation is that the hollow is due to the impact of a meteorite which lies buried at some depth

---

[1] *Rec. G.S.I.* vol. xxxviii., pt. 2, 1909.

at the site, or to an engulfment produced by the sinking of the surface within a circular fracture into a cavern in the lavas.

COASTS

The coasts of India are comparatively regular and uniform, there being but few creeks, inlets, or promontories of any magnitude. It is only on the Malabar coast that there are seen a number of lakes, lagoons or back-waters which form a noteworthy feature of that coast. These back-waters, *e.g.* the *Kayals* of Kerala, are shallow lagoons or inlets of the sea lying parallel to the coast-line. They form an important physical as well as economic feature of the Malabar coast, affording facilities for inland water-communication. The silts brought by the recurring monsoon floods support large forests and plantations along their shores. At some places, especially along the tidal estuaries, deltaic fronts, or salt-marshes, there are the remarkable mangrove-swamps lining the coasts. The whole sea-board is surrounded by a narrow submarine ledge or platform, the "plain of marine denudation", where the sea is very shallow, the soundings being much less than 100 fathoms. This shelf is of greater breadth on the Malabar coast and on the Arakan coast than on the Coromandel coast. From these low shelving plains the sea-bed suddenly deepens, both towards the Bay of Bengal and the Arabian Sea, to a mean depth of 2,000 fathoms in the former and 3,000 fathoms in the latter sea. The seas are not of any great geological antiquity, both having originated in the earth-movements of the Cretaceous or early Tertiary times as bays or arms of the Indian Ocean overspreading foundered areas of a large southern continent (Gondwanaland), which, in the Mesozoic ages, connected India with Africa and with Australia. As will be seen later, both these seas had long inlets or gulfs penetrating far to the north during the Cretaceous and Eocene—in the case of the former sea, to Assam (this Assam gulf ran parallel with the Burma gulf of the same sea), and in the case of the Arabian Sea to north Punjab and beyond to Simla. The coast-line in front of the deltas of the Indus and Ganges is greatly changeable owing to the constant struggle between the sea-ward growth of the delta and the erosion of the waves, the formation of lagoons, lakes and sand-bars. Extensive mangrove-swamps are a feature of these coasts. The coast of Sind forms part of the plain of marine denudation, with the sea hardly a few fathoms deep.

The Malabar coast is fronted by a broad continental shelf which stretches in a straight line from Cape Comorin to Karachi at a depth of less than 100 fathoms. It suddenly plunges to 1,100 fathoms to a deep submarine valley, separating the shelf from the broad and irregular submarine ridge that stretches intermittently from lat. 15° to the Laccadive, Minicoy and Maldive islands to Chagos. These islands are the unsubmerged peaks of the ridge which rises steeply from the sea bottom.

It has long been the belief of geologists that the escarpment of the Western Ghats, parallel with the Malabar coast, has been formed by scarp-faulting, while Blanford considered the Mekran coast of Baluchistan to be largely shaped by an E.W. fault. The south-east coast of Arabia and the Somaliland coast as far south as Zanzibar are likewise believed to be determined by scarp-faults. The whole of the north border of the Arabian Sea is thus surrounded by a series of steep fractures believed to be of Pliocene or even later age.

The first researches on the submarine topography of the Arabian Sea, conducted by the Murray Expedition of 1933-4, have revealed some further interesting facts. These have shown that there are intermittent submerged ridges, 3,050 m. high, some 100 km. from the Mekran coast. Two parallel ridges, separated by a deep rift valley, 2,000 fathoms below the surface of the sea (extension of the present valley of the Indus?), starting from Karachi extend up to the Gulf of Oman. The axes of these ridges are probably in tectonic continuation with the Kirthar range of Sind composed of Eocene and Oligocene rocks. The *Murray Ridge* is the name given to the innermost of these. Its southward continuation is the *Carlsberg Ridge*, which extends a great distance south-south-eastwards towards Chagos Island and thence as a broad submarine range to lat. 46° S.—the *Mid-Indian Ridge*.

It is generally believed that there is a remarkable similarity between the topographic highs and lows on the floor of the Arabian Sea and the region of the great Rift Valleys of East Africa.[1]

Important geodetic data obtained by Colonel E.A. Glennie suggest that the Laccadive archipelago, prolonged northwards by a chain of shoals, is on a continuation of the axes of the Aravalli mountains. The islands of the seas are continental islands, with the exception of the group of coral islands, the Maldives and the Laccadives, which are atolls or barrier-reefs, reared on shallow submarine banks, the unsubmerged, elevated points of the ancient continent. Barren Island and Narcondam are volcanic islands east of the Andamans. The low level and smooth contours of the tract of country which lies in front of the S.E. coast below the Mahanadi suggest that it was a submarine plain which has emerged from the waters at a comparatively late date. Behind this coastal belt are the gneissic highlands of the mainland—the Eastern Ghats —which are marked by a more varied relief and rugged topography. Between these two lies the old shore-line.

---

[1] R.B. Seymour Sewell, Geographic and Oceanographic Research in Indian Waters, *Memoirs, As. Soc. Beng.* vol. ix. pp. 1-7, 1935.

The Arakan coast of the Bay of Bengal, with its numerous drowned valleys and deep inlets, owes its features to recent depression. The numerous islands of this coast as well as of the Malay archipelago and the East Indies are regarded as only the unsubmerged portions of a once continuous stretch of land from Akyab to Australia.

The topography of the bottom of the Bay of Bengal is comparatively simple. The recent surveys by the International Indian Ocean Expedition, however, have revealed a 1,000 km.-long and up to 40 km.-wide valley running along the crest of a submerged range extending from the Andaman Sea to the north tip of Sumatra, analogous to the mid-ocean ridge traversing the Atlantic. A number of submerged cañons, as much as 1,300 metres deep, have also been observed during these surveys traversing the floor of the bay between Orissa and Tamilnadu. These may be due to the eroding action of strong periodic turbidity currents. Much new light is shed on the topography, geology and other features of the Indian ocean floor by the explorations of the joint 1961-65 International Indian Ocean Expedition.[1]

On the east (Coromandel) coast, the continental shelf is much narrower. It is however certain that the island of Sri Lanka is a part of the Madras mainland, severed only in sub-Recent times, and separated by a submerged platform which is barely 5 fathoms deep. Outside the Indo-Sri Lanka strait the coastal shelf plunges to 1,000-1,600 fathoms. This coast of India has been invaded by the sea again and again from the Jurassic to mid-Pliocene times, but the broad outlines of the coast were determined in the Cretaceous. The coastal shelf broadens considerably and shallows to the south of the Ganges delta, because of the heavy alluvial deposits of the Ganges and the Brahmaputra.

---

[1] The seismically active mid-ocean ridge system running into the Gulf of Aden and joining the East African Rift Valley, the occurrence of a seismic ridge and the micro-continents like the Seychelles Bank, the Chagos-Laccadive ridge with volcanic foundation associated with the Deccan Traps, the Indus and the Ganges zone with abnormal thickness of sediments, the confirmation of the northern movement of India by the recent concept of sea floor spreading, the likely occurrence of heavy metals along the rift zone of Central Indian Ocean Ridge, submerged terraces on the western shelf, the presence of a major fault along the western shelf edge, submarine canyons on the eastern continental shelf, etc., are some of the outstanding results of the International Indian Ocean Expedition. The reader may refer to the UNESCO publication titled "International Indian Ocean Expedition, Collected Reprints, Volumes I to VIII (1965-1972);" "Structure of the Indian Ocean," Volume IV, Part 2, Wiley Inter-Science, 1970; "Proceedings of the Symposium on Indian Ocean," Bull. N.I.S.I., No. 38, Pt. 1, 1968, and later publications.

VOLCANOES

Barren Island volcano—There are no living or active volcanoes anywhere in the Indian region. The Malay branch of the

Fig. 3—The Volcano of Barren Island in the Bay of Bengal.
(H. F. Blanford.)

line of living volcanoes—the Sunda chain—if prolonged to the north, would connect a few dormant or extinct volcanoes belonging to this region. Of these the most important is the now dormant volcano of Barren Island (Fig. 3) in the Bay of Bengal, to the east of the Andaman Islands, 12° 15′ N. lat.; 93° 54′ E. long. What is now seen of it is a mere truncated remnant of a once much larger cone—its *basal wreck* or caldera. It consists of an outer amphitheatre, about 3 km. in diameter, breached at one or two places, the remains of the old cone, surrounding an inner, much smaller, but symmetrical cone, composed of regularly bedded lava-sheets of comparatively recent eruption. At the summit of this newer cone is a crater, about 300 m. above the level of the sea. But the part of the volcano seen above the waters is quite an insignificant part of its whole volume. The base of the cone lies some thousands of metres below the surface of the sea.

The last time it was observed to be in eruption was early in the nineteenth century; since then it has been dormant, but sublimations of sulphur on the walls of the crater point to a mild solfataric phase into which the volcano has declined. F. R. Mallet, of the Geological Survey of India, has given a complete account of Barren Island in *Memoir*, vol. xxi. pt. 4, 1885.[1]

---

[1] Captain Blair has described an eruption of Barren Island in 1795. Glowing cinders and volcanic blocks up to some tonnes in weight were discharged from the crater at the top of the new cone, which was also ejecting enormous clouds

Narcondam. Popa—Another volcano, along the same line, is that of the island of Narcondam, a craterless volcano composed wholly of andesitic lavas. From the amount of denudation that the cone has undergone it appears to be an old extinct volcano. The third example is the volcano of Popa, a large centrally situated cone composed of trachytes, ashes, and volcanic breccia, situated about fifty miles north-east of the oil-field of Yenangyaung. This is also extinct now, the cone is much weathered, and the crater is only preserved in part. From the fact that some volcanic matter is found interstratified in the surrounding strata belonging to the Irrawaddy group, it seems that this volcano must have been in an active condition as far back as the Pliocene.

Koh-i-Sultan—One more volcano, within the Indian region, but far on its western border, is the large extinct volcano of Koh-i-Sultan in the Nushki desert of western Baluchistan.

There are some unverified records of a number of living and dormant volcanoes in Central Tibet and in the Kuen-Lun range of mountains to its north. None of these, however, have been proved to be active recently, although reports about the eruption of some of these having been witnessed by Tibetan travellers from a distance have been current.

Among the volcanic phenomena of recent age may also be included the crateriform lake of Lonar. Though its origin through impact of a meteorite is possible, many still uphold its connection with sub-Recent volcanic action. Some more recent volcanic phenomena may be mentioned. (i) An evanescent volcanic eruption in 1756 off the Pondicherry coast which threw up large volumes of ashes and pumice and built up, at the site, an island two miles long while the eruption lasted. This island was soon reduced and eroded by the sea waves. (ii) The fairly wide distribution of fragments and pebbles of pumice on the coasts of Sri Lanka (notably the east coast)—an island totally devoid of land volcanic eruptive rocks—may be due to some unrecorded recent or sub-Recent eruption under coastal waters. Fishermen have reported the occurrence of a band of pumice running for some miles parallel with the coast north of Trincomalee.

Mud-Volcanoes

Distribution of Mud-Volcanoes—We must here consider a curious phenomenon—what was once regarded as a decadent phase of volcanic action, but which has no connection whatever with

---

of gases and vapours. Another observer, in 1803, witnessed a series of explosions at the crater at intervals of every ten minutes, throwing out masses of dense black gases and vapours with great violence to considerable heights.

PLATE III. MUD VOLCANO. ONE OF THE LARGEST, MINBU, BURMA.

Photo. E. H. Pascoe. (*Geol. Survey of India*, vol. xl.)

vulcanicity.[1] In the Irrawaddy valley and Arakan coast of Burma and the Mekran coast of Baluchistan, there occur groups of small and, more rarely, large cones of dried mud; from small holes ("craters") at the top there are discharged hydrocarbon gases (principally marsh-gas), muddy saline water, and often traces of petroleum. These conical mounds, known as mud-volcanoes, occur in great numbers in the Ramri and Cheduba islands on the Arakan coast, the majority being about six to nine metres high although some are much higher. Near Minbu in Burma the cones reach about twelve metres, but in the dry climate of Baluchistan some are nearly 90 metres high. The great majority of mud-volcanoes are associated with a very gentle flow of muddy water, but in exceptional cases the mud-volcanoes are subject to occasional outbursts of great violence, fragments of the country rock being thrown out with force; the friction may even be sufficient to ignite the accompanying hydrocarbon gases.

Association with Petroleum—The gas, which is the prime cause of the mud-volcanoes, has the same origin as petroleum, and not only do many of the mud-volcanoes exude small quantities of petroleum, but a large number are in close proximity to small oil-fields or to seepages of petroleum. Most of the mud-volcanoes are near the crests of anticlinal folds or on lines of faulting. In the Yenangyaung oil-field of Burma there have been observed veins of dried mud penetrating the Miocene strata; these veins represent the channels supplying mud to mud-volcanoes that have long since disappeared. The mud is derived from the disintegration of shales of Tertiary age lying beneath the surface in close association with the gas-bearing strata.

EARTHQUAKES

The earthquake zone of India—Few earthquakes have visited the Peninsula since historic times; but those that have shaken the extra-Peninsula form a long catalogue.[2] It is a well-authenticated generalisation that the majority of the Indian earthquakes have originated from the great plains of India, or from the peripheral mountain-ranges to their North, West and East. No less than 74 destructive earthquakes have been recorded in this part of India in the last two centuries.

Of the great Indian earthquakes recorded in history the best-known are: Delhi, 1720; Calcutta, 1737; Eastern Bengal and the Arakan coast, 1762; Kutch, 1819; Kashmir, 1885; Bengal, 1885; Assam, 1897; Kangra, 1905; North Bihar, 1934; Baluchistan,

---

[1] J. Coggin Brown, *Rec. G.S.I.* vol. xxxvii. pt. 3, 1900; Sir E. H. Pascoe, *Mem. G.S.I.* vol. xl. pt. 1, 1912.

[2] Oldham, List of Indian Earthquakes, *Mem. G.S.I.* vol. xix. pt. 3, 1883.

1935; Mekran, 1945; and N.E. Assam, 1950. All of these, in the sites of their origins, agree with the above statement.

The area noted above is the zone of weakness and strain implied by the severe crumpling of the rock-beds in the elevation of the Himalayas within very recent times, which has, therefore, not yet attained stability or quiescence. It is also according to some authorities a belt of *underload*, its rocks being about 18 per cent. lighter than normal rocks. It falls within the great earthquake belt which traverses the earth from east to west.

The Assam Earthquake—On the 12th June, 1897, there occurred in Assam, heralded by a roar of extraordinary loudness, one of the most disastrous earthquakes of the world on record, the disturbed area bounded by the isoseismal of V or VI being no less than 4,140,000 sq. km. Shillong, with the surrounding country of 380,500 sq. km, was laid waste in less than one minute, all communications were destroyed, the plains riddled with rents and flooded and the hill-sides were scarred by gigantic landslips. The seismic motion was a complicated undulatory movement of the ground, the vertical component of which must have been high, for stones on the roads of Shillong were tossed in the air "like peas on a drum". The maximum amplitude of horizontal vibration was as much as 17.5 cm. the period being one second. Wide, gaping earth-fissures opened out in all directions in the alluvial plains, from which issued innumerable jets of water and sand, like fountains, spouting up to 1 or 1-3 metres in the air. Beds of rivers, tanks and even wells were ridged up, or filled, by the outpouring sand, thus greatly disturbing the drainage system of the land and causing extensive flooding. Over a wide area encircling the epicentre, the mountains precipitated landslips of unusual dimensions, which further obstructed the drainage.

The main shock was succeeded by hundreds of after-shocks during the first month, felt all over the shaken area. These shocks originated in a large number of shifting foci, scattered over the main epicentral tract in a fitful manner, certain districts registering far more shocks than others.

Of great significance geologically are the concomitant structural changes produced on the surface of the ground, such as fault-scarps and fractures, local changes of level, compression of the ground, and slight changes in the heights of hills. The most important fault-scarp ran parallel with the Chidrang river for 20 km., with a vertical throw varying from 30 cm. to 11 m. producing a number of water-falls and as many as thirty lakes in the course of the river.

R. D. Oldham, the author of a valuable memoir on this earthquake, has stated that the complex phenomena of this quake and the occurrence of many maxima of intensity are inconsistent with a simple or single fault-dislocation. He believes that there were numerous foci, or centres of disturbance, situated over a tract

PHYSICAL FEATURES 39

320 km. long and 80 km. wide. The original disruption starting in a thrust fault initiated numerous sympathetic shocks along branch-faults. The after-shocks were closely connected with the subsequent movements of these faults and served in some degree to locate them.

Oldham has computed the velocity of the earth-waves as about 3 km. per second and the depth of origin of the main shock at only 8 km. or even less.[1]

The Kangra Earthquake—The earthquake took place on the early morning of the 4th April, 1905. The shock, which was felt over the whole of India north of the Tapti valley, was characterised by exceptional violence and destructiveness along two linear tracts between Kangra and Kulu, and between Mussoorie and Dehra Dun. These were the *epifocal* tracts. The destruction grew less and less in severity as the distance from them increased, but the area that was perceptibly shaken, and which is encompassed by the *isoseist* of Intensity II of the Rossi-Forel scale included such distant places as Afghanistan, Quetta, Sind, Gujarat, the Tapti valley, Puri and the Ganges delta. The *centra* or the foci of the original concussion, or blow, were linear, corresponding to the two linear epicentra, Kangra-Kulu and Mussoorie-Dehra Dun, or regions which were directly above and in which the vibrations had a large vertical component. The *isoseists*, or curves of equal intensity, were hence ellipsoidal.

The *velocity* of the quake was difficult to judge, because of the absence of any accurate time-records at the different outlying places. But from a number of observations, the mean velocity of the earth-wave is deduced to be nearly 3.089 km. per second.

Middlemiss does not support the view that earthquakes of great severity originate near the surface in a complex network of faults and fractures. He ascribes to the present earthquake a deep-seated origin, and calculates, from Dutton's formula for deducing the depth of focus, a depth varying from 34 to 64 km.

The main shock was sudden, with only a few premonitory warnings, but the *after-shocks*, of moderate to slight intensity, which succeeded it for weeks and months, were several hundred in number. During the whole of 1906 the number of after-shocks was from ten to thirty a month. In 1907 they decreased in number, but scarcely in intensity. In the succeeding years the number of shocks grew fewer till they gradually disappeared.

The *geological effects* of the earthquake were not very marked. There were the usual disturbances of streams, springs, and canals; a number of landslips and rock-falls took place, also a few slight alterations in the level of some stations and hill-tops (*e.g.*, Dehra Dun and the Siwalik hills showed a rise of about 30 cm. relatively

---

[1] R. D. Oldham, *Memoirs, G.S.I.* vol xxix., 1899.

to Mussoorie). No true fissures of dislocations were, however, seen. In the above respects this earthquake offers a marked contrast to the Assam quake of 1897, where the geological results were of a more serious description and more permanent in their effect.

With regard to the *cause* of the earthquake, there is no doubt that it was a tectonic quake. Middlemiss is of opinion that it was due to a slipping of one of the walls, or change of strain, of a fault parallel to the "Main Boundary Fault" of the outer Himalayas at two points. Just where the two epicentra lie are two very well-defined "bays" or inpushings of the younger Tertiary rocks into the older rocks of the Himalayas, showing much packing and folding of the strata. Relief was sought from this compression by a slight sinking of one side of the fault.[1]

The Bihar Earthquake—On the afternoon of 15th January, 1934, North Bihar and Nepal were shaken by an earthquake of high intensity. Within three minutes the cities of Monghyr and Bhatgaon (Nepal) were in ruins, and towns so far apart as Kathmandu, Patna and Darjeeling were strewn with débris of many public and private buildings. Houses in Purnea and Sitamarhi tilted and sank under the ground, and sand and water were emitted from countless fissures in the ground opened on either side of the Ganges. The intensity of the main shock was so great that the recording apparatuses of the majority of the seismographs were thrown out of action, while the shocks were recorded at seismological stations as far away as Pasadena, Leningrad and Tokyo. The area enveloped by the Isoseist of Intensity II was roughly 4,900,000 sq. km. The main epicentra, where the intensity reached the degree of X, were three: (1) Motihari-Madhubani, (2) Kathmandu and (3) Monghyr. 28,500 sq. km. of the Ganges basin were riddled by fissures and sand-vents which ejected large volumes of water and sand, flooding the cultivated country and killing the standing crops. The total loss of human life is estimated at more than 12,000.

The effects of the earthquake on the general configuration and drainage of the country, alterations of level, fault-scarps, landslips, etc., were not so marked as in the Assam quake of 1897. The period and amplitude of vibrations and the maximum acceleration of the earth-wave were likewise not so remarkable.

Estimates of the depth of focus on the various standard methods of calculation vary largely, but it is probable that the movements responsible for the shock may have been along a highly inclined fracture or fractures.

With regard to the *cause*, there is some agreement that this earthquake was not primarily caused by displacements along the Himalayan Boundary Faults or thrusts, but that a more probable source

---

[1] *Memoirs G.S.I.*, vol. xxxviii., 1910.

of disturbance lay in the folded and fractured zone of the crust underneath the Gangetic Basin—a geosynclinal depression, the bottom of which must conceivably be under great strain.[1]

Baluchistan (Quetta) Earthquake—This seismic disaster, though comparatively local in incidence, brought unusual destruction of life and property on the town of Quetta on the night of 31st May, 1935. In a few moments this large military station was converted into a graveyard entombing 20,000 people. The epicentral tract is calculated to be only about 110 km. long and 26 km. broad, between Quetta and Kalat, away from which the intensity of damage rapidly decreased. The area over which the shock was felt, enclosed by Isoseists of Intensity IV and V, was 26,000 sq. km. which, considering the extraordinary destruction caused at the epicentre, is unusually small. From this fact, as also from the one that the intensity of the quake, as judged by the distribution of the damage, fell off rapidly from the epicentre, it was evident that the focus of origin of this earthquake could not be very deep-seated.

Extensive rock-falls took place from the limestone cliffs around Quetta and the ground, where composed of alluvium or loose soil, was fissured by a network of cracks. There were however no marked upheavals on the sides of the cracks, which were mainly superficial.

The earthquake was of the tectonic kind, though no connection has been established between this (or the less severe previous quake of 1931) and the various faults that have been noted in this region of severely compressed and looped fold-axes. The mountains of the Quetta region form a deep re-entrant angle, their tectonic axes being as it were festooned around a pivot near Quetta. The strain on the rock-folds arising from such a structure is probably responsible for the well-known seismic instability of this part of Baluchistan.[2]

Mekran Coast—In November 1945 an earthquake of some intensity, with its epicentre near the Mekran coast, 260 km. northwest of Karachi, took place, accompanied by a violent tidal wave and eruptions of mud. The earthquake-wave was of such intensity that it was recorded in Australia and the accompanying tidal wave, which reached a height of 12 m. at some Mekran ports, caused great damage all along the coast. Even at Bombay the tidal wave, 2 m. high, swept the coast and washed away a number of people. Large eruption from a submarine mud volcano led to the appearance of an island a few miles off the coast. This earthquake appears to be of a tectonic nature from the large area that was affected, being

---

[1] *Records G.S.I.* vol. lxviii. pt. 2, 1934. *Mem. G.S.I.* vol. lxxiii., 1938.

[2] W. D. West, *Records G.S.I.* vol. lxix. pt 2, 1935, and *Memoirs G.S.I.* vol. lxvii. pt. 1, 1934.

connected with the great Mekran coast fault.

North-east Assam—On the evening of 15th August, 1950, north-east Assam was shaken by an earthquake of high intensity comparable in some respects with the 1897 disaster. The area suffering most extensive damage in life and property was 39,000 square km. including the districts of Lakhimpur, Sibsagar and Sadiya, while the area of less damage, encompassed by Isoseist VIII, was nearly 200,000 square km. The earthquake was accompanied by all the usual surface effects—huge fissures discharging sand and water, subsidence of the ground in some areas and elevation of other tracts altering the drainage of the country and causing extensive floods. A few days later these floods were greatly accentuated by the bursting of dams of numerous temporary lakes created by landslides on the Dihang, Subansiri and other tributaries of the Brahmaputra. Landslips of great size scarred the ranges on the north-east of Assam disrupting the drainage of innumerable streams, inundations of which swept the countryside for months after the quake. The epicentre of the earthquake, determined by the seismographic recordings in India and other countries, was about 320 km. north of Sadiya in mountainous country on the north-east border of Assam. More damage to life and property was caused by river floods than by the earthquake. Changes in the main drainage lines of north-east Assam including that of the Brahmaputra have been reported.

Koyna—The Koyna earthquake of December 11, 1967, with epicentre about 320 km. south of Bombay, was the first recorded major tectonic earthquake in the Deccan Peninsular Shield (peripheral area). Considerable damage to life and property was caused, mostly around the epicentral tract in the vicinity of Koyna, though the large dam and reservoir at Koyna remained unharmed. The epicentre (17° 22′N, 73° 44′E) lies about 55 km. east of the Malabar coastline, the site of the great Malabar Fault, from Kutch to Cape Comorin, of Mio-Pliocene age, which disrupted this coastline, throwing down a large slice of W. India into the sea. The steep scarp-face of the Western Ghats suggests a parallel satellite fault and the Koyna earthquake may be ascribed to a slippage along this fault-plane. Pending collection of data by detailed geological mapping, the depth of focus is provisionally regarded at about 8 kilometres and the intensity of the quake 6.5 on the Richter Scale. The area shaken encompassed over 181,300 sq. km. in an ellipse stretching N.–S. from Surat to Bellari. The main shock was followed by a large number of after-shocks, nearly 200 being recorded at Poona in 24 hours.

Local Alterations of Level

Elevation of the peninsular tableland—Few hypogene disturbances have interfered with the stability of the Peninsula as a continental land-mass for an immense length of geological time, but there have been a few minor movements of secular upheaval and depression along the coasts within past as well as recent times. Of these, the most important is that connected with the slight but appreciable elevation of the Peninsula, exposing portions of the plain of marine denudation as a shelf or platform round its coasts, the west as well as the east. Raised beaches are found at altitudes varying from 30 to 45 m. at many places round the coasts of India; a common type of raised beach is the littoral concrete, composed of an agglutinated mass of gravel and sand with shells and coral fragments; while marine shells are found at several places some distance inland, and at a height far above the level of the tides. The steep face of the Sahyadri mountains, looking like a line of sea-cliffs, and their approximate parallelism to the coast lead to the inference that the escarpment is a result of a recent elevation of the Ghats from the sea and subsequent sea-action modified by subaerial denudation. Marine and estuarine deposits of post-Tertiary age are met with on a large scale towards the southern extremity of the Peninsula.[1]

Local alterations—Besides these evidences of a rather prominent uplift of the Peninsula, there are also proofs of minor, more local alterations of level, both of elevation and depression, within sub-Recent and pre-historic times. The existence of beds of lignite and peat in the Ganges delta, the peat deposits below the surface near Pondicherry, the submerged forest discovered on the eastern coast of the island of Bombay, etc., are proofs of slow movements of depression. Evidences of upheaval are furnished by the exposure of some coral reefs along the coasts, low-level raised beaches on various parts of the Ghats, and recent marine accumulations above the present level of the sea.

Submerged forest of Bombay. Alterations of level in Kutch—The submerged forest of Bombay is nearly 3.6 m. below low-water mark and 9 m. below high-water; here a number of tree-stumps are seen with their roots *in situ* embedded in the old soil.[2] On the Tinnevelli coast a similar forest or fragment of the old land surface, half an acre in extent, is seen slightly below high-water mark. Further evidence to the same effect is supplied by the thick bed of lignite found at Pondicherry 73 m. below ground level, and the layers of vegetable débris in the Ganges delta. About twenty miles from the coast of Mekran the sea deepens suddenly to a great hollow. This is thought to be due to the submergence

---

[1] E. Vredenburg, Pleistocene Movement in the Indian Peninsula, *Rec. G.S.I.* vol. xxxiii, 1906.

[2] *Rec. G.S.I.* vol. xi. pt. 4, 1878. Also T.D. La Touche, *Rec. G.S.I.*, vol. xlix. pt. 4, 1919.

of a cliff formerly lying on the coast. The recent subsidence, in 1819, of the western border of the Rann of Kutch under the sea, accompanied by the elevation of a large tract of land (the *Allah Bund*), is the most striking event of its kind recorded in India, and was witnessed by the whole of the local population. Here an extent consisting of roughly 5,180 square km. in area was suddenly depressed to a depth of from 3.6 to 4.5 m., and the whole tract converted into an inland sea. The fort of Sindree, which stood on the shores, the scene of many a battle recorded in history, was also submerged underneath the waters, and only a single turret of that fort remained, for many years, exposed above the sea. As an accompaniment of the same movements, another area, about 1,550 square km. was simultaneously elevated several metres above the plains, into a mound which was appropriately designated by the people the "Allah Bund" (built of God). The elevated tract of land known as the Madhupur jungle, near Dacca, is believed to have been upheaved as much as 30 m. in quite recent times. This upheaval caused the deflection of the Brahmaputra river eastward into Sylhet, away from the Ganges valley. Since this change the Brahmaputra has again gradually changed its bed to the west.

Even within historic times the Rann of Kutch was a gulf of the sea, with surrounding coast towns, a few recognisable relics of which yet exist. The gulf was gradually silted up, a process aided no doubt by a slow elevation of its floor, and eventually converted into a low-lying tract of land, which at the present day is alternately a dry saline desert for a part of the year, and a shallow swamp for the other part.

The branching *fjords*, or deep, narrow inlets of the sea, in the Andaman and Nicobar islands in the Bay of Bengal, point to a submergence of these islands within late geological times, by which their inland valleys were "drowned" in their lower parts. Good examples of drowned valleys occur on the Arakan coast, which, with its numerous estuaries and inlets proceeding inland from a submarine shelf, gives proof of recent submergence along the whole stretch of country from Akyab to Indonesia. In some of the creeks of Saurashtra near Porbander, on the other hand, oyster-shells were found at several places and at levels much above the present height of the tides, while barnacles and *serpulae* were found at levels not now reached by the highest tides. In Sind a number of oyster-banks have been seen several feet above high-water mark. Oyster-shells discovered lately at Calcutta likewise point to a slight local rise of the eastern coast.

Himalayas yet in a state of tension—It is the belief of some geologists that appreciable changes of level have recently taken place, and are still taking place, in the Himalayas, and that although the loftiest mountains of the world, they have not yet attained their maximum elevation but are still rising. That alterations of level have lately taken place is clear from a number of circumstances. Many of the rivers bear incontrovertible proofs of recent rejuvenation, due to the uplift of their water-shed. Another fact, suggesting the same inference, is the frequency and violence of earthquakes in the Himalayas and in the depressed tract lying at their foot. By far the largest number of disastrous Indian earthquakes have occurred, as already remarked, along these tracts. They indicate that the strata under the Himalayas are in a state of tension and are not yet settled down to their equilibrium plane. Relief is therefore sought by the subsidence of some tracts and the elevation of others.

ISOSTASY

India is particularly favourably circumstanced for the study of geodesy (the science of surveying and measuring large areas of the earth). Its triangular shape provides, from the foot of the Himalayas to Kanya Kumari, a stretch of 2,735 km. of land over one meridian. Again the deformation of the *geoid* (the shape, or as it is called, the figure of the earth) in India is such that in no other part of the world has the direction of gravity been found to undergo such abnormal variations as have been detected by the Survey of India in North India and by the Russian surveyors north of the Pamirs in Ferghana. According to Burrard, in no other country in the world does a surface of liquid at rest deviate so much from the horizontal. It was in India that it was discovered that a deficiency of matter underlies that vast pile of superficial matter, the Himalayas; that, on the other hand, a chain of dense matter runs hidden to the south of the Indo-Gangetic plains; and that seaward deflections of the pendulum, rather than towards the Ghats, prevail round the Deccan coasts. These discoveries led to the formulation of the theory of *mountain compensation* in about 1854 by G. B. Airy and the Rev. J. H. Pratt, Archdeacon of Calcutta, a theory which was subsequently elaborated and expanded by C. E. Dutton into the doctrine of *Isostasy*. This simple hypothesis, which has had a great vogue, particularly in America, implies a certain amount of hydrostatic balance between the different segments of the earth's crust, and an adjustment between the surface topographic relief and the arrangement of density in the sub-crust, so that above each region of less density there will be a bulge, while over tracts of greater density there will be a hollow—the former will be the continents, plateaus and mountains, the latter the ocean-basins. The excess of material over the portions of the earth above sea-level will thus be compensated for by a defect of density in the underlying material, the continents and mountains being floated because they are composed of relatively light material. Similarly the floor of the ocean will be depressed because it is composed of unusually dense rocky substratum. If an extra load is imposed on any part of the surface, *e.g.* ice-sheets during a glacial epoch, it must sink under it, while regions exposed to prolonged denudation must rise until equilibrium is established. The depth at which isostatic compensation is supposed to be complete is found, in the United States of America, according to the calculation by Hayford and Bowie of the U. S. Geodetic Survey, to be about 76 miles (122.3 km.) In India it is difficult to arrive at any such definite figure, for isostatic conditions must evidently be different in the Peninsula, a region of high geological antiquity, from those of the extra-Peninsular mountain region, which has undergone very recent orographic movements of the crust. In the former area isostatic balance must be more perfect than in the Himalayas and in the great plains of the north.

Plumb-line and pendulum observations at Dehra Dun have shown that the "topographic deflection," *i.e.* that due to the calculated visible mass of the Himalayas to the north, is 215 cm. but the true observed deflection is only 77.5 cm. For Murree the figures are 112.5 cm. and 30 cm. respectively; while for Kaliana, near Meerut, which is only 80 km. from the foot of the Himalayas, the observed deflection is only 2.5 cm. whereas it ought to be 145 cm. These observations prove that the Himalayas are largely compensated, though not fully; for the differences between the observed deflections and the theoretical, even under the assumption of isostatic compensation, are too great. The outer and middle Himalayas are found to be under-compensated, while the central ranges appear to be over-compensated.

On the Indo-Gangetic plains the deflections are invariably to the south and not towards the Himalayas. This southerly deflection increases as far as lat. 23° N., to the south of which the plumb-line deflects to the north. These discrepant data have been explained by Burrard by assuming that there exists underneath the plains a chain of dense rock, from Orissa north-westwards through Jabalpur into Kalat—an assumption which is borne out by gravity measurements of recent years.

Although measurements of gravity and deviations of the vertical, as carried out by the Geodetic Survey of India during the last few decades, broadly confirm the main postulates of the theory of isostasy, this theory is found to be inadequate in explaining the large anomalies of gravity which exist in India, even when there are no surface features present to account for them. For the main relief features of India, although a certain degree of compensation does exist, there are serious anomalies between the theoretical and observed values of the direction and force of gravity, which remain to be accounted for. For instance the gravimetric surveys have definitely proved belts of excess of density and of defects of density in North India which are not represented by any surface deeps or heights.

It is these data and their interpretation to establish a satisfactory correlation with surface geology that have drawn attention to the insufficiency of the isostatic theory as pronounced by Hayford. The gravimetric surveys have now definitely proved a deep-seated belt of excess of density underneath the plains—the *Hidden Range* of Burrard extending north-west and south-east of Jabalpur from Karachi to Orissa. To the north and south of this are belts of defects of density. This irregular variation of density is inconsistent with isostasy, which postulates that underlying excesses or defects of gravity must be reflected in surface deeps or heights. The gravity measurement work carried out by the Survey of India during the last few years, from a large number of stations scattered over India, has enabled them to explain provisionally the numerous gravity anomalies by assuming a series of upward bulges or downward warps or troughs in the sub-crust which may be taken to be some 16 to 32 km. below the surface. These crustal warps elevate or depress the denser, more basic layers of the sub-crust, the *Sima*, which underlie the lighter more acidic rocks of the surface crust—*Sial*—above or below their equilibrium plane. The theory is still in a stage of discussion but it promises to explain the residual anomalies in the force of gravity that are so commonly observed in India.

It appears that India as a whole is an area of defective density. Gravity in India is in deficit by an amount of material that is measured approximately by a stratum of rock 183 m. in thickness, spread over an area of 5,180,000 square km.[1]

As a result of gravimeter surveys made in the course of prospecting for oil, the distribution of gravity anomalies is known in some detail over large parts of north-eastern India and Burma, and the evidence has been analysed by P. Evans and W. Crompton. A line of maximum gravity has been traced from the extreme north of Burma southwards to continue by way of Narcondam and Barren Island to a similar line in Sumatra and Java; a gravity minimum lies under the Arakan Yoma and connects with a line running through the Andamans to islands off the south-west coast of Sumatra.[2]

---

[1] E. A. Glennie, Gravity Anomalies and the Structure of the Earth's Crust, *Surv. of Ind.* Prof. Pap. 27, 1932; H. J. Couchman, Progress of Geodesy in India, *Proc. Nat. Inst. Sc. Ind.* vol. iii, 1937.

[2] P. Evans and W. Crompton, Geological Factors in Gravity Interpretation illustrated by Evidence from India and Burma, *Q.J. Geol. Soc.* cii., 1946.

DENUDATION

Monsoonic alternations—Among the physical features of India a brief notice of the various denudational processes in operation in the country at the present time must be included. Inasmuch as climate is an important determining factor in the denudation of a region, the peculiar features which the climate of India possesses require consideration. The most unique feature in the meteorology of India is the monsoonic alternations of wet and dry weather. The division of the year into a wet half, from May to October, the period of the moist, vapour-laden winds from the south-west (from the Bay of Bengal and the Arabian Sea) towards Tibet and the heated tracts to the north, and the dry half, from November to April, the period of the retreating dry winds blowing from the north-east, has a preponderating influence on the character and rate of the subaerial denudation of the surface of the country.

Lateritic regolith—The intensity of the influence exercised by this dominating factor in the atmospheric circulation of the Indian region will be realised when the extent and thickness of the peculiar surface formation, *laterite*, is considered. Laterite is a form of regolith highly peculiar to India, and covers the whole expanse of the Peninsula from the Ganges valley to Kanya Kumari; it is believed by most authorities to have resulted from the subaerial alteration of its surface rocks under the alternately dry and humid (*i.e.* monsoonic) weather of India. Other characteristic products of weathering of the surface rocks *in situ* in the Peninsula are the Red Soil of Tamilnadu and that capping the gneissic tracts of the Deccan generally, and the Black Soil (*regur*),[1] which covers also large tracts of country in South India. The *Reh* efflorescences of the plains of North India and the formation of *nitre* in some soils should also be noted in this connection.

General character of denudation sub-tropical. Desert-erosion in Rajasthan—If this factor is excluded, the general atmospheric weathering or denudation of India is that characteristic of the tropical or sub-tropical zone of the earth. This, however, is a very general statement of the case. Within the borders of India every variety of climate is met with, from the torrid heat of the vast inland plains of the Punjab and North-east Baluchistan and upland plateaus (like Ladakh) to the Arctic cold of the higher ranges of the Himalayas; and from the reeking tropical forests of the coastal tracts of the Peninsula to the desert regions of Sind, Punjab and Rajasthan. Rock *disintegration* is the predominant process in the one area, rock *decomposition* in the other. The student can easily imagine the intensity of frost-action in the Himalayan highlands and the comparative mildness of the other agents

---

[1] The subject of soils of India is treated in Chapter XXVI, p. 473.

of erosion in that area, such as rapid alternations of heat and cold, chemical action, etc., and the vigorous chemical and mechanical erosion of the tropical monsoon-swept parts of the Peninsula, the denudation of some parts of which partakes of the character of that prevailing in the equatorial belt of the earth. In the desert tracts of Rajasthan, Sind and Baluchistan, mechanical disintegration due to the prevalent drought with its great extremes of heat and cold, the powerful insolation and wind-action, is dominant, to the exclusion of other agents of change. In this belt the action of the powerful summer-winds and dust-storms which blow for about two months preceding the summer must result in the transport of vast quantities of fine detritus, the prolonged accumulation of which has been the cause of the wide-spread loëss deposits of N. W. India. The transporting power of winds in the drier regions of India is enormous. Thousands of tons of dust and fine sand and silt are carried by the upper currents of winds for distances of hundreds of kilometres and dropped where their velocity decreases. Considerable erosion of the surface and of the soil-caps results in this manner in some Punjab and Rajasthan tracts. Rajasthan affords a noteworthy example of the evolution of desert topography within comparatively recent geological times. It also affords excellent illustrations of the geological action of winds in modifying the surface-features of a country. (See Sand-dunes, *Bhur* lands, etc.) This change has been brought about by the great dryness that has overcome this region since Pleistocene times, leading to the intensity of aeolian action on its surface.

Denudation by rivers—The geological work of Indian rivers calls for a few remarks. Some experiments by Everest prove that the Ganges conveys annually to the Bay of Bengal, at a conservative estimate, more than 356,000,000 tonnes of sand and clay—an average of over 900,000 tonnes of silt a day. There are some rivers of India whose waters are more silt-laden than those of the Ganges for many days of the year. The solid matter suspended in the Indus waters and discharged below Hyderabad, in Sind, is roughly estimated at 1,000,000 tonnes daily. The Brahmaputra carries down more silt than the Indus or Ganges. To the mechanically transported débris must be added the invisible amounts of chemically dissolved matter in the waters of the rivers. Exact measurements of these have not been made, but analyses of average samples of river-water show that the amounts of salts, *e.g.* sulphate and carbonate of calcium, silica, and the salts of Na, Mg, Fe, etc., removed from the land to the sea in solution by a river such as the Narmada or the Jhelum run into several millions of cubic feet per annum. There are wide fluctuations in the saline contents of river waters draining different rock terrains, from less than 50 to over 400 parts per million. The salinity of the Mahanadi river rising in the region of Archaean crystalline rocks, near Cuttack, is found to amount to 86 parts in a million parts of water.

Peculiarity of river-erosion in India—The Indian rivers accomplish an incredible amount of erosion during the wet half of the year, transporting to the sea an enormous load of silt, in swollen muddy streams. A stream in flood-time accomplishes a hundred times the work it performs in the normal seasons. If the same amount of rainfall, therefore, were evenly distributed throughout the year, the denudation would be far less in amount.

Their floods—The Himalayan streams and rivers are specially noted for their floods of extraordinary severity in the spring and monsoon seasons. This arises from the absence in the Indian rivers of lakes which exercise a restraining influence on the number, violence and duration of river-floods. Several of the Indus floods are noted in history, the most recent and best remembered being those of 1841 and 1858. Drew[1] gives a graphic account of the 1841 flood, when, after a period of unusually low level of the waters in the winter and spring of that year, the river, all of a sudden, descended in a black, mighty torrent that in a few minutes tore and swept away everything in its course, including a whole Sikh army that had encamped on its banks below Attock with its tents, baggage and artillery. The cause of this flood is attributed to a landslip in the narrow, gorge-like part of the river in Gilgit, which blocked up the water and converted the basin of the river above it into a lake fifty-six kilometres long and some hundreds of metres in depth. The sudden bursting of the barrier by the constantly increasing pressure of the water on it after the spring thaw is supposed to have caused the inundation.

Many mountain channels are known to have been dammed back by the precipitation of a whole hillside across them. In 1893 in Garhwal, the Alaknanda, a tributary of the Ganges, was similarly blocked by the fall of a hillside, and was converted into a lake at Gohna. The lake spread in extent and steadily rose in height for several months, till the waters ultimately surmounted the obstacle and caused a severe flood by the sudden draining of a large part of the lake.[2] A similar flood is recorded of the Sutlej in 1819. The shoulder of a mountain gave way in the deep gorge of the river, some 32 kilometres north-west of Simla, damming up the river to a height of 122 m., and producing the usual devastating flood when the obstruction burst. The formation of a lake, 150 m. deep and 24 to 32 km. around, in the Shyok river of Baltistan, by the interposition of the snout of the Chong Kundun glacier across the valley, successively in the springs of the years 1924, 1927 and 1930 is a recent instance. The bursting of the glacier

---

[1] *Jammu and Kashmir Territories*, London, 1875.
[2] *Rec. G.S.I.* vol. xxvii. pt. 2, 1894.

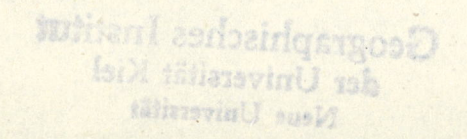

barrier made the Indus at Attock, situated 1,127 km. downstream from the Shyok dam, rise in flood at each occasion.

The increased volume of water, combined with the high velocity of the rivers in flood-time, multiplies their erosive and transporting power to an inconceivable extent, and boulders and blocks, several metres in diameter, are rolled along their beds, and carried in this manner to distances of 80 or even 160 km. from their sources, causing much injury to the banks and wear and tear to the beds of the channels.

Late Changes in the Drainage Systems of North India

Many and great have been the changes in the chief drainage lines of North India since late Tertiary times[1]—changes in fact which have produced a complete reversal of the directions of flow of the chief rivers of North India. The formation of the long thin belt of Siwalik deposits along the foot of the Himalayas from Assam, through Kumaon and the Punjab to Sind, widening steadily in its westward extension, is now ascribed to the flood-plain deposits of a great north-west-flowing river lying south of and parallel with the Himalayan chain from Assam to the furthest north-west corner of the Punjab, and then flowing southwards to meet the gradually receding Miocene sea of Sind and Punjab. This river has been named the "Siwalik River" by Pilgrim and the "Indobrahm" by Pascoe, from the combined discharge of the Brahmaputra, Ganges and Indus which it carried at one time. This old river is believed to be the successor of the narrow strip of the sea—the remnant of the Himalayan sea left after the main uplift of those mountains—as the latter gradually withdrew, through the encroachment of the delta of the replacing river, from Naini Tal, Solon, Muzaffarabad and Attock to Sind. The Nummulitic limestone deposits of these localities testify to the extent and boundary of the Eocene gulf. The final extinction of this gulf, which once stretched from Assam to Sind, left behind it a wide river-basin in which were laid down the thick series of Murree and Siwalik deposits during the interval between the early Miocene and the end of the Pliocene. Post-Siwalik movements in the N.W. Punjab brought about a dismemberment of this river-system, which hitherto had flowed from the head-waters in Assam, through the whole breadth of India, to the Potwar and thence to the receding head of the Sind Gulf, into three subsidiary systems: (1) the present Indus from north-west Hazara; (2) the five Punjab tributary

---

[1] E. H. Pascoe, The Indobrahm, *Quart. Journ. Geol. Soc.*, vol. lxxv. pp. 138–155 (1919); G. E. Pilgrim, The Siwalik River, *Journ. Asiat. Soc. Beng.*, vol. xv. (New Series), pp. 81–99 (1919); M. S. Krishnan and N. K. Aiyengar, Did the Indobrahm River exist? *Rec. G.S.I.* vol. lxxv. paper 6, 1940.

rivers of the Indus; (3) the rivers belonging to the Ganges system which finally took a south-easterly course.

The elevation of the Potwar into a plateau converted the north-west section of the main river into a separate independent drainage basin, with the Sutlej as its most easterly tributary. Hitherto the main river had travelled to its confluence with the Indus along a track which was a north-western prolongation of the present course of the Jumna, thence *via* the present bed of the Soan to the Indus. After these elevatory movements and separation of the north-west section, the remaining upper portion of the main channel was subjected to a process of reversal of flow, its water being forced back by the Punjab elevations to seek an outlet into the Bay of Bengal along the now aggraded, more or less levelled sub-montane plains. In this reversal of the old drainage Pascoe assigns the chief share to a process of river-capture by head-erosion of the tributaries. The competence of the agency of river-capture alone in accomplishing this far-reaching change is debatable and differential earth-movement as the chief contributory cause is suggested, aided by the recently levelled and uniformly graded drainage-lines on the surface of these wide plains.

The severed upper part of the Siwalik River became the modern Ganges, it having in course of time captured the transversely running Jumna and converted it into its own affluent. The transverse Himalayan rivers, *e.g.* the Alaknanda, Karnali, Gandak and Kosi, which are really among the oldest water-courses of North India, continued to discharge their waters into this new river, irrespective of its ultimate destination, whether it was the Arabian Sea or the Bay of Bengal. During sub-Recent times some interchange took place between the easterly affluents of the Indus and the westerly tributaries of the Jumna by minor shiftings of the watershed, now to one side, now to the other. There are both physical and historical grounds for the belief that the Jumna during early historic times discharged into the Indus system, through the now neglected bed of the Saraswati river of Hindu traditions, its present course to Prayag being of late acquisition.

The Punjab portions of the present Jhelum, Chenab, Ravi, Beas and Sutlej have originated after the uplift of the topmost stage of the Siwalik system and subsequent to the severance of the Indus from the Ganges. The Potwar Plateau-building movements could not but have rejuvenated the small rivulets of the southern Punjab, which until then were discharging into the lower Indus. The vigorous head-erosion resulting from this impetus enabled them to capture, one bit after the other, that portion of the Siwalik River which crossed the Potwar on its westerly course to the Indus. Ultimately, the head-waters joining up with the youthful torrents descending from the mountains, these rivers grew much in volume and formed these five important rivers of the province, having their sources in the snows of the Great Himalaya Range and deriving their waters from as far east as the Manasarowar lake on the Kailas Range. The western portion of the broad but now deserted channel of the main river, after these mutilating operations, has been occupied to-day by the puny, insignificant stream of the Soan, a river out of all harmony with its great basin and the enormous extent of the fluviatile deposits with which it is choked.

The Himalayas are undergoing a very active phase of subaerial erosion; being a zone of recent folding and fracture, their disintegration is proceeding at a more rapid rate than is the case with older earth-features of greater geological stability. The plains of India and the Ganges delta are a fair measure of the amount of matter worn down from a section of the Himalayas since the Pliocene period. Landslips, soil-creep, breaking off of enormous blocks from the mountain-tops, are phenomena familiar to visitors to these mountains. The denudation in the dense forests of the hill-slopes in the Eastern Himalayas recalls that of the tropical lands in its intensity and character.

REFERENCES

H. B. Medlicott and W. T. Blanford, *Manual of the Geology of India*, vol. i., 1887, Introduction.

Sir S. Burrard and A. M. Heron, *The Geography and Geology of the Himalaya Mountains*, Second Edition, 1934.

Records of the Geological Survey of India, vol. xxxv. pts. 3 and 4; vol. xl. Pt. 1; vol. xliv. pt. 4; vol. lxiii. pt. 2, 1930, Glaciers of the Himalayas.

Physical Atlas of Asia, W. & A. K. Johnston.

The Bathy-Orographical Map of India, W. & A. K. Johnston.

W. H. Hobbs, *Earth-Features and their Meaning*, 1912 (Macmillan).

Mt. Everest Expeditions: Publications by, 1921-33.

Sven Hedin, *Southern Tibet*, vols. i–ix., Stockholm, 1917.

G. Dainelli, *Italian Expedition to the Himalaya and Karakoram* (1913-14), vols. i–xiii., Bologna, 1923-35.

R.D. Oldham, The Evolution of Indian Geography, *Geographical Journal*, London, March, 1894; Support of the Mountains of Central Asia, *Rec. G.S.I.*, vol. xlix., 1918.

J. D. Wiseman and R. B. Sewell, The Floor of the Arabian Sea, *Geol. Mag.* vol. 74 and 75, 1937 and 1938.—The John Murray Expedition, *Scientific Reports*, vol. i., 1936.

A. Cunningham, Ancient Geography of India, London, 1871.

R. B. Sewell, Geographic and Oceanographic Researches in Indian Waters, *Mem. As. Soc. Beng.*, vol. ix., 1935.

O. H. K. Spate, *India and Pakistan, Regional Geography* (Methuen), London, 1953.

Vening, Meinesz, *Gravity Expeditions at Sea*, 1923-32, vol. ii.

F. Drew, Jammu and Kashmir Territories, E. Stanford, London, 1875.

J. W. Gregory, *Structure of Asia*, London, 1924.

E. A. Glennie, *Gravity Anomalis and The Figure of the Earth*, Surv. of Ind., Prof. Pap., 1940.

H. J. Couchman, *Progress of Geodesy in India*, Nat. Inst. Sc. Ind., 1937.

CHAPTER II

Stratigraphy of India—Introductory

Correlation of Indian formations with those of the world—An outstanding difficulty in the study of the geology of India is the difficulty of correlating accurately the various Indian systems and series of rocks with the different divisions of the European stratigraphical scale, which is accepted as the standard for the world. The difficulty becomes much greater when there is a total absence of any kind of fossil evidence, as in the enormous rock-systems of the Peninsula or in the outer zone of the Himalayas, in which case the determination of the geological horizon is left to the more or less arbitrary and unreliable tests of lithological composition, structure, and the degree of metamorphism acquired by the rocks. These tests are admittedly unsatisfactory, but they are the only ones available for fixing the homology of the vast pre-Cambrian formations of the Peninsula, which form such an important feature of the pre-Palaeozoic geology of India.

The basis of stratigraphy is the determination of the natural order of superposition of strata; until the exact original succession of deposits in a stratified series is ascertained no correlation of strata at different localities is possible. It is the function of stratigraphy to discover and arrange the sedimentary deposits of the earth's crust in the order of their age, so that each originally older bed is lower in position than the next newer one. Apart from the complications introduced by folding and faulting, which makes the application of the principle of superposition difficult, there is the difficulty arising from frequent lateral variations of sedimentary strata. A sandstone or limestone lying between two shale beds may thicken or thin out until the whole series becomes a group of sandstones or limestones or shales.

The discovery by William Smith, at the end of the eighteenth century, that groups of strata are characterised by the preservation in them of particular fossil organisms and can be identified by them, laid the foundation of historical geology. In establishing correlations of formations in distant areas the following criteria are employed:

1. The order of superposition.
2. Fossil organisms.
3. Lithological characters.

4. Stratigraphical continuity.
5. Unconformities.
6. Degree of metamorphism.
7. Tectonic and structural disturbance.

There is no question, of course, of establishing any absolute contemporaneity between the rock-systems of India and those of Europe, because neither lithological correspondence nor even identity of fossils is proof of the synchronous origin of two rock-areas so far apart. Biological facts prove that the evolution of life has not progressed uniformly or in a simple straightforward direction all over the globe in the past, but that in different geographical provinces the succession of life-forms has been marked by widely varying rates of evolution due to physical differences existing between them, and that the process of distribution of species from the centre of their origin is very slow and variable. The idea, therefore, of contemporaneity is not to be entertained in geological deposits of two distant areas, even when there is a perfect similarity in their fossil contents.

What is essential is that the rock-records of India, discovered in the various parts of the country, should be arranged in the order of their superposition, *i.e.* in a chronological sequence. They should be classified with the help of local breaks in their sequence, or by the evidence of their organic remains, and named according to some local terminology. The different outcrops should then be correlated among themselves. The last and the most important step is to correlate these, on the evidence of their contained fossils, of failing that, on lithological grounds, to some equivalent division or divisions of the standard scale of stratigraphy worked out from the fossiliferous rock-records of the world.

In illustration of the above it may be remarked that the Carboniferous system of Europe is characterised by the presence of certain types of fossils and by the absence of others. If in any part of India a series of strata is found, containing a suite of organisms in which many of the *genera* and a few of the *species* can be recognised as identical with the above, then the series of strata thus marked off is correlated with the Carboniferous system of Europe, though on account of local peculiarities and variations, the system is often designated by a local name. It is not of much significance whether they were or were not deposited simultaneously, so long as they point to the same epoch in the history of life upon the globe; and since the history of the development of life upon the earth, in other words, the order of appearance of the successive life-forms, has been proved to be broadly uniform in all parts of the earth, there is some unity between these two rock-groups. As a substitute for geological synchronism Prof. Huxley introduced the term *Homotaxis*, meaning "Similarity of arrangement", and implying a corresponding position in the geological series.

The different "facies" of Indian formations—It often happens that one and the same geological formation in the different districts is composed of different types of deposits, *e.g.* in one district it is composed wholly of massive limestones, and in another of clays and sandstones. These divergent types of deposits are spoken of as belonging to different *facies*, *e.g.* a calcareous facies, argillaceous facies, arenaceous facies. etc. There may also be different facies of fauna, just as much as facies of rock-deposits and the facies is then distinguished after the chief element or character of its fauna, *e.g.* coralline facies, littoral facies, etc. Such is often the case with the rock-formations of India. From the vastness of its area and the prevalence of different physical conditions at the various centres of sedimentation, rocks of the same system or age are represented by two or more widely different facies, one coastal, another deep-water, a third terrestrial, and sometimes even a fourth, volcanic. The most conspicuous example of this is the Gondwana system of the Peninsula and its homotaxial equivalents. The former is an immense system of fresh-water and subaerially deposited rocks, ranging in age from Upper Carboniferous to Upper Jurassic, whose fossils are ferns and conifers, fishes and reptiles. Rocks of the same age, in the Himalayas, are marine limestones and calcareous shales of great thickness, and containing deep-sea organisms like *Lamellibranchs*, *Cephalopods*, *Crinoids*, etc., from the testimony of which they are grouped into Upper Carboniferous, Permian, Triassic and Jurassic systems. In the Salt-Range these same systems often exhibit a coastal facies of deposits like clays and sandstones, with littoral organisms, alternating with limestones.

In this connection it must be clearly recognised how these deposits, which are homotaxial and more or less the time-equivalents of one another, should come to differ in their fossil contents. The reason is obvious. For not only are marine organisms widely different from land animals and plants, but the littoral species that inhabit the sandy or muddy bottoms of the coasts are different from those pelagic and abyssal organisms that find a congenial habitat in the clearer waters of the sea and at great distances from land. Again, the animal life of the seas of the past ages was not uniform, but it was distributed according to much the same laws as those that govern the distribution of the marine biological provinces of to-day. The fossils entombed in some formations are of markedly local or provincial affinities. Provincialisation of faunas arises from various causes—the dependence of organisms on their environments, their isolation, or from relative preponderance or absence of competing species, or from physical barriers to migration of species. Pelagic, or free-swimming, members of a fauna attain a wider horizontal or geographical distribution than bottom-living forms. It thus arises that the fossils present in a series of deposits are not a *function* only of the period when the deposits were laid down, but, as Lyell says, are a "function of three variables", *viz.* (1) the geological period at which the rocks were formed, (2) the zoological or botanical provinces in which the locality was situated, and (3) the physical conditions prevalent at the time, *e.g.* depth, salinity and muddiness of water, temperature, character of the sea-bottom, currents, etc.

A new aid to stratigraphy that has come into vogue may be just mentioned. The discovery that uranium and thorium break up into other elements through atomic disintegration, producing as a final residuum lead, the change taking place at a definite and measurable rate, has placed in the hands of the geologist a new weapon for the determination of the age of the great azoic pre-Cambrian systems.

In recent years the method of age determination of rocks by investigation of lead ratio and helium-ratio in radio-active minerals has been greatly reinforced by new methods based on the radio-active isotopes of K^{40} and Rb^{87} (the potassium-argon and rubidium-strontium ratio methods). The scale of ages of the main geological systems now arrived at from these data is :

	Million years		Million years
Third glaciation	0.3	Base of Cretaceous	135
Pleistocene	1.0	Base of Jurassic	180
Base of Eocene	70	Base of Trisassic	225
Base of Permian	270	Base of Silurian	440
Base of Carboniferous	350	Base of Ordovician	500
Base of Devonian	400	Base of Cambrian	600

(From Arthur Holmes, 1959)

The chief geological provinces of India—Geographically as well as geologically India is one single well-defined unit. Though a peninsula of the Eurasian continent, it is structurally marked off from it and has ever since the Upper Carboniferous period pursued its own geological evolution as a separate entity. In the geological study of any part of this sub-continent, therefore, this aspect should be constantly kept in view, for political divisions and boundaries have no significance in the structure and formational lines of a land-mass. For this reason the region of India is dealt with as a geographical and not a political unit and includes Pakistan, Burma and Sri Lanka.

The following are parts of the Indian region which contain one or the other section of the geological record in some degree of fullness. These isolated fragmentary records from different areas when pieced together compose the geological history of India; each area, therefore, needs careful study.

1. The Salt-Range (Pakistan).

This range of mountains is a widely explored region of India. It was one of the earliest parts of India to attract the notice of the geologists, both on account of its easily accessible position as well as for the conspicuous manner in which most of the geological systems are displayed in its precipices and defiles. Over and above its stratigraphic and palaeontological results, the Salt-Range illustrates a number of phenomena of dynamical and tectonic geology.

2. The Himalayas.

As mentioned in the first chapter, a broad zone of sedimentary strata lies to the north of the Himalayas, behind its central axis, occupying a large part of Tibet. This is known as the Tibetan zone of the Himalayas. This zone of marine sediments contains one of the most perfect developments of the geological record seen in the world, comprising in it all the periods of earth-history from the Cambrian to the Eocene. It is almost certain that this belt of sediments extends the whole length of the Himalayan chain, from Hazara and Kashmir to the furthest eastern extremity; but so far only two portions of it have been surveyed in some detail, the one the north-west portion—the Kashmir Himalayas—and the other the mountains of the central Himalayas of the Simla region, especially the Spiti valley, and the northern parts of Kumaon and Garhwal.

(i) North-West Himalayas.

This area includes Hazara, Kashmir, the Pir Panjal, and the ranges of the inner Himalayas. A very complete sequence of marine Palaeozoic and Mesozoic rocks is met with in the inner zone of the mountains, while a complete sequence of Tertiary development is seen in the outer, Jammu hills. The Kashmir basin, lying between the Zanskar and the Panjal ranges, contains the most fully developed Palaeozoic system seen in any part of India. For this reason, and because of the easily accessible nature of the formations to parties of students, in a country which climatically forms one of the best parts of India, the geology of Kashmir is treated in some detail.

(ii) Central Himalayas.

Many eminent explorers have unravelled the geology of these mountains since the early thirties of the last century, and parts of this region, such as Spiti, form the classic ground of Indian geology.

The central Himalayas include the Simla hills, Spiti, Kumaon and Garhwal provinces. The great plateau of Tibet ends in the northern parts of these areas in a series of gigantic south-facing escarpments, wherein the stratigraphy of the northern or Tibetan zone of the Himalyas, referred to above, is typically displayed. The Spiti basin is the best known for its fossil wealth as well as for the completeness of the stratigraphic succession from the Cambrian to Cretaceous. The systems of Kashmir are on a north-west continuation of the strike of the Spiti basin. Much detailed work has been done of late years in the Simla-Chakrata area.

3. Sind (Pakistan).

Sind possesses a highly fossiliferous marine Cretaceous and Tertiary record. The hills of the Sind-Baluchistan frontier contain the best-developed Tertiary sequence, which is recognised as a type for the rest of India.

4. Rajasthan.

Besides the development of a very full sedimentary record, divided into three pre-Palaeozoic systems of Archaean-Dharwar age and an interesting facies of the Vindhyan system in the Aravalli, range, Western Rajasthan contains a few isolated outcrops of marine Mesozoic and early Tertiary strata underneath the Pleistocene desert sand, which has concealed by far the greater part of the solid geology.

5. Burma and Baluchistan.

These two countries, at either extremity of the extra-Peninsular area, contain a large section of the stratified marine geological record which helps to fill up the gaps in the Indian sequence. Many of these formations are again highly fossiliferous, and afford good ground for comparison with their Indian congeners. Within the geographical term "India" is now included all these regions which are regarded as its natural physical extensions on its two borders—Pakistan, Afghanistan and Baluchistan on the west and Burma on the east. The student of Indian geology is therefore expected to know of the principal rock-formations of Baluchistan and Burma.

6. Coastal System of India.

Along the eastern coast of the Peninsula and to a less extent on the Mekran coast, there is a strip of marine sediment of Mesozoic, Tertiary or Quaternary ages, in more or less connected patches—the records of several successive "marine transgressions" on the coasts.

7. Peninsular India.

As must be clear from what we have seen regarding its physical history in the first chapter, the Deccan peninsula is a part of India which contains a remarkably full Archaean and pre-Cambrian sequence and a most imperfectly developed post-Cambrian geological record. The Palaeozoic group is unrepresented but for the fluviatile Permian formations; the Mesozoic era has a fairly full record, but except as regards the Cretaceous it is preponderatingly made up of fluviatile, terrestrial and volcanic accumulations; while the Tertiary is almost unrepresented except by the coastal Tertiaries and the partly Eocene lavas forming the Deccan Traps.

The student of Indian geology should first familiarise himself with the representatives of the various geological systems that are found in these provinces of India and correlated to the principal divisions of the European sequence.

The idea of a geological system is not confined to a summary of facts regarding its rocks and fossils. These are the dry bones of the science; they must be clothed with flesh and blood, by comparing the processes and actions which prevailed when they were formed with those which are taking place before our eyes in the world of to-day. A sand-grain or a pebble of the rocks is not a mere particle of inanimate matter, but is a *word* or a *phrase* in the history of the earth, and has much to tell of a long chain of natural operations which were concerned in its formation. Similarly, a fossil shell is not a mere chance relic of an animal that once lived, but a valuable *document* whose preservation is to be reckoned an important event in the history of the earth. That mollusc to which the shell belonged was the heir to a long line of ancestors and itself was the progenitor of a long line of descendants. Its fossil shell marks a definite stage in the evolution of life on earth that was reached at the time of its existence, which definite period of time it has helped to register. Often it tells much more than this, of the geography and climate of the epoch, of its contemporaries and its rival species. In this way, by a judicious use of the imagination, is the bare skeleton given a form and clothed; the geological records then cease to be an unintelligible mass of facts, a burden to memory, and become a living story of the various stages of the earth's evolution.

In reading stratigraphical geology the student should remind himself to take note of the illustrations of the principles of dynamical and tectonic gelogy, of which every page of historical geology is full. Many of the facts of dynamical and structural geology find a pertinent illustration in the part they play in the structure or history of a particular country or district. The problems of crust-deformations, of vulcanicity, of the variations, migrations and extinctions of life-forms with the passage of time, and a host of other minor questions that are inscribed in the pages of the rock-register, must be thought over and interpreted with the clue that modern agencies in the earth's dynamics furnish.

The following table gives the standard STAGES in which the world's stratified GEOLOGICAL RECORD is divided.

RECENT	Present-day alluvium, etc.
PLEISTOCENE	{ Younger alluvium { Older alluvium
PLIOCENE	{ Villefranchian { Astian { Plaisancian
MIOCENE	{ Pontian { Sarmatian { Tortonian } Vindobonian { Helvetian } { Burdigalian { Aquitanian
OLIGOCENE	{ Chattian { Rupelian (Stampian) { Lattorfian (Sannoisian)
EOCENE	{ Ludian } Priabonian { Bartonian } { Auversian { Lutetian (Parisian) { Ypresian (Cuisian) { Sparnacian { Thanetian { Montian
CRETACEOUS	{ Danian { Maestrichtian { Campanian { Senonian { Turonian { Cenomanian { Albian { Aptian { Barremian } { Hauterivian } Neocomian { Valanginian }

STRATIGRAPHY OF INDIA

JURASSIC
- OOLITE
  - Tithonian (Purbeck)
  - Portlandian
  - Kimmeridgian
  - Sequanian (Lusitanian)
  - Oxfordian
  - Callovian
  - Bathonian
  - Bajocian
  - Aalenian
- LIAS
  - Toarcian
  - Charmouthian
  - Sinemurian
  - Hettangian

TRIASSIC
- Rhaetic
- Noric
- Carnic
- Ladinic
- Anisic
- Scythian

PERMIAN
- Thuringian (Zechsrein)
- Saxonian (Punjabian)
- Artinskian

CARBONIFEROUS
- Stephanian (Uralian)
- Westphalian ⎫
- Namurian   ⎭ (Moscovian)
- Dinantian (Avonian)

DEVONIAN
- Famennian
- Frasnian
- Givetian
- Eifelian
- Coblenzian
- Gedinnian

SILURIAN
- Downtonian
- Clunian (Ludlow)
- Salopian (Wenlock)
- Valentian (Llandovery)

ORDOVICIAN
- Ashgillian ⎫
- Caradocian ⎬ Bala
- Llandeilian ⎭
- Llanvirnian
- Arenig

CAMBRIAN - - $\begin{cases} \text{Shumardia Series (Tremadocian)} \\ \text{Olenus Series} \\ \text{Paradoxides Series} \\ \text{Olenellus Series} \end{cases}$

REFERENCES

Sir T. H. Holland, *Imperial Gazetteer of India*, vol. i. chapter ii., 1907.

Marr, *Principles of Stratigraphic Geology* (C.U. Press).

Holland and Tipper, *Mem. G.S.I.* vol. xliii. pt. 1, Indian Geological Terminology, 1913 and (Second Edition) vol. li. pt. 1, 1926.

A. W. Grabau, *Principles of Stratigraphy* (Seiler), 1924.

W. H. Twenhofel, *Treatise on Sedimentation* (Williams & Wilkins), 1926.

Lexicon of International Stratigraphy, vol. iii., Asia. *Internat. Geol. Congress*, Paris, 1956.

TABLE OF THE GEOLOGICAL FORMATIONS OF INDIA.

	Peninsula.	Himalayas.	Salt-Range.	Other Areas.	Age.
Aryan Group	Newer alluvium of the deltas; newer raised-beaches; coral banks.	Modern river-deposits. Dry deltas, fans, etc.	Blown sand, loess; travertine, etc.	Newer alluvium—*Khadar* of the Indus and the Ganges.	Recent.
	Cave-deposits of Kurnool. Older alluvium of the Narmada, Godavari, etc.; *Palaeolithic gravels*; low-level *laterite*, *Porbander* sandstone; raised-beaches; sand dunes; loess; desert sands of Rajasthan and Kutch. **Upper Cuddalore** sandstone.	**Ice Age.** Glacial moraines; perched blocks, etc.; *Upper Karewas* of Kashmir; old high-level alluvia of the Sutlej, etc. River-terraces.	Loess deposits. Travertine masses.	The Indo-Gangetic Alluvium—*Bhangar*, Loess of Baluchistan.	Pleistocene.
	Laterite (high-level) of the Peninsula.	**Siwalik System.** *Upper Siwalik.*	**Siwalik System.** *Upper Siwalik.*	Sind and Baluchistan. *Dibing Series.* **Manchar System** of Sind.	Pliocene.
	Ossiferous conglomerate of Perim island.	*Middle Siwalik.*	*Middle Siwalik.*	Assam. *Dupi Tila Series.* **Mekran System** of Baluchistan.	Upper Miocene.
	Miocene of Puri and Baripada. **Cuddalore Series** of the east coast (part); Tertiary of Quilon.	*Lower Siwalik* or *Nahan*	*Lower Siwalik.*	*Tipam Series.* **Gaj Series** of Sind.	Middle Miocene.
	Gaj Series of Kutch.	*Murree* Series of Punjab Himalayas; *Kasauli* and *Dagshai* Series of Simla Himalayas. *Fatehjang* beds.	Upper Murree (in eastern part).	*Surma Series.* *Flysch of Baluchistan.* *Bugti* beds of Baluchistan.	**Irrawaddy System.** **Pegu System.** *Upper Pegu.* Lower Miocene.
	Nari Series of Kutch; *Dwarka* beds of Saurashtra.	Intrusive granites, etc., in the core of the Himalayas.		**Nari Series** of Sind.	*Lower Pegu.* Oligocene.

TABLE OF THE GEOLOGICAL FORMATIONS OF INDIA—*Continued.*

PENINSULA.	HIMALAYAS.	SALT-RANGE.	OTHER AREAS.	AGE.
Nummulitics of Surat and Broach; Lower Tertiaries of Kutch; nummulitic limestone of Rajasthan and Saurashtra.	*Subathu* and *Chharat* Series of Outer Himalayas; Lower Tertiaries of the Inner Himalayas, Indus Valley and Hundes.	Nummulitic limestone, with shale and coal.	*Barail* Series of Assam. **Kirthar Series** (Nummulitic limestone) of Assam (*Jaintia* Series), Burma, Sind and Baluchistan.	*Upper Eocene.*
Laki Series of Bikaner. Eocene of Pondicherry.	Coal-bearing beds of Jammu. *Laki* Series of Kohat.	*Laki* Series.	**Laki Series** of Sind and Baluchistan.	
Ranikot Series	Hill Limestone of Hazara and Punjab. *Ranikot* Series.	*Ranikot* Series.	**Ranikot Series** of Sind. Paunggyi conglomerate of Burma.	*Lower Eocene.*
Deccan Trap.	**Cretaceous of Central Himalayas.** Plutonic and volcanic rocks of Astor and Dras. Flysch of C. Himalayas.	Cretaceous of Trans-Indus Salt-Range—Chichali.	*Disang* Series of Assam. Intrusive granite, gabbros, and serpentine of Baluchistan and Burma. *Cardita beaumonti* beds.	*Danian.*
South-East Coast Cretaceous. Niniyur Stage. Ariyalur Stage. *Lameta* Series. Trichinopoly Stage. *Bagh* beds. Utatur Stage.	*Chikkim* Series of North Himalayas.	Gault of Kala Chitta.	*Pab* sandstone of Sind and Baluchistan.	*Cenomanian.* *Wealden.*

ARYAN GROUP (*contd.*)

GEOLOGICAL FORMATIONS OF INDIA

PENINSULA.	HIMALAYAS.	SALT-RANGE.	OTHER AREAS.	AGE.
Himatnagar sandstone.	Giumal sandstone of Spiti and Hazara. Volcanic series and Orbitolina limestone of Burzil and Astor.		Cretaceous of Assam and Burma. Parb limestone of Baluchistan.	
Marine Umia beds. **Upper Gondwana System.** Umia Series.	**Jurassic of the Himalayas.** Spiti shales. Kioto lime- { Tagling Stage. stone { Para Stage. (Megalodon limestone.) Tal Series of Garhwal. Oolite of Hazara. Jurassic of Banihal	Upper and Middle Jurassic of the Trans-Indus Salt-Range. Oolite of Kala Chitta.	Massive limestone of Baluchistan (Oolite). Black crinoidal limestone of Baluchistan (Lias). Saighan Series of Afghanistan. Namyau beds of Northern Shan States. Tabbowa beds of Sri Lanka. Loi-am coal-beds.	Jurassic.
Marine Jurassic of Kutch. Katrol Series. Chari Series. Patcham Series.				
Rajmahal Series. Jabalpur Series. Kota Series.				
Marine beds in the East Coast Gondwanas.				
Middle Gondwana System. Maleri Series. Parsora Stage.	**Trias of the Himalayas.** Upper Trias.	**Trias of Salt-Range.** Trias of Kala Chitta.	Upper Trias of Baluchistan and Napeng beds of Burma.	Triassic.
Mahadev Series.		Middle Trias.		
Panchet Series.	Middle Trias. KROLS Lower Trias (Otoceras zone).	Lower Trias (Ceratite beds).		

ARYAN GROUP (contd).

TABLE OF THE GEOLOGICAL FORMATIONS OF INDIA—*Continued.*

	PENINSULA.	HIMALAYAS.	SALT-RANGE.	OTHER AREAS.	AGE.
ARYAN GROUP (*contd.*).	**Lower Gondwana System.** Damuda Series. { *Raniganj* Stage. *Ironstone* shales stage. *Barakar* Stage. *Productus* bed of Umaria. } Talchir Series. { *Karharbari* Stage. *Talchir* Stage. (Talchir boulder-bed.) }	**Permian of the Himalayas.** *Zewan* Series of Kashmir and the *Productus* shales of Spiti and the central Himalayas. *Sirban* limestone and *Krol* Series of Kashmir & Simla. Gondwanas of Himalayas. *Gangamopteris* beds of Kashmir. Panjal Volcanics. Infra-Krol of Gharwal. *Blaini* and *Tanaḳki* boulder-beds.	**Productus limestone.** *Productus* limestone. Speckled sandstone. Conularia beds. Boulder-bed.	*Plateau* limestone— Upper part of the Northern Shan States. *Subansiri* beds of Assam	Permian. Permo-Carboniferous. Upper Carboniferous.
DRAVIDIAN GROUP		**Carboniferous of Spiti and Kashmir.** *Po* Series of Spiti and *Fenestella* shales of Kashmir. *Lipak* Series of Spiti. *Syringothyris* limestone of Kashmir.		*Plateau* limestone— Middle part, of the Northern Shan States. *Fusulina* limestone of Baluchistan. *Plateau* limestone— Lower part of the Northern Shan States. *Sariko* Series of Chitral and Pamir.	Middle Carboniferous. Lower Carboniferous.

Peninsula.	Himalayas.	Salt-Range.	Other Areas.	Age
	Muth Series of Spiti and Kashmir. Devonian of Chitral. *Jaunsar* and *Lower Tanawal* Series.		**Devonian of Burma.** Crystalline limestones of Padaukpin. *Wetwin* shales.	*Devonian.*
	Silurian of Spiti and Kashmir. Fossiliferous Silurian beds of Spiti and Kashmir.		*Zebingyi* Series. *Namshim sandstone.*	*Silurian.*
	Ordovician of Spiti and Kashmir. Fossiliferous Ordovician beds in Central Himalayas and in Kashmir.		*Naung Kangyi* Series of Northern Shan States.	*Ordovician.*
? Upper *Vindhyans* of **Central India.**	**Cambrian of Spiti and Kashmir.** *Haimanta* System of Central Himalayas. Cambrian of N.W. Kashmir.	**Cambrian of Salt-Range** *Salt-pseudomorph* shales. *Magnesian* sandstone. *Neobolus* beds. Purple sandstone. Salt marl and gypsum.	*Chaung Magyi* Series of Northern Shan States.	*Cambrian.*

Dravidian Group (*Contd.*).

TABLE OF THE GEOLOGICAL FORMATIONS OF INDIA—Continued.

	PENINSULA.	HIMALAYAS.	SALT-RANGE	OTHER AREAS.	AGE
PURANA GROUP	**Vindhyan System.** Upper { *Bhander Series.* *Rawab Series.* *Kaimur Series.* Lower { *Semri Series: Karnool Series; Malani rhyolites and Jalor and Sirwana granites.* **Cuddapah System: Delhi System.** Upper Cuddapah Lower Cuddapah : *Raialo Series.*	*Attock* slates of Peshawar and Hazara. *Dogra* slate of Kashmir. *Baxa* Series of Eastern Himalayas. *Simla slate and Deoban* Series of central Himalayas.		*Miju* Series of Assam.	Torridonian. Algonkian.
ARCHAEAN GROUP GNEISSES AND GRANITES	**Archaean.** **Dharwar System : Aravalli System** *Iron-ore Series. Gondite Series.* *Sausar Series. Kodurite Series.* { *Charnockite Series of S. India* *Bundelkand gneiss.* *Bengal gneiss and schistose gneisses of the Peninsula.*	*Vaikrita* Series of Spiti; *Jutogh* and *Salkhala* Series of Punjab Himalayas ; *Daling* Series of Eastern Himalayas. "Central gneiss" in part. Basement gneiss, granulite and schists.		*Shillong* Series. Crystalline limestones, etc. of Burma. *Mergui* Series of Burma. Archaean fundamental gneiss and intrusive granites of Burma, Assam, Sri Lanka, Baluchistan, etc.	Huronian. Lewisian.	Archaean.

TABLE OF THE GEOLOGICAL FORMATIONS OF KASHMIR—SIMLA

Kashmir.	Hazara.	Spiti-Simla-Garhwal.	Age.
Sub-Recent—river-terraces, low-level alluvia of Jhelum and Indus; pebble-beds. Recent moraines. Upper Karewas—later moraines. Upper Siwalik ⎫ Middle Siwalik ⎬ 4,880 m. ancient moraines. Lower Karewas. Lower Siwalik ⎭	Sub-Recent—fluviatile, lacustrine and glacial alluvia.	Sub-Recent—alluvial terraces of the Sutlej.	Pleistocene.
		Glacial moraines. Hundes alhuvium.	Pliocene.
	Siwalik series.	Siwaliks of the foot-hills.	Miocene.
Murree series > 2,100 m.	Murree series.	Kasauli series. Dagshai series.	? Oligocene.
Eocene—Nummulitics of Pir Panjal and Outer hills; Volcanics of Dras, Ladakh and ? Burzil.	Nummulitics { Laki series. Ranikot series. } 600 m.	Nummulitics of Hundes. Subathu of the Outer Himalayas. Flysch series.	Eocene.
Chikkim series—*Orbitolina* limestone and Volcanics of Burzil.	Chikkim series, 3 m.	Chikkim series.	Cretaceous.
Spiti shales.	Giumal series, 90 m. Spiti shales, 30 m.	Giumal series. Spiti shales. Tagling limestone.	Jurassic.
Megalodon (Kioto) limestone. Jurassic of Banihal and ? Baltal.	Lower Jurassic (Oolite).	Para limestone.	
Trias { Upper, 1,525 m. Middle, 360 m. Lower, 90 m. }	Upper Trias, 300 m.	Upper Trias. Middle Trias. Lower Trias.	Triassic.
Zewan series, 240 m. Lower Gondwanas—	Panjal Trap > 1,500 m.	Tal series.	Permian.

Aryan Era · Panjal Trap.

TABLE OF THE GEOLOGICAL FORMATIONS OF KASHMIR–SIMLA—Continued.

	Kashmir.	Hazara.	Spiti-Simla-Garhwal.	Age.
Aryan Era	*Gangamopteris* beds, 240 m. Upper Tanawals of Pir Panjal (many thousand m.) Agglomeratic slates. ? Sirban limestone of Uri and Riasi > 450 m. Agglomeratic slates, 1,525 m. Fenestella series > 600 m.	Agglomeratic Slate. Panjal Trap. Sirban limestone 600 m. Speckled sandstones. Tanakki boulder-beds. Upper Tanawal (Lr. Gondwana)	Productus shales series. Productus conglomerate. Blaini boulder-beds / Infra-Krol. Po series. / Krol series / Infra-Krol.	Upper Carboniferous. Middle Carboniferous. Lower Carboniferous.
Dravidian	Syringothyris limestone, 900 m. Muth quartzite, 900 m. Silurian of Lidar and Sind. Ordovician of Hundawar and Lidar. Upper Cambrian of Hundawar, 1,525 m. Lower Cambrian of Shamsh Abari, 900 m.	Lower Tanawal.	Lipak series. Muth quartzite. Silurian. Ordovician. Haimanta system. Deoban series. / Jaunsar series.	Devonian. Silurian. Ordovician. Cambrian. ? Lower Cambrian.
Purana	Dogra slate, 1,525 m. passing into Lr. Cambrian	Hazara and Attock slate.	Simla slate.	
Archaean Pre-Cambrian	Salkhala series (many thousand metres). Gneisses and granulites with intrusive granite and basic plutonics.	Salkhala series. Ortho- and para-gneisses and schists with acid and basic intrusives.	Jutogh and Chail series. Vaikrita series. Gneisses and schists with intrusives. ? Chor granite.	Algonkian. Archaean.

CHAPTER III

The Archaean System—Gneisses and Schists

Introduction—The oldest rocks of the earth's crust that have been found at the bottom of the stratified deposits, in all countries of the world, exhibit similar characters regarding their structure as well as their composition. They form the core of all the great mountain-chains of the world and the foundations of all its great ancient plateaus. They are all *azoic*, thoroughly crystalline, extremely contorted and faulted, are largely intruded by plutonic intrusions, and generally have a well-defined foliated structure. These conditions have imparted to the Archaean rocks such an extreme complexity of characters and relations that the system is often known by the names of the "Fundamental Complex", the "Basement Complex", etc. (Fig. 4.)

The way in which the Archaean crystalline rocks have originated is not well understood yet, and various modes of formation have been ascribed to these rocks. (1) Some are believed to represent, in part at least, the first-formed crust of the earth by the consolidation of the gaseous or molten planet. (2) Some are believed to be the earliest sediments formed under conditions of the atmosphere and of the oceans in many respects different from those existing at later dates, and afterwards subjected to an extreme degree of thermal and regional metamorphism. (3) Some are thought to be the result of the bodily deformation or metamorphism of large plutonic igneous masses under great earth-movements or stresses. (4) Some are believed to be the result of the consolidation of an original heterogeneous magma erupted successively in the crust (*cf*. the banded granites and gabbros).

Distribution—The crystalline metamorphosed sediments and gneissic rocks of the Archaean system form an enormous extent of the surface of India. By far the largest part of the Peninsula, the central and southern, is occupied by this ancient crystalline complex. To the north-east they occupy wide areas in Orissa, Assam, Central Provinces (now Madhya Pradesh) and Chota Nagpur. Towards the north the same rocks are exposed in an extensive outcrop covering the whole of Bundelkhand; while to the north-west they found in a number of isolated outcrops, extending from north of Baroda to a long distance in the Aravallis and Rajasthan (Dharwar System).

In the extra-Peninsula, gneisses and crystalline rocks are again exposed along the whole length of the Himalayas, forming the

(about 1/15 natural size)

FIG. 4—Diagram showing contortion in the Archaean gneiss of Bangalore.

bulk of the high ranges and the backbone of the mountain-system. This *crystalline axis* runs as a broad central zone from the westernmost Kashmir ranges to the eastern extremity in Burma. The eastern part of the Himalayas, from Nepal eastwards, has not been explored, with the exception of Sikkim, but it is certain that the crystalline zone is quite continuous. It is a matter of great uncertainty, however, what part of the great gneissic complex of the Himalayas (designated as the "Central" or "Fundamental" gneiss) represents the Archaean system, because much of it is now ascertained to be highly metamorphosed granites or other intrusives of the late Mesozoic or even Tertiary ages.

THE ARCHAEAN SYSTEM

A fairly broad crystalline zone, similar to the gneisses and metamorphosed sediments of the Peninsula, constitutes almost the whole framework of the island of Sri Lanka which in fact is a part of the Deccan Peninsula, but recently separated by a shallow sea. The gneisses and schists of Archaean-Dharwar affinities reappear in Burma as a broad belt, after crossing over north-east from Assam; this belt runs along Burma from north to south, constituting the so-called *Martaban* system[1] of the southern or Tenasserim division, and the *Mogok gneiss* of N. Burma.

Petrology of the Archaean system—Over all these areas of many hundred thousand square km. the most common Archaean rock is *gneiss*—a rock which in mineral composition may vary from granite to gabbro, but which possesses a constant, more or less foliated or banded structure, designated as *gneissic*. This characteristic banded or streaky character may be either due to an alternation of bands or layers of the different constituent minerals of the rock, or to the association of layers of rocks of varying mineral composition. At many places the gneiss appears to be a mere intrusive granite, exhibiting clearly intrusive relations to its neighbours. The gneiss, again, frequently shows great lack of uniformity either of composition or of structure, and varies from place to place. At times it is very finely foliated, with folia of exceeding thinness alternating with one another; at other times there is hardly any foliation or schistosity at all, the rock looking perfectly granitoid in appearance. The texture also varies between wide limits, from a coarse holocrystalline rock, with individual phenocrysts as large as 2.5 or 5 cm. to almost a felsite with a texture so fine that the rock appears quite homogeneous to the eye.

The constituent minerals of the commoner types of the Archaean gneiss are: orthoclase, oligoclase or microcline, quartz, muscovite, biotite, and hornblende with a variable amount of accessory minerals and some secondary or alteration products, the tourmaline, apatite, magnetite, zircon, chlorite, epidote and kaolin. Orthoclase is the most abundant constituent, and gives the characteristic pink or white colour to the rock. Plagioclase is subordinate in amount; quartz also is present in variable quantities; hornblende and biotite are the most usual ferro-magnesian constituents, and give rise to the hornblende- and biotite-gneisses, which are the most prevalent rocks of the central ranges over wide tracts of the Himalayas. Tourmaline is an essential constituent of some gneisses of the Himalayas. Chlorite occurs as a secondary product, replacing either hornblende or biotite. Less frequent

---

[1] Recent work shows that the *Martaban* gneisses are probably largely Mesozoic granites.

minerals, and occurring either in the main mass or in the pegmatite veins that cross them, are apatite, epidote, garnets, cordierite, scapolite, wollastonite, beryl, tremolite, actinolite, jadeite, corundum, sillimanite, kyanite, together with spinels, ilmenite, rutile, graphite, iron ore, etc. Besides the composition of the gneiss being very variable over wide areas, almost all gradations are to be seen, from thoroughly acid to intermediate and basic composition (granite-gneiss, syenite-gneiss, diorite-gneiss, gabbro-gneiss).

By the disappearance of the felspars the gneisses pass into schists, which are the next most abundant components of the Archaean of India. The schists are for the most part thoroughly crystalline, mica-, hornblende-, talc-, chlorite-, epidote-, sillimanite- and graphite-schists. Mica-schists are the most common, and are often garnetiferous. Less common rocks of the Archaean of India, and occurring separately or as interbedded lenses or bands in the main complex, are slates, phyllites, granulites, crystalline limestones (marbles), dolomites, graphite, iron-ores, and some other mineral masses. The gneisses and schists are further traversed by an extensive system of basic trap-dykes of dioritic or doleritic or ultrabasic composition.

The Archaeans evince a generally high grade of regional or dynamic metamorphism, due to the three or four periods of diastrophism (mountain-building movements) which they experienced and the widespread igneous activity of this and subsequent periods.

But there are, in the Peninsula, areas remarkably immune from these disturbances, which show a feeble (the epi-grade) metamorphism, characterised by the prevalence of rock-types such as phyllites or schists with such minerals as talc, chlorite and epidote. From this there is a progressive rise in grade of metamorphism (the meso-grade), characterised by the presence of mica-, hornblende-, garnet-, and staurolite-schists and gneisses, to the high grade of plutonic metamorphism (kata-grade), characteristic of the deeper, more loaded zones of the crust, in which such dense and compact minerals as pyroxene, cordierite, graphite, garnets, sillimanite are developed in rock-masses like granulites and eclogites.

But the Archaean group of India, as of the other countries of the world, is far more complex in its constitution than is expressed by the above few simple statements. In it, though several distinct petrological elements have been recognised, yet their relations are so very intimate that separation of these is very difficult or impossible. Among these gneisses and schists those which, by reason of their chemical and mineralogical composition, are believed to be the highly deformed and metamorphosed equivalents of plutonic igneous masses of later ages are known as ortho-gneisses or ortho-schists, while others that suggest the characters of highly altered sediments deposited in the ancient seas are known

THE ARCHAEAN SYSTEM 77

as para-gneisses or para-schists, the Dharwars; a third kind again is also distinguished, which, according to some authors, may be the original first-formed crust of the earth. It thus appears that the Indian Archaean representatives do not belong to any one petrological system, but are a "complex" of several factors: (1) an ancient fundamental basement complex into which (2) a series of plutonic rocks are intruded, granite bathyliths and some varieties of Bundelkhand gneisses, while there is (3) a factor representing highly metamorphosed schistose sediments, the para-gneisses and schists, which probably are mainly of Dharwar age, and are generally younger than the (1) gneisses.

Petrological types—Included in the Archaean gneisses and schists there are some interesting petrological types, discovered during the progress of the Indian Geological Survey, which the student should know. Some of these are described below:

Granite.	Of Karnataka, North Arcot, Tamilnadu Rajasthan, etc.
Augite-granite.	Of Salem.
Augite-syenite (Laurvikite).	
Nepheline-syenite. Elaeolite-syenites and their pegmatites.	Of Coimbatore, Vishakapatnam, Kishengarh (Rajasthan) and Junagadh (Saurashtra). These are a group of intermediate plutonic rocks foliated among the gneisses. Among their normal essential minerals are *calcite* and *graphite* in a quite fresh state. The pegmatites of the elaeolite-syenite of Kishengarh[1] contain large crystals of beautiful blue *sodalite* with *sphene*, garnet, etc. as accessories.
Elaeolite-syenite. Augite-syenite. Corundum-syenite.	Of the Coimbatore district, constitute the so-called *Sivamalai* series of Holland. These are genetically related rocks, all derived from a common highly aluminous magma.
1. Charnockite. 2. Augite-norite. 3. Norite. 4. Hyperite. 5. Olivine-norite. 6. Pyroxenite. 7. Anorthosite.	Of Tamilnadu and Bengal,[1] are acid, intermediate basic and ultra-basic members respectively of a highly differentiated series of holo-crystalline, granitoid, hypersthene-bearing rocks of the Peninsula, distinguished by Holland and named by him Charnockite series from Job Char-

---

[1] For the soda-bearing syenite suite of the N.E. end of the Aravallis, see Heron, *Rec. G.S.I.* vol. lxvi. pt. 2, 1924.

8. Granulite.
9. Garnetiferous-leptynite.
10. Pyroxene-diorite.
11. Seapolite-diorite.

Khondalite
(Quartz+sillimanite
+garnet+graphite).

Named after the Khonds of Orissa, occurs in Orissa, Madhya Pradesh, etc.; are light-coloured richly garnetiferous gneisses and schists characterised by the abundance of the mineral sillimanite and the presence of graphite. They are regarded as para-gneisses and schists.[1] The Khondalite group is well-developed also in Kerala and Sri Lanka where these rocks are the carriers of the deposits of graphite. This group is of much wider geographical prevalence than has been so far recognised.

nock, the founder of Calcutta. (See Charnockite series below.)

1. Gondite
(Quartz+manganese-garnet+rhodonite).
2. Rhodonite rock.

Named from the Gonds of Madhya Pradesh by Dr. L. Fermor. These are a series of metamorphosed rocks belonging to the Archaean and Dharwar systems and largely composed of quartz, spessartite, rhodonite and other manganese silicates. These rocks are supposed to be the product of the dynamic metamorphism of manganiferous clays and sands deposited during Dharwar times. On the chemical alteration of the manganese silicates so produced, these rocks have yielded the abundant manganese-ores of the Dharwar system.

Kodurite
(Orthoclase
+manganese-garnet
+apatite).

From Kodur in Vishakapatnam district. These are a group of plutonic rocks, associated with the Khondalite series and possibly of hybrid origin. The normal type, or *Kodurite* proper, has the composition noted above, and is a basic plutonic rock classified with Shonkinites, but there are acid as well as ultra-basic varieties of the series like the *spandite-rock, manganese-pyroxenite*, containing manganese-garnet,-amphibole, -pyroxene, -sphene, etc. at one end, and quartz-orthoclase rock and quartz-kodurite at

---

[1] *Mem. G.S.I.* vol. xxxiii. pt. 3, 1902.

THE ARCHAEAN SYSTEM

the other. These rocks also have yielded manganese-ores of economic value by chemical alteration.

Calc-gneiss, calciphyres and crystalline limestones.
The first two of these are highly calcareous rocks which are found associated with the Archaean rocks of Madhya Pradesh and some other localities in India. They are a series of granulite-like rocks with an unusually high preponderance of lime silicates, diopside, hornblende, labradorite, epidote, garnet, sphene, and similar alumino-calcareous silicates. From such a composition, they are believed to be para-gneisses, *i.e.* formed by the metamorphism of a pre-existing calcareous and argillaceous series of sediments.

The oxidation by meteoric agencies of these series has given rise to the crystalline limestones, the third class of rocks mentioned in the heading. These are very intimately associated with the two former rocks in Madhya Pradesh and in Burma. The abundant lime and magnesian silicates of these gneisses have been altered by percolating waters, carrying dissolved CO_2, into calcite and magnesite. Besides the crystalline limestones and dolomites of Madhya Pradesh, the celebrated ruby-limestone associated with the Mogok gneiss of Burma is another example. The origin of these limestones was a puzzle because they could not be explained on the supposition of their being of either sedimentary, organic or chemical deposition.[1]

Quartz-haematite-schist (Jaspilite).
Quartz-magnetite-schist.
Composed of quartz and haematite or quartz and magnetite. These are of very common prevalence in many parts of South India, especially among the Dharwar schists. The iron-ore and jasper or quartz are generally in very intimate association arranged in thin layers or folia.

Cordierite-gneiss.
Of Tamilnadu, Bastar, etc., is a contact-metamorphosed, basic aluminous sediment

---
[1] Fermor, *Rec. G.S.I.* vol. xxxiii. pt. 3, 1906.

	with high magnesia content. In the more metamorphosed types anthophyllite, sillimanite and garnet are frequently developed. Acid plagioclase and quartz, also biotite, are often present in these gneisses.
Andalusite(sillimanite, kyanite, chiastolite) -schist.	Of Tamilnadu and Madhya Pradesh, representing aluminous sediments highly metamorphosed by granitic intrusions.
"Streaky gneisses."	So called on account of the arrangement of the leucocratic and melanocratic components of the rock in parallel streaks and bands. It is a composite rock and the origin of the structure is due in many cases to the *lit-par-lit* injection of an acid aplitic material along the foliation-planes of a schistose melanocratic country-rock, giving rise to a migmatite.
Felspathic gneiss.	Generally composed of an acid plagioclase, subordinate microcline, small flakes of biotite and muscovite, and quartz. Often a para-gneiss, it represents a thoroughly recrystallised aluminous sediment, the metamorphism being due to granitic intrusions.
Pegmatites.	These coarse-grained differentiates of igneous rocks, especially acidic ones, are widely distributed in the Archaean complex of India. They occur chiefly as veins and dykes intersecting the older rocks, and sometimes as segregation-patches in the body of the rock of which they are differentiates. The acid granite-pegmatites sometimes attain large dimensions and in Nellore, Hazaribagh, Gaya, Rajasthan and Ajmer-Merwara have been found to contain many rare-earth minerals and mica deposits of economic importance.
Peridotite (Olivine+femic minerals). Dunite (essentially olivine). Saxonite (Olivine+rhombic	These ultra-basic rocks, though not widely spread, are of importance because of their association with minerals of economic uses. They are the source of chromite in India and of serpentine and magnesite. The chromite occurs as bed-like veins and scattered grains in these rocks.

THE ARCHAEAN SYSTEM

pyroxenes).	Among the well-known occurrences are those of Salem, some districts in Karnataka, Singhbhum, Hindu Bagh in Baluchistan and Dras in Kashmir.
Amphibolites.	Of widespread distribution in the Peninsula and extra-Peninsula; these rocks consist essentially of tremolite, actinolite, or some other amphibole, with varying amounts of plagioclase. Quartz, epidote and garnet are often present. They are products of the metamorphism of basic igneous masses, tuffs, or sediments.
Quartzite. Ortho-quartz.	Common in the Archaean and Dharwar and the older Purana systems is a granulose, recrystallised metamorphic rock, composed essentially of quartz with or without sericite, rutile, or other accessory constituents. It may be derived from original siliceous sediments or from quartz-reefs and vein-quartz. In the absence of stratification planes, ripple marks and other sedimentary characters, it is difficult to distinguish sedimentary from igneous quartzites.
Phyllites.	Very widespread in the Archaean and Dharwar systems. Often markedly graphitic and interbedded with crystalline limestones. The Himalayan Dharwars are especially characterised by the prevalence of graphitic phyllite and schist (Salkhala and Jutogh series). Passing into mica-, sillimanite-, or staurolite-schist by further metamorphism.

Groups—The gneissic Archaean rocks of India are generally described under the following three areal groups, each of which in its respective area has some well-defined types:

1. *Bengal gneiss*—Highly foliated, heterogeneous, schistose gneisses and schists, of Bengal, Bihar, Orissa, Carnatic and large tracts of the Peninsula.
2. *Bundelkhand gneiss*—Massive, granitoid gneisses of Bundelkhand and some parts of the Peninsula. This gneiss is regarded as intrusive into the former.
3. *Charnockite series*—*Nilgiri gneiss*—Massive, eruptive, dark-coloured hypersthene-granitoid gneisses of South India.

1. **BENGAL GNEISS** is very finely foliated, of heterogeneous composition, the different schistose planes being characterised by material of different composition. The gneiss is closely associated with schists of various composition. The gneiss is often dioritic, owing to the larger proportion of the plagioclase present. Numerous intercalated beds of limestones, dolomites, hornblende-rock, epidote-rock, corundum-rock, etc. occur among the gneiss. There is an abundance of accessory minerals, contained both in the rock itself and in the accessory beds associated with it, such as magnetite, ilmenite, schorl, garnet, calcite, lepidolite, beryl, apatite, epidote, corundum, micas, and sphene. In all the above characters the rocks commonly designated Bengal gneiss differ strikingly from those commonly named Bundelkhand gneiss, in which there are no accessory constituents, and but few associated schists.

The weathering of some parts of the gneiss of North Bengal is very peculiar; it gives rise to semi-circular, dome-like hills, or ellipsoidal masses, by the exfoliating of the rock in regularly circular scales. From this peculiarity the gneiss has received the name of *Dome gneiss*.

The gneiss in some places of Bengal closely resembles an intrusive granite with well-marked zones of contact-metamorphism in the surrounding gneisses and schists in which it appears to have intruded. Its plutonic nature is further shown in its containing local segregations (autoliths) and inclusions of foreign rock-fragments (xenoliths).

Types of Bengal gneiss—Besides the foregoing varieties some other petrological types are distinguished in the Bengal gneiss, the most noted being the *Sillimanite-gneiss* and *Sillimanite-schist* of Orissa, known as Khondalites (from the Khond inhabitants of Orissa). These give clear evidence of being metamorphosed sediments (para-schists) and are discussed in the next chapter. A large part of the schistose and garnetiferous gneiss of South India, commonly designated "Fundamental gneiss" or "Peninsular gneiss", belongs appropriately to this division. The Bengal gneiss *facies* is revealed in the gneisses of Bihar, Manbhum and Rewah, and some other parts of the Peninsula also. The *Carnatic* and *Salem* gneisses are examples. Carnatic gneiss is schistose, including micaceous, talcose, and hornblendic schists. The well-known mica-bearing schists of Nellore, which support the mica mines of the district, belong to the facies of the Bengal gneisses. The schistose type of Bengal gneiss is regarded as probably the oldest member of the Archaean Complex. The *Peninsular* gneiss of Karnataka, covering 64,750 square km., is now believed to be a granitic gneiss intrusive into an older Dharwar complex. What have been called the *Closepet* and *Champion* gneisses are also later granites intruded in the same basement complex. Recent work in the Archaean Com-

PLATE IV. BELLARY GRANITE, GNEISS COUNTRY, HAMPI.

plex of South India has shown that many of the fine-grained gneissic rocks are actually granitoid phases of recrystallised pre-existing formations and do not represent the crushed or foliated phases of true eruptive granites. (*Records M.G.D.*, Vol. 42, 1944)

2. BUNDELKHAND GNEISS. Bundelkhand gneiss occurs in the type area of Bundelkhand. It looks a typical pink granite in hand specimens, the foliation being very rude, if at all developed. In its field relations, the Bundelkhand gneiss differs from ordinary intrusive granite only in the enormous area which it occupies. Indeed, it may be regarded as a granite intruded into older gneisses, large patches of which it has remelted. Schists are associated with the gneisses very sparingly, *e.g.* hornblende-, talc- and chlorite-schists. No interbedded marbles or dolomites or quartzites occur in the Bundelkhand gneiss, nor is there any development of accessory minerals in the mass of the rock or in the pegmatite-veins. Bundelkhand gneiss is traversed by extensive dykes and sills of a coarse-grained diorite, which persist for long distances. It is also traversed by a large number of coarse pegmatite-veins as in a boss of granite. Quartz-veins or reefs (the ultra-acid modification of the pegmatite-veins), of great length, run as long, narrow, serrated walls, intersecting each other in all directions, giving to the landscapes of the country a peculiar feature. They intersect the drainage-courses of the district and are the cause of the numerous small lakes of Bundelkhand, whose formation consequently requires no further explanation.

This type of gneiss is also met with in the Peninsula at several localities, and is recognised there under various names — *Balaghat gneiss* (also named *Bellary gneiss*), *Hosur gneiss*, *Arcot gneiss*, *Cuddapah gneiss*, etc. The oldest basement gneiss of some parts of Rajasthan belongs to this system. The rock is quarried extensively for use as a building-stone, and has in the past contributed material of excellent quality for the building of numerous temples and other edifices of South India.

3. CHARNOCKITE SERIES. This is the name given to a series of granitoid rocks of South India, occurring among the older Archaean gneisses and schists of the Peninsula. These rocks are of wide prevalence in Tamilnadu, and constitute its chief hill-masses—the Nilgiris, Palnis, Shevaroys, etc. They are medium to coarse-grained, dark-coloured, basic holocrystalline granitoid gneisses, possessing such a distinctive assemblage of petrological characters and mineral composition that they are easily distinguished from the other Archaean rocks of the Peninsula. This group includes many varieties and forms which are modifications of a central type (the Charnockite proper), but these different varieties exhibit a distinct "consanguinity" or family relationship to each other. From this circumstance the Charnockite gneisses of South

India afford a very good instance of a *petrographical province* within the Indian region. The name Charnockite which was originally given by the discoverer of these rocks, Sir T. H. Holland, to the type-rock present near Madras was, therefore, extended by him (Charnockite series) to include all the more or less closely related varieties occurring in various parts of Tamilnadu and other parts of the Peninsula.

Petrological characters—The mineralogical characters which give to these rocks their distinctive characters are the almost constant presence of the rhombic pyroxene, hypersthene or enstatite, and a high proportion of the dark ferro-magnesian compounds which impart to the rock its usual dark colours. The ordinary constituents of the rock include blue-coloured quartz, plagioclases, augite, hornblende and biotite with zircon, iron-ores and graphite as accessories. Garnets are of very common occurrence. The presence, in different proportions, of the above constituents imparts to the different varieties a composition varying from an acid or intermediate hypersthene-granite (Charnockite proper) through all gradations of increasing basicity to that of the ultra-basic felsparless rocks, pyroxenites. The specific gravity and silica content range from 2.67 and 75 per cent respectively, in the normal hypersthene-granite, to 3.03 and 52 per cent in the norites and hyperites. In the pyroxenites the specific gravity rises to 3.37, corresponding to a fall in silica to 48 per cent. These ultra-basic types occur only locally as small lenses or bands in the more acid and commoner types.

That the Charnockites are of the nature of igneous plutonic rocks, intruded into the other Archaean rock-masses, is believed by some workers to be established by their field relations and possession of features which are regarded as evidence of magmatic segregation and differentiation, protruding of dykes and apophyses into surrounding rocks and well-defined contact-metamorphic aureoles at junctions with other rocks.

More recent studies of charnockite rocks and the discovery of similar suites of rocks in other parts of the world, for example Antarctica, West Africa, Uganda and West Australia, have for some years past raised an interesting controversy with regard to the origin of charnockites. Vredenburg suggested in 1918 that the Indian charnockites were not plutonic gneisses but were metamorphosed Dharwars. Several other workers have also expressed the view that the charnockites are intensely metamorphosed rocks, and owe their common characteristics to an intense metamorphic impress upon rocks that were originally non-hypersthenic, so that the presence of hypersthene does not necessarily prove they were all genetically related to each other. The belief is growing that many of the characters of charnockite are

PLATE V. BANDED PORPHYRITIC GNEISS (YOUNGER ARCHAEAN), NAKTA NALA, CHHINDWARA DISTRICT.
Geol. Survey of India, Records, xliii. pl. 1.

to be ascribed to plutonic metamorphism at high temperature and pressure at great depths of the crust (Kata-zone). B. Rama Rao, from the evidence provided by a study of the Mysore rocks, suggests that the charnockites are essentially reconstructed rocks resulting from recrystallisation of rocks that were in part of ultimate sedimentary origin and in part of ultimate igneous origin (*Bull. No. 18, M.G.D.* 1945).

The Charnockite series is mainly confined to Tamilnadu and Southern India extending as far as Sri Lanka; a few of its types, *viz. anorthosite*, a rock principally composed of labradorite felspar, and *olivine-norites*, are found in Bengal near Raniganj.

Archaean of the Himalayas—As already said, the bulk of the high ranges of the Himalayas forming the central or Himalayan zone proper is formed of crystalline or metamorphic rocks, like granites, granulites, gneisses, phyllites, and schists. The high snow-peaks of the central axis extending from Nanga Parbat on the Indus to Namcha Barwa on the Brahmaputra have a substratum composed of these rocks. In this complex, known formerly as the Central gneiss, from its occupying the central axis of the mountain-chain from one extremity to the other, the representatives of the Archaean gneisses of the Peninsula are to be found. It is, however, now known for certain, by the researches of General McMahon and later investigators, that much of the gneiss is of intrusive origin, and, therefore, of very much younger age. It is found intrusive into the Panjal Volcanic series of Permian age in the Pir Panjal and elsewhere; into the Jurassic in Chitral; into the Cretaceous *Orbitolina*-bearing beds in the Burzil valley of Kashmir; and into the Eocene of Eastern Tibet. These granites have passed into gneisses by assuming a foliated structure, while the Archaean gneiss proper has assumed the aspect of granites, owing to the high degree of dynamic metamorphism. It is again quite probable that a certain proportion of the central gneiss is to be attributed to highly metamorphosed ancient (Purana) sediments. It is therefore difficult to separate in this complex the constituent elements of the Archaean gneiss from gneissose granite or from the metamorphosed sediments of later age. The postulated Archaean age of the Himalayan granite of most localities, especially in the Kashmir and Hazara areas, remains to be proved.

The old view that the Central Himalayan axis is wholly composed of granite or gneiss, which also build the high peaks of the range, is not supported by the findings of the many Himalayan expeditions of recent years. These have shown that most of the peaks on the axial range of the Central Himalaya, Nepal and further east in Bhutan and Sikkim are composed of altered or metamorphosed sedimentary strata — slates, quartzites and crystalline limestones. The peak of Nanga Parbat in Kashmir is also similarly built. Orthogneiss and granite no doubt build the substratum of these mountains, but the peaks rising above them often are of stratified, even of fossiliferous, sediments, *e.g.* the peak of Mt. Everest.

There is reason to believe that the gneiss and granite in the vicinity of the central axis and around the majority of the high peaks of the Himalayas belong to the intrusive category rather than to the old Archaean foundation; they probably mark zones of special elevation connected with the welling up of acidic magma at certain points at the time of the uplift of the mountains.

The sedimentary Archaean complex of the Himalayas is dealt with in the next chapter.

The Crystalline Complex of Kashmir

This would be the best place to describe briefly the so-called "fundamental gneiss" with intrusive granite of the Kashmir Himalayas, though it is now clearly recognised that only a part of it is of Archaean age.

"Fundamental Gneiss" with intrusive granites—Crystalline rocks, granites, gneisses and schists, occupy large areas of the N.W. Himalayas of Kashmir and Simla, to the north of the Middle Ranges, forming the core of the Dhauladhar, the Zanskar and the ranges beyond in Ladakh and Baltistan. These rocks were all regarded as igneous and called "Central Gneiss" by Stoliczka and were taken to be Archaean in age. Later investigations have proved that much of this gneiss, as is the case with that of the Himalayas as a whole, is not of Archaean age, but is of intrusive origin and has invaded rocks of various ages at a number of different geological periods. Also a considerable part of this crystalline complex has now been found to be of pre-Cambrian metamorphic sedimentary origin, forming the basement on which all the subsequent geological formations rest. The latter have been distinguished as the *Salkhala series* in the Kashmir-Hazara area and the *Jutogh series* in the Simla-Chakrata area. Some affinity of these series with the Dharwars of Rajasthan and Singhbhum is apparent; while it is difficult or impossible to demarcate the areas of truly Archaean gneiss from the widespread later intrusive granites, the distinction of the sedimentary Archaeans from the fundamental gneisses and the intrusives is in general recognisable in many cases. The three elements of the great basement complex of the Himalayas are thus mixed up and may best be described at this place: (1) the metamorphosed sedimentary Archaeans, (2) intrusive granite and gneisses of later periods, and (3) remnants of Archaean granites, granulites, ortho-gneisses and schists. The presence of the latter can be inferred from the occurrence of granite pebbles and boulders, beds of arkose and of the widespread quartzites in the Palaeozoic sediments. The true Archaean gneisses have often assumed a coarse granitoid aspect, while owing to extreme dynamic metamorphism the very much younger intrusive granites have developed a gneissic structure. Foliation thus is not a criterion of age.

Petrology—Three kinds of granite have been recognised in this Archaean complex: biotite-granite, hornblende-granite and tourmaline-granite. Of these the most prevalent is the biotite-gneiss or granite, the one showing a quick transition to the other. The composition is acidic; pink orthoclase is rare, so also is muscovite; the bulk of the gneiss is made up of milk-white orthoclase, acid plagioclases with quartz and a conspicuous amount of biotite, arranged in schistose or lenticular manner, foliation being fine, or coarse, or absent altogether. This rock is the most prevalent Himalayan gneiss from Kashmir to Assam. It is often porphyritic, with orthoclase phenocrysts as much as 5-10 cm. across, giving rise to an apparent augen structure. Accessory minerals are not common, except garnet and tourmaline. Hornblende-gneiss is much less common, but it has a very similar structure and composition, the biotite being replaced by hornblende. Both the gneisses are traversed by veins of intrusive tourmaline-granite varying from 30 cm. to 6 or more metres in breadth, which in some cases penetrate the surrounding sedimentary strata as well. These pegmatite and aplite veins have a greater diversity of mineral composition than their hosts, often carrying such accessories as microcline, oligoclase, rock-crystal, garnet, tourmaline (schorl as well as the coloured transparent varieties rubellite and indicolite), muscovite, beryl (aquamarine), fluorspar, actinolite, corundum.

Next to the gneisses the most frequent rock is biotite-schist, passing into fine, silky schists, chlorite-talc-, hornblende-, and muscovite-schists, and migmatite, composite rock produced by injection of granite in schists.

These rocks are abundantly traversed by dykes, stocks and masses of basic intrusives such as dolerite, epidiorite, gabbro, pyroxenite, etc.

Distribution—With regard to the distribution of the gneissic rocks in the area, the main crystalline development is in the north and north-east portions, in the Zanskar range and the region beyond it, in Gilgit, Baltistan and Ladakh, while in the ranges to the south of the valley they play but a subordinate part. The core of the Dhauladhar range is formed of these rocks, but they are not a very conspicuous component of the Pir Panjal range, where they occur in a number of minor intrusions. The trans-Jhelum continuation of this range, known as the Kaz Nag, has a larger development of the crystalline core. A broad area of Kishtwar is also occupied by these rocks which continue in force eastwards to beyond the valley of the Sutlej. It is from the circumstance of the prominent development of the crystalline core in the Zanskar range, in continuity with the central Himalayan axis, that the range is regarded as the principal continuation of the Great Himalaya chain, after its bifurcation at Kangra. The other branch, the Pir Panjal, is regarded only as a minor offshoot. North of the Zanskar the outcrop of the crystalline series becomes very wide, encompassing almost the whole of the region up to the Karakoram, with the exception of a few sedimentary tracts in central and south-east Ladakh. The largest occurrence of hornblende-granite is in the mountains between Astor and Deosai. Its post-Cretaceous age is definitely proved by its intrusive contact with *Orbitolina* limestone at the Burzil Pass (4,270 m.)[1] Tourmaline-granite is of relatively subordinate occurrence in dykes and pegmatite veins.

REFERENCES

R. D. Oldham, *Geology of India*, 2nd Edition, chapter ii, 1893.

Sir T. H. Holland, Charnockite Series, *Mem. G.S.I.* vol. xxviii. pt. 2, 1900; Sivamalai Series, *Mem. G.S.I.* vol. xxx. pt. 3, 1901.

R. B. Foote, Geology of Bellary District, Madras Presidency, *Mem. G.S.I.* vol. xxv., 1896.

C. A. McMahon, Microscopic Structure of some Himalayan Granites, *Rec. G.S.I.* vol. xvii. pt. 2, 1884.

Sir L. L. Fermor, Correlation of the Archaean of Peninsular India, *Mem. G.S.I.* vol. lxx. pt. 1, 1936.

Records of Mysore State Geological Department; all of these deal with the rocks described in this chapter.

C. S. Pichamuthu, *The Charnockite Problem*, Mysore Geologists Association, 1953.

A. M. Heron, Geology of Central Rajputana, *Mem. G.S.I.* vol. lxxix., 1953.

---

[1] D. N. Wadia, Cretaceous Volcanic Series of Astor-Deosai, Kashmir, *Rec. G.S.I.* vol. lxxii. pt. 2, 1937.

CHAPTER IV

Archaean System (*Continued*)

THE DHARWAR SYSTEM

Introduction—In this chapter are described the most ancient metamorphosed sedimentary rock-systems of India, as old as, and in some cases older than, the basement gneisses and schists described in the last chapter. These sedimentary Archaeans are grouped under the name of Dharwar System, but the difference of name does not denote difference of systematic position. According to the commonly received interpretation, during the Archaean era the meteoric conditions of the earth appear to have been changing gradually. We may suppose that the decreasing temperature, due to continual radiation, condensed most of the vapours that were held in the thick primitive atmosphere and precipitated them on the earth's surface. The condensed vapours collected into the hollows and corrugations of the lithosphere, and thus gave rise to the first-formed ocean. Further loss of heat produced condensation in the original bulk of the planet, and as the outer crust had to accommodate itself to the steady diminution of the interior, the first-formed wrinkles and inequalities became more and more accentuated. The oceans became deeper, and the land-masses, the skeletons of the first continents, rose more and more above the general surface. The outlines of the seas and continents being thus established, the geological agents of denudation entered upon their work. The weathering of the pristine Archaean gneisses and schists yielded the earliest sediments which were deposited on the bed of the sea, and formed the oldest sedimentary strata, known in the geology of India as the Dharwar System. They are often so metamorphosed into schists and gneisses that they are indistinguishable from the primitive gneisses and schists. In fact, at several localities indubitable Dharwar sediments are found to be older than some orthogneisses and schists with which they are associated. The above is only a partial definition of the term Dharwar system, whose exact limits and relations with respect to the Archaean igneous rocks are not yet fully understood. In the present chapter the term Dharwar System is used as synonymous with metamorphosed Archaean sediments, and including all the schistose series below the eparchaean unconformity.

These sedimentary strata appear to rest over the gneisses at some places with an unconformity, while at others they are largely interbedded with them, and in some cases are of undoubtedly older age than some of the gneisses. Although, for the greater part at least, of undoubted sedimentary origin, the Dharwar strata are altogether unfossiliferous, a circumstance to be explained as much by their extremely early age, when no organic beings peopled the earth, as by the great degree of metamorphism they have undergone. The complex foldings of the crust in which these rocks have been involved have obliterated nearly all traces of their sedimentary nature, and have given to them a thoroughly crystalline and schistose structure, hardly to be distinguished from the underlying gneisses and schists. They are besides extensively intruded by granitic bosses and veins and sheets, and by an extensive system of dolerite dykes, thus rendering these rock-masses still more difficult of identification.

All these circumstances have led to the sedimentary nature of the Dharwar rocks of several areas, notably of Karnataka, being doubted by some geologists who regard the bedded schists, limestones and conglomerates as of igneous origin, the conglomerates having resulted from the autoclastic crushing of quartz-veins and plutonic dykes. Field work in Karnataka has indicated that many of the subjacent gneisses have intrusive relations towards what were previously included among the Dharwars and are therefore younger than them. But such is not universally the case, and during the last few years the sedimentary nature of many terrains of Dharwar rocks has been demonstrated beyond doubt.[1]

Of late there has been a tendency to discard the term Dharwar and to designate this system by the name Archaean. The use of the term Dharwar to embrace all the great sedimentary systems, either associated with or resting upon the fundamental basement gneisses of India, and separated from the overlying Purana systems by a pronounced eparchaean unconformity, seems appropriate and has the sanction of long usage. In one of the best-studied Archaean provinces of India Dr. A. M. Heron has proved at least two, and possibly three, great cycles of Archaean sedimentary deposits, separated by important unconformities, denoting periods of diastrophism, erosion and peneplanation, overlying the Bundelkhand gneiss. These clastic Archaean rock-formations of great thickness and extent, reposing over an older gneissic floor, need a distinguishing term to separate them from the igneous Archaeans.

The lithology of the Dharwars—The rocks of this system possess the most diverse lithological characters, being a complex

---

[1] B. Rama Rao, *Records, Mysore Geol. Dept.* vol. xxxiv., 1936.

THE DHARWAR SYSTEM 91

of all kinds of rocks—clastic sediments, chemically precipitated rocks, volcanic and plutonic rocks—all of which generally show an intense degree of metamorphism. The principal types have been described in the last chapter (pp. 77-76). No other system furnishes such excellent material for the study of the various aspects and degrees of rock metamorphism. The rocks are often highly metalliferous, containing ores of iron and manganese, occasionally also of copper, lead, and gold. The bulk of the rocks of the system is formed of phyllites, schists, and slates. There are hornblende-, chlorite-, haematite- and magnetite-schists, felspathic schists; quartzites and highly altered volcanic rocks, *e.g.* rhyolites and andesites turned into hornblende-schists; abundant and widespread granitic intrusions; crystalline limestones and marbles; serpentinous marbles; steatite masses; beds of brilliantly coloured and ribboned jaspers; and massive beds of iron and manganese oxides.

Plutonic intrusions—The plutonic intrusions assumed to be of Dharwar age are copious and of varied characters; they have given rise to some interesting rock-types, some of which have already been described in the last chapter, *viz*. nepheline-syenites of Rajasthan, differentiated into the elaeolite-syenite and sodalite-syenite of Kishengarh, which carry the beautiful mineral, sodalite. Many of the granites of the Dharwar system are tourmaline-granites; among other intrusives are the quartz-porphyry of Rajasthan, and the dunites of Salem. The pegmatite-veins intersecting some of the plutonics are often very coarse, and, especially when they cut through mica-schists, bear extremely large crystals of muscovite, the cleavage sheets of which are of great commercial value. Such is particularly the case with the mica-schists of Hazaribagh, Nellore, and parts of Rajasthan, where a large quantity of mica is quarried. Besides muscovite, the pegmatites carry several other beautifully crystallised rare minerals, *e.g.* molybdenite, columbite, pitchblende, gadolinite, torbernite, beryl, allanite, samarskite, etc.

Here must also be considered the curious group of manganiferous crystalline limestones of Nagpur and Chhindwara districts of Madhya Pradesh, originating by metasomatism of gneisses, containing such minerals as piedmontite (Mn-epidote), spessartite (Mn-garnet), with Mn-pyroxene, -amphibole, -sphene, etc., which have given rise, on subsequent alteration, to some quantity of manganese ores. As mentioned on pp. 78-79 these crystalline limestones are assigned a curious mode of origin. Fermor has shown them to be due to the metasomatic replacement of Archaean calc-gneisses and calciphyres, which in turn were themselves the product of the regional metamorphism of highly calcareous and manganiferous sediments.[1]

Another peculiar rock is the *flexible sandstone* of Jind (Kaliana).

---
[1] *Rec. G.S.I.*, vol. xxxiii. pt. 3, 1906.

The rock was originally formed from the decomposition of the gneisses, and had a certain proportion of felspar grains in it. On the subsequent decomposition of the felspar grains the rock became a mass of loosely interlocking grains of quartz, with wide interspaces around them, which allow a certain amount of flexibility in the stone.

Outcrops of the Dharwar rocks—One important peculiarity regarding the mode of occurrence of the Dharwar rocks—as of generally all other occurrences of the oldest sediments that have survived up to the present—is that they occur in narrow elongated synclinal outcrops among the gneissic Archaeans—as outliers in them. This tectonic peculiarity is due to the fact that only those portions of the Dharwar beds that were involved in the troughs of deep synclinal folds and have, consequently, received a great deal of compression, are preserved, the limbs of the synclines, together with their connecting anticlinal tops, having been planed down by the weathering of ages.

Distribution of the Dharwars—The Dharwarian rocks are very closely associated with the gneisses and schists, described in the last chapter, in many parts of the Peninsula. The principal exposures in the Peninsula are: (1) Southern Deccan, including the type-area of Dharwar and Bellary and the greater part of Karnataka extending southwards to the Nilgiris, Madurai and Sri Lanka (2) the Dharwar areas of Carnatic, Chota Nagpur, Jabalpur, Nagpur, etc., with those of Bihar, Rewah and Hazaribagh; (3) the Aravalli region, extending as far northwards as Jaipur, and in its southern extremity including north Gujarat. In the extra-Peninsula the Dharwar system is well represented in the Himalayas, both in the central and northern zones, as well as in the Shillong plateau of the Assam ranges.

In the following pages some important developments of Archaean rocks of the Dharwar facies met with in six regions are described.

1. DHARWAR-KARNATAKA (*the Type-area*). The rocks occur in a number of narrow elongated bands, the bottoms of old synclines, extending from the southern margin of the Deccan traps to the Kaveri. The general dip of the strata is towards the middle of the bands. The constituent rocks are hornblende-, chlorite-, talc-schists, together with slates, quartzite, and conglomerates and very characteristic brilliantly banded cherts; these rocks are associated with various types of ortho-gneisses and schists and lavas of dioritic composition. The Dharwar slates exhibit all the intermediate stages of metamorphism (anamorphism) into schists, *viz*. unaltered slates, chiastolite-slates, phyllites and mica-schists. Numerous quartz-veins or reefs traverse the Dharwar rocks of these areas. Some of those are auriferous and contain enough disseminated gold to support some gold-fields. The principal

gold-mining centre in India, the Kolar fields in Karnataka, is situated on the outcrops of some of these quartz-veins or reefs.

The Dharwar System is very well developed in Karnataka where it forms three belts and several narrow strips and stringers covering an area of 15,540 square km. It has been intensively studied by the State Geological Department.[1] The older group of geologists, led by Dr. W. F. Smeeth, held the view that the system was entirely an igneous formation and contained no clearly recognisable sediments. They regarded all the crystalline schists in the Dharwars, and even the types like conglomerates, quartzites and limestones found there, as having originated from severe crushing and extreme alteration and modification of various types of acid and basic igneous rocks. All the conglomerates were held to be autoclastic in origin; the quartzites were regarded as the crushed phases of felsites and rhyolites, or vein-quartz; and the limestones as highly calcified phases of decomposed basic volcanic rocks, or metasomatic replacements of schistose acid igneous rocks.

Intensive field investigations in recent years have revealed, however, at several places in Karnataka, remnants of current bedding, ripple-marks, graded bedding and similar other structures which, though ill-preserved, afford undoubted proofs of sedimentation. Several of the types of crystalline schists chemically analysed in recent years also indicate clearly their sedimentary origin. The various types of crystalline rocks in the Dharwar System of Karnataka thus have been grouped as under:

(a) Volcanic rocks consisting of altered acid and basic lava flows, sills, sheets and dykes, tuffs and agglomerates.

(b) The crystalline schists and granulites which form much the larger portion of the Dharwars of Karnataka are made up of several types classified as chlorite-schists, mica-schists, hornblende-schists and tremolite-actinolite-schists, together with granulitic schists containing kyanite, sillimanite, staurolite, cordierite, graphite, garnet and corundum. Chemical examination of these rocks has shown that the dark hornblende-schists and tremolite-actinolite-schists are of igneous origin, and some mica-schists, chlorite-schists and the granulitic schists containing kyanite, staurolite and other highly aluminous minerals are evidently of sedimentary origin.

(c) Deformed Sediments—These form a comparatively small proportion but are specially interesting on account of the prolonged controversy they have raised regarding their origin. They form conglomerates, quartzites, ferruginous quartzites, phyllites

---

[1] Sampat Iyengar, *Acid Rocks of Mysore*, *Seventh Indian Science Congress Proceedings*, Calcutta, 1921.

and limestones. Most of these have proved to be definitely sedimentary in origin (*Mysore Geol. Dept. Bull.* No. 17, 1940).

(d) **Basic and Ultrabasic Intrusives**—In the central and northern parts of Karnataka, masses of coarse diorites and epidiorites are found intruding the schists. In the southern parts of the State, pyroxenites and peridotites are intruded into the basement schists.

Since the sedimentary origin of the conglomerates was established, they have been observed to fall into two well-marked series occupying different horizons. These two sets of basal conglomerates within the Dharwar system of Karnataka have enabled them to be classified into three sections forming upper, middle and lower divisions. According to Rama Rao, the lower division forms mainly a complex of volcanic rocks; the middle division is essentially a sedimentary group with volcanic material and intruded plutonic rocks; the upper division is also sedimentary, composed of ferruginous clays and silts, quartzites and conglomerates. It is a peculiarity of the Karnataka Dharwars that the grade of metamorphism shows a progressive increase to the south. The three-fold division mentioned above is, therefore, recognisable clearly only in the northern parts of the State.

Correlation of the Karnataka Dharwars—The correlation of the crystalline schists of Karnataka with those found in other widely separated Dharwar areas of India (given in a table on page 109) is yet provisional. The *Aravalli system* of Rajasthan with its southern continuation, the *Champaner series* of Gujarat; the *Sausar, Sakoli* and *Chilpi Ghat* series of Madhya Pradesh; the *Iron-Ore series* and *Older Metamorphics* of Bihar and Orissa; and the *Khondalites* of the Eastern Ghats and Sri Lanka are all more or less similar in their general lithological characters and metamorphic grade to the crystalline schists of the Dharwar system of Southern India. It is, however, to be recognised that most of these schistose series fall into more than one well-defined division separated by profound unconformities covering long ranges of time.

2. RAJASTHAN (Rajputana). Rocks which may be regarded as belonging to the Dharwarian group occupy a wide surface extent of Rajasthan, constituting the vast system of pre-Cambrian sediments designated as the *Aravalli system*. The results of a comprehensive study of this ancient sedimentary system, which is separated from the oldest Purana system by a hiatus, represented by one or two profound unconformities, have become available.[1] The relations of the Aravalli system in the different parts of Rajasthan are shown in the annexed table (page 96).

Aravalli mountains—The type rocks are exposed in a very large

---

[1] A. M. Heron, Synopsis of Pre-Vindhyan Geology of Rajputana, *Trans. Nat. Inst. Sc. Ind.* vol. i. No. 2, 1935.

outcrop in the Aravalli range of Rajasthan. This, the most ancient mountain-chain of India, came into existence at the close of the Dharwar era, when the sediments that were deposited in the seas of that age were ridged up by an upheaval of an orogenic nature. Since then the Aravalli mountains remained the principal feature in the geography of India for many ages, performing all the functions of a great mountain-chain and contributing their sediments to many deposits of later ages. Evidence exists that this mountain-chain received renewed upheavals during the early Palaeozoic and was of far greater proportions in past times, and that it stretched from the Deccan to perhaps beyond the limits of the Himalayas.

The Aravalli range, marking the site of one of the oldest geosynclines of the world, is still the most distinct mountain range of the Indian Peninsula, with summits of 1,200 to 1,500 m. It was peneplaned in pre-Cretaceous times but has been rejuvenated in the central part, large tracts of western Rajasthan remaining a peneplain. Structurally it is a closely plicated synclinorium of rocks of the Aravalli and Delhi systems, the latter forming the core of the fold for some 800 km. from Delhi to Idar in a N.E.-S.W. direction. Though the north-west flank of the synclinorium is a straight line, there is no evidence of a fault there. The curving east boundary of the fold, on the other hand, marks the line of the Great Boundary Fault of Rajasthan, which brings the Vindhyans against Aravallis and Bundhelkhand gneiss.

Aravalli system—The Dharwarian rocks of the Aravalli region form a long and wide synclinorium in the basement schistose gneisses of Rajasthan, constricted in the middle. Heron has classified these rock-groups into two great pre-Cambrian systems separated by a profound regional unconformity—the lower division forming the *Aravalli system* and the upper forming the *Raialo series*.

The lower, Aravalli system, is a vast formation, aggregating over 3,000 m. in vertical extent, composed of basal quartzites, conglomerates, shales, slates, phyllites and composite gneisses. It rests with a great erosional unconformity on the finely schistose and banded gneiss (Bundelkhand gneiss). Its metamorphism is variable, and there are exposures of almost unaltered Archaean shales in one part of the outcrop and such highly metamorphosed rocks as hornblende-schists and schistose conglomerates in another. The schists include numerous secondary aluminous and calcareous silicates, *e.g.* andalusite, sillimanite, staurolite, and a great many garnets. At a few localities the Aravallis include lodes of copper, lead and zinc, with traces of nickel and cobalt. Granite and amphibolite have intruded at many places into the slates and phyllites in the form of veins, attended with offshoots of quartz-veins and pegmatites. *Lit-par-lit* injections of granite in slaty rocks have given rise to composite gneisses.

TABLE OF THE GEOLOGICAL FORMATIONS OF RAJASTHAN

	JODHPUR.	MEWAR, AJMER-MERWARA.	CHITOR.	JAIPUR.	ALWAR.
Delhi System.		Calc-gneisses. Calc-schists. Biotite-schists. Quartzites. Basement arkose grits. Garnetiferous biotite-schists.	Jiran sandstone. Sawa shales and grit.	Ajabgarh series. Alwar series.	Ajabgarh series. Hornstone breccia. Kushalgarh limestone. Alwar series.
Raialo Series.	Raialo (Makrana) marble, limestones of Ras.	Raialo (Rajanagar) marble. Local basal grit.	Raialo (Bhagwanpura) limestone.		Raialo limestone. Raialo quartzite.
	Unconformity not seen.				
Aravalli System.	Shales (Sojat). Schists of Godwar.	Phyllites, cherty limestones, quartzites and composite gneisses. Basal quartzites, grits and local conglomerates. Local thick volcanic series.	Khardeola and Kanoj grits, Badesar quartzites. Vague unconformity. Ranthambhor quartzites. Shales and cherty limestones. Basal quartzites and grits.	Quartzites and schists of Baonli-Awan ridge and Bechun, Biana and Lalsot hills. Volcanics of Basi. Schists of Rajmahal.	Limestones and schists of Baswa and Rajgarh. Quartzite and conglomerate of Rewasa.
	Unconformity not seen.	Banded gneissic complex.	Bundelkhand gneiss.	Gneissic granite of Karela and Ganor.	
	Grey homogeneous gneiss.				

Great Boundary Fault separates Chitor column between Jiran sandstone/Binota shales and the lower units.

Raialo series. Delhi system—The Raialo series comes above the Aravallis with a pronounced unconformity. This series is rich in crystalline limestones, associated with quartzites, grits and schistose rocks. The famous Makrana marbles, the source of the material for the celebrated Moghul buildings of Delhi and Agra, are a product of this rock-series. The Raialos are succeeded in the northern part of the Aravallis, after another great unconformity, by the system of quartzites, grits and schistose rocks constituting the famous Ridge of the city of Delhi. These form the *Delhi system*. The Delhi system is now regarded as of Cuddapah age and is described on p. 116, Chapter V. On a possible prolongation of the Aravalli strike to the interior of the plains of the Punjab, a few small straggling outliers of the same rock-series are found, composed of ferruginous quartzite and slate, together with a great development of rhyolitic lavas (Malani rhyolites, p. 124). These outliers constitute the low, deeply weathered hills known as Kirana and Sangla, lying between the Jhelum and the Chenab.[1]

Features of great interest in the study of metamorphism are brought to light in the survey of the ancient sedimentary systems of Rajasthan. Schistose and banded gneisses in the Aravallis have been traced along the strike into rocks still in the condition of practically unaltered shales and slates. By the injection of granite, sedimentary rocks have been converted into banded composite gneisses on a large scale, which may easily be mistaken for orthogneisses. Comparatively newer sediments, *e.g.* of the Delhi system, occurring in the centre of the synclinorium of the Aravalli strata, evince a higher grade of metamorphism and tectonic deformation than the Aravallis on which they rest with a great hiatus. This anomalous metamorphism of a newer series is explained as due to the fact that the Delhi strata have been buried more deeply in their synclinal roots and therefore subjected to more intense pressures and intrusive action than the underlying Aravallis which flank the Delhis.

Dr. Heron has observed that the Aravallis of Rajasthan are analogous to, if not contemporaneous with, the Dharwars of South India, and has suggested a very general correlation of these with the Dharwars of Madhya Pradesh and Chhota Nagpur, and the Mergui series of Burma.

One further outlier of the Aravalli series, but this time to the south-west extremity of its strike, is found in the vicinity of Baroda on the site of the ancient city of Champaner. It overspreads a large area of northern Gujarat and is known as the *Champaner series*. The component rocks are quartzites, conglomerates, slates and limestones, all highly metamorphosed. A

---

[1] *Rec. G.S.I.* vol. xliii. pt. 3, 1913.

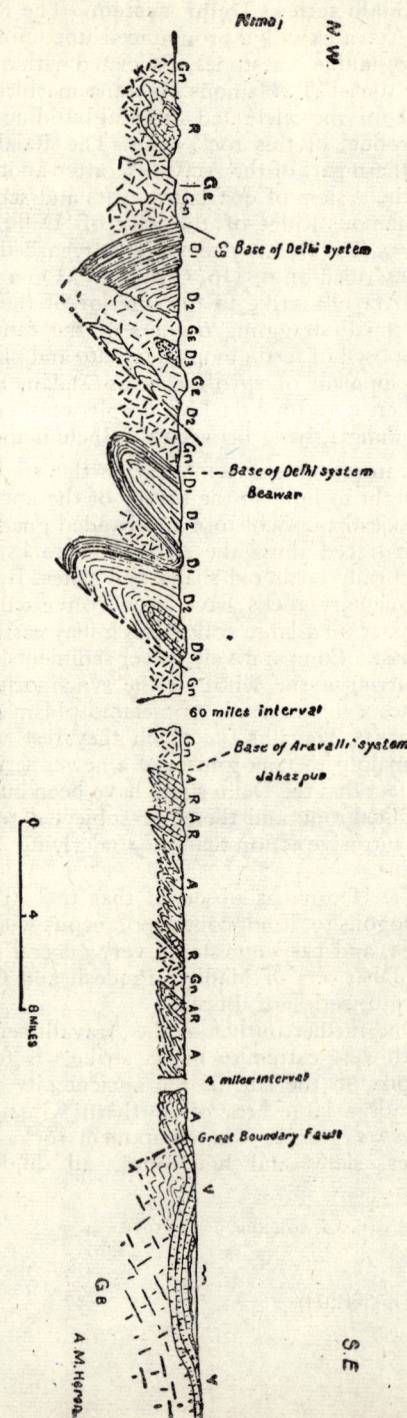

Fig. 5—Section across the Aravalli Range to the Vindhyan Plateau showing the Peneplaned Synclinorium of the most ancient mountain range of India.

1. Gn, Pre-Aravalli gneisses. GB, Bundelkhand gneiss.
2. A, Aravalli System (Schists).
3. R, Raialo series.
4. D1, }
   D2, } Delhi System
   D3, }
   {Biotite-schists with basal Conglomerate.
   Calc-schists.
   Calc-gneisses and Limestone.
5. GE, Erinpura granite.
6. V, Vindhyan System.

Note: In this figure and in subsequent figures where the mile or foot scale is given, 1 mile=1.609 kilometres. 1 foot=0.3048 metre.

green and mottled marble of exquisite beauty is quarried from these rocks near Motipura.

3. ASSAM. The *Shillong series* which occurs within the Assam hills is a group of parallel deposits which may be mentioned at this place. It is a widely developed formation, consisting of a thick series of quartzites, slates and schists, with masses of granitic intrusions and basic interbedded traps. The Shillong series is for the greater part of its extent overlain by horizontally bedded Cretaceous sandstones.

4. MADHYA PRADESH (Central Provinces). The Dharwarian system covers large connected areas within Madhya Pradesh and Bihar, spreading over Balaghat, Nagpur and Jabalpur districts, and over Hazaribagh and Rewah. In these areas it possesses a highly characteristic metalliferous facies of deposits which has attracted a great deal of attention lately on account of the ores of manganese and iron associated with it. The lithology of the Dharwars in these exposures is very varying, but each outcrop possesses a sufficient variety of its peculiar rock-types to reveal the identity of the system. The Dharwarian rocks of the Nagpur, Chhindwara and Bhandara districts of Madhya Pradesh have been named the *Sausar series*. They consist of granulites, calciphyres, dolomitic marble in lenticular association with mica-sillimanite-quartz-schists, diopsidites, hornblende-schist, etc. These rocks carry important economic deposits of manganese-ores. The Sausar series has been subdivided into stages which have a wide geographical extent in Madhya Pradesh and can therefore be correlated in distant outcrops of the series. The series is largely of aqueous sedimentation, but subsequently it has been metamorphosed and invaded by acid and basic plutonic rock-masses. The *Sakoli series* of the more southern portions of Madhya Pradesh, consisting of less altered slates, chlorite-schists, jaspilites and haematitic quartzites, is probably an upward extension of the Sausars. In the Balaghat district, and probably some other districts, the local representatives of the Dharwars are distinguished as the *Chilpi series*, from the Chilpi Ghat; these rocks include a great thickness of highly disturbed slates and phyllites, with quartzite and basic trappean intrusions. In Jabalpur the outcrop is distinguished by the occurrences of perfectly crystalline dolomitic limestones. The famous "marble-rocks" of Jabalpur in the Narmada gorge belong to this system.[1] In other parts of Madhya

---

[1] A series of Dharwar marbles and Deccan traps dissected into a number of magnificent dazzling white steeps, through which the Narmada, after its fall (*Dhurandhar*), runs for about two miles in a defile that is barely twenty yards in width.

Pradesh and in Rewah, and also some places in the Gujarat State (Panch Mahals),[1] etc., the exposures are distinguished by a richly manganiferous facies, containing large deposits of workable manganese-ores. Sir L. Fermor has given the name *Gondite series* to these rocks, because of their containing, as their characteristic member, a spessartite-quartz-rock, to which he has given the name of *Gondite* (p. 78). Besides spessartite, the rock contains many other manganese silicates; it is the decomposition of these manganese silicates that has given rise to the enormous deposits of manganese-ores in these occurrences of the Dharwar system.

Manganiferous series in Dharwar system. Gondite series—The origin of these rocks is interesting. According to Fermor they have originated from the metamorphism of sediments deposited during Dharwar times which were originally partly mechanical clays and sands, and partly chemical precipitates—chiefly of manganese oxides. The same metamorphic agencies that have converted the former into slates, phyllites and quartzites have altered the latter into crystalline manganese oxides, when pure, and into a number of manganese silicates where the original precipitates were mixed with clayey or sandy impurities.

Outcrops of the Gondite series are typically developed in the Balaghat, Chhindwara, and Nagpur districts of Madhya Pradesh and a few localities in Maharashtra and Gujarat, Central India and in Banswara in Rajasthan. The same authority regards the manganese deposits of Tamilnadu as due to the alteration of a series of plutonic intrusions (belonging to the *Kodurite series*) which may be of hybrid origin and due to the incorporation in acid intrusives of manganese ore-bodies of the Gondite type. The Kodurite series is typically developed in the Vizianagram area of the Vishakapatnam district of Tamilnadu.

5. SINGHBHUM—ORISSA. The next important area of Dharwar development is in Bihar-Orissa. In north Bihar, Dharwar rocks are met with in the Ranchi, Hazaribagh and Gaya districts. This area contains the well-known mica-fields of N. India. A more geologically interesting development is in South Bihar, where a large area extending from Gangpur through Singhbhum to the Mayurbhanj region covered by the Dharwars, has been studied in detail by H. C. Jones, J. A. Dunn and M. S. Krishnan. A widely sweeping zone of thrust, more or less E.-W. in direction, separates a comparatively unmetamorphosed tract to the south from a heavily metamorphosed tract on the north.

The chief interest of the Singhbhum Dharwars is in their enclosing a thick group of ferruginous sediments.

---

[1] The manganese-ores of the Panch Mahals occur in the south extension of the Aravalli system (Champaner series).

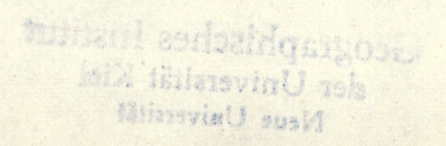

PLATE VI. "MARBLE ROCKS" (DOLOMITE MARBLE), JABALPUR.

This area contains the following sequence of Archaean sediments. It consists essentially of a series of iron-bearing sediments—phyllites, tuffs, lavas, quartzites, and limestones, designated as the *Iron-ore series*—resting unconformably on an older metamorphic series. The age of the Iron-ore series is regarded as Upper Dharwar:

Iron-ore Series	⎧ Shales, phyllites, tuffs with lava-flows. ⎪ Phyllites, quartzites, limestones with tuffs and lavas. ⎨ Banded haematite-quartzites and iron-ores. ⎪ Shales and phyllites with sandstones and limestones. ⎩ Sandstones, conglomerates.

Unconformity

? *Sausar Series*	{ *Gangpur Series* — schists, crystalline limestones, phyllites with Mn-ore bodies.

Unconformity

Older Metamorphics — hornblende-schists, mica-schists and quartzites.

The Iron-ore series is economically the most important (p.441), containing interbedded ore-bodies of large dimensions, estimated to yield a total of over three thousand million tonnes of high-grade iron-ore. In its petrogenesis the series is believed to be akin to the other well-known pre-Cambrian iron-bearing formations of the world, *e.g.* the Lake Superior deposits of the U.S.A. and those of Brazil. The question of the ultimate source of the iron oxides and the exact processes which segregated them here on such an immense scale yet awaits solution. Indian geologists generally regard these ores as, in the main, marine chemical precipitates in the form of oxides, carbonates and silicates. Some secondary changes and replacement have taken place subsequent to their deposition, but it is not believed that organic agencies such as algae or bacteria have helped in the precipitation of the iron. It is possible, however, that no single mode of origin applies to all the occurrences. While the larger deposits of iron-ore, such as those of Singhbhum or Keonjhar, may be sedimentary, there are other deposits belonging to the series which have probably originated by a process of metasomatic replacement under terrestrial conditions, in a period of marked volcanic activity.

The ores occur as massive beds and lenses of ferric oxides, soft powdery haematite, and as banded or ribboned haematite-quartzite or jasper, from which the free ore is liberated by the leaching out of the interlaminated silica. There is a considerable amount of igneous volcanic action in this area, witnessed by the bosses of Singhbhum and Bonai granite, by masses of ultrabasic intrusives and by lava-flows and tuffs. The basic intrusives have given origin to the chromite, asbestos and steatite of Singhbhum.

102   GEOLOGY OF INDIA

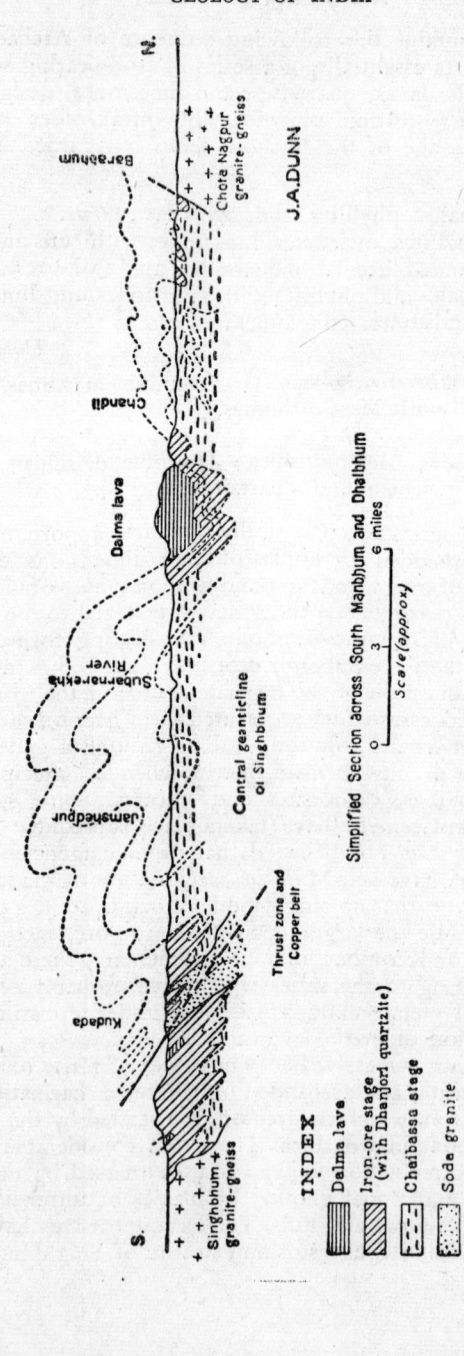

FIG. 6—Section across the Singhbhum Anticlinorium, Chota Nagpur

J. A. Dunn has studied the *Iron-ore series* and associated Archaeans of Singhbhum in detail together with various problems of economic minerals, petrogenesis and ore-genesis which these rocks present. Dunn recognises no earlier rocks, sedimentary or igneous, in Singhbhum than the Iron-ore series, which is a group of phyllites, shales and quartzites, overlain by tuffs and basic lavas, the more strongly folded portions of which show every grade of metamorphism from schists of the epizone to gneissic rocks of the hypozone. M. S. Krishnan has published a memoir in which he discusses the correlation of the Archaeans of the area to the south-west. He distinguishes a group of basal phyllites, overlying the Iron-ore series, which he regards as of Upper Dharwarian age. In Sir L. L. Fermor's scheme of correlation between the Archaeans of different parts of India he uses the Gonditic rocks, with marbles as confirmatory evidence, as a datum-line on the assumption that the manganese-ores and marbles mark one single stage of deposition in the Archaeans. On this ground he correlates the Gangpur series with the Sausars and the Iron-ore series with the Sakolis.

In the Bastar area (south-east of Madhya Pradesh), H. Crookshank differentiates three Archaean groups—andalusite gneiss and quartz-schists, a group of haematitic-quartzites, and quartzite. A part at least of the Bastar sequence is probably correlated to the Iron-ore series of Singhbhum.

L. A. N. Iyer has recently studied the gneissose granite of Bengal. Much of this granite is foliated, but coarse-grained and porphyritic types are present. Clear evidence of intrusion into schists of Dharwarian type is furnished by the hybrid injection-gneisses produced. Tourmalinisation of the schists is a feature of the intrusion.

Overlying the Iron-ore series are altered basalts and associated sub-aerial volcanic products—*Dalma traps*.[1] Within the Iron-ore series also there are dykes and sills of igneous ultra-basic rocks, dunites, peridotites and saxonites, generally serpentinised and at places carrying lodes of chromite ore.

Manganese-ores of Dharwar system—Almost the whole of the *Manganese-ores* annually produced in India is derived directly or indirectly from the Dharwar rocks. With regard to their geological relations Dr. Fermor has divided the ore bodies into three classes.

(1) *Deposits connected with the intrusive rock, Kodurite*, a basic plutonic rock, possessing an exceptional mineralogical composition, in being unusually rich in manganese silicates like manganese-garnets, rhodonite, and manganese-pyroxenes and -amphiboles.

---

[1] J. A. Dunn, Origin of Iron Ores in Singhbhum, *Econ. Geol.* vol. xxx. p. 643, (1935); *Mem. G.S.I.* vol. lxix. pt. 1, 1937.

The ores of the Vishakapatnam district have resulted from the meteoric alteration of these manganese silicates, while the felspar has altered into masses of lithomarge and chert, the other products being wad, ochres, etc. The ore-bodies resulting in this manner are of course of extremely irregular form and dimensions, and the grade of the ore is low.

(2) *Deposits contained in the Gondite series* are developed in Madhya Pradesh, Central India, the Panch Mahals, etc. As already described, the Gondite rocks were originally elastic sediments, including precipitates of manganese oxides like those of iron oxides enclosed in the sedimentary rocks of various ages. Their dynamic or regional metamorphism has given rise to crystallised ores of manganese, like braunite, hausmannite, hollandite, etc. The resulting ore-bodies are large and well-bedded, following the strike of the enclosing rocks, indicating that they have had the same origin as the latter. Sometimes, as in Chhindwara and Nagpur the manganese-ores are found in the crystalline limestone and calc-gneisses associated with the other Dharwar rocks. In addition to the ores psilomelane, braunite, hollandite, the crystalline limestone contains usually piedmontite (the manganese-epidote). The Gondite deposits yield by far the largest part of the economically important manganese-ores.

(3) *Lateritic deposits* are due to metasomatic surface replacement of Dharwar slates and schists by manganese-bearing solutions. These ores occur in Singhbhum, Jabalpur, Bellary, etc. They are irregular in distribution, occurring as caps on the outcrops of the Dharwar rocks, as is evident from the peculiar nature of their origin.

These ore-deposits have brought to light some new mineral species and beautiful crystallised varieties of already recognised manganese minerals. They are: *Vredenburgite*, *Sitaparite*—manganese and iron oxides; *Hollandite* and *Beldongrite* are manganates; *Winchite* is a blue manganese-amphibole, and *Blanfordite* a pleochroic manganese-pyroxene; *Spandite* is a manganese-garnet, intermediate in composition between spessartite and andradite; *Grandite* is similarly a "hybrid" of grossularite and andradite; *Alurgite* is a pink-coloured manganese-mica.[1]

6. THE HIMALAYAS. Rocks belonging to this, the oldest sedimentary system, occur in a more or less continuous band between the central axis of the higher Himalayas and the outer ranges. They occupy tracts of North Hazara, Kohistan, Gilgit, Ladakh and the Mid-Himalayan ranges from Garhwal to Assam. They are closely associated with the Central gneiss and also at places with the younger Puranas, to which they are distinctly unconformable in the less disturbed areas. They consist

---

[1] Fermor, *Mem. G.S.I.* vol. xxxvii., 1909.

of slates, phyllites (often graphitic), schists, migmalite, quartzites and crystalline limestones. They have been named *Salkhala series* in the Kashmir are.., *Jutogh series* in the Simla area, and *Daling series* in eastern Himalayas. The gneissification of these rocks at some places and the wide prevalence of later intrusive granites, especially in the central axial range of the Himalayas, make it difficult to separate from this complex any remnants of the Archaean gneisses. The Great Himalaya range, west of Ladakh, is largely composed of the Salkhalas converted into para-gneiss, the Nanga Parbat (8,119 m.) massif being almost wholly built of this, with intrusive biotite-gneiss and hornblende-granite rocks. South of this range the Salkhalas show a steadily decreasing grade of metamorphism, clearly revealing their sedimentary characters. Some of the rock-elements present in them show remarkable resemblance to the Dharwars of Rajasthan and Singhbhum; and it appears probable that the Great Himalaya range represents the basements of the old Peninsular Archaeans on which the Tethyan sediments were laid down in the Himalayan geosyncline. It thus denotes the protaxis of the Himalayas.

Dynamic metamorphism generally of a high grade is evident in the series, but all types of rocks are met with from dense compact carbonaceous slates and finely crystalline limestone to adinole-like beds, micaceous, garnetiferous and graphitic schists, saccharoidal marble, calc-schist and gneisses. From the Indus to Garhwal a chain of massive porphyritic biotite-gneiss intrusions occurs in these ancient sediments. The Salkhala sediments have been subjected to an intense granitisation at places, in Kaghan, in the ranges north of the Kishenganga, and in the Nanga Parbat area. The argillaceous components have been converted by the injection of magma to biotite-gneisses; while the calcareous and dolomitic members are changed into dark hornblende- and garnet-gneisses. A host of secondary minerals have resulted from metamorphic action—phlogopite, actinolite, epidote, zoisite, sphene, idocrase, tourmaline, beryl, etc. Elsewhere the metamorphism is of a curiously subdued type, and the Salkhala slates are then scarcely distinguishable from some Dogra slates.

The prominent peak of Nanga Parbat, Mt. Diyamir, 8,119 m. the culminating point of the Punjab Himalaya, is composed almost entirely of finely schistose biotite-gneiss, a para-gneiss, with interbedded marble, graphite-schists, etc., of Salkhala age. Through this para-gneissic complex are intruded sheets and bosses of gneissose granite of two later periods.[1]

There are no Archaean outcrops between the Aravallis and the Punjab Himalayas, except the few straggling hillocks of Kirana

---

[1] Wadia, Geology of Nanga Parbat and parts of Gilgit District, *Rec. G.S.I.*, vol. lxvi. pt. 2, 1932.

and Sangla, which probably are the unburied peaks of a branch of the Aravallis, buried under the alluvium of the Punjab.

Principal areas of Himalayan Dharwars—Different exposures of Himalayan Archaeans have received different names, according to the localities of their distribution. In the district of Spiti, the equivalents of the Dharwars are known as the *Vaikrita series*. On the south there occur more extensive exposures of metamorphosed rocks of distinctly older age than Cambrian. A part of these may be regarded as Dharwar in age, but owing to the complicated folding and inversions of the strata it is not easy to distinguish the representatives of the Dharwars from younger sediments, much less to correlate and group together the widely-separated outcrops of these formations in the different parts of the Himalayas. One of the most important occurrences of these ancient sediments is in the neighbourhood of Simla, covering large tracts to its east and west, which was previously known under the general name of the *Simla system*. Recent investigations have enabled this comprehensive system to be differentiated: the basal part, named the *Jutogh series*, being referred to Dharwar age, while a newer series coming unconformably over it is of Purana or still newer age—*Simla slate* series. The Jutoghs are a series of carbonaceous slates, limestones and dolomites, quartzites and schists, possessing a high order of metamorphism. Intervening between the Jutoghs and the Simla slates are a group of light grey schistose slates and talcose quartzites which have been named the *Chail series*. The *Chor granite*, prominently exposed in the Chor peak, has intruded on these formations. The Chails show thrust-fault relations to the series above and below.

Simla—The tectonics of the Simla area are of great interest. Pilgrim and West have proved that the highly metamorphosed Jutoghs now resting on top of the practically unaltered Simla slates at Simla are not in their normal position, but have been inverted and thrust southward, from their original position in the central axis of the Himalayas, along a horizontal plane of thrust that has travelled for many miles. The effects of denudation on this overthrust sheet of the Jutoghs is to leave isolated outliers, "klippen", of older rocks capping the summits of the Jakko and Jutogh hills, while the main body of these mountains is built of younger rocks.

Central and Eastern Himalayas—In the eastern Himalayas, a series of schists of the same formation constitutes the *Daling series*. The Daling series extends along the Tista valley into Sikkim and thence to Bhutan, consisting of much-contorted slates and chloritic and sericitic phyllites with hornblende-schists and quartzites. Some lodes of copper are associated with these rocks which are commercially workable in Sikkim. A large part of the Mid. Himalayas of Kumaon and Nepal is covered by a zone of metamorphosed rocks identical with the Dalings, leaving little doubt about their continuity with the Sikkim Dalings. The Daling series of this region includes an overlying group of schists copiously injected with gneiss—*Darjeeling series*, which was formerly regarded as an older series overthrust on the Daling. It has been shown now that the apparently higher grade of the metamorphism of the Darjeeling group is not due to greater age, but to the inter-laminated intrusive granite of a later date. The Daling and Darjeeling series of Nepal, Sikkim, and Bhutan are homotaxial with the Jutoghs and Salkhalas of Simla and Kashmir; the Himalayan Archaeans of the intervening Garhwal area being composed of similar elements—highly metamorphised slates, migmatites, granulites and schists, containing such kata-metamorphic minerals as garnet, sillimanite and staurolite. Granite intrusions of several periods are prevalent in this wide, monotonously uniform belt of azoic rocks, which separate the central axial Himalayan zone from the Tertiaries of the Outer or Sub-Himalaya zone. Further to the east, in the yet imperfectly surveyed region of the Baxa Duars, Miris, Abor and Mishmi country of the Assam Himalayas, a more or less similar assemblage of metamorphosed rocks is recognised by its containing several characteristic types of the Dharwars of the Peninsula.

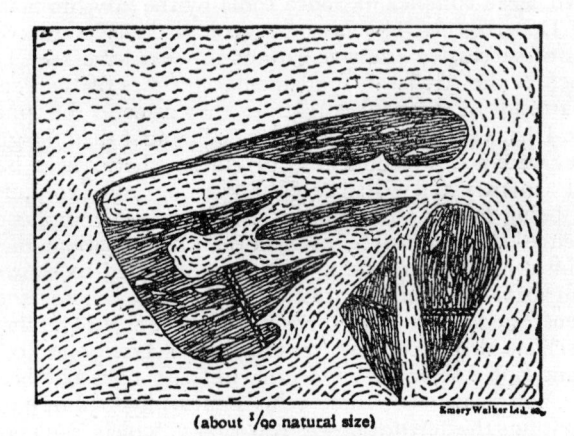

FIG. 7.—Diagram showing the relation of Dharwar schists with the gneisses.

(After Sampat Iyengar, *Rec. M.G.D.* vol. xi.)

With the exception of copper occurring at some places in Sikkim and Garhwal and some gemstones in Kashmir, the Himalayan Dharwars are barren of economic mineral deposits, unlike their peninsular representatives which carry many ores and deposits of industrial minerals. The metallogenic phase of the Dharwars of the South is, so far as present knowledge goes, missing from this structurally largest and potentially most promising belt of the Himalayas.

Homotaxis of the Dharwar system—With regard to the age of the Dharwar rocks, there is no doubt that they are far older than the Cambrian, separated therefrom by an immense interval of geological time represented by three or possibly four vast cycles of deposition, mountain-building and base-levelling. With regard to their lower limit, they are so closely associated and intermixed with the Archaean gneisses at certain places that they leave no doubt that some of the gneisses are younger than some of the Dharwar schists. From their field-relations, and from the circumstance of a widespread unconformity separating the Dharwars from all younger formations, Sir T. H. Holland has grouped them along with the Archaean. There is no parallel system of deposits comparable to the Dharwars in England or many parts of Europe, but the Dharwars show a degree of affinity with the Huronian rocks of America in their stratigraphic position and their petrological constitution.

A very careful and detailed investigation has been made in the great Archaean complex of South India by the Mysore State Geological Department. The Mysore geologists have unravelled a number of successive eruptive groups in what have been hitherto described as the Archaean fundamental gneisses of the Peninsula, and as a result of these investigations they came to the conclusion that the Dharwar schists were *all* decidedly older than the gneisses; that they were not of sedimentary origin as hitherto held, but were certainly in part and possibly entirely of igneous volcanic derivation, being in fact strictly basic lava-flows metamorphosed into hornblende- and chloritic schists. In their field-relations the Dharwar schists have again and again been observed to show a distinct intrusive contact towards the invading gneisses, and have been penetrated by the latter times without number. The characters of the schists also, according to these observers, point to an igneous and not a sedimentary origin, for they have not been able to trace any passage of these schists into phyllites or unaltered slates within the territories of Karnataka which encompass an area of nearly 77,700 square km. On the other hand, they show a gradual transition into epidiorites or hornblende-rocks. Many of the Dharwar conglomerates, likewise, are believed to be of crushed, *autoclastic*, origin. Fig. 7 gives an idea of the nature of the association of the two rock-groups. These views have been to a considerable extent modified as the result of later work by the State geologists.

The subject is one of the major controversies of Indian geology,

TABLE OF CORRELATIONS OF DHARWAR FORMATIONS

Mysore.	Singhbhum.	Rajasthan.	Eastern Ghats.	Madhya Pradesh.	Assam.	N.W. Himalayas.
Closepet Granite	Granite and "Dome Gneiss"	Alkali granites and syenites	Bellary, Hosur and Arcot Granite	Granite	Granite intrusions	Kazi Nag and Chor Granite?
Charnockite			Charnockite			
Peninsular gneiss	Bengal gneiss		Granite-gneiss	Granite-gneisses		Chail series
Sedimentary Dharwars	Iron-ore series Gangpur series	Raialo series: Aravalli system: Champaner series	Bastar Iron-ore series: Khondalites and Kodurites	Sakoli, Sausar and Chilpi series: Gondites	Shillong series Calc-gneiss	Salkhala series Jutogh series
Older gneisses	Older gneiss	Bundelkhand gneiss Banded gneisses		Older gneisses and schists		Archaean schists

but the prolonged study of the South Indian crystalline complex, by members of the Indian and Mysore State Geological Surveys, extending from 1902, has helped to clear it considerably. Present opinion tends to support the Mysore view in so far as the age of the main body of the Dharwars is concerned, though work in extra-Mysore areas equally supports the older views as regards the sedimentary nature and origin of a portion of these rock-bodies, there being little doubt about the detrital nature of the phyllites and quartzites.

The following general scheme of classification of the Archaeans of India,

4. The Charnockite and Bundelkhand Gneisses, with intrusions such as Peridotites, Granites and Syenites;
3. Re-melted masses of the Basement Gneiss, now constituting much of the schistose and garnetiferous Bengal and Peninsular Gneisses which include some para-gneisses and schists;
2. Dharwar sediments and contemporaneous lavas, also Khondalites;
1. The oldest Basement Gneisses representing, in part at least, the primitive crust of the earth,

adopted by Sir Lewis Fermor in 1919, is now amplified by the sub-division of the Archaean foundation of the Peninsula into 15 distinct provinces, based largely on their petrological characters. The Archaean terrain of India is first broadly divided into two regions, the *Charnockitic* and the *non-Charnockitic*; these major regions are further subdivided into a number of provinces, grouped under (1) Iron-ore provinces, (2) Manganese-ore-marble provinces, and (3) Igneous provinces, based on their compositional differences. In establishing these divisions and their correlations in different parts of the Indian Peninsula, Fermor uses the following criteria:

1. Stratigraphic sequence.
2. Structural relationships—unconformities, periods of folding, etc.
3. Relationship to igneous intrusives.
4. Associated ore-deposits of epigenetic origin.
5. Lithological composition.
6. Chemical composition.
7. Grade of metamorphism.
8. Lead and helium ratios[1].

Economics—The Dharwar system carries the principal ore deposits of the country, *e.g.* those of gold, manganese, iron, chromium, copper, tungsten, lead, etc. These with their associated rocks are also rich in such industrially useful products as mica, corundum,

---

[1] *Memoirs G.S.I.* vol. lxx. pt. 1, 1936.

etc.; the lithium minerals, lepidolite and spodumene; titanium and thorium minerals; rare valuable minerals like pitchblende, monazite and columbite, etc.; and a few gems and semi-precious stones like ruby, sapphire, emerald, aquamarine, beryl, chryso-beryl, zircon, spinels, garnets, tourmalines, amethyst, rock-crystal, etc. This system is also rich in its resources of building materials, e.g. granites, marbles, ornamental building stones, and roofing slates. The famous marbles of which the best specimens of ancient Indian architecture are built are a product of the Dharwar system.

Chronological Classification of the Archaeans

No system of classification of the oldest rock formations of India has met with general acceptance. However the application of U-Pb & Rb-Sr ratio methods of age measurement by Prof. A. Holmes and others enables us to arrive at the following sequence of the Archaean formations of India:

    Delhi system : Cuddapah system
        (Huronian) 750 m.y.
―――――――――――――――――――――Eparchaean unconformity
Cycle of intrusive gneisses and granites of
    the Peninsula
        (Satpura orogeny) 1000 m.y.
―――――――――――――――――――――Unconformity
    Aravalli system
        (E. Ghats & Rajasthan
        orogeny) 1600 m.y.
―――――――――――――――――――――Unconformity
    Dharwar system
        2400–1800 m.y.
―――――――――――――――――――――Eruptive unconformity
Fundamental gneisses and schists
    (Mysore gneisses: Bundelkhand
    gneiss) 2500 m.y.

REFERENCES

L. L. Fermor, *Mem. G.S.I.* vol. xxxvii., 1909; *J.A.S.B.* vol. xv. (New Series), 1919; *Mem. G.S.I.* vol, lxx., 1936.

A. M. Heron, Geology of Rajputana, *Mem. G.S.I.* vol. xlv., 1917 and 1922; and *Mem.* vol. lxviii. pt. 1., 1936; *Rec.* vol. liv. pt. 4, 1922.

Sir T. H. Holland and G. H. Tipper, *Mem. G.S.I.* xliii. pt. 1, 1913, and (Second Edition) *Mem.* li. pt. 1, 1926, The Dharwar System.

J. A. Dunn, Mineral Deposits of Eastern Singhbhum, *Mem. G.S.I.* vol. lxix. pt. 1, 1937; Geology of North Singhbhum, *Mem. G.S.I.* vol. liv., 1929.

H. C. Jones, Iron-Ore Deposits of Bihar and Orissa, *Mem. G.S.I.*, vol. lxiii. pt. 2, 1933.

R. B. Foote, Dharwar System in South India, *Rec. G.S.I.*, vol. xx. and xxi., 1888–1889; *Mem. G.S.I.*, vol. xx. pt. 1, 1883.

W. King, Geology of Trichinopoly, Salem, etc. *Mem. G.S.I.* vol. iv. pt. 2, 1864.

G. E. Pilgrim and W. D. West, Structure of the Simla Rocks, *Mem. G.S.I.* vol. liii., 1928.

D. N. Wadia, Geology of Nanga Parbat and parts of Gilgit, *Records G.S.I.* vol. lxvi. pt. 2, 1932.

B. Rama Rao, *Records, Mysore Geol. Dept.*, vol. xxxiv., 1936; *M.G.D. Bull.* 17, 1940.

M. S. Krishnan, Geology of the Gangpur State, *Mem. G.S.I.* vol. lxxi., 1937.

CHAPTER V

The Cuddapah System

Introduction—The closing of the Dharwar era must have witnessed earth-movements on a very extensive scale, which folded the Dharwar sediments into complicated wrinkles, creating a number of mountain-ranges, the most prominent among them being the mountain-chain of the Aravallis. No such powerful crustal deformation, of an equal degree of magnitude, seems to have occurred since then in the Peninsula, since all the succeeding systems show less and less disturbance of the original lines of stratification and of their internal structures, till, at the end of the Vindhyan era, all orogenic forces almost disappeared from this part of the earth.

Cuddapah system—A vast interval of time elapsed before the next rock-system began to be deposited, during which a great extent of Dharwar land, together with its mountains and plateaus, was cut down to the base-level by a cycle of erosion. For it is on the deeply denuded edges of the Dharwar rocks that the basement strata of the present formation rest. This formation is known as the Cuddapah system, from the occurrence of the most typical, and first-studied, outcrops of these rocks in the district of Cuddapah in the middle of Tamilnadu. The Cuddapah is a series of formations or systems, rather than a single system, it being composed of a number of more or less *parallel* series or groups of ancient sedimentary strata, each of the thickness and proportions of a geological system by itself. They rest, with a great unconformity, at some places on the Dharwars and at other places on the gneisses and schists, and themselves underlie with another unconformity the immediately succeeding Vindhyan system of Central India.

Lithology of the Cuddapahs—This system is mainly composed of much indurated and compacted shales, slates, quartzites, and limestones. The shales have acquired a *slaty cleavage*, but beyond that there is no further metamorphism into phyllites or schists; such secondary minerals as mica, chlorite, andalusite, staurolite, garnets, etc. have not been developed in them; nor are the limestones recrystallised into marbles, as in the Dharwar rocks. Quartzites, which are the most common rocks of the system, are metamorphosed sandstones, the metamorphism consisting of the introduction and deposition of secondary silica, in crystalline continuity with the rolled quartz-grains of the original sandstone. Contemporaneous volcanic action prevailed on a large scale during the lower half of the system, the records of which are left in a series of bedded traps (lava-flows) and tuff-beds. (See Fig. 8.) Besides the above rocks, the Lower Cuddapahs contain brilliantly coloured and banded cherts and jaspers and some interstratified

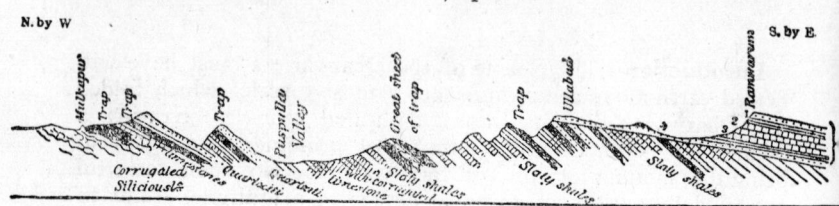

FIG. 8.—Sketch section illustrating the relation of Cuddapah and Kurnool rocks (marked K).
After King, *Mem. G.S.I.*, vol. viii, 1872.

iron- and manganese-ores, very much like those of the Dharwar system. In these two peculiarities, most noticeable in the lower part, the Lower Cuddapahs therefore resemble the Dharwar system; while the upper half, in its unmetamorphosed shales and limestones, shows a close resemblance to the overlying Vindhyan rocks.

On account of the absence of any violent tectonic disturbance of the Peninsula during later ages, the Cuddapah rocks have in general low angles of dip, except towards the eastern coast, where they form a part of the Eastern Ghats (the Yellaconda range of hills), and where consequently they have been subjected to much plication and over-thrust. To account for the enormous thickness of the Cuddapah sediments, which amounts to more than 6,000 m. in the aggregate, of slates and quartzites, it is necessary to suppose that a slow and quiet submergence of the surface was in progress all through their deposition, which lowered the basins of sedimentation as fast as they were filled.

Absence of fossils in the Cuddapahs—The entire series of Cuddapah rocks is totally unfossiliferous, no sign of life being met with in these vast piles of marine sediments. This looks quite inexplicable, since not only are the rocks true clastic sediments, and not chemical precipitates, laid down on the floor of the sea and very well fitted to contain and preserve some relics of the life inhabiting the seas, but also all mechanical disturbances and chemical changes, which usually obliterate such relics, are absent from them. It cannot again be surmised that life had not originated in this part of the world, since in formations immediately

THE CUDDAPAH SYSTEM

subsequent to the Cuddapahs, and in areas not very remote from them, we find evidence of fossil organisms, which, though the earliest animals to be discovered, are by no means the simplest or the most primitive. The geological record is in many respects imperfect, but in none more imperfect than this—its failure to register the first beginnings of life, by far the most important event in the history of the earth.

Classification—The Cuddapah system is divided into two sections, an upper and a lower, separated by a great unconformity. Each of these divisions consists of several well-defined series, whose stratigraphic relations to each other, however, are not definitely established, and which may be quite parallel or homotaxial to each other instead of successional. The Cuddapahs of the S.E. are represented by the *Delhi system* and *Gwalior series* in the north of India.

Kurnool series (Lr. Vindhyan)
Unconformity.

Upper Cuddapah
- *Kistna series*—slates and quartzites—*Kaladgi series*.
- 600 m.   3,355 m.
- *Nallamalai series* { *Cumbum slates*.
- 1,037 m.   *Bairenkonda quartzites*.

Unconformity.

Lower Cuddapah
- *Cheyair series*—shales and quartzites—*Bijawar series*.
- 3,200 m.
- *Papaghani series* { *Vaimpalli slates*. *Delhi system*.
- 1,372 m.   *Gulcheru quartzites*—*Gwalior series*

Unconformity.

Archaean and Dharwarian.

Distribution—A large development of these rocks occurs in the type area of Cuddapah district of Andhra. The outcrop is of an irregular crescent shape, the concave part of which faces the coast, the opposite side abutting on the gneisses. Another large development of the same system lies in the Chhatisgarh locality of Madhya Pradesh. A few isolated exposures occur in the intervening Godavari and Pranhita valleys, in the Vishakapatnam, Bastar and Kalahandi districts, and in the Singhbhum and Keonjhar districts further north (*Kolhan series*). A contemporaneous system of strata, compressed in the tight geosynclinal orogen, occurs in Rajasthan as the present Aravalli chain, the shrunken roots of a once great mountain-range.

The Lower Cuddapah—*The Papaghani series*. The lowest member is named from the Papaghani river. The bottom beds are sandstones followed by shales and slates, with a few limestone layers in the shales. Contemporaneous lava-flows, with intrusions of the same magma in the form of dykes and sills, are common; in the latter case, where the invading rock comes in contact with limestones, these are found to be converted into marbles, serpentines, and talc.

Economically the slate and limestone series (*Vaimpalli slates*) are of importance, because considerable deposits of barytes and asbestos occur in these rocks and their associated basaltic sills.[1]

The Delhi system—The Delhi system of strata referred to in the last chapter is probably of Lower Cuddapah age, though in its intense structural disturbance and degree of folding it departs from the general tectonic features of this system. It appears to be a locally specialised type of the Cuddapahs, owing its structural disturbance to local orogenic flexures and also to the intrusion of large bodies of granite and amphibolite. The Delhi system occupies a large extent of E. Rajasthan country extending from Delhi to Idar in constricted, sorely eroded synclinal bands in the centre of the great Aravalli synclinorium, its fullest development being found in the main Rajasthan geosyncline of Ajmer-Merwara and the Mewar area. The *Alwar quartzites*, which constitute a prominent part of the system, are quartzites, grits and flagstones. The Delhi system is intruded by a varied series of basic rocks and by a series of granite bosses and laccolites, with their related group of pegmatites and aplites (*Erinpura granite*), covering a large area to the west of the Aravalli range. The *Idar granite* (granite, microgranite and granophyre) occurs in a number of scattered masses at the south extremity of the outcrop of the Delhi system. The Delhi system, which may be taken as marking the commencement of the Purana Era, is characterised by a great variety and abundance of igneous intrusions and by an intenser grade of metamorphism than that observed in the older Aravallis (Archaeans). This circumstance is explained by the fact that the Delhis were buried more deeply in the roots of the synclinorium than the older Aravalli rocks, which form the flanks of the fold and have thus escaped severe metamorphism. The Purana Era in Rajasthan was one of igneous and orogenic activity, localised and more or less confined to the Aravalli mountains. Over the whole of this area the Delhi system exhibits violent unconformity with the Aravallis at its base, while towards the newer Vindhyan terrain to the east its relations are those of a great boundary fault, with a throw of over 1,500 m. Dr. A. M. Heron has classified the Delhi system as follows:

Semri series (Lr. Vindhyan) of Chitor

Unconformity.

Delhi System { *Ajabgarh series*: biotite-schist, phyllites, quartzites and impure biotitic limestones and calciphyres - - - - } 1,525 m.
Hornstone breccia - - Of variable thickness.

---

[1] A. L. Coulson, *Mem. G.S.I.* vol. xliv., 1934.

THE CUDDAPAH SYSTEM 117

Delhi   ⎧ *Kushalgarh limestone* - - - -   457 m.
System   ⎨ *Alwar series* : quartzites, arkose, conglome- ⎫ 3,050 m.
            ⎩   rates and mica-schists with bedded lavas   ⎭ 3,965 m.
Unconformity. 〜〜〜〜〜〜〜〜〜〜〜〜〜〜〜
Raialo   ⎧ Raialo limestones and marble.
Series   ⎩ Raialo quartzites.

The Bijawar series—The upper division of the Lower Cuddapah is more widely developed, and occurs extensively at Bijawar, Cheyair, Gwalior, etc. The *Bijawar series* is composed of cherty limestones, siliceous hornstones and ferruginous sandstones, haematite beds, and quartzites, resting unconformably on the gneisses. But the most distinctive character of the Bijawar series is the presence in it of abundant products of contemporaneous volcanic action—ash-beds, lava-flows and sills of a basic augite-andesite or basalt, now resting as a number of interbedded green traps. The dykes of these lavas that have penetrated the older formations are supposed to be the parent-rock of the diamonds of India. The celebrated "Golconda" diamonds were mostly derived from a conglomerate mainly composed of the rolled pebbles of these dykes. V. S. Dubey has reported a "diamondiferous plug" (a post-Bijawar trap dyke intrusive into the Bijawars) in the Rewah conglomerates of the Panna region.[1] Small diamonds are found in the matrix of this rock, which may be found to correspond to the "diamond pipes" of Kimberley, the prolific source of South African diamonds. Wherever the andesitic lava of the Bijawar series is subjected to folding and compression, it has altered into an epidiorite.

An exposure of very similar character, occurring in the valley of the Cheyair river, is known as the *Cheyair series*, while the one at Gwalior, on which the town of Gwalior stands, forms the *Gwalior series*. In the latter series there is a very conspicuous development of unmetamorphosed ferruginous shales, jaspers, porcellanites, and hornstones, associated with the andesitic or basaltic lavas of the Bijawar type. The porcellanite and lydite-like rocks appear to have originated from the effects of contact-metamorphism on argillaceous strata, while the preponderance of hornstones, cherts and other siliceous rocks points to the presence of solfataric action, connected with the volcanic activity of the period. Solfataras or hot siliceous springs come into existence during the declining stages of volcanoes; they precipitate large quantities of silica on the surface, likewise bringing about a good deal of *silicification* of the previously existing rocks by chemical replacement (metasomatism) in the underlying rocks. The lower division of the Gwalior series, resting upon the basement gneiss,

---

[1] *Ind. Sc. Cong. Proc.*, pt. iii., 1948.

is known as the *Par*, and the upper is designated the *Morar series*. Dr. Heron regards the Gwalior series as an isolated outcrop of unmetamorphosed Aravalli series, which owe their horizontality and absence of metamorphism to their distance from the main axes of folding of the Aravalli range and their protection by the resistant mass of Bundelkhand gneiss upon which they rest.[1]

An outlier formed of identical rocks is seen in the valley of the Pranhita, and is named *Penganga beds*. It must be understood that the reason for giving these different local names to the different occurrences of what may ultimately prove to be the same division of the Lower Cuddapah is the uncertainty, which is always present in the case of unfossiliferous strata, of correlating them with one another in the absence of any positive evidence.

No indubitable fossil remains have been discovered in the Cuddapahs except some discoid tabular or columnar forms in limestones belonging to various horizons including *Kaladgi* series. These bodies, known as *Stromatoliths*, possessing finely lamellar structure from 2.5 cm. to several metres in size, are believed to be marine *algae* remains, though no cellular structure is preserved.

The Upper Cuddapahs—The Upper Cuddapahs rest unconformably over the rocks last described at a number of places. The most important development is in the type area of the Cuddapah basin, where it has received the name of the *Nallamalai series*, from the Nallamalai range of hills in which it is found. The component rocks of the *Nallamalai series* are quartzites (Bairenkonda quartzites) in the lower part, and indurated shales and slates (Cumbum slates) in the upper. In the limestone beds that occur intercalated with the shales there is found an ore of lead, galena.

The Upper Cuddapahs of Chhattisgarh, occupying large areas in the upper Mahanadi valley, in Drug, Rajpur, Bilaspur and Sambulpur districts, constitute two series, *Rajpur series*, of shales and limestones, 600 m. thick and *Chandarpur series* of quartzitic sandstones, 300 m., resting unconformably over the Dharwars. They are composed of comparatively less distorted gently folded rocks and for that reason the two series were taken to belong to the Kurnool, or Lr. Vindhyan horizon rather than to the Cuddapah.

The Kaladgi series—The *Kaladgi series*, another member of the same system, is several thousand metres of quartzites, limestones, shales, conglomerates and breccias, occupying the country between Belgaum and Kaladgi in the Bijapur district. Towards the west they disappear under the basalts of Deccan Trap age. The upper part includes some haematite-schists, which include sometimes so much haematite as to constitute a workable ore of iron. Besides

---

[1] *Mem. G.S.I.* vol. lxviii. pt. 1, 1936.

the above there are other localities where rocks of the Upper Cuddapah horizon occur, *viz*. in the Krishna valley (the *Kistna series*), in the Godavari valley (the *Pakhal series*, of 2,285 metres of quartzites, slates and flinty limestone), and in Rewah. C. Mahadevan suggests that the Pakhals are really much older, belonging to the Dharwar system, and comparable with the *Gangpur series* of Orissa, or with the less metamorphosed outcrops of *Khondalites*. It is also possible that a part of the Kaladgi series, the part occurring in Ratnagiri district, heavily intruded by acid and basic rocks, is likewise of Dharwar age (L. A. N. Iyer).

Economics—The economic importance of the Cuddapah rocks lies in some iron and manganese ores, interbedded with the shales and slates. Numerous workable deposits of barytes and asbestos occur among the Papaghanis in the Cuddapah and Kurnool Districts of Tamilnadu (p. 463). Other products of some use are variegated marbles, steatite, and the bright-coloured jaspers and cherts, which are used, when polished, in interior decoration and inlaid work, as in the old Moghul buildings. The Delhi system contains some lodes of metallic compounds. Most of the copper-ores and some cobalt and nickel ores known in Rajasthan are associated with rocks of the Delhi system.

Stratigraphic position—The stratigraphic relations of the Cuddapahs prove that they are far younger than the Dharwars. On the other hand, their thoroughly azoic nature, and the moderate degree of metamorphism they have undergone, show that the Cuddapahs are older than the Vindhyans. In their lithological characters they show much resemblance to the pre-Cambrian Algonkian system of North America. The Cuddapah basin has, during late years, received much attention regarding its stratigraphy, tectonics, geomorphology and effects granite intrusives.

It is difficult to correlate with the Cuddapahs the Himalayan representatives of the Peninsular Puranas occurring to the south of the crystalline axis of the Range (p. 129). It seems probable, however, that the older members of the Attock, Dogra and Simla Slate series, as well as the thick succession of strata intervening between the latter and the underlying Salkhalas and Jutoghs, which has received local names, e.g. the *Chail*, *Chandpur* and *Mandhali series* in the Chakrata-Garhwal area of the Mid. Himalayas and the less metamorphosed members of the *Vaikrita* and the *Haimanta* system of the Spiti area north of the Central Himalayan axis, may be of Cuddapah affinities. The age of the Mandhali series is now held in some doubt. It is found resting upon the Deoban limestones (p. 128) and comprises a group of boulder-conglomerate beds, thought to be the equivalent of the Blaini beds of Simla (p. 216) of much newer age.

REFERENCES

W. King, Kadapah and Karnul Formations in Madras Presidency, *Mem. G.S.I.* vol. viii. pt. 1, 1872.

R. B. Foote, Geology of Madras, *Mem. G.S.I.* vol. x. pt. 1, 1873; Geology of Southern Mahratta Country, *Mem. G.S.I.* vol. xii., 1876.

A. M. Heron, Geology of Central Rajputana, *Mem. G.S.I.* vol. lxxix., 1953; and Geology of South-Eastern Mewar, *Mem. G.S.I.* vol. lxviii. pt. 1, 1936; Correlation of the Cuddapah and Kurnool formations, *Jour. Hyd. Geol. Surv.*, vol. v. pt. 1, 1947.

C. Mahadevan, Some Aspects of the Puranas of South India, *Ind. Sc. Cong. Proc.*, pt. 2, 1949.

A. Holmes, Pre-Cambrian Formations of India, *Geol. Mag.* vol. 87-3, 1950.

P. Lake, Basic eruptive rocks of the Cuddapah area, *Rec. G.S.I.* vol. 23, 1890.

CHAPTER VI

The Vindhyan System

Extent and thickness—The Vindhyan system is a vast stratified formation of sandstones, shales and limestones encompassing a thickness of over 4,270 m., developed principally in the central Indian highlands which form the dividing ridge between Hindustan proper and the Deccan, known as the Vindhya mountains. They occupy a large extent of the country —a stretch of over 103,600 square km.—from Sasaram and Rohtas in Western Bihar to Chitorgarh on the Aravallis, with the exception of a central tract in Bundelkhand; while a large area of Vindhyan rocks is covered by the Deccan trap. The outcrop has its maximum breadth in the country between Agra and Neemuch.

Rocks. Structural features—The Vindhyan system is composed of two distinct facies of deposits, one marine, calcareous and argillaceous, characteristically developed in the lower part, and the other almost exclusively arenaceous, of fluviatile, or estuarine deposition, forming the upper portion. The shale, limestone and sandstone strata show very little structural displacement or disturbance of their primeval characters; they have preserved almost their original horizontality of deposition over wide areas; the rocks show no evidence of metamorphism, as one is led to expect from their extreme age, beyond induration or compacting. The shales have not developed cleavage nor have the limestones undergone any degree of crystallisation. The only locality where the Vindhyan strata show any marked structural disturbance is along the south-east edge of the Aravalli country, where they have been affected by folding and overthrust due to the crust-movements which succeeded their deposition, and their internal mineral structure considerably altered, especially in the case of the freestones which have become quartzites. The epeirogenic upheaval which lifted up the Vindhyan deposits from the floor of the sea to form a continental land-area was the last serious earth-movement recorded in the history of the Peninsula, no other disturbance of a similar nature having ever affected its stability as a land-mass during the long series of geological ages that we have yet to review. The Peninsula has remained an impassive solid block of the lithosphere, unsusceptible to any folding or plication, and only affected at its fringes by slight movements of secular upheaval and depression.

The Vindhyan sandstones throughout their thickness give evidence of shallow-water deposition in their oft-recurring ripplemarked and sun-cracked surfaces, and in their conspicuous current-bedding or diagonal lamination, characters which point to the shallow agitated water of the coast, near the mouths of rivers, and the constantly changing velocity and direction of its currents.

Life during the Vindhyan Age—Except for a few obscure traces of animal and vegetable life occasionally discernible in the Vindhyan system, and such plausible evidences of the existence of life as are furnished by the presence of thick limestone strata and beds of carbonaceous shales, glauconitic sandstones, and some lenticles of bright coaly matter (vitrain), occurring at the base of the Kaimurs at Japla, this vast pile of sandstones, shales and limestones is characterised by an almost total absence of recognisable organic remains. The only fossils that have been hitherto discovered in these rocks are small carbonised, horny discs, 1–3 mm., which are believed to belong definitely to some fossil organism; these have been found embedded in black shales at the base of the Kaimur series (*Suket shales*) by Mr. H. C. Jones, near Rampura, Central India. But the specimens are too imperfectly preserved for specific or even generic determination and have been variously identified by palaeontologists as minute horny valves of primitive brachiopods, possessing affinities with *Acrothele* or *Neobolus*, and also as *algal* plant remains. Stromatoliths of algal origin have been discovered from various Vindhyan horizons. *Fucoid* markings, belonging to indistinguishable *thallophytic* plants, are usually seen on the ripple-marked and sun-cracked surfaces of sandstones and shales. The age of the Vindhyan system is thus uncertain, though it is probable that the topmost part of the system may represent a basal Cambrian horizon. The striking lithological similarity of the Upper Vindhyans with the *Purple sandstone* of the Salt-Range Cambrian is suggestive in this respect.

Classification—The Vindhyan system has been divided into the Lower and Upper divisions of very unequal proportions, but justified by an unconformity between the two parts, quite apparent at some places and non-existent at others, and also by a sharp lithological contrast between the lower and upper portions of the system.

The Lower Vindhyans show tectonic deformation by folding movements, while the Upper Vindhyans are generally lying in undisturbed horizontal strata.

PLATE VII. UPPER REWAH SANDSTONE, RAHUTGARH, SANGOR DISTRICT.

THE VINDHYAN SYSTEM

Series.		Stages.
Upper Vindhyan	Bhander.	Upper Bhander sandstone. Sirbu shales. Lower Bhander sandstone. Bhander limestone.
		Conglomerate-bed.
	Rewah.	Upper Rewah sandstone. Jhiri shales. Lower Rewah sandstone. Panna shales.
		Conglomerate-bed.
	Kaimur.	Upper Kaimur sandstone. Kaimur conglomerate. Bijaigarh shales. Lower Kaimur sandstone. Suket shales.

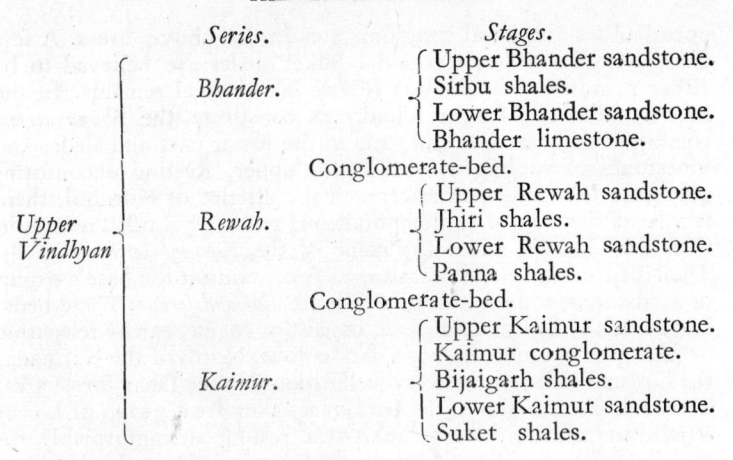

Lower Vindhyan—
   Semri Series. Kurnool Series. Bhima Series.
   Malani Series of rhyolites and tuffs.
   Granite bosses of Jalor and Siwana.

Distribution of the Lower Vindhyan—The most typical, and at the same time the most conspicuous, development of the system is along the great series of escarpments of the Vindhyan range, north of the Narmada Valley, particularly in Malwa and Bundelkhand in Central India, from which the system takes its name. The lower division is well displayed in the Son valley, in Chhatisgarh and in the valley of the Bhima. The Lower Vindhyans of the Son valley have been the subject of a detailed study by J. B. Auden which throws light on conditions of sedimentation, palaeogeography, climate and the question of the prevalence of life at the time. He groups together 900 metres of limestones, shales and sandstones with interbedded porcellanites (silicified ash and tuffs), glauconitic sandstones, and intrusive dolerites into the *Semri series*, which conformably underlies the Kaimur series of the Upper Vindhyan. There are conglomerates, epiclastic breccias, and pebble-beds in the Semris, which show the great variability and instability of physical conditions of the period, in contrast with the striking uniformity of deposition which persisted all through the Upper Vindhyan. The *Semri series*, or its equivalents, are found in the Son valley, Karauli State (Rajasthan) and at Chitor. The uppermost stage, known as the *Rohtas stage*, of 150–215 metres, is composed of limestones and shales which support the cement industry of the Son valley. Its equivalents are the *Suket shales* in Chitor, and the *Tirohan limestone* in Karauli overlain by beds of *Tirohan Breccia*. The Rohtas stage is underlain by olive shales, glauconitic beds, porcellanites and basal conglomerates in the above areas. A few discoid bodies occurring in the Suket shales are believed to be either primitive brachiopods (*Fermoria*) or algal remains. In the Bhima valley the Lower Vindhyans constitute the *Bhima series*, composed of quartzites and grits in the lower part and shales and limestones of varying colours in the upper. Resting unconformably over the Cuddapah system, in the district of Kurnool, there is a large outcrop of contemporaneous rocks, about 360 metres in thickness, known under the name of the *Kurnool series* (Fig. 8). The Kurnool series is interesting as it contains at the base a group of sandstones, some bands of which are *diamondiferous*. These beds, known as the *Banaganapalli beds*, consist of coarse, earthy felspathic or ferruginous sandstones of a dark colour. North of the Narmada, the Lower Vindhyans are very well exposed in the Dhar forest area. The *Sullavai sandstones* of the Godavari valley are a group of Lower Vindhyan sandstones and quartzites resting unconformably on the *Pakhal quartzites*. Contemporaneous in age with the Kurnools is the great thickness of limestones, shales and quartzites, constituting the *Palnad series* of Hyderabad and adjoining areas. The composition of all these occurrences shows local variations in the rock-types, but in the main conforms to the argillaceous and calcareous nature of the system. Some of the limestones show a concretionary structure, the concentric layers exhibiting different colours and giving to the polished rock a beautiful marble-like appearance. The limestones of the Lower Vindhyan formation are extensively drawn upon for burning as well as for building purposes. The Rohtas limestone of the Shahabad district is especially valuable for lime and cement manufacture, and is largely quarried.

The Vindhyans of Rajasthan. The Malani series—The unique sequence of Archaean and Purana sedimentary deposition in the Rajasthan synclinorium came to an end with the Vindhyan period. A large development of Vindhyans is seen on the east flank of the Aravallis and a lesser one, in detached outcrops, in the desert regions to its west. The Lower Vindhyan rocks of Western Rajasthan deserve special notice. Rocks which may be correlated to this system show there a very much altered facies, being composed of a group of rhyolitic lavas with abundant pyroclastic material, resting unconformably on the Aravalli schists. This volcanic series is known as the *Malani series*, from the district of that name (near Jodhpur in Rajasthan). The Malani rhyolites cover some thousands of square km. around Jodhpur. They are partly glassy, much devitrified, amygdaloidal lavas largely interstratified with tuffs and volcanic breccia. The lavas vary in acidity from rhyolites to quartz-andesites. In the majority of cases they have undergone such an amount of devitrification that they appear almost as felsite, the glassy ground-mass having completely disappeared. An outcrop of the Malani series composed of felsitic rhyolites and tuffs occurs, remote from the Aravallis, in the plains of Northern India, in the Kirana hills in the Punjab, small highly eroded outliers of the Aravalli chain.[1] In the Vindhyan terrain of S.E. Mewar the Malani volcanic and the Semri series are represented by a group of limestones, shales and sandstones with breccias and conglomerates.

Connected with these lava-flows, as their subterranean plutonic roots or magma-reservoirs which supplied the materials of the eruptions, are bosses of granite, laid bare by denudation in some parts of Rajasthan. Two varieties of granite are recognised in them—one, hornblende-biotite-granite (*Jalor granite*), and the other, hornblende-granite (*Siwana granite*). The latter boss shows distinctly intrusive relations to both the Malani series and the Aravalli schists; it rises to a height of nearly 900 m. above sea-level.

With the Vindhyan era, the most important chapter in the geological history of Rajasthan came to a close. Deposits of some Mesozoic and Eocene systems are found only in a few scattered outliers in Eastern Rajasthan, for the most part concealed under the desert sands. The tectonics of Rajasthan is of great interest as revealing the structure of the part of the Indian foreland whose northern promontory, the "Punjab wedge", has played such a part in moulding the orientation of the Himalayan, and according to Mushketov, also of the Pamir and Ferghana ranges. The main period of crustal deformation and igneous activity in Rajasthan was the Purana Era. The orogenic activity was localised and more or less confined to the Aravalli belt from north of Delhi to Gujarat, so that outside this orogenic zone the rocks, even though so ancient, are unmetamorphosed.

Meaning of "Lower" and "Upper" Vindhyans—The Lower Vindhyan is separated from the Upper by an unconformity that is very apparent in the north but which tends to disappear in the south areas of Mewar, Chitor and the Son valley. This signifies that earth-movements supervened after the deposition of the Lower Vindhyan sediments which elevated them into land in the Aravalli area of the north and put a stop to further sedimentation in these areas. When, after re-submergence, deposition was renewed, an interval of time had elapsed, during which the former set of conditions disappeared, and the mountains and highlands which yielded the detritus changed completely. Such earth-movements, causing cessation of deposition in a particular area, with a change in the physical conditions, are at the root of stratigraphic divisions. Smaller and more local breaks in the continuity of a

---

[1] *Rec. G.S.I.* vol. xliii. pt. 3, 1913.

stratified succession have led to its further subdivision into *series* and *stages*, while profounder changes, accompanied by more pronounced alterations of land and sea, affecting the inter-continental and inter-sea migrations of life inhabiting them, determine the limit between *system* and *system*.

Upper Vindhyan—In their type-area, north of the Narmada, the Upper Vindhyan sandstones consist of three well-marked divisions (series):

Bhander series - - $\begin{cases} \text{Upper Bhander sandstone,} \\ \text{Sirbu shales.} \\ \text{Lower Bhander sandstone.} \\ \text{Bhander limestone.} \\ \text{Ganurgarh shales.} \end{cases}$

Diamondiferous beds.

Rewah series - - $\begin{cases} \text{Upper Rewah sandstone.} \\ \text{Jhiri shales.} \\ \text{Lower Rewah sandstone.} \\ \text{Panna shales.} \end{cases}$

Diamondiferous beds.

Kaimur series - - $\begin{cases} \text{Upper Kaimur sandstone.} \\ \text{Kaimur conglomerate.} \\ \text{Bijaigarh shales.} \\ \text{Lower Kaimur sandstone.} \\ \text{Suket shales.} \end{cases}$

The East India Railway from Katni to Allahabad runs through the heart of the Vindhyan country, and thence up to Dehri-on-Son passes along its north-eastern margin, without ever leaving sight of the outcrops of horizontally bedded red or buff sandstones. Another Vindhyan province lies in central India, on the eastern borders of the Aravalli chain. This country is also crossed by the railway from Jhalrapatan to Bharatpur, which almost constantly keeps within sight of, or actually meets, a series of illustrative outcrops of the system. Prevalence of arid, continental conditions in the Upper Vindhyan times is suggested by the perfect rounding of quartz-grains in the majority of the sandstones, and also by the prevailing red and brown colours of the sediments and by the occasional presence of gypsum in the Bhander shales.

The junction of the Upper Vindhyans with the older rocks of the Aravallis, at their north-east extremity, reveals an extremely long fault of great throw, which has brought the undisturbed, almost horizontal strata of the Vindhyan sandstone (*Bhander series*) in contact with the highly folded and foliated schists of the Aravallis. This great fault, which has a throw of 1,525 m., is roughly parallel with the course of the river Chambal and can be traced from the western limit of the outcrop to as far north as Agra, a distance of 800 km. It is possible that this junction is not of the nature of an ordinary fracture or dislocation, but marks the approximate *limit of deposition* of the younger Vindhyan sandstone

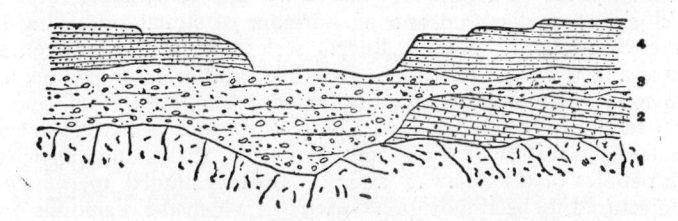

FIG. 9.—Section showing relation between Gwalior series and rocks of the Vindhyan system (after Oldham).
4. Vindhyan (Kaimur) sandstone.
3. Kaimur conglomerate.
2. Gwalior series (Par sandstone).
1. Bundelkhand gneiss.

against the foot of the Aravallis which was modified subsequently by faulting and thrusting. The fault, therefore, is of the nature of a "Boundary Fault", which recalls the much better known case of the junction of the younger with the older Tertiaries of the Himalayas. (See Siwalik System, Chapter XX, pp. 336–7.)

Vindhyan sandstones—Sandstones are by far the most common rocks throughout this division with the exception of the lower Bhander series, which is for the greater part calcareous. The sandstones are of a uniformly fine grain, preserving their uniformity of texture and composition unchanged for long distances. The colours are variegated shades of red, yellow or buff, or grey, while they are often mottled or speckled, owing to the variable dissemination of the colouring matter, or to its removal by deoxidation. The Kaimur as well as the Bhander sandstones are fine-textured, soft, easily workable stones of a deep red tint, passing now and then into softer shades of great beauty. These sandstones are available for easy quarrying in any quantity in all the localities mentioned. No other rock-formation of India possesses such an assemblage of characters, rendering it so eminently suitable for building or architectural work. When thinly stratified, the rock yields flags and slabs for paving and roofing purposes; when the bedding is coarse, the rock is of the nature of freestone, and large blocks and columns can be cut out of it for use in a number of building and architectural applications.[1]

Shales are sparsely developed in the Upper Vindhyan division,

---
[1] See Chapter XXVI—Building Stones, p. 424.

and are of local occurrence only. They are often carbonaceous. At other times they are siliceous or calcareous. They are distinguished under various names, such as Bijaigarh shale, Panna shale, Jhiri shale, etc., from their localities.

Economics—The Upper Vindhyans are remarkable for their enclosing two diamond-bearing horizons of strata, one lying between the Kaimur and the Rewah series, the other between the latter and the Bhander series. The historically famous Panna and Golconda diamonds were mined from these beds, from one or two small productive patches. The country-rock is a conglomerate containing water-worn pebbles of older rocks, among which are pebbles of the Bijawar andesite already alluded to, which is conjectured to be the original matrix in which the diamonds once crystallised. The Vindhyan system is not possessed of any metalliferous deposits, but is rich in resources of building materials, which furnish an unlimited measure of excellent and durable freestones, flagstones, ornamental stones, and large quantities of limestones for the manufacture of lime and cements. The Bhander stage has yielded materials for the building of some of the finest specimens of Indian architecture. The famous *stupas* of Sanchi and Sarnath, the Moghul palaces and mosques of Delhi and Agra, and the modern government edifices of New Delhi are built of Vindhyan sandstones. The economic aspects of the Vindhyan rocks are dealt with in the chapter on Economic Geology.

Himalayan Vindhyans—The extra-Peninsular representatives of the Vindhyans, and probably also of the Cuddapahs, are surmised to be largely present in the belt of unfossiliferous sedimentary rocks that lies between the crystalline rocks of the central and the younger rocks of the outer Himalayas. It is a question how far they are homotaxial with the Vindhyans, or with the Raialos or the Delhis of Rajasthan. They are designated by various names in the different parts of the mountains. Near Peshawar they form a large outcrop of dark slates (the *Attock slates*), with a few limestones and sandstones here and there, permeated with trappean intrusions; in Hazara also there is a large outcrop of black unfossiliferous slates. A prominent belt of slates and associated rocks occurs in the south-west flank of the Pir Panjal and Dhauladhar ranges of the Kashmir Himalaya. This series has been named the *Dogra slates*. The Dogra slates are unconformably overlain by a great thickness of unfossiliferous sediments—the *Tanawal series*. In the Simla area the Vindhyans are probably recognisable in a thick series of dark unaltered slates and micaceous sandstones under the name of *Simla slates*. The Simla slates are succeeded after a pronounced hiatus, by a group of banded slates, sandstones and pebbly quartzites, named the *Jaunsar series*. The Tanawals and Jaunsars are in all probability representatives of the Lower and Mid-Palaeozoic fossiliferous formations described in the following two chapters. North of Chakrata, rocks of this age, forming the peak of Deoban, are known as the *Deoban series*. They consist of extremely compact grey dolomite and limestones with cherty concretions. Near Darjeeling, the Western Duars and the foot-hills of Bhutan, they constitute the *Baxa series* of quartzites, slates and dolomites occurring in bands between the Daling outcrop and the Gondwana strips of the eastern sub-Himalayas. All the Vindhyan rocks of the Himalayas are distinguished from the Vindhyans of the Peninsula by the scanty development in them of the arenaceous facies and the predominance of argillaceous elements; also, as is quite obvious, they are much folded, compressed and inverted by being involved in the severe flexures of the mountains. As a rule these older rocks overlie the younger members of the sub-Himalayan zone along a plane of overthrust—this being the most persistent feature of the structure of the Outer Himalayas from the Punjab to Assam (see p. 392).

The relation of the Himalayan unfossiliferous system to the Peninsular Puranas—It is the belief of the Indian Geological Survey, first promulgated by Sir T. H. Holland, that these old unfossiliferous formations developed on the south of the central Himalayan axis, representing the Dharwar, Cuddapah and Vindhyan systems of the Peninsula, are only the northern outliers or prolongations of the respective Peninsular systems, which were once continuous and connected before the Himalayan area became demarcated from the Peninsula by the upheaval of the Himalayan chain and the concomitant formation of the deep Indo-Gangetic depression. During these movements the extra-Peninsular extensions of the Dharwar, Cuddapah and Vindhyan systems were caught up in the Himalayan system of flexures, while their "Peninsular congeners" were left undisturbed. The belief receives strong confirmation from the fact that on the northern side of the central axis, *i.e* the Tibetan, there is an altogether different sequence of strata from that occurring on the Indian side, being composed of marine fossiliferous sediments of almost every geological age from the Cambrian to the Eocene. This total difference in the facies of the deposits of the two sides of the chain suggests the prevalence of altogether different physical and geographical conditions in them, and indicates that the two areas (Tibet and India) were from the earliest times separate and underwent altogether different geological histories.

Homotaxis—With regard to the homotaxis of the Vindhyan system there exists some difference of opinion. From its lithological agreement with the fossiliferous Cambrian of the Salt-Range, Vredenburg has considered it to be Cambrian in age, while Sir T. H. Holland regarded all the unfossiliferous Peninsular formations resting above the Archaean-Dharwar complex as pre-Cambrian, occupying much the same position as the Torridon sandstone of Scotland overlying the Lewisian gneisses, and grouped them in his *Purana group*. The Purana group of this eminent author includes the unmetamorphosed but more or less disturbed and folded rock-system that intervenes between the crystalline Archaean and the fossiliferous younger systems of the Peninsula. The Purana group thus forms a sort of transition between the foliated and the highly metamorphosed Dharwar and Archaean gneisses and the fossiliferous Palaeozoic strata. It includes the major part of what, in the early days of Indian geology, was called the *Transition System*. The discovery of the few undoubted organic remains and stromatoliths suggestive of the action of life, both in the Lower and Upper Vindhyan, now lifts this rock-system from the pre-Cambrian to an indefinite horizon in the Cambrian. Future discoveries of fossils may prove that the upper part of the apparently barren Puranas of parts of the Himalayas is really Lower Palaeozoic, and owes its generally unfossiliferous character to accidental circumstances.

We have seen in Chapter IV that the same author has linked the Dharwar with the Archaean system, recognising, in the unconformity that separates the former from the Puranas, a far wider significance and more extensive lapse of time than in that which separates the Archaean from the Dharwars.

	Recent ↑	
	Productus Series and Talchir Series (Upper Carboniferous and Permian).	Aryan.
Fossiliferous.	~~~~~~~~~~~~~~~~	Palaeozoic unconformity.
	Po Series (Lower to Middle Carboniferous). ↑ *Haimanta System (Cambrian).*	Dravidian.
	~~ ↑	Post-Vindhyan break.
	Vindhyan System. ↑ *Cuddapah System.*	Purana.
Unfossiliferous.	~~~~~~~~~~~~~~~~	Eparchaean unconformity.
	Dharwar System and Archaean System.	Archaean.

The table above shows in outline the scheme of classification of the Indian formations adopted by the Geological Survey of India. The classification of the post-Purana systems is based upon the recognition of the two most profound breaks in the continuity of that series of deposits. These breaks or "lost intervals" have a fundamental meaning in the geological history of India; they denote periods of great crust-movements and erosion, and mark the commencement of new eras of life and sedimentation. The first break was subsequent to the Vindhyans, and is universally observed in both the Peninsula and the extra-Peninsula. The other is a somewhat less pronounced break at the base of the Permian in the extra-Peninsula. In all the other areas of India, the post-Vindhyan break is the most momentous and universal, and comprehends a long cycle of unchronicled ages from the Vindhyan to the Permo-Carboniferous.

REFERENCES

R. D. Oldham, Geology of Son Valley, Jabalpur, etc., *Mem. G.S.I.* vol xxx. pt. 1, 1900.

T. H. D. La Touche, Geology of Western Rajputana, *Mem. G.S.I.* vol. xxxv. pt. 1, 1902.

H. B. Medlicott, Vindhyan Rocks of Bundelkhand, *Mem. G.S.I.* vol. ii. pt. 1, 1859.

F. R. Mallet, Vindhyans of Central Provinces, *Mem. G.S.I.* vol. vii., pt. 1, 1869.

A. M. Heron, *Geography* and *Geology of the Himalaya*, pt. 4 (Second Edition), 1934.

Sir T. H. Holland and G. H. Tipper, *Mem. G.S.I.* vol. xliii. pt. 1, 1913 and (Second Edition) vol. li. pt. 1, 1926, Archaean—Dharwar—Purana.

A. M. Heron, Geology of S. E. Rajputana, *Mem. G.S.I.* vol. xlv. pt. 2, 1922; *Mem.* vol, lxviii. pt. 1, 1936, vol. lxxix, 1953.

J. B. Auden, Vindhyans of the Son Valley, *Mem. G.S.I.* vol lxii. pt. 2, 1933.

E. Vredenburg, Geology of the Panna State Diamondiferous deposits, *Rec. G.S.I.*, vol. xxxiii. pt. 4, 1906.

W. King, Kadapah and Kurnoof formations of Madras, *Mem. G.S.I.*, vol. viii., 1872.

CHAPTER VII

The Cambrian System

The Cambrian of India—Marine fossiliferous rocks of Cambrian age are found in a thick series of strata at three places in the extra-Peninsula, each of which deserves a separate description. The first and the most easily accessible locality is the Salt-Range in the north-west Punjab; the second is the remote district of Spiti in the northern Himalayas, in the district of Kangra, beyond the crystalline axis of the Himalayas. The third area is the Baramula district of Kashmir. These rocks contain well-preserved fossils, and hence their age is no longer a matter of conjecture or hypothesis, as was the case with the Peninsular formation last dealt with.

[*The Salt-Range*—The Salt-Range is the most important locality in India for the study of physical as well as stratigraphical geology. Since very early times it has attracted the attention of geologists, not only because it contains a very large portion of the fossiliferous stratified record of the Indian region, but because of the easily accessible nature of the deposits and the clearness with which the various geological formations are exposed in its hills. Besides the stratigraphical and palaeontological interest, there is inscribed in its barren cliffs and dried gullies such a wealth of geodynamical and tectonic illustrations that this imposing line of hills can fitly be called a field-museum of geology. The Salt-Range is a continuous range of low, flat-topped mountains rising abruptly out of the flat Punjab plains. The range extends from long. 74° to 71° E. with an approximately east-west strike, from the Jhelum westwards, through the Indus, to a long distance beyond, undergoing where it crosses the Indus a deep bend of the strike to the south-west. In structural, stratigraphical as well as physiographic features the Salt-Range offers a contrast to the north-western portion of the Himalayas, which rise hardly 80 km. farther north, although the two ranges are clearly allied in their geological systems and in their orogenic phases. The prominent structural peculiarity of the Salt-Range is the more or less level plateau-top, ending abruptly on the one side in a long line of steep escarpments and cliffs overlooking the Punjab, and on the other northern side inclining gently towards and merging into the high Potwar plains, which represent a synclinal trough between the Salt-Range and the

THE CAMBRIAN SYSTEM

Rawalpindi foot-hills, filled up by Tertiary deposits. The general dip of the strata is in the north direction, from one end of the range to the other. Thus, it is on the north border that the youngest Tertiary rocks of the mountains are seen, inclining away from the steep escarpment, while it is in these steep escarpments that the oldest Palaeozoic formations are exposed. The line of high precipitous cliffs is intersected by a number of deep gullies and ravines, some of them deserving the name of cañons, affording sections which distinctly reveal the inner architecture of the range, as well as the details of its stratigraphy. There is little vegetation or covering of decomposed rock or soil to hide the details of these sections. Extensive heaps of talus or scree-deposits are seen all along the southern foot of the range at the base of the bold bare cliffs.

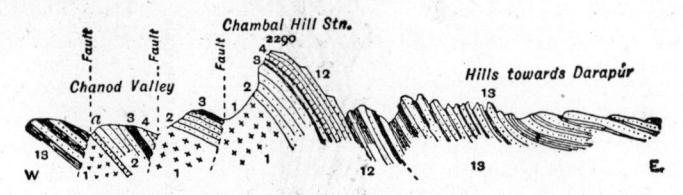

FIG. 10.—Section illustrating the general structure of the Salt-Range (Block-faults). Section over Chambal Hill (East).

12-13 Siwalik sandstones and clays (Upper Tertiary).
4. Magnesian sandstone.
3. *Neobolus* beds.
2. Purple sandstone.
a. Dolomite bed in Salt-marl.
1. Salt-marl and gypsum.

Wynne, *Mem.*, *G.S.I.*, vol. xiv.

The entire length of the range is faulted in a most characteristic fashion by a number of transverse dip-faults into well-marked blocks (block-structure), (Fig. 10). These clean-cut faulted blocks are so conspicuous to one who looks at the range from the plains that they can be separated out, and the main elements of their composition recognised, from great distances. At many places the faults are of the *reversed* type, sometimes intensified into thrust-planes, which have introduced a great deal of complication into the structure and stratigraphy of the area. (See Figs. 10, 11, 20 and 21).

The name Salt-Range is aptly derived from the circumstance that its lowest exposed rock contains large beds or lenses of pure common salt, all throughout its extent. In this way an immense quantity of rock-salt is embedded and available for extraction in many parts of these mountains.]

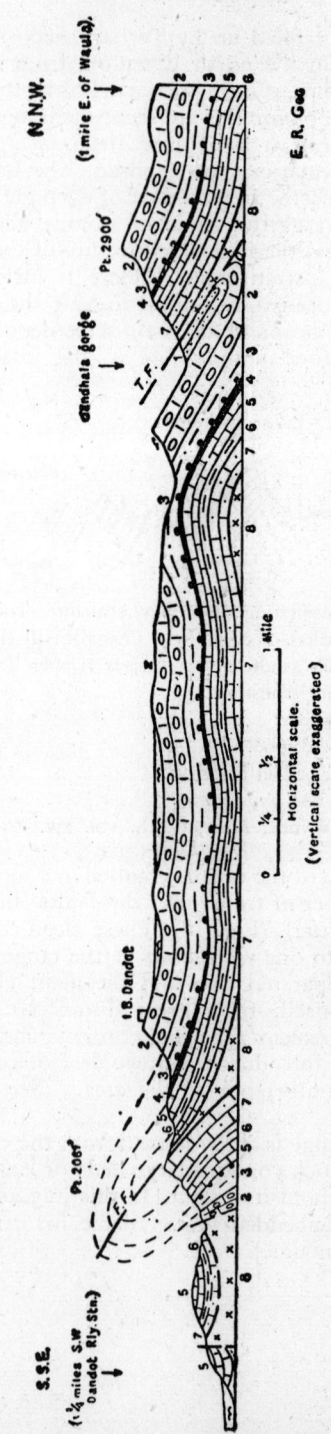

Fig. 11—Section across the Dandot scarp from Khewra to Gandhala. Salt-Range

THE CAMBRIAN SYSTEM

The Salt-Range Cambrian—At the eastern extremity of the Salt-Range a thick stratified series of rocks occurs in a conformable sequence. They are subdivided into the following groups in the order of superposition (Fig. 11):

Salt-pseudomorph shales: 137 m.	Bright red or green flaggy argillaceous beds, with cubic clay pseudomorphs of salt-crystals.
Magnesian sandstone: 76 m.	Laminated white or cream-coloured sandstones, often dolomitic.
Neobolus shales: 30 m.	Grey or dark-coloured shales containing brachiopods, trilobites, gastropods, etc.
Purple sandstone: 137 m.	Dark red or purplish-brown well-bedded sandstones with maroon-coloured shales at the base.
Saline series: 457 m.?	Stiff clay or marl, mainly dark red and vermilion, with abundant gypsum and salt, and thin beds of dolomite.

The Saline Series—The age of the lowest group, composed of *salt-marl*, gypseous marl, salt, gypsum, and dolomite, presents a difficult problem which has long been one of the major controversies of Indian geology. The boundary between the Saline series and the overlying Purple sandstone is much disturbed and is undoubtedly not a regular one. This fact has been interpreted in different ways; one view is that this disturbed boundary is merely the result of differential movement between two very different types of rock—the very "competent" Purple sandstones, and the soft, plastic, and "incompetent" beds of the Saline series; another interpretation stresses the effects of solution of saline material and suggests that this has led to the severe disturbance and brecciation noticeable wherever the Saline series is in contact with other rocks. A widely different interpretation has been put forward by several geologists and is supported by recent work. It is that the apparently infra-Cambrian position of the Saline series is due to a large overthrust and that the salt-marl and associated beds are really of Eocene age. B. Sahni has found micro-fossils of angiosperm plants embedded in the salt, gypsum and associated rocks from different outcrops of the Saline series. About the indigenous nature of these micro-fossils, however, some doubt has been expressed. E. R. Gee has established that the large masses of gypsum in the western part of the main Salt-Range—where it borders on the

Indus valley—are of Laki (Eocene) age, and although the age of the gypsum and salt of the central part of the range cannot be directly established in the same way, it seems a reasonable assumption that it is of the same age as the gypsum and associated beds a short distance further north-west. Gee, on the other hand, has found evidence which is regarded by many geologists as establishing the Cambrian age of the Saline series. The Talchir boulder-bed, which rests unconformably on the Cambrian, when traced W.N.W. from Khewra is seen to lie on successively lower members of the Cambrian succession, passing from the Salt-pseudomorph beds at Khewra to the Magnesian sandstone, the Neobolus beds and the Purple sandstone, and thence on to the Saline series near Sakesar. The contact appears to be an ordinary sedimentary junction and pebbles of rocks from the Saline series occur in the basal Talchir conglomerate. If this reading of the section is accepted, it follows that the Saline series is pre-Carboniferous at least. Also the Cambrian view derives some support from the Joya Mair bore near Chakwal, where a deep oil boring passed from the Purple sandstone to the Saline series at 2,683 metres. It has however been suggested that this may be an intrusive contact and not a sedimentary junction, and that the evidence does not necessarily imply a Cambrian age for the Saline series.

[Near Khewra, the accumulation of gypsum and rock-salt is on a large scale. At the Mayo Salt Mines, at Khewra, there is a mass of nearly pure crystalline salt of a light pink colour, interbedded with some seams of impure red earthy salt (*Kalar*), of the total thickness of 90 m. Above this is another bed of the thickness of 75 m. The upper deposit is not so pure as the lower, for it contains more intercalations of Kalar and is associated with other salts, *viz*. calcium sulphate and magnesium, potassium, and calcium chlorides, in greater proportions. The lateral extension of the saltbeds appears to be very great, amounting to several square km. in area, and there is thus a very large supply of salt from the Khewra deposits. To this must be added the salt contained in the red marl at other parts of the range, and worked in several smaller mines. The associated gypsum occurs in large masses and also in smaller beds; it exhibits an irregular bedding and varies greatly in purity and in degree of hydration, passing at times into anhydrite.

The origin of the *salt-marl* is not known with certainty. Oldham suggested that it is an alteration product of pre-existing sediments by the action of acid vapours and solutions. Christie has brought forward evidence to show that the salt and gypsum were formed by the evaporation of sea-water in inland or enclosed basins which were intermittently cut off from the main ocean by barriers. The red saline earth or Kalar seams are held to indicate the last stage of the desiccation of the sea-bed; the occurrence of potassium salts mentioned below, just underneath the Kalar, is pointed to as further evidence in support of the evaporation theory; for, in a seabasin undergoing desiccation, the salts of potassium are the last to be precipitated, after nearly 98 per cent of the water has evaporated. It is argued that the stratification-planes which were originally present, both in the enclosing marl and in the salt, have been obliterated subsequently by superficial agencies as well as by the effects of compression and earth-movements on a soft plastic substance like the marl.

There is no doubt that although much of the Saline series outcrop is devoid of clear stratification, other parts show the clearest disposition of the different components of the Saline series into distinct beds which are of sedimentary origin. This is particularly shown by the dolomites and shales associated with the red marl and also by the bands of gypsum and salt. This prominent stratification shows that hypotheses based on an "igneous" or "intrusive" origin are inapplicable, and that the Saline series is in the main of sedimentary origin. Nevertheless, the discovery in Kohat and in the north-west end of the Salt-Range shows that the gypsum is—at least in part—an alteration product of limestones. The intimate association of limestones and shales with the gypsum in the Salt-Range is closely paralleled in Kohat.

Economics—The economic importance of the salt deposits is great, as they produce about 150,000 tonnes of salt per year. Besides the chloride of sodium, there are found other salts, of use in agriculture and industries. Of the latter the salts of *Potassium* (Sylvite, Kainite, Blödite and Langbeinite), which occur in seams underlying beds of red earthy salts (Kalar), are the most important. Magnesium salts are Epsomite and Kieserite.]

The Purple sandstone—Overlying the salt-marl, but in a most irregular and mechanically disturbed manner, is a series of purple or red-coloured sandstones. The junction-plane between the two series of strata is so discordant that the marl appears to have intruded itself into the lower beds of the Purple sandstone. The Purple sandstone is a red or purple-coloured series of sandstone beds. It is a shallow-water deposit, as can be seen from the frequency of oblique lamination, ripple-marks and sun-cracks, and such surface marks as rain-prints, worm-burrows, fucoid impressions, etc. The lower beds are argillaceous, being known as the "Maroon shales," gradually becoming more arenaceous at the top. Worm-tracks and fucoid marks are the only signs of life in these rocks.

Neobolus beds—This stage is succeeded by the most important beds of the system, a group of dark micaceous shales with white dolomitic layers known as the *Neobolus beds*, from their containing the fossil brachiopod *Neobolus*. Other fossils are *Discinolepis*,

Schizopholis, Lakhmina, Lingula, Orthis, Conocephalites, Redlichia (a trilobite resembling *Olenellus*) and the probable pteropod *Hyolithes*. The brachiopods and trilobites resemble those of the Cambrian of Europe, and hence the Neobolus beds stamp the whole connected series of deposits as Cambrian. This division of the Cambrian of the Salt-Range is well displayed in the hill surmounted by the old Khusak fortress in the neighbourhood of Khewra.

Magnesian sandstone stage—Overlying the Neobolus beds is the Magnesian sandstone stage, a sandstone whose matrix is dolomitic and imparts to the rock its white or cream colour. There are also some beds of dolomite, among which are a few oolitic or pisolitic bands. Some of the beds in this group are very finely laminated; sometimes a hundred laminae can be counted in the thickness of an inch. When showing oblique lamination and minor faulting in hand-specimens, they form prize specimens in a student's collection. The only fossil contained in these rocks is *Stenotheca*, a lower Cambrian mollusc, besides a few unrecognisable fucoid and annelid markings.

Salt-pseudomorph shales—The Salt-pseudomorph shales are bright red and variegated shales with thin-bedded sandstones. The name of the group is derived from the numerous pseudomorphic casts of large perfect crystals of rock-salt very prominently seen on the shale-partings. It is evident that these strata were formed on a gently shelving shore which was laid bare at each retreating tide. In the pools of salt-water on the bare beach crystals of salt would be formed by evaporation, which would be covered up by the sediments brought by the next tide. The cavities left by their subsequent dissolution would be filled up by infiltrated clay.

Trans-Indus Cambrian—In the west of the Salt-Range, in the trans-Indus area, the Cambrian beds are seen near Saiduwali in the Kirri-Khasor range. The lowest beds are the Purple sandstones of the Salt-Range succession but higher in the sequence there are massive gypsum, dolomite, and bituminous shales; the facies thus differs somewhat in lithology from the corresponding beds in the upper part of the Cambrian sequence of the Salt-Range.

CAMBRIAN OF SPITI

In the Spiti valley[1] lying amid the north-eastern ranges of the Kangra district, and in some adjoining parts of the central Himalayas, a nearly complete sequence of fossiliferous Palaeozoic and

---

[1] The Spiti river is a tributary of the river Sutlej, running N.W.–S.E. in a tract of mountains which form the boundary between the N.E. Punjab and Tibet (Lat. $32°\ 10'$ N., Long. $78°$ E.).

THE CAMBRIAN SYSTEM 139

Mesozoic strata is laid bare, in which representatives of all the geological systems, from Cambrian to Eocene, have been worked out in detail by a number of geologists since the middle of the last century.

The Spiti area, the classic ground of Indian geology, which will recur often in the following pages, is in general a broad synclinal basin (a *Geosyncline*) which contains the stratified deposits of the old Himalayan sea, representative of the ages during which it occupied the northern Himalayas and Tibet.

The axis of the syncline is north-west-south-east, in conformity with the trend of the Himalayas. The youngest Mesozoic formations are, obviously, exposed in the central part of the basin, while the successively older ones are laid bare on the flanks, the oldest, Cambrian being the outermost, *i.e.* towards the Punjab. The dip of the latter formations is northerly in the main, *i.e.* towards the interior. All these formations are fossiliferous, the fossils being the means of a very precise correlation of these systems with those of Europe. The student should consult Dr. Hayden's memoir on the geology of Spiti.[1] Hayden's researches have contributed a great deal in elucidating the Palaeozoic geology of this region.

The Cambrian of Spiti. Cambrian fossils—The Cambrian of Spiti rests over the highly metamorphosed pre-Cambrian series of schists (the Vaikrita series), which in turn are underlain by what have been regarded as the Archaean gneisses. There is a great thickness of highly folded and disturbed sedimentary strata comprising the whole of the Cambrian system—Lower, Middle and Upper. The system has been named *Haimanta*, from its occurrence in high snow-capped peaks. The component rocks are principally argillaceous and siliceous rocks such as slates and quartzites; the latter occupy the base, followed by red and black slates, with much enclosed haematite in the former and carbonaceous matter in the latter. At the top are again siliceous slates and shales interbedded with dolomite. The upper portion of the group, constituting a thickness of some 366 m. is fossiliferous. A fairly abundant Cambrian fauna has been discovered in it, of which trilobites form the chief element. The following are the leading genera: *Olenus*, *Agnostus*, *Microdiscus*, *Ptychoparia* (many species) and *Dicellocephalus*. Among the other fossils are the brachiopods *Lingulella*, *Obolus* and *Obolella*, and a few crinoids and gastropods (*Bellerophon*). The species of the above-named genera of fossils show clear affinities with the European Cambrian forms.

The most complete development of these strata is exposed in the valley of the Parahio, a tributary of the Spiti river. (See Fig. 14. p. 149).

---

[1] *Mem. G.S.I.*, vol. xxxvi. pt. 1, 1909.

Autoclastic conglomerates—Some conglomerate layers among the slates are of interest because of their uncommon mode of origin. They are not ordinary clastic conglomerates of sedimentary derivation, but, according to Dr. Hayden, they are of "autoclastic" origin, *i.e.* they were produced by the crushing of veins of quartz into more or less rounded fragments or lenticles scattered in a fine-grained micaceous matrix, this latter having been formed from the slates.

CAMBRIAN OF KASHMIR

Fossiliferous Cambrian rocks are developed on a large scale in the mountains of the Baramula district of Kashmir to the north of the Jhelum, forming a broad-irregular band on the north limb of the Palaeozoic basin of Hundawar.

Dogra slates—Underlying the fossiliferous Cambrian of Kashmir conformably, and at some localities showing a transitional passage into it, there is a thick zone of slaty rocks—argillaceous cleavage slates, with generally oblique cleavage, with thin sandy or quartzitic partings, often ripple-marked. They are quite unfossiliferous and their exact horizon, whether Purana or possibly Lower Cambrian, is uncertain. Lithologically identical groups occur in Hazara and Simla, recognised as the *Hazara slates* and *Simla slates*.

The Dogra slates occupy long belts in the Pir Panjal (where they are associated with a great thickness of contemporaneous basic trap), the Kishenganga valley and in Hazara.

Basins of Palaeozoic rocks—Fossiliferous Palaeozoic rocks of Kashmir occupy elongated ellipse-shaped patches of the country north of the alluvial part of the valley, stretching from north-west of Hundawar to the south-east end of the Kashmir sedimentary "basin", where it merges into the Spiti basin. The Lidar valley development is the more typical. The long axis of this ellipse, north-west to south-east, corresponds to the axis of a broad anticlinal flexure, in which the whole series of Palaeozoic rocks is folded. Denudation has exposed, in the central part of this anticlinal, a broad oval outcrop of the most ancient fossiliferous rocks of Kashmir—the Cambrian and Ordovician—flanked on its two sides successively by thinner bands of the younger formations, Silurian, Devonian and Carboniferous (see Pl. VIII). A similar section is exposed in the Basmai anticline of the Sind valley between Sonamarg and Kolahoi. Palaeozoic rocks, especially of the younger systems, are also conspicuous in the Vihi district, in east Karnah, and, to a less degree, in the Pir Panjal, while the great series of volcanic rocks of Upper Carboniferous age are quite ubiquitous in their distribution over the whole area of Kashmir, forming the main mass of the Panjal range and of the mountains

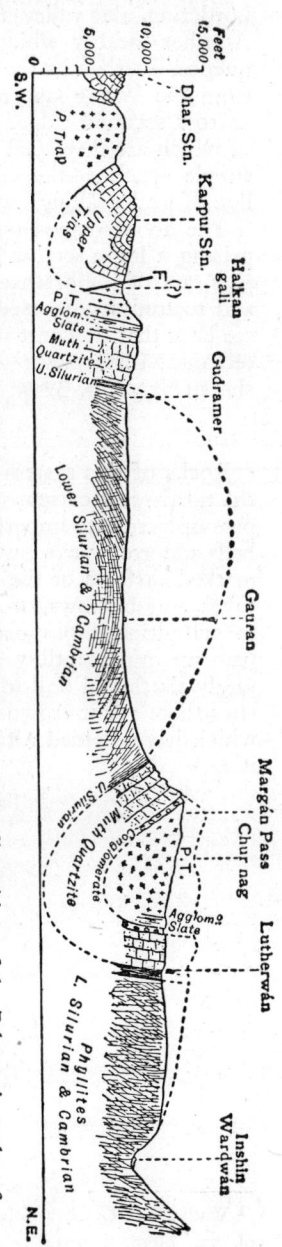

FIG. 12.—General section, Naubug Valley, Margan Pass and Wardwan to show the disposition of the Palaeozoic rocks of Kashmir. (Middlemiss, *Rec. G.S.I.* vol. xl. pt. 3.)

Note—The Lower Silurian of this figure is Ordovician of present-day nomenclature.

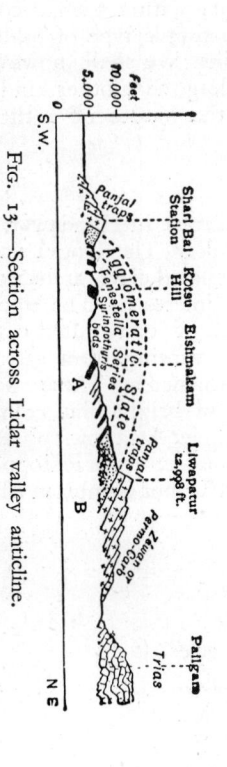

FIG. 13.—Section across Lidar valley anticline. (Middlemiss, *Rec. Geological Survey of India,* vol. xl. pt. 3.)

bordering the valley to the north-west, north and north-east. Another locality which epitomises a part of the Palaeozoic sequence, overlain by the Trias, is the large synclinal basin extending from the Wular lake to Tithwal. The fold is traversed by the narrow serrated ridge, the Shamsh Abari, in the steep precipices of which are displayed fine sections of the Palaeozoic folded in a simple syncline, the crest of the syncline (4,238 m.) building a line of peaks falling away in bare rock-faces of thousands of feet.

The above-named outcrops of Palaeozoic rocks, besides comprising a large section of geological history within a small compass, are of importance in illustrating the simple type of folding and tectonics witnessed in these mountains. We shall, however, see later that this part of Kashmir has undergone another kind of tectonic disturbance—displacement of the nature of a thrust sheet (*Nappe*, p. 394).

Cambrian

Rocks of this system cover an extensive tract in Hundawar, at the north-west extremity of the Kashmir valley. The Dogra slates pass upward into imperfectly cleaved and foliated clays, arenaceous beds and greywackes, with a few lenticular limestones. The ripple-marked surfaces of the strata are often full of convoluted casts, tubes and burrows of tubicolous *Vermes*, varying from threads to cylindrical pipes reaching 5 cm. in diameter. These beds pass up imperceptibly into massive clays of bright blue colour, sandy slates and oolitic or pisolitic limestones. At a few sporadic sites there occur crowds of trilobites and obolaceous brachiopods, which have yielded a fauna of Middle and Upper Cambrian affinities:[1]

Trilobites :
Agnostus.
Microdiscus.
Conocoryphe, 3 species.
Tonkinella, 2 species.
Anomocare, 6 species.
Chaungia, 3 species.
Solenopleura, 2 species.
Blountia.
Ptychoparia.
Hundwarella, 2 species.
Saukia.

---

[1] Wadia, *Rec. G.S.I.* vol. lxviii. pt. 2, 1934; Cowper Reed, *Pal. Indica*, N.S. vol. xxi. Mem. 2, 1934.

Brachiopods:
Obolus.
Lingulella.
Acrothele.
Botsfordia.
Lingulepis.

Pteropod:
Hyolithes.
Crinoid :
Eocystites.
Sponge :
Hazelia.

The most noteworthy feature of this fauna, according to Dr. Cowper Reed, is its strictly provincial character, showing no affinities with the adjacent Cambrian life-provinces of the Salt-Range, Spiti, or the Persian Gulf. Many of the sixteen genera of trilobites found in this area and all the species are new. The whole fauna thus is markedly endemic, having no relationship with adjacent Indian or neighbouring extra-Indian provinces. On the other hand, the Kashmir Cambrian fauna exhibits affinities with the Cambrian of Indo-China.

No good Cambrian fauna has been found in the Lidar, Sind, or Vihi area, where the fossiliferous Silurian exhibits a conformable passage downwards into a thick group of knotted, crudely foliated slates and arenaceous beds, greywackes, etc. In the Wardwan valley the same rocks reappear by a synclinal bending underneath the younger strata of the intervening ground between it and the Lidar. Here the Cambrian slates have a phyllitic or schistose aspect owing to contact metamorphism by granitic intrusions. In the Banihal valley also the Cambrians show a considerable amount of foliation; beyond annelid markings and indistinct pteropod shells no determinable fossils have been found.

As we have to turn often to the Himalayas for study of the successive marine formations, Palaeozoic to Tertiary, a few notes of historic and general information on the stratigraphic sequences worked out in parts of the Himalayas (Kashmir-Hazara, Simla-Garhwal) that have been more explored geologically than others are given here.

Stratigraphy of Kashmir

R. Lydekker in the eighties of the last century made a geological survey of Kashmir. His results were published in *Memoirs* of the Geological Survey of India (vol. xxii., 1883). Lydekker in his preliminary survey grouped all the stratified formations of Kashmir into three broad divisions—the Panjal, the Zanskar and the Tertiary groups—the homotaxial relations of whose constituent series and systems were not clearly distinguished because of the absence of satisfactory fossil evidence. Middlemiss worked in the same field from 1908–1917. Middlemiss's researches have revealed a series of fossiliferous strata in different parts of the province, belonging to various divisions of the Palaeozoic and the

Mesozoic, which have enabled him to make a more perfect classification of the Kashmir record. Thus he has resolved what was formerly one comprehensive group, the Panjal system, which encompassed almost the whole of the Palaeozoic sequence, into no less than seven well-defined systems or series, the representatives of the Cambrian, Ordovician, Silurian, Devonian, Carboniferous and Permian, and the homotaxial equivalents of those of the classic ground of Spiti.

Of the Mesozoic systems, the Trias is the best and most fully developed; the Jurassic and Cretaceous outcrops are few and mostly confined to the mountains of Ladakh which have scarcely been systematically surveyed by geologists. All the Tertiary systems are fully represented in the outer mountains and have been studied by a number of workers.

The broad outlines of the stratigraphy of Hazara and North-West Kashmir are similar; these two regions form one more or less continuous sedimentary terrain, though now isolated by the deep knee-bend of the mountains across the Muzaffarabad promontory of the foreland. A great regional unconformity encompassing the period from the top of the Silurian to the Middle Carboniferous is a distinctive feature of this north-west province. The south-east part of Kashmir has a continuous Palaeozoic record similar to that of Spiti.

The account given of the successive geological systems of Kashmir, in the following chapters, is deduced from the writings of Lydekker, Middlemiss and Wadia. For more detailed information with regard to the whole of the Palaeozoic group and the Triassic system, the student should consult original publicatons, *Rec. G.S.I.* vol. xl. part 3; 1910, and vol. lxviii. part 2, 1934. For the remaining systems, and the tectonics of Kashmir, the present writer's work should be consulted.[1]

Large Unsurveyed Areas of the Himalayas

In spite of the large blanks still existing on the geological map of the Himalayas, representing nearly three quarters of the total area of the Himalayas, there has been during recent years a considerable advance in our knowledge of the geology of these mountains, their stratigraphic and structural plan. Except for the immediate neighbourhood of Mount Evere t, geologically reconnoitred by successive Mount Everest expeditions, only Spiti, Hazara-Kashmir and Simla-Chakrata, have been mapped in some detail; Garhwal and Kumaon have had reconnaissance surveys by Heim and Gansser and the officials of the Geological Survey of India and some

---

[1] D. N. Wadia, *Mem. G.S.I.* vol. li. pt. 2, 1928; *Rec. G.S.I.* vol. lxv. pt. 2, 1931, vol. lxvi., pt. 2, 1932, and vol. lxxii., pt. 2, 1937.

THE CAMBRIAN SYSTEM

Indian universities.[1] While the entire block of Assam Himalaya, except for areas in Sikkim and Bhutan, still remains blank, Nepal has lately been geologically surveyed in some detail. The results so far obtained in stratigraphy and tectonics disclose a unity of structure and constitution for the whole of this mountain system from the Indus to the Brahmaputra. Data are slowly accumulating which tend to show that the baffling complexity of structure and diversity from area to area of the Alps, though encountered in a few local patches, are not met with in the same degree in the Himalayas, a fact which, if substantiated by further work, will enable a complete synthesis of Himalayan geology and orogeny to be built up in the near future.

Himalayan Stratigraphy—Kumaon Sector

Medlicott in the Kumaon and Middlemiss in the Kashmir Himalayas in 1910 securely laid the foundations of the stratigraphy of the Himalayas. The total absence of fossils in the Simla-Himalaya introduces great difficulty and uncertainty in correlating even the broad divisions of strata, but of late years, careful study of relative metamorphism and structural relations of thrust-planes and unconformities has enabled the natural order of superposition of strata to be established more or less in parallel with the fossiliferous systems of Kashmir. The Himalayas have been divided into three longitudinal stratigraphical zones: an outer or *Sub-Himalayan zone*, composed of Tertiary rocks; a central or *Himalayan zone*, composed of crystallines and unfossiliferous slaty sediments constituting the bulk of the Central ranges; and a northern or *Tibetan zone* composed of fossiliferous marine sediments ranging from Cambrian to Eocene. It is probable that the middle Himalayan zone denotes the central geanticline within the main Himalayan geosyncline. In the Eastern Himalaya, this geanticlinal axis (approximately following the line of the Great Himalaya range) lies close to what was the southern shore of the Tethys, with the result that almost the whole of the Tibetan zone is to the north of the range, leaving but little Palaeozoic and Mesozoic sediments on the Indian side of the axis. In the Western Himalaya of Kashmir and Hazara, however, the axis lies well to the north of what was the margin of the Tethys, so that the Tibetan zone is not confined to the north of the Great Himalayan range but is found in detached patches on both sides of the axis. The unfossiliferous sedimentary systems of Simla-Nainital may be regarded as detached outliers of the Tibetan formations of Spiti, though their total lack of fossils is still an inexplicable circumstance. Perhaps it may be explained

---

[1] A. Heim and A. Gansser, *Geological Observations in the Kumaon Himalayas*, Zürich, 1939.

as due to their being laid down in distant outlying basins of the main Sea to the north.

In this sector of the Himalayas, both the Palaeozoic and Mesozoic record is highly imperfect and scrappy; almost the whole of the Palaeozoic with the exception of the Up. Carboniferous and the Permian—known as the *Blaini, Infra-Krol* and *Krol series*, is missing, save for some locally developed rock groups of uncertain age in scattered inland basins, which are recognised today as the *Nagathat, Jaunsar, Deoban series*, etc. The Mesozoic is also absent, but for a few patches of marine Jurassic strata, belonging to the Tethyan system of deposits, as can be judged from some contained fossils. Only the Tertiary sequence is preserved in full, composed of over 15,250 m. of marine, estuarine and fresh-water deposits building the Outer Himalayan ranges and the foot-hills.

The existence of a Palaeozoic sea of undefined boundaries, occupying large areas of the present middle and outer Himalayas, is inferred from the presence of vast thicknesses of unfossiliferous sediments that have been designated the *Tanawal series* in Kashmir, as *Jaunsar, Nagathat* and *Deoban series* in the Garhwal area. This ancient Himalayan sea, the predecessor of the *Tethys* of later (Up. Carboniferous) period, came into existence through the depression of the old pre-Palaeozoic land surface of the Deccan mainland, covered under the mantle of the Jutogh, Chail and Simla slates formations, the northern extensions of Peninsular Puranas.

REFERENCES

A. B. Wynne, Geology of the Salt-Range, *Mem. G.S.I.* vol. xiv., 1878.

C. S. Middlemiss, *Rec. G.S.I.* vol. xxiv. Pt. 1, 1891.

H. H. Hayden, Geology of Spiti, *Mem. G.S.I.* vol. xxvi. pt. 1, 1904.

D. N. Wadia, Cambrian-Trias Sequence of N. W. Kashmir, *Rec. G.S.I.* vol. lxviii. pt. 2, 1934.

F. R. C. Reed, The Cambrian Fossils of Spiti, *Palaeontologia Indica*, Series XV. vol. ii. mem. 1, 1910; Cambrian and Ordovician Fossils of Kashmir, *Pal. Ind.* New Series, vol. xxi. mem. 2, 1934.

C. L. Griesbach, Geology of the Central Himalayas, *Mem. G.S.I.* vol. xxiii., 1891.

M. Stuart, Origin and History of the rock-salt deposits of Punjab and Kohat, *Rec. G.S.I.* vol. 1. pt. 1., 1919.

CHAPTER VIII

The Ordovician, Silurian, Devonian and Lower and Middle Carboniferous Systems

Introduction—These great groups of Palaeozoic strata do not occur at all in the Peninsular part of India, while their occurrences in the extra-Peninsular area are also, with one exception, outside the geographical limits of India proper, and confined to the northernmost borders of the Himalayas and to Upper Burma. In the Peninsula there exists, between the Vindhyan and the next overlying (Upper Carboniferous) deposits, a great hiatus arising from a persistent epeirogenic uplift of the country during the ages that followed the deposition of the Vindhyan sediments. The absence from India of these formations, constituting nearly three-fourths of the Palaeozoic history of the earth, is quite noteworthy, as it imparts to the Indian geological record, especially of the Peninsula, a very imperfect and fragmentary character. The Himalayan occurrences of these rock-groups, referred to above, are restricted also to the northernmost or Tibetan zone of the Himalayas, where a broad belt of marine fossiliferous sedimentary rocks extends from the western extremity, Hazara and Kashmir, through Spiti, Garhwal and Kumaon to Nepal, Sikkim and Bhutan, in which representatives of almost all the rock-systems from Cambrian to Eocene are recognised.

1. SPITI AREA

Ordovician and Silurian—Overlying the Haimanta system in all parts of Spiti there is a thick series of red quartzites and grits, underlain by conglomerates and passing upwards into shales with bands of limestone and dolomite. The accompanying table shows the relations of the Ordovician and Silurian of Spiti with the overlying and underlying formations (see Fig. 14):

Devonian.	Muth Quartzite.	
Silurian.	Grey coloured siliceous limestones. Coral limestones. Shaly limestones with brachiopods, corals and gastropods.	
Ordovician.	Hard grey dolomitic limestones. Dark and grey limestones with *cystidea*, brachiopods and trilobites. Shales and flaggy sandstones and quartzites. Thick mass of pink or red quartzite, gritty unfossiliferous coarse conglomerates.	610 m.
Cambrian.	Haimanta black shales and slates.	

The lower, arenaceous, beds are unfossiliferous, but the upper shaly and calcareous portion has yielded numerous fossil brachiopods, cystids, crinoids, corals and trilobites. Of these the most important genera are: (Trilobites) *Cheirurus, Illaenus, Asaphus, Calymene* and *Bronteus*; (Brachiopods) *Orthis, Strophomena, Leptaena, Atrypa, Pentamerus* (?); (Corals) *Favosites, Halysites, Cyathophyllum, Syringopora* and *Chaetetes*; (Hydrozoa) *Stromatopora*; (Gastropods) *Bellerophon* and *Pleurotomaria*; (Cystids) *Pyrocystites* and *Craterina*. The above-named genera bear close zoological relations to those obtained from the Palaeozoic of England and Europe, a relationship which extends also to many of their species, a certain number of them being common to both regions.

Devonian—Resting over the Silurian beds is a thick series of white hard quartzite, which is mostly unfossiliferous and whose age therefore, whether Upper Silurian or Devonian, was uptil lately uncertain. This quartzite is known as the *Muth quartzite* from its occurrence very conspicuously in a broad belt forming the entire south margin of the Spiti synclinorium and its typical development in the Muth Pass in Spiti. The Muth quartzite is a highly conspicuous formation well recognized by its smooth white quartzite outcrops from N. Kumaon in the east to Kashmir. It was regarded by some as partly Silurian and partly Devonian. However, its Devonian age is now nearly settled by the discovery of a typical Devonian fauna of brachiopods, trilobites and fish from the 'Muth quartzite' series in Kashmir, which, in its stratigraphic position as well as lithologic assemblage, shows complete identity with the Spiti series. The Muth quartzites, together with an overlying group of hard siliceous limestone, some 90 m. in thickness in the neighbouring locality of Bashahr, may be taken to represent in part at

THE SILURIAN AND DEVONIAN SYSTEMS

least the Devonian Age in the Himalayas.[1]

Carboniferous. Lipak series—The Muth quartzite is overlain by a thick series of limestones and quartzites more than 600 m. in thickness. The limestones are hard, dark-coloured and splintery. They are, however, very prolific in fossils, the fossiliferous bands alternating with white and grey barren quartzites. This series is known as the *Lipak series*, from a typical outcrop in the Lipak valley in the eastern part of Spiti. The fossils are characteristic Lower Carboniferous organisms belonging to such genera as : (Brachiopods) *Productus* (spp. *cora* and *semireticulatus*), *Chonetes*, *Athyris* (sp. *roysii*), *Syringothyris* (sp. *cuspidata*), *Spirifer*, *Reticularia*; (Lamellibranchs) *Conocardium*, *Aviculopecten*; the Carboniferous trilobite *Phillipsia*; (Cephalopods) *Orthoceras* and *Platyceras*; (Gastropods) *Euomphalus*, *Conularia*, *Pleurotomaria*; (Crustacea) *Estheria*; fish-teeth, etc.

The Po series—The Lipak series is succeeded, in the same continuous sequence, by a group of dark-coloured shales and quartzites constituting what is known as the *Po series*. (See Fig. 22.) The lower division is for the most part composed of black shales, traversed by intrusive

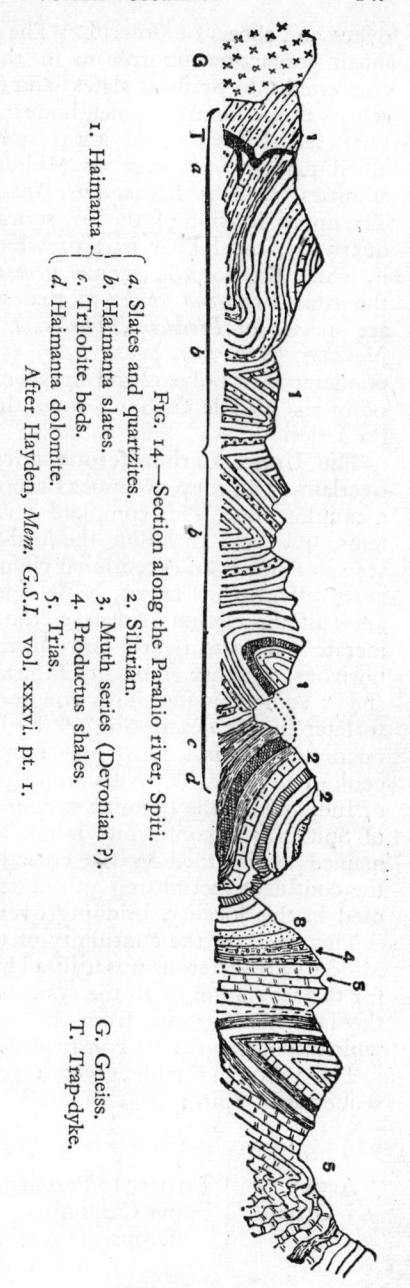

[1] V. J. Gupta, 1967

Fig. 14.—Section along the Parahio river, Spiti.

1. Haimanta $\begin{cases} a. \text{ Slates and quartzites.} \\ b. \text{ Haimanta slates.} \\ c. \text{ Trilobite beds.} \\ d. \text{ Haimanta dolomite.} \end{cases}$
2. Silurian.
3. Muth series (Devonian ?).
4. Productus shales.
5. Trias.

After Hayden, *Mem. G.S.I.* vol. xxxvi, pt. 1.

G. Gneiss.
T. Trap-dyke.

dykes and sheets of dolerite. The intruded rock has induced much contact-metamorphism in the shales, some of which are converted into pyritous slates and even into garnetiferous micaschists in the immediate neighbourhood of the igneous rock. The unaltered shales contain impressions of the leaves of ferns and allied plants, of Lower or Middle Carboniferous (Moscovian) affinities, such as *Rhacopteris*, *Sphenopteridium*, *Sphenopteris*, etc. The upper division of the Po series is composed of shales and quartzites, the higher part of which contains marine organisms in which the polyzoan genus *Fenestella* preponderates, and gives the name *Fenestella shales* to that subdivision. The other fossils are species of *Productus*, *Dielasma*, *Spirigera*, *Reticularia*, *Spirifer*, *Nautilus*, *Orthoceras*, *Protoretepora* (sp. *ampla*), etc. From the preponderance of polyzoa and the species of brachiopods characteristic of the Middle Carboniferous, the latter age is ascribed to the Po series.

The Upper Carboniferous unconformity—The Po series is overlain by a group of Upper Carboniferous strata beginning with a conglomerate. This complete development of the Palaezoic systems, upto and including the Mid-Carboniferous, which we have seen in Spiti, is an exceptional circumstance and confined to some parts only, for in Hazara, N.W. Kashmir, Simla and several other areas of the central Himalaya, the Upper Carboniferous conglomerate is seen to overlie unconformably formations of far lower horizons, whether Haimanta, Silurian or Muth, all the intervening stages being missing. This conglomerate, which will be referred to later in our description of the Upper Carboniferous and Permian systems, is a most important horizon, a *datum-line*, in the geology of India. It covers an unconformity universal in all parts of India where the Permian system is seen. In this particular area of Spiti this unconformity is not apparent, because this area remained undisturbed by the crustal readjustments of the rest of the continent, permitting an uninterrupted sedimentation to proceed in this locality, bridging over the gap.

This break in the continuity of the deposits at the top of the Middle Carboniferous was utilised by Sir T. H. Holland as the basis for the separation of all the systems below it (collectively forming the Dravidian group) from the remaining systems of later ages which come above it, constituting the great Aryan group.

The following table gives a general view of the Palaeozoic sequence in Spiti :

Aryan Group. { Tertiary to Permian.
Upper Carboniferous. { Basement conglomerate.

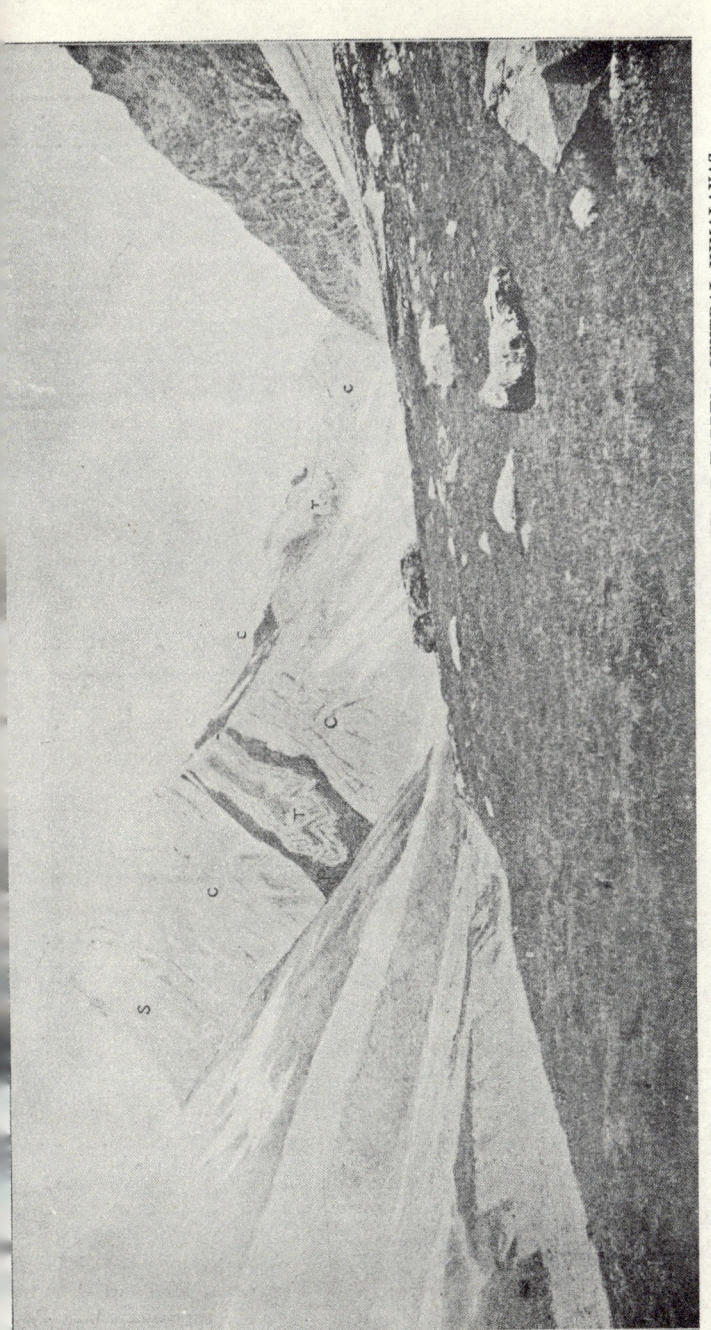

PLATE VIII. OVERFOLDING OF THE PALAEOZOIC ROCKS, UPPER LIDAR VALLEY, CENTRAL HIMALAYAS.

S. Silurian quartzites, shales and coral limestones. C. Limestones and quartzites of the Lipak and Po series; the latter are seen in the form of a synclinal in which lie the highly crumpled strata of the black Productus shales (P), and the Lower Trias of the *Otoceras* zone (T). Notice the smooth, worn surface of the cliff polished by a glacier. *Geological Survey of India, Mem.* vol. xxiii.

THE SILURIAN AND DEVONIAN SYSTEMS 151

Slight unconformity.

Dravidian Group.	Middle Carboniferous.	Po Series, 600 m.	Fenestella shales. Shales and quartzite with plants (*Culm*).
	Lower Carboniferous.	Lipak Series, 600 m.	Shales and limestones with *Syringothyris*, *Spirifer*, etc.
	Devonian.	Muth quartzite and limestone, 240 m.	
	Silurian and Ordovician.	Quartzites, shales and coral limestone, etc. 600 m.	
	Cambrian.	Haimanta slates and quartzites with dolomite, 1,220–1,520 m.	

Purana Group.   Pre-Cambrian.   Vaikrita series of schists and phyllites.

2. KASHMIR

A stratified series, in many respects identical with the above sequence in Spiti, is developed in Kashmir in a "basin" of sediments which lies on a direct north-west continuation of the strike of the Spiti basin, the only instance within the limits of India of a continuous and conformable well-developed Palaeozoic succession. In this there is a very perfect succession of all the primary stratigraphical systems—Cambrian, Ordovician and Silurian, Devonian, Carboniferous and Permian—conformably overlying the unfossiliferous slate series (Dogra slates) of basal Cambrian or late Purana age. In the Lidar Valley of Kashmir, Middlemiss has proved a continuous succession of fossiliferous Palaeozoic strata from Ordovician to Permian (see pp. 205–214). The following table shows the section up to Middle Carboniferous:

Fenestella Series
600 m.   Middle Carboniferous.
Syringothyris limestone
300 m.   Lower Carboniferous.
Muth quartzites
900 m.   ? Devonian.
Silurian and Ordovician
30 m.

In North-West Kashmir later work has shown a very pronounced stratigraphic break between the Muth quartzites and Upper Carboniferous. At many localities the Ordovician and Silurian also are not developed, and the Cambrian comes to be covered by the basal beds of the Upper Carboniferous volcanic series of deposits.

Unfossiliferous representatives, however, of what are believed to be continental types of the older Palaeozoic systems, are observed in parts of Hazara, Kashmir and the Simla Himalayas under the name of *Tanawal system* in the former and *Jaunsars* in the latter.

Ordovician—The Ordovician, in one part of Kashmir, is found to contain *Didymograptus* and other fragmentary graptolites. In the north limb of the Shamsh Abari syncline near Trehgam (Hundawar Tehsil) a series of sandy ferruginous slates, quartzose greywackes and limestones occur conformably above the Upper Cambrian in a synclinal warping of the latter, and here the Ordovician is recognised by the presence of some species of *Orthis*, among them *O.* cf. *calligramma* Dalm., and other Orthid and Strophomenid brachiopods, *Leptelloidea*, crinoid stem-joints, etc. Fragments of proparian trilobites (? *Cheirurus*) are common. The limestones, though frequently crowded with organic fragments, have yielded no recognisable fossils. The *Trehgam beds* pass up into the Silurian, small patches of which occur on either limb of the main syncline underlying the Muth Quartzites.

Detailed work in the thick group intervening between the Cambrian and the Muth Quartzites in the core of the Basmai anticline of the Sind valley is likely to bring to light some further outcrops of the Ordovician.

Silurian. Distribution—Round the oval expanse of the core of the Lidar anticline there runs a thin but continuous band of unmistakable Silurian strata, from which well-preserved Silurian organisms have been obtained. These rocks are continuously met with on the north-east side of the anticlinal from the neighbourhood of Eishmakam in the Lidar valley to Lutherwan in the Wardwan valley. On the south-west flank the outcrops are not as continuous, being hidden under the recent alluvium of the Lidar and Arpat streams and their tributaries.

Rocks—Lithologically the strata bear close resemblance to the underlying Cambrian and Ordovician, being composed of sandy shales or shaly sandstones with impure yellow limestones, but they are distinguished by the presence of a well-preserved suite of fossil organisms. Limestones and calcareous rocks are less common than in the corresponding rocks of Spiti. The aggregate thickness of the fossil-bearing Silurian strata is only 30 m. but the organisms preserved in them leave no doubt of their age, thus denoting a highly valued geological horizon in India. They offer one of the few instances, in the whole of the Indian region, where a well-defined Silurian fauna occurs. The presumed Silurian rocks of the Shamsh Abari area are of much greater thickness, but they are obscurely fossiliferous or unfossiliferous over wide stretches, and their age is inferred from their superposition on the Upper Cambrian or their conformable position underneath the Muth quartzites. A fossiliferous Silurian horizon exists in the Central Himalayas above the Haimanta system of Spiti and in the neighbouring area of Kumaon and Garhwal; another example is in the Shan States of Upper Burma. The occurrence of Silurian rocks is suspected, on strong lithological grounds, in Poonch and in Chitral, but no index fossil has been obtained from these localities hitherto, and their definite correlation is a matter of doubt.

Fossils[1] —Several species e.g. *Orthis* and other Brachiopods *Laptaena*, *Strophodonta*, *Atrypa*, *Meristella*, *Crania*, *Strophomena*, *Conchidium* with *Monograptus* (graptolites) are the principal fossils.

Of Trilobites the following genera occur *Calymene*, *Illaenus*, *Phacops*, *Acidaspis*, *Encrinurus*, *Beyrichia*.

The Cephalopods are represented by *Orthoceras* and *Cyrtoceras*. Some corals occur, including *Alveolites*, *Petraia* or *Lindstroemia*.

The absence from this fauna of the well-known Silurian corals, *Favosites*, *Heliolites*, *Cyathophyllum*, *Syringopora*, etc., which are present in the homotaxial deposits of Spiti, is noteworthy. The evidence of the other fossils, however, points to a similarity between these two deposits, a correspondence borne out by all other subsequent formations.

Devonian. Occurrence—The Devonian of Kashmir comes conformably on the group last described. Its outcrop follows the outcrop of the Silurian in normal stratigraphic order and is coextensive with the latter. Devonian strata are well seen on both the flanks of the Lidar anticlinal as thin bands; they are also well exposed in the Wardwan district, where their re-appearance is due to a synclinal folding.

An even band of hard, snow-white quartzites, 300-600 m. thick, follows the hair-pin loop of the pitching tip of Cambrian and Silurian outcrops in the Shamsh Abari syncline. It makes a regular even belt lying between the Cambro-Silurian and the outcrop of the next succeeding series, the Panjal Volcanic series.

The rocks regarded as Devonian are a great thickness of massive white quartzite. This rock, both in its composition and texture as well as in its stratigraphic relations to the rocks below and above it, exactly resembles the Muth quartzite of Spiti and Kumaon. As in Spiti, these massive beds of quartzite, reaching the thickness of 900 m. at places, were considered devoid of any fossil remains. The inference of their age, therefore, was solely

---

[1] Cowper Reed, Silurian Fossils from Kashmir. *Rec. G.S.I.*, vol. xlii, pt. 1,1912.

based on their stratigraphic position between fossiliferous Upper Silurian beds below and fossil-bearing Carboniferous beds above. The recent discovery of a well-preserved suite of typical Devonian fossils—corals, brachiopods, trilobites, lamellibranch conodonts, and fish by V. J. Gupta, Punjab University, from a locality in the Anantnag district, Kashmir, now definitely fixes its systematic position. In this Muth faunal assemblage, corals brachiopods are most prolific in species which bear a general assemblage to the fauna of Padaukpin limestone of the Burma Devonian. Outcrops of the Muth series are easily detected by the prominent escarpments and cliffs which it forms, due to the harder and more compact quartzites resisting the action of the denuding agencies better than the underlying slates.

Lower Carboniferous. Syringothyris Limestone Series. Distribution—Next in the order of superposition is a series of limestone strata lying conformably over the Muth quartzites. The outcrop of this limestone forms a thin band bordering the northwest half of the ellipse we are considering; it cannot be traced further eastwards, being to a great extent hidden under superficial deposits, such as river alluvia. It has also suffered greatly by the overlapping of the Panjal traps, which approach it from the north by successively overlapping the younger series. The present series is well exposed at Eishmakam and Kotsu, which are good localities for collecting fossils.

Outcrops of the *Syringothyris* limestone of considerable thickness, 600–900 m., are observed in the Banihal valley of the Pir Panjal, unconformably overlying the Cambrian. In the Sind valley, narrower bands of this limestone conformably overlie the Muth quartzites. Both these outcrops have suffered through the overlap of the Panjal volcanics.

Lower Carboniferous fossils—The rocks composing the Lower Carboniferous of Kashmir are thin-bedded flaggy limestones of a grey colour with clay or quartzite partings which occasionally assume large bulk. The maximum thickness is over 900 m. The calcareous consitution of this series readily distinguishes it from the older series, which are devoid of strata of limestone. The limestones are crowded with fossils principally belonging to the brachiopod class. The most frequently occurring brachiopod, which characterises the series, is *Syringothyris cuspidata*. This is a valuable index fossil, being also very typical of the Lipak series of Spiti. *Chonetes* is found in large numbers, together with many species of *Productus*, of which the species of *P. cora* is the most common, while *P. scabriculus* and *P. reticulatus* are not so abundant. *Athyris*, *Derbya* and *Rhynchonella* are among other brachiopods.

The age of the Syringothyris limestone series is determined by that of the Lipak series, with which it shows exact parallelism.

From the association of *Syringothyris cuspidata* with species of Trilobites (*Phillipsia*), regarded as Lower Carboniferous, in the Lipak group of Spiti, Hayden has ascribed to that group a Lower Carboniferous horizon.

Middle (?) Carboniferous Fenestella Shales—Overlying the upper beds of the Syringothyris limestone there comes some thickness of unfossiliferous quartzites and shales before the first beds of the characteristic Fenestella-bearing strata begin. These intermediate beds in their composition are allied to the upper group—the Fenestella shales to be presently described—but since they contain no fossils proper to that series, they are regarded as "passage beds" between the two series.

In distribution this group is even more restricted than the last described, being confined only to the north-west part of the ellipse of the Palaeozoic anticline of the Lidar and to some outcrops near Banihal and Budil in the Pir Panjal. To the south-west the series is totally missing, having been obliterated by the overlap of the Panjal lavas. In the Banihal anticline a broad band of *Fenestella shales* series conformably overlies and surrounds the outcrop of the Syringothyris limestone and reaches over 900 m. in vertical extent. Its relations with the overlying volcanic agglomeratic slate are perfectly conformable and even transitional, some of the black shales being crowded with pyroclastic and glassy débris, crystals of felspar, quartz, etc.

In the Hundawar basin this series, in common with the Syringothyris limestone, is absent, the Muth quartzites here being overlain by the Panjal Volcanic series. It is also absent from the Sind Valley.

Lithology—Lithologically the Fenestella shales are a great thickness (more than 600 m.) of thickly bedded quartzites interstratified with black shales, sandy or micaceous, and thick, coarse conglomerates. The shales are more prevalent at the base, becoming scarce at the middle and top. The shales are the only fossil-bearing horizons in the series, being rich repositories of fossil polyzoa—*Fenestella*, which gives the name to the series—brachiopods, corals and lamellibranchs.

The following is a characteristic section seen at Lehindajjar:

	Panjal agglomerate-slates.	Upper Carboniferous.
Fenestella shales.	Uppermost Fenestella shales, not thick. Unfossiliferous quartzites and shales, 150–200 m.	
	Black sandy shales with *Fenestella*, 30 m.	
	Quartzite, 20 m. (approx.)	Middle Carboniferous (?)
	Greyish shaly sandstone, obscure fossils, 60 m. (approx.)	

	Dark shales full of *Fenestella*, corals, brachiopods, lamellibranchs 45 m.
Fenestella shales. Contd.	Quartzite, 30 m.
	Sandy shales, full of *Productus* and other fossils, 150 m. (approx.)
	Base not seen.

Fauna—The most abundant fossils are casts of species of *Fenestella*, whose fan-shaped zoaria are preserved in great perfection. Brachiopods are also abundant in number as well as in species; the most commonly occurring are *Spirifer* (*S. middlemissii* and *S. varuna*), *Productus undatus*, *P. cora*, *P. lidarensis*, *P. spitiensis*, *P. scabriculus*, *Dielasma*, *Uncinella*, *Aulosteges*, *Camarophoria*, *Rhynchonella*. The lamellibranchs are *Modiola* and *Aviculopecten*. Pygida of *Phillipsia* occur. Besides *Fenestella* another polyzoon, though very rare, is *Protoretepora*.

Age of the Fenestella series—The fauna of the Fenestella series possesses, according to Dr. Diener, strikingly individual characters of its own. Many of the fossil forms are quite special to it, bearing no relations to any definite Carboniferous horizon.[1] For this reason their stratigraphic position is dubitable, between Lower and Upper.

The disposition of the outcrop of the Fenestella shales reveals the existence of a dip-fault traversing it along the Lidar basin. The fault is not important, but its effect upon the outcrop on the two banks of the river is quite illustrative. The exposure on the left bank lies much higher up the river than the right-bank outcrop. This is in consequence of a lateral shift (heave) produced by a fault cutting across the strike of the beds.

3. NEPAL, SIKKIM, BHUTAN

Palaeozoic strata, from Ordovician to Permo-Carboniferous with their characteristic fossils, are found developed in restricted and interrupted chain of outcrops in the northern ranges of Nepal, Sikkim and Bhutan. They form part of the Tibetan Zone of marine geosynclinical formations of the Himalayas that are more fully developed in Spiti and Kashmir. Many of the high peaks of Nepal and Bhutan, including Everest, have a capping of these rocks. These Himalayan areas, however, have so far not been surveyed in any detail. East of Bhutan, the entire length of Assam Himalayas has remained unexplored.

4. CHITRAL (HINDUKUSH)

In the valley of Chitral river, at the north-west frontier, Devonian

---

[1] Diener, *Pal. Indica, N.S.*, vol. v, mem. 2, 1915.

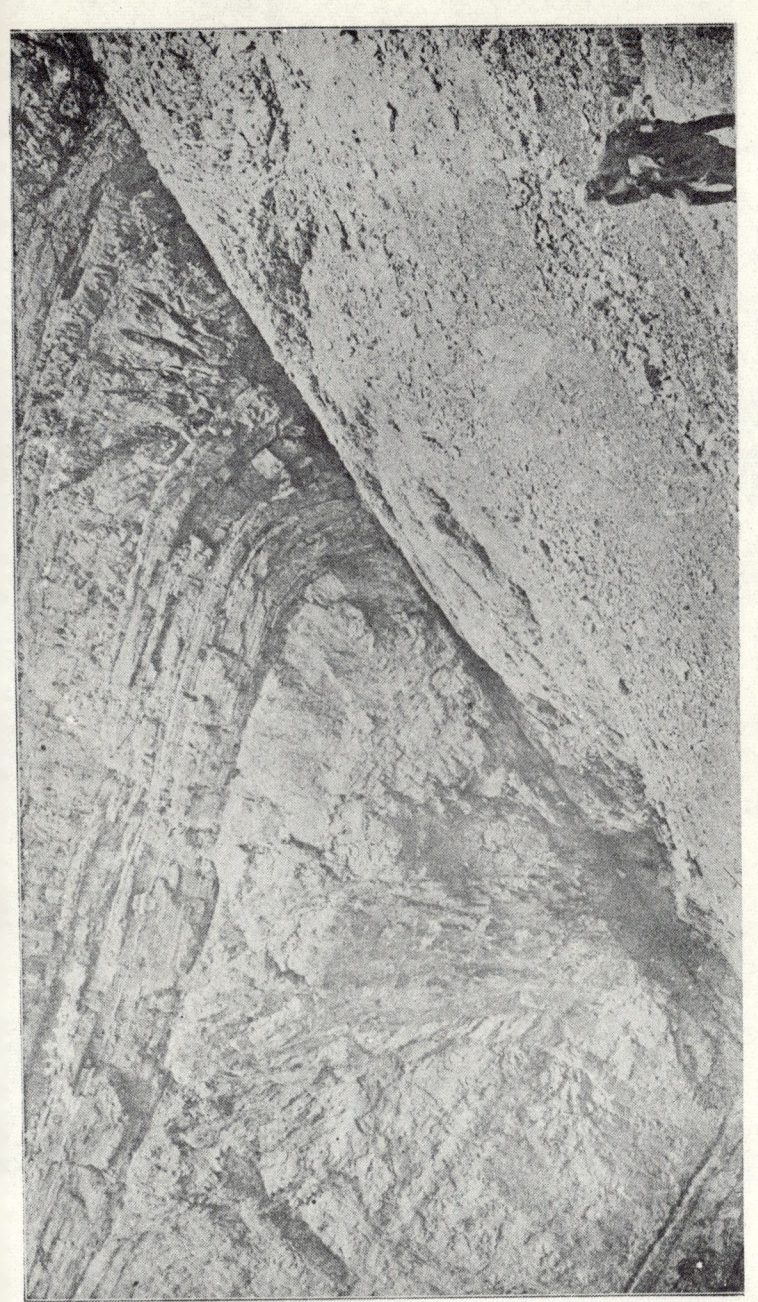

PLATE IX. REVERSED FAULT IN CARBONIFEROUS ROCKS, LEBUNG PASS, CENTRAL HIMALAYAS.

The Permian Productus shales (much disintegrated) are overlain by Carboniferous quartzites. (*Geol. Survey of India, Mem.* vol. xxiii.)

strata are found containing corals and brachiopods, *Favosites*, *Cyathophyllum*, *Orthis*, *Athyris*, *Atrypa*, *Spirifer*, showing a conformable sequence from Lower Devonian to the *Fusulina* limestone of Upper Carboniferous age. The structure in these mountains is highly complicated, and the Devonian is as a rule thrown against a Cretaceous or Lower Tertiary conglomerate (Reshun conglomerate) by a great fault. The Carboniferous occurs in well-marked bands and embodies the *Chitral slates* and *Sarikol shales* besides *Fusulina* limestone and some *Bellerophon* beds. Lithologically the Devonian of Chitral is a thick series of limestones overlying a series of older Palaeozoic strata, quartzites, red sandstones and conglomerates, in which are to be recognised the probable equivalents of the Muth quartzite and the Upper Silurian horizons of the better-known areas.[1]

5. BURMA

(Northern Shan States.)

But a much more perfect development of marine Palaeozoic rocks is found in the eastern extremity of the extra-Peninsula, in the Shan States of Upper Burma, in which the Indian Geological Survey have worked out a succession of faunas, revealing a continuous history of the life and deposits of the Palaeozoic group from Ordovician to Permian. The Shan States of Burma are a solitary instance, with the exception of Spiti and Kashmir, within the

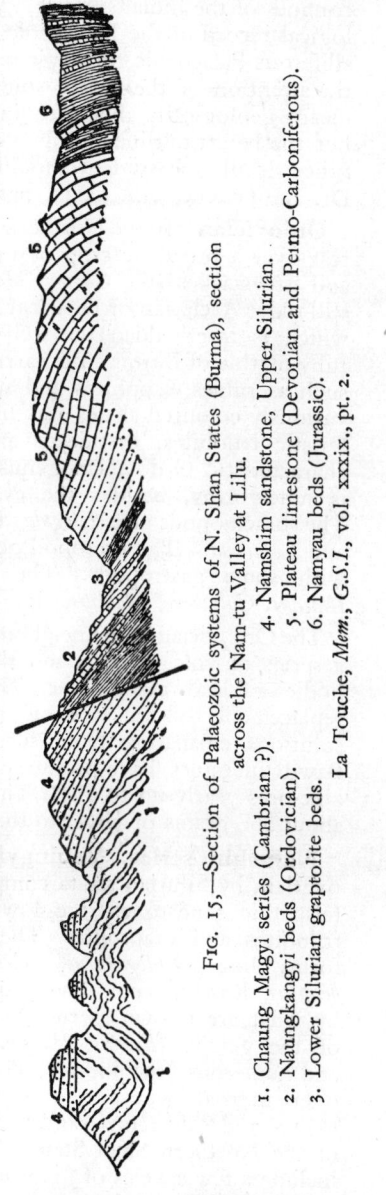

FIG. 15.—Section of Palaeozoic systems of N. Shan States (Burma), section across the Nam-tu Valley at Lilu.
1. Chaung Magyi series (Cambrian?). 2. Naungkangyi beds (Ordovician). 3. Lower Silurian graptolite beds. 4. Namshim sandstone, Upper Silurian. 5. Plateau limestone (Devonian and Permo-Carboniferous). 6. Namyau beds (Jurassic).
La Touche, *Mem.*, G.S.I. vol. xxxix., pt. 2.

---

[1] H.H. Hayden, *Rec. G.S.I.* vol. xlv., pt. 4, 1915; G.H. Tipper, *Rec. G.S.I.* vol. lv, p. 38, and vol, lvi. pp. 44–48, 1924.

confines of the Indian region, which possesses a complete geological record of the Palaeozoic era. The extreme rarity of fossiliferous Palaeozoic rock-systems in the Indian Peninsula compels the attention of the Indian student to this distant, though by no means geologically alien, province for study. We can here give but the barest outline of this very interesting development. For fuller details the student should consult the original *Memoir* by Dr. La Touche, vol. xxxix. part 2, 1913.

Ordovician—In the Northern Shan States, Ordovician exposures rest over a broad outcrop of unfossiliferous Cambrian quartzites and greywackes (the *Chaung Magyi* beds). These in turn overlie still older Archaean or Dharwar gneisses (the Mogok gneiss), with which is interbedded the well-known crystalline limestone (the ruby-marble of Burma), the carrier of a number of precious stones, such as rubies, sapphires and spinels. The Ordovician rocks are variously coloured shales and limestones containing the characteristic trilobites, cystideans and brachiopods of that age. The characteristic Ordovician genus of stemmed cystid, *Aristocystis*, is noteworthy, besides the cystids *Caryocrinus* and *Heliocrinus*. The brachiopods are *Lingula, Orthis, Strophomena, Plectambonites* and *Leptaena*. The pteropod genus *Hyolithes* is present, together with some gastropods. The trilobites are *Ampyx, Asaphus, Illaenus, Calymene, Phacops*, etc.

The Ordovician in the neighbourhood of Bawdwin is underlain by a series of volcanic tuffs and rhyolites which carry valuable orebodies of lead, zinc, silver. The Bawdwin ores are metasomatic replacements of the volcanic rocks, brought about by thermal solutions emanating from surrounding granite intrusions. The Bawdwin mines have produced some 465,000 tonnes of lead- and zinc-ores yearly up to 1941. The ores are argentiferous and carry about 567 grams of silver to the tonne of ore.

Namshim series. Zebingyi series—The Ordovician beds are overlain by Silurian strata composed of a series of quartzites and felspathic sandstones, the lower beds of which contain many trilobites and graptolites. The graptolites include characteristic forms like *Diplograptus, Climacograptus, Monograptus, Cyrtograptus, Rastrites*, etc. The graptolite-bearing beds are succeeded by what are known as the *Namshim series*, containing trilobites of the genera *Illaenus, Encrinurus, Calymene, Phacops, Cheirurus*, and numerous brachiopods. The Namshim sandstones are in turn overlain by a newer series of fossiliferous, soft yellow and grey limestones and calcareous sandstones, constituting the *Zebingyi series* of the Northern Shan States. The fossils of the Zebingyi series include a few species of graptolites of the type-genus *Monograptus*, together with cephalopods and trilobites (*Phacops* and *Dalmanites*), possessing affinities somewhat newer than the Wenlock limestone of England. These fossils indicate an uppermost Silurian age of the enclosing strata. The Zebingyi stage is thus to be regarded as forming the passage-beds between the Silurian and the overlying Devonian.

Silurian fauna of Burma—The Silurian fossils obtained from both the Namshim and Zebingyi horizons of the Shan States are:
Brachiopods—*Lingula, Leptaena, Orthothetis, Strophomena, Orthis, Pentamerus, Atrypa, Spirifer, Meristina.*
Lamellibranchs—*Pterinea, Modiolopsis, Glassia, Dualina, Conocardium.*
Gastropods—*Tentaculites.*
Cephalopods—Many species of *Orthoceras.*
Numerous broken stems of crinoids.
Rugose coral—*Lindstroemia.*
Worm borings and tubes.
Trilobites—*Illaenus, Proetus, Encrinurus, Calymene, Cheirurus, Phacops, Dalmanites,* and fragments of many other trilobites.

During the nineteen-thirties geological work in Burma established the existence of a more or less parallel series of fossiliferous Ordovician and Silurian in the Southern Shan States, comparable with those of the Northern Shan States through the help of a rich graptolite and brachiopod fauna.[1] The graptolites have established the Valentian and Salopian horizons of the Silurian.

Devonian—The Devonian is represented by a series of crystalline dolomites and limestones of Padaukpin, which have yielded a very rich assemblage of Devonian fossil, till lately the only undoubted occurrence of Devonian fauna met with hitherto in the Indian region. The fossils are very numerous and belong to all kinds of life of the period—corals, brachiopods, lamellibranchs, gastropods, cystids, crinoids, polyzoa, crustacea, etc.

Devonian fauna—The Devonian fauna of Burma :
Corals—*Calceola* (sp. *sandalina,* the characteristic Devonian coral), *Cyathophyllum, Cystiphyllum, Alveolites, Zaphrentis, Heliolites, Pachypora,* etc.
Polyzoa—*Fenestrapora, Hemitrypa, Polypora.*
Brachiopods—*Orthis, Atrypa, Pentamerus, Chonetes, Spirifer, Cyrtina, Merista, Meristella,* etc.
Lamellibranchs—*Conocardium, Avicula.*
Gastropods—*Loxonema, Pleurotomaria, Murchisonia, Euomphalus, Bellerophon.*
Cephalopods—*Anarcestes.*

---
[1] V. P. Sondhi and J. Coggin Brown : *Rec. G.S.I.,* vols. lxvi, pt. 2, 1932, and lxvii, pt. 2, 1933.

Trilobites—*Phacops*, etc.
Crinoids—*Cupressocrinus, Taxocrinus, Hexacrinus.*

The Wetwin slates—The limestone and dolomite are followed by an argillaceous series of yellow-coloured shales and slates of Upper Devonian age, known as the *Wetwin slates*, also fossiliferous, and containing *Lingula, Athyris, Chonetes, Janeia, Nucula* and *Bellerophon* as the commonest fossils. With the Wetwin slates are associated fine crystalline dolomites and limestones with remains of corals and foraminifers.

Carboniferous and Permo-Carboniferous—The Devonian is succeeded, in the same locality and in one continuous succession, by a great development of limestones and dolomites belonging to the Lower and Upper Carboniferous and Permian systems, which on account of their forming (together with the Devonian limestones) the plateau country of the Northern Shan States have been collectively known as the *Plateau limestone*. The limestones, which are extensively crushed and brecciated, vary from pure limestones through dolomitic limestones to pure dolomites. There are foraminiferal limestones (Fusulina limestone, from the preponderance of *Fusulina* in it, a rock-building *foraminifer* highly peculiar to this age in many parts of the world). The fossils of the upper portion of the Plateau limestone very closely correspond in facies with those of the Productus limestone of the Salt-Range (Chapter XI) of Permian age. (See Figs. 15 and 24.) In the Southern Shan States, where the Plateau limestone covers vast expanses of the plateau country, it has been divided into Lower (Devonian and Lower Carboniferous) and Upper (Carboniferous and Permian) on lithological differences, supported by some measure of palaeontological evidence. The supposed Devonian part of the limestone is generally a white or grey dolomite, extensively brecciated, and in the main unfossiliferous; while the upper part is more calcareous and contains a fauna showing affinities with the Productus fauna of India.

The faunas throughout the whole series of strata following the Wetwin shales are closely related and are stamped with the same general facies. The Lower Carboniferous forms are not separable from the Upper, nor are these from the Permian. For this reason the two groups of Carboniferous and Permian rocks are described under the name of *Anthracolithic* group, a grouping which was applicable to the Permo-Carboniferous rocks of some other parts of India as well, before their fossil faunas were differentiated.

The foregoing facts are summarised in the following table of geological formations of the Shan States, Upper Burma:

THE SILURIAN AND DEVONIAN SYSTEMS

	Burma.	Other parts.
Rhaetic.	Napeng beds.	
Permo-Carboniferous (Anthracolithic) System.	Upper Plateau limestone. *Fusulina* and *Productus* limestones. Partly dolomitic and brecciated. In the main unfossiliferous.	Productus limestone of the Salt-Range. Productus shales of Spiti and Zewan beds of Kashmir.
Devonian System.	Crystalline dolomites and limestones, much crushed, with *Calceola sandalina*, *Phacops*, *Pentamerus*, etc. (of Padaukpin), forming the plateau country. Wetwin shales with *Chonetes* and a very rich Devonian fauna (Eifelian).	Muth Series and Devonian of Chitral.
Silurian System.	Zebingyi beds, blue and grey flaggy limestones with *Graptolites*, *Tentaculites*, *Orthoceras*. Namshim sandstones, quartzose and felspathic sandstones, soft marls, and limestones with *Orthoceras*, *Trilobites*, etc.	Silurian of Spiti and Kashmir.
Ordovician System.	Nyaungbaw beds, brown limestones with shales containing Upper Ordovician fossils. Naungkangyi beds, yellow or purple shales with thick limestones. Cystids, *Orthis*, *Strophomena*, *Trilobites*.	Ordovician of Kashmir.
Cambrian System.	Chaung Magyi beds, thick quartzites, slaty shales and greywackes: unfossiliferous.	Haimanta of Spiti and Cambrian of N.W. Kashmir and the Salt-Range.

11 (45-36/1974)

	Burma.	Other parts.
Archaean System.	Mogok gneiss, gneiss and interbanded crystalline limestones with intrusive granites.	Peninsular gneisses.

Physical changes at the end of the Dravidian era—With the advent of the Upper Carboniferous, the second great era of the geological time-scale in India ended. Before we pass on to the description of the succeeding rock-groups we have to consider a great revolution in the physical geography of India at this epoch, whereby profound changes were brought about in the relative distribution of land and sea. The readjustments that followed these crust-movements brought under sedimentation large areas of India which hitherto had been exposed land-masses. An immense tract of India, now forming the northern zone of the Himalayas, was covered by the waters of a sea which invaded it from the west, and overspread North India, Tibet and a great part of China. This sea, the great Tethys of geologists, was the ancient central or *mediterranean* ocean which encircled almost the whole earth at this period in its history, and divided the continents of the northern hemisphere from the southern hemisphere. It retained its hold over the Himalayas for the whole length of the Mesozoic era, and gave rise, in the geosynclinal trough that was forming at its floor, to a system of deposits which recorded a continuous history of the ages between Permian and Eocene. This long cycle of sedimentation constitutes the second and last marine period of the Himalayan area.

During this interval the Peninsula of India underwent a different cycle of geological events. The Upper Carboniferous movements interrupted its long unbroken quiescence since the Vindhyan. Although the circumstances of its being a horst-like segment of the crust gave it immunity from deformations of a compressional or orogenic kind, yet it was susceptible to another class of crust-movements, characteristic of such land-masses. These manifested themselves in tensional cracks and in the subsidence of large linear tracts in various parts of the country between more or less vertical fissures of dislocation in the earth (block type of earth-movements), which eventually resulted in the formation of chains of basin-shaped depressions on the old gneissic land. These basins received the drainage of the surrounding country and began to be filled by its fluviatile and lacustrine débris. As the sediments accumulated, the loaded basins subsided more and more, and subsidence and sedimentation going on *pari passu*, there resulted thick deposits of fresh-water and subaerial sediments several thousand metres in vertical extent and entombing among them many relics of the terrestrial plants and animals of the time. These records, therefore, have preserved to us the history of the land-surface of the Indian continent, as the zone of marine sediments, accumulated in the geosynclinal of the Northern Himalayas, has that of the oceans. Thus a double facies is recognisable in the two deposition-areas of India in the systems that follow— a marine type in the extra-Peninsula and a fresh-water and subaerial type in the peninsula.

REFERENCES

H. H. Hayden, Geology of Spiti, *Mem. G.S.I.* vol. xxxvi., pt. 1, 1904.

C. L. Griesbach, Geology of Central Himalayas, *Mem. G.S.I.*, vol. xxiii., 1891.

T. H. D. La Touche, Geology of the Northern Shan States, *Mem. G.S.I.*, vol. xxxix. pt. 2, 1913.

F. R. C. Reed, *Palaeontologia Indica*, New Series, vol. ii. mem. 3 and mem. 5, 1906–8 and vol. xxi. mem. 3, 1936; Series XV. vol. vii., mem. 2, 1912, and New Series, vol. vi. mem. 2, 1922.

H. L. Chhibber, *Geology of Burma* (Macmillan), 1935.

C. S. Middlemiss, Silurian-Trias Sequence of South-East Kashmir, *Rec. G.S.I.* vol. xl, pt. 3, 1910.

A. Heim and A. Gansser, *Rocks and Structure of Kumaon*, Schw. Naturf. Gesell., Zürich, 1939.

A. B. Wynne, Trans-Indus Salt Range, *Mem. G.S.I.*, vols. xi. and xvii., 1880.

V. J. Gupta, *New Fossil Finds from the Lr. Palaeozoic of Kashmir and Spiti*, Punjab University and other publications, 1966–68.

CHAPTER IX

The Gondwana System

General. The Ancient Gondwanaland—Rocks of later age than Vindhyan in the Peninsula of India belong to a most characteristic system of land-deposits, which range in age from the Upper Carboniferous, through the greater part of the Mesozoic era, upto the end of the Jurassic. As mentioned in the last chapter, their deposition on the surface of the ancient continent commenced with the new era, the Aryan era. This enormous system of continental deposits, in spite of some local unconformities, forms one vast conformable and connected sequence from the bottom to the top. It is distinguished in the geology of India as the Gondwana system, from the ancient Gond kingdoms south, of the Narmada, where the formation was first known. Investigations in other parts of the world, *viz.* in South Africa, Madagascar, Australia and even South America, have brought to light a parallel group of continental formations, exhibiting much the same physical as well as organic characters. From the above circumstance, which in itself is adequate evidence, as well as from the additional proofs that are furnished by important palaeontological discoveries in the Jurassic and Cretaceous systems of India, Africa and Patagonia, it is argued by many eminent geologists that land-connection existed between these distant regions across what is now the Indian Ocean, either through one continuous southern continent, or through a series of land-bridges or isthmian links, which extended from South America to India, and united within the same borders the Malay Archipelago and Australia. The presence of land connections in the southern world for a long succession of ages, which permitted an unrestricted migration of its animal and plant inhabitants within its confines, is indicated by another very telling circumstance. It is the effect of such a continent on the character and distribution of the living fauna and flora of India and Africa of the present day. Zoologists have traced unmistakable affinities between the living lower vertebrate fauna of India and that of Central Africa and Madagascar, relationships which could never have subsisted if the two regions had always been apart, and had each pursued its own independent course of evolution. From data obtained from the distribution of fossil Cretaceous reptiles, especially the Sauropods, Prof. Von Huene suggests a distinct land-connection through *Lemuria* (the name given to the Indo-Madagascar continent) to South America. According to this authority, the Cretaceous dinosaurs of Madhya Pradesh belonged to the same faunistic province as Madagascar, and there is a great similarity in the fauna of the latter with that of Patagonia, Brazil, and Uruguay. These facts point to unrestricted inter-migration of land animals over a vast southern continent. The northern frontier of this continent was approximately co-extensive with central chain of the Himalayas and was washed by the waters of the Tethys.

The evidence, from which the above conclusion regarding an Indo-African land connection is drawn, is so weighty and so many-sided that the differences of opinion that exist among geologists appertain only to the mode of continuity of the land and the details of its geography, the main conclusion being accepted as one of the settled facts in the geology of this part of the world. The subaerial deposits formed by the rivers of this continent during the long series of ages are preserved in a number of isolated basins throughout its area, indicating a general uniformity and kinship of life and conditions on its surface. The term Gondwana system has been consequently extended to include all these formations, while the name of *Gondwanaland* is given to this Mesozoic Indo-African-American continent or archipelago. The Gondwanaland, called into existence by the great crust-movements at the beginning of this epoch, persisted as a very prominent feature in ancient geography till the commencement of the Cainozoic age, when, collaterally with other physical revolutions in India, large segments of it drifted away, or subsided, permanently, under the ocean, to form what are now the Bay of Bengal, the Arabian Sea, etc., thus isolating the Peninsula of India.

The *Gondwana system* is in many respects a unique formation. Its homogeneity from top to bottom, the fidelity with which it has preserved the history of the *land*-surface of a large segment of the earth for such a vast measure of time, the peculiar mode of its deposition in slowly sinking faulted troughs into which the rivers of the Gondwana country poured their detritus, and the preservation of valuable coal-measures lying undisturbed among them stamp these rocks with a striking individuality among the geological systems of India.

The geotectonic relations of the Gondwana rocks—The most important fact regarding the Gondwana system is its *mode of origin*. The formation of thousands of metres of river and stream deposits in definite linear tracts cannot be explained on any other supposition than the one already briefly alluded to. It is suggested that the mountain-building and other crustal movements of an earlier date, such, for instance, as the rejuvenation of the Aravalli and the Eastern Ghat ranges, had their reaction now in the subsidence of large blocks of the country to the equilibrium-plane, between vertical or slightly inclined normal faults in the crust. These depressions naturally became the gathering-grounds for the detritus of the land, for the drainage system must soon have betaken itself to the new configuration. The continually increasing load of the sediments that were poured into the basins caused them to sink relatively to the surrounding Archaean or Vindhyan country from which the sediments were derived, and thus gave rise to a continuation of the same conditions without interruption.

Although in a general way the Gondwanas were deposited in faulted depressions which have a general correspondence to the present disposition of their outcrops, it should not be supposed that in every case these outcrops imply the original fault-bound basin. Some of the boundary faults may be of post-Gondwana age. The original limits of deposition of the individual beds now found in these basins may not correspond in every case to the present ourcrops. The strike of these faults delimiting the Gondwana basins is E.-W. in the Bengal-Bihar area and N.W.-S.E. in the Mahanadi and Godavari Valleys. The down-throws of the main bounding faults are generally unequal in amount, *e.g.* on the south side of the Damodar valley basins the throw is much greater than on the north margin; the basins on the Godavari and Mahanadi have subsided much more on their N.E. margins than on the S.W. It is this circumstance that has determined the prevailing dip of Gondwana strata to the south in the former area and to the N.E. in the latter. Minor cross or oblique faults are also seen in the basins; these have afforded channels for the later igneous intrusions.

It is this sinking of the loaded troughs among the Archaean crystalline rocks that has tended to preserve the Gondwana rocks from removal by surface denudation, to which they would certainly have been otherwise subject. The more or less vertical faulting did not disturb the original horizontal stratification of the deposits beyond imparting to them minor warping, or a slight tilt now to one direction, now to the other, while it made for their preservation during all the subsequent ages. As almost all the coal of India is derived from the coal-seams enclosed in the Gondwana rocks, this circumstance is of great economic importance to India, since to it we owe not only their preservation from erosion, but their immunity from all crushing or folding which would have destroyed their commercial value by making the extraction of the coal difficult and costly.

Their fluviatile origin—The *fluviatile nature* of the Gondwana deposits is proved not only by the large number of the enclosed terrestrial plants, crustaceans, insects, fishes, amphibians, reptiles, etc., and by the total absence of the marine molluscs, corals and crinoids, but also by the character and nature of the very detritus itself, which gives conclusive evidence of deposition in broad river-valleys and basins. The rapid alternations of coarse- and fine-grained sandstones, and the numerous local variations met with in the rocks, point to a depositing agency which was liable to constant fluctuations in its velocity and current. Such an agency is river water. Further evidence is supplied by the other characters commonly observed in the alluvial deposits of river valleys, such as the frequency of false-bedding, the existence of several local unconformities due to what is known as "contemporaneous erosion" by a current of unususal velocity removing the previously deposited sediment, the intercalations of finely laminated clays among coarsely stratified sandstones, etc.

It is probable that in a few instances the deposits were laid down in lakes and not in river-basins, *e.g.* the fine silty shales of the *Talchir stage* at the bottom of the system. The distinctive character of the *lacustrine* deposits is that the coarser deposits are confined to the margin of the lake or basin, from which there is a gradation towards the centre where only the finest silts were precipitated. Breccias, conglomerates and grits mark the boundary of ancient lakes, while finely laminated sandstones and clays are found in the middle of the basins. This is frequently observed in deposits belonging to the Talchir series.

Climatic vicissitudes—The Gondwana system is of interest in bearing the marks of several *changes of climate* in its rocks. The boulder-bed at its base tells us of the cold of a Glacial Age at the commencement of the period, an inference that is corroborated, and at the same time much extended in its application, by the presence of boulder-beds at the same horizon in such widely separated sites as Hazara, Kashmir, Simla, Salt-Range, Rajasthan, Madhya Pradesh and Orissa. This Upper Carboniferous glacial epoch is a well-established fact not only in India, but in other parts of Gondwanaland, *e.g.* in Australia and South Africa. The thick coal-seams in the strata of the succeeding epoch, pointing to superabundance of vegetation, suggest a much warmer climate. This is followed by another cold cycle in the next series (the *Panchet*), the evidence for which is contained in the presence of undecomposed felspar grains among the clastic sediments. The last-mentioned fact proves the existence of ice among the agents of denudation, by which the crystalline rocks of the surface were *disintegrated* by frost-action and not *decomposed* as in normal climates. The thick red Middle Gondwana sandstones succeeding the Panchet beds indicate arid desert conditions during a somewhat later period, a conclusion warranted by the prevalence in them of so much ferruginous matter coupled with the almost total absence of vegetation.

Life of the period—The *organic remains* entombed in the sediments of the Lower Gondwana division are predominantly plants, members of the *Gangamopteris* and *Glossopteris* flora; they were succeeded by characteristic Middle and Upper Gondwana floras. These floras are numerous and of great biological interest, as furnishing the natural history of the large continent; but they do not help us in fixing the homotaxis of the different divisions of the system, in terms of the standard stratigraphical scale, with other parts of the world. The palaeontological value of terrestrial and fresh-water fossil organisms is limited, as they do not furnish a continuous and connected history of their evolution, nor is the geographical distribution of their species wide enough, as is the case with the marine molluscs, echinoderms, etc. Plant fossils are abundant, and are of service in enabling the different groups of exposures to be subdivided and correlated *among themselves* with some degree of minuteness. The lower Gondwanas contain numerous pteridosperms, ferns and equisetums; the middle part of the system contains a fairly well differentiated invertebrate as well as vertebrate fauna of crustacea, insects, fish, amphibia, and crocodilian and dinosaurian reptiles, besides plants, while in the upper division there is again a rich assemblage of fossil plants, now chiefly of the higher vegetable sub-kingdom (spermaphyta), cycads and conifers, with fish and other vertebrate remains.

A succession of distinct *floras* has been worked out from the shale and sandstone beds of the various Gondwana divisions by palaeobotanists, and distinguished as the *Talchir*, *Damuda*, *Raniganj*, *Rajmahal*, *Jabalpur* flora, etc., each possessing some individual characteristic of its own.

The isolation of Gondwanaland from the northern world in the Permo-Carboniferous is indicated by its individualised fossil

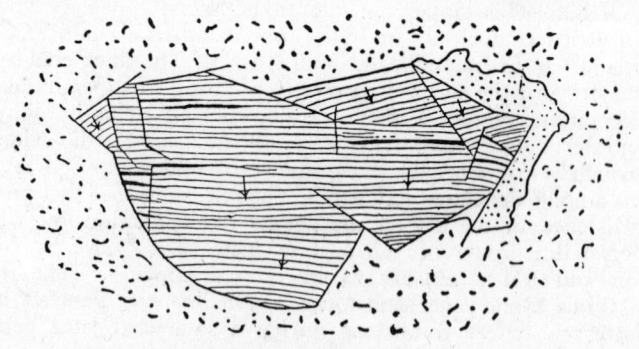

FIG. 16.—Sketch map of typical Gondwana outcrop

floras. The contemporaries of the *Gangamopteris* and *Glossopteris* flora were the *Gigantopteris* flora of North America and China, the *Angaraland* flora of Siberia and North Europe, and the *Lepidodendron* and *Sigillaria* flora characteristic of West European Carboniferous coal-measures, all unlike each other. The last-named flora was the direct descendant of the old *Rhacopteris* flora which pervaded the whole world in the beginning of the Carboniferous. While there are a few forms common to the Gondwana and the Angaraland floras (probably due to some island or isthmus connection across the Tethys via Kashmir and the Pamirs, during the Talchir-Damuda period), there is nothing in common between the Gondwana and West European floras, or the China and Indo-China floras. The diversification from the uniform *Rhacopteris* flora was brought about by the great geographical revolutions of the Hercynian epoch—an epoch of great earth-movements preceding the Upper Carboniferous.

Land-bridge between Gondwanaland and Angaraland—The idea of a northward migration of Gondwanaland plants to the northern continent of Angaraland was suggested sixty years ago by the discovery of some marked affinities of Russian and Siberian fossil plants with the *Glossopteris* flora of India. Zalessky advocated the view of intermigration, and suggested an isthmus connecting the two continents across the Himalayan sea. Field work in Kashmir has proved that during the greater part of the Silurian-Middle Carboniferous interval dry land existed in N.W. Punjab, Salt-Range, Hazara, and Kashmir, to as far north as the Pamirs. In all these areas the Cambrian, Ordovician or Silurian strata are overlain by the Upper Carboniferous, commencing with the Panjal Volcanic series, with a pronounced and widespread regional unconformity. This mid-Palaeozoic land-mass of Kashmir must have functioned as a land-bridge between the two continents before the Upper Carboniferous age. Even during the Upper Carboniferous, the Kashmir part of the Tethys must have been studded with an archipelago of volcanic islands which may well have permitted an interchange of land plants.

Both the supporters of Wegener's theory of Continental Drift and its opponents have looked for evidence in support of their respective views in the later geological history of the different units of Gondwanaland. The separation of the now discrete units of the once continuous southern continent of the Palaeozoic (Pangea) was brought about, according to one view, by the drifting away (*i.e.* north-easterly drift) of India from Africa ; and by the fragmentation and foundering of large segments of the land under the oceans, according to the other.

Palaeontological facts clearly show that the Indian Mesozoic systems, from the Trias to the Danian stage of the Cretaceous, are more closely related to those of Madagascar and South Africa than to Europe. Only at the end of the Cretaceous does the fauna enclosed in the Infra- and Inter-trappean beds show relationships to Sind, Persia and further west. In an important paper on the geographical relations of Gondwanaland, the eminent American geologists, Schuchert and Bailey Willis, present geological and biogeographic evidence which strongly supports the existence of land-bridges or isthmuses of the nature of Cordilleras, rather than a continuous land-mass, connecting Brazil, Africa and India, from the pre-Cambrian to the end of the Cretaceous. A. L. du Toit, on the other hand, supports the hypothesis of continental drift in a paper on the geological comparison of the sedimentary sequence in South America with that in South Africa.[1]

Distribution of the Gondwana rocks—Outcrops of the Gondwana system are scattered in a number of more or less isolated basins (see Figs. 16 and 17) lying in the older rocks of the Peninsula along certain very definite lines, which follow approximately (though not always) the courses of some of the existing rivers of the Peninsula. Three large tracts in the Peninsula can be marked out as prominent Gondwana areas : (1) a large linear tract in Bengal along the valley of the Damodar river, with a considerable area in the Rajmahal hills ; (2) an extensive outcrop in Madhya Pradesh prolonged to the south-east in a belt approximately following the Mahanadi valley ; (3) a series of more or less connected troughs forming an elongated band along the Godavari river from near Nagpur to the head of its delta. Besides these main areas, outliers of the Upper Gondwana rocks occur in Saurashtra, Kutch, Western Rajasthan and, the most important of all, along the east coast. Similar rocks, containing typical Upper Gondwana

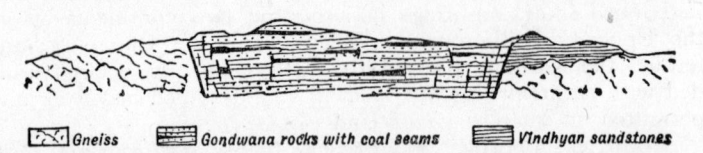

Fig. 17.—Tectonic relations of the Gondwana rocks. Vertical scale exaggerated.

cycads and ferns, are found in Sri Lanka in two small faulted basins. The Gondwana system, however, is not confined to the Peninsular part of India only, since we find outliers of the Lower Gondwanas to the north of the Peninsula on the other side of the Indo-Gangetic alluvium, at such distant centres as the Punjab Salt-Range,

---
[1] See also his *Our Wandering Continents*, London, 1937.

Shekh Budin hills, Hazara, Afghanistan, Kashmir, Nepal, Sikkim, Bhutan,[1] Assam, and the Abor country.[2]

From what has been said regarding their mode of origin and their geotectonic relations with the older rocks into which they have been faulted, the above manner of disposition of the Gondwana outcrops will easily be apparent. It also follows that the boundaries of the outcrops are sharply marked off on all sides, and that there is a zone of somewhat disturbed and fractured rock along the boundary while the main body of the rocks is comparatively undisturbed. These are actually observed facts, since the Gondwana strata never show any folding or plication, the only disturbance being a gentle inclination or dipping, usually to the south but sometimes to the north and north-east. The extra-Peninsular occurrences, on the other hand, have been much folded and compressed, along with the other rocks, and as a consequence the sandstones, shales and coal-seams have been metamorphosed into quartzites, slates and carbonaceous (graphitic) schists. These extra-Peninsular occurrences are of interest as indicating the limit of the northern extension of the Gondwana continent and the spread of its peculiar flora and fauna.

Classification—The system is classified into three principal divisions, the Lower, Middle, and Upper, corresponding in a general way respectively to the Permian, Triassic and Jurassic of Europe. The following tables show the division of the principal sections into series and stages, their distribution in the different Gondwana areas and the names by which they are recognised in these areas:

I. *Broad Correlation of the Gondwana System of India with equivalent deposits of other parts of the Southern Hemisphere.* [*C. S. Fox.*]

INDIA.	SOUTH AFRICA.	S.E. AUSTRALIA	S. AMERICA.	AGE.
Jabalpur group.				Jurassic.
Rajmahal group.	Stormberg Series.			
Panchet group.	Beaufort Series.	Hawkesbury Series.	Santa Catharina System.	Triassic.
Damuda group. (Coal Measures)	Ecca Series. (Coal Measures)	Maitland Series. (Upper Coal Measures)		Permian.
Talchir Series. (Glacial)	Dwyka Series. (Glacial)	Murree Series. (Lower Coal Measures and Glacial)	Rio Tubaro Series. (Glacial)	Upper Carboniferous.

(India column bracketed as **Gondwana**; South Africa column bracketed as **Karoo**.)

[1] *Rec. G.S.I.* vol. xxxiv. pt. 1, 1906.   [2] *Rec. G.S.I.* vol. xlii. pt. 4, 1912.

1. LOWER GONDWANA SYSTEM

Talchir series—The lowest beds of the Lower Gondwana are known as the Talchir series, from their first recognition in the Talchir district of Orissa. The series is divided into two stages, of which the lower, the *Talchir stage*, has a wide geographical prevalence, and is present in all the localities where Gondwana rocks are found, from the Rajmahal hills to the Godavari and from Raniganj to Nagpur. The group is quite homogeneous and uniform in composition over all these areas, and thus constitutes a valuable stratigraphical horizon. The component rocks (90–125 m. thick) are green laminated shales and soft fine sandstones. The sandstones contain *undecomposed* felspar grains, a fact which suggests the prevalence of land-ice and the disruptive action of frost. Glacial conditions are, however, more clearly indicated by a boulder-bed also of very wide prevalence in all the Gondwana areas, containing the characteristically glaciated, striated and faceted blocks of rock brought from afar and embedded in a fine silt-like matrix. The presence of this matrix suggests a *fluvioglacial* agency of transport and deposition rather than glacial. The boulders and blocks were transported in floating blocks of ice, and dropped in the Talchir basins, in which the deposition of fine silt was going on. Proofs of similar glacial conditions at this stage exist in many other parts of India, *viz*. the Aravallis, Rajasthan, Salt-Range, Hazara and Simla. The Aravallis in the north and the Eastern Ghats in the south-east were, it appears, the chief gathering-grounds for the snow-fields at this time, from which the glaciers radiated out in all directions. Many parts of the southern hemisphere, as shown in the table on page 171, experienced glacial conditions at this period. Boulder-conglomerates (tillites) homotaxial with the Talchir stage occur in South Africa (Dwyka series), south-east Australia (Murree series) and in South America (Itarara boulder-beds).

Talchir fossils—Fossils are few in the Talchir stage, the lower beds being quite unfossiliferous, while only a few remains of terrestrial organisms are contained in the upper sandstones ; there are impressions of the fronds of the most typical of the Lower Gondwana seed-ferns *Gangamopteris*, and *Glossopteris* with its characteristic stem named *Vertebraria* ; also spores of various shapes have been found on some fertile fronds ; wings of insects, worm-tracks, etc. are the only signs of animal life. The Talchir stage is succeeded by a group of coal bearing strata known as the *Karharbari stage*, 150–180 m. in thickness, also of wide geographical prevalence. The rocks are grits, conglomerates, felspathic sandstones and a few shales, containing seams of coal. Plant fossils are numerous, the majority of them belonging to genera of

II. Table of Correlation of the Series and Stages of the Gondwana System in different parts of Peninsular India.

System	Series	Stages	Damodar Valley	Rajmahal	Son & Mahanadi Valleys	Satpura	Godavari Valley	East Coast	Kutch	Age
Up. Gondwana		Umia.	—	—	—	—	—	—	Umia.³	Lower Cretaceous.
		Jabalpur. Rajmahal.	Rajmahal.	Rajmahal.	Jabalpur. (Athgarh sandstone). Chicharia.	Jabalpur. Changan.	Chikiala.	Tripetty, Pavalur, etc. Raghavapuram,² Sripermatur, etc.	Jabalpur.	Upper Jurassic.
		Kota.¹	—	Dubrajpur sandstone.	—	—	Kota.	Golapilli and Budavada² stages.	—	Lias.
		Maleri	—	—	Tiki.	Bagra. Denwa.	Maleri.	—	—	Keuper and Rhaetic. Muschelkalk
Mid. Gondwana		Mahadev (or Pachmarhi).	Durgapur.	—	Parsora.	Pachmarhi.	—	—	—	Bunter.
		Panchet.	Panchet.	—	Daigaon.	Panchet.	Panchet or Mangli.	—	—	Upper Permian.
Lr. Gondwana		Damuda.	{ Raniganj. Ironstone shales (Barren measures). Barakar.	Barakar.	Himgir. Barakar.	Bijori. Motur. Barakar. Umaria marine beds.	Kamthi. Barakar.	Chintalpudi sandstone;	—	Middle Permian.
		Talchir.	{ Karharbari. Talchir.	Talchir.	Karharbari. Talchir.	Karharbari. Talchir.	Talchir.	—	—	Upper Carboniferous.

¹ The relationship of the *Kota* and *Rajmahal* stages is uncertain; possibly the Kota beds are younger than the Rajmahal.
² These beds are now assigned to the Lower Cretaceous (see p. 190).
³ The lower Umia beds are uppermost Jurassic (see pp. 191 and 249).

unknown affinity, provisionally referred to the class of seed-ferns (Pteridosperms). The chief genera are:

(Pteridosperms) *Gangamopteris*—several species—this genus being represented at its best in the Karharbari stage, *Glossopteris* and its stem *Vertebraria, Gondwanidium* (formerly known as *Neuropteridium*).

(Cordaitales) *Noeggerathiopsis, Euryphyllum.*
(Equisetales) *Schizoneura.*
(Incertae) *Buriada, Ottokaria, Arberia.*

Besides, there occur the seed-like bodies *Samaropsis* and *Cordai carpus*, as well as scales with an entire or lacerated margin.

Damuda series—The Talchir series is succeeded by the second division of the Lower Gondwanas, the *Damuda series*, the most important portion of the Gondwana system. Where fully developed, as in the Damuda area of Bengal, the series is divided into three stages, in the descending order :

>Raniganj—1,500 m.
>Ironstone shales (Barren measures)—400 m.
>Barakar—600 m.

Of these the Barakar stage, named from the Barakar branch of the Damodar river, alone is of wide distribution among the Gondwana basins outside Bengal, *viz.* in the Satpura and the Mahanadi and Godavari valleys ; the middle and upper members are missing from most of them, being restricted chiefly to the type-area of the Damodar valley. The *Barakar stage* rests conformably upon the Talchir series, and consists of coarse, soft, usually white, massive sandstones and shales with coal-seams. The Barakars contain a large quantity of coal in thick coal-seams, though the quality of the coal is variable. The percentage of carbon is sometimes so low that the coal passes into mere carbonaceous shale by the large admixture of clay. It is usually composed of alternating bright and dull layers.[1] The coal is often spheroidal, *i.e.* it breaks up into ball-like masses. The *Ironstone shales* are a great thickness of carbonaceous shales with concretions (Sphaerosiderites) of impure iron carbonate and oxides. They have yielded much ore of iron formerly used in the blast-furnaces of Bengal. This stage, which is about 600 m. thick, also known under the name of *Barren measures*, from its total lack of coal seams, consists mostly of sandstones and carbonaceous shales. The stage is met with in the Jharia and Karanpura coalfields but when followed westwards it merges into the overlying Raniganj series. The group is of a most inconstant thickness and appears only at a few localities in the Damuda area, being altogether missing from the

---

[1] C. S. Fox, Natural History of Indian Coal, *Mem. G.S.I.* lvii., 1931.

rest of the Gondwana areas. This is succeeded by the *Raniganj stage* of the Damuda series, named from the important mining town of Bengal. The Raniganj stage is composed of massive, false-bedded, coarse and fine sandstones and red, brown and black shales, with numerous interbedded coal-seams. The sandstones are felspathic, but the felspar in them is all decomposed, *i.e.* kaolinised. The coal is abundant and of good quality as a fuel, with a percentage of fixed carbon generally above 55.

Igneous rocks of Damuda coal-measures—Many of the coal-fields of the Damodar valley, especially those of the eastern part, are invaded by dykes and sills of dolerite and of an ultra-basic rock which have wrought much destruction in the coal-seams by the contact-metamorphism they have induced. The invading rock is a mica-peridotite, containing a large quantity of apatite in the Damodar valley, and dolerite or basalt (Deccan trap) in the Satpura and Rewah areas. The peridotite has intruded in the form of dykes and then spread itself out in wide horizontal sheets or sills. Another intrusive rock is a dolerite, whose dykes are thicker, but they are fewer and are attended with less widespread destruction of coal than the former.

Effects of contact-metamorphism—The coal is converted into coke, and its economic utility destroyed. The reciprocal effects of contact-metamorphism on the peridotite as well as the coal are very instructive to observe. The peridotite has turned into a pale earthy and friable mass with bronze-coloured scales of mica in it, but without any other trace of its former crystalline structure. On the other hand, the coal has coked or even burnt out, becoming light and cindery, and at places it has developed prismatic structure.

The Damuda flora—The Damuda fossils are nearly all plants. The flora is chiefly cryptogamic, associated with only a few spermaphytes. It is exceedingly rich in Pteridosperm leaves of the net-veined type, the genus *Glossopteris* here attaining its maximum development, while *Gangamopteris* is on the decline. The following are the most important genera:

(Pteridosperms)—*Glossopteris* with *Vertebraria*, at least nine species, several of them confined to the Raniganj stage, *Gangamopteris, Belemnopteris, Merianopteris, Sphenopteris, Pecopteris, Palaeovittaria.*

(? Ginkgoales)—*Rhipidopsis.*
(Cordaitales)—*Noeggerathiopsis, Dadoxylon.*
(Cycadophyta)—*Taeniopteris, Pseudoctenis.*
(Filicales)—*Cladophlebis.*
(Equisetales)—*Schizoneura, Phyllotheca.*
(Sphenophyllales)—*Sphenophyllum.*

(Lycopodiales)—? *Bothrodendron*.

(Incertae)—*Barakaria, Dictyopteridium*, scales, seeds including *Samaropsis* and *Cordaicarpus*.

The animals include *Estheria*, Labyrinthodonts and some Fishes.

The Damuda series of other areas—In the Satpura area the Damuda series is represented, in its Barakar and Raniganj stages, by about 3,050 metres of sandstone and shale, constituting what are known as the *Barakar, Motur* and *Bijori stages*, respectively of this province. The Mohpani and the Pench valley coal-fields of the Satpura region belong to the Barakar stage of this series. In the strata of the last-named stage, at Bijori, there occur bones and other remains of a Labyrinthodont (*Gondwanosaurus*). Other fossils include scales and teeth of ganoid fishes, and seed-ferns and equisetums identical with those of Bihar. It is quite probable that large expanses of the Lower Gondwana rocks are buried under the basalts of the Satpuras, and they must have contained, and possibly still contain, some valuable coal-seams.

Another area of the Peninsula where the Damuda series is recognised, though greatly reduced and with a somewhat altered facies, is in the Godavari valley, where a long but narrow band of Lower Gondwana rocks stretches from the old coal-field of Warora to the neighbourhood of Rajahmundry. The Barakar stage of the Damuda series prevails in these outcrops which bear the coal-fields of Warora, Singareni, Ballarpur, etc.

One more outcrop of the Damuda group is seen in the Rewah region, Vindhya Pradesh, which at one or two places has workable coal-seams, *e.g.* in the Umaria field. The division of the Lower Gondwana exposed in this field also is the Barakar. The Raniganj stage is represented in Nagpur, Chanda and the Wardha Valley by the *Kamthi beds* by the *Bijori stage* in the Satpuras; by the *Pali beds* in South Rewah; and the *Himgir beds* in the Mahanadi Valley.

Homotaxis of the Damuda and Talchir series—Few problems in the geology of India have aroused greater controversy than the problem of the lower age limit of the Gondwana system. The Talchir series has been referred, by different authors, to almost every stratigraphic position from Lower Carboniferous to Trias. The discovery, however, of a Lower Gondwana horizon in Kashmir, bearing the eminently characteristic genera *Gangamopteris* and *Glossopteris* overlying the Upper Carboniferous and underlying marine fossiliferous strata of undoubtedly Permian age, has settled the question beyond doubt. A similar occurrence of Lower Gondwana plants has been noted in the Lower and Middle Productus limestone of the Salt-Range, the marine fossils of which point to Lower and Middle Permian affinities. The Upper Carboniferous, or Permo-Carboniferous, age attributed to the Talchir glacial horizon by this circumstance is quite in keeping with the

PLATE X. BARRIER OF COAL ACROSS KARARIA NALA. *Photo. L. L. Fermor.* (*Geol. Survey of India, Mem.* vol. xli.)

The elephant is standing on the parting between seams 4 and 5.

internal evidence that is furnished by the Talchir and Damuda floras, as well as by the fish and labyrinthodont remains of Bijori. The occurrence of *Eurydesma cordatum* and the typical Lower Gondwana fossil plants *Gangamopteris* and *Glossopteris* in sandstones directly overlying the glacial boulder-bed of the Salt-Range, and considerably below the horizon of the Lower Productus limestone containing the Fusulinid, *Parafusulina kataensis* (a Permian form), places the boulder-bed at the top of the Moscovian or at most in the Uralian. Over 150 m. of sandstones and shales containing intercalations bearing fossil fronds of genera belonging to the *Glossopteris* flora of the Talchir horizon separate the two zones. The eminent American palaeontologist, Professor Charles Schuchert, has, however, ascribed a definitely Permian (Lower to Middle) age to the Talchir glacial epoch.

Further positive evidence leading to the same inference is supplied by the Lower Gondwanas of Victoria and New South Wales, Australia. Here, *Gangamopteris* and other plant-bearing beds of undoubted Gondwana facies, underlain by a glacial deposit identical with the Talchir boulder-bed, are found interstratified with marine beds which contain an Upper Carboniferous fauna with *Eurydesma*, resembling that of the Speckled sandstone group of the Salt-Range.

Economics—The Damuda series contains a great store of mineral wealth in its coal-measures, and forms, economically, one of the most productive horizons in the geology of India. It contains the most valuable and best worked coal-fields of the country. The mining operations required for the extraction of coal from these rocks are comparatively simple and easy because of the immunity of the Gondwana rocks from all folding or plication. Also, mining in India is not so dangerous, on account of the less common association of highly explosive gases (marsh-gas or "firedamp") with the coal, as compared with European coal-fields. There are, however, special difficulties associated with the working of thick seams, and fires and subsidences in mines have proved very troublesome.

Although coal occurs in India in later geological formations also —*e.g.* in the Tertiary of Assam, Punjab, Rajasthan and Tamilnadu and in the Jurassic rocks in Kutch and Kalabagh—the Damuda series is the principal source of Indian coal, contributing over 80 per cent of the total Indian production, now about 70 million tonnes per annum. The principal coal-fields are: Bihar and adjoining area: Raniganj, Jharia, Giridih, Bokaro, Karanpura and Rampur; Madhya Pradesh: Pench Valley, Singrauli and Korea; Orissa: Talcher; Hyderabad: Singareni, Tandur. The Raniganj coal-field covers an area of 1,500–1,600 sq. km. containing many seams of good coal with interbedded iron-stones. The thickness of individual seams of good coal is great, 12-15 metres with occasional seams 25 metres thick. The annual output of coal from Raniganj collieries is 19 to 20 million tonnes. The Jharia field is smaller in area but has at present the largest output, nearly 20 to 21 million tonnes per annum. It is the most important coal-field of India, with the largest resources in coking coal. The coal of the Jharia-field belongs to the Barakar stage. It has less moisture and a greater proportion of fixed carbon than that of the Raniganj stage, which has more water and greater volatile content. The coal-fields of Karanpura and Bokaro, to the west of Jharia, contain thick seams of valuable coal, only a small portion of which is coking coal. The Damodar Valley fields yield nearly 80 per cent of the total coal in India. Of the remaining, those of Pench Valley, Singrauli and Korea are the more important. In general, Gondwana coal is bituminous, good to moderate quality steam and gas-coal, with 11 to 20 per cent ash and 6,000 to 8,500 calorific value. Anthracite is rare, confined to coal-fields lying in compressed mountain areas.

India's resources in coking coal are almost confined to the Jharia field and parts of the Giridih and Bokaro fields. The coal of the Giridih field is free from phosphorus and is, therefore, of value in the manufacture of ferro-manganese. Fuel research experiments during the last several years have indicated the possibility of refining high-volatile, high-ash coals by washing and flotation; by this means, and by blending of suitable varieties, some non-coking coals can be used for coke making.

Besides coal, iron, fire-clay, kaolin, and terra-cotta clays occur in considerable quantities in Bihar and Madhya Pradesh. The Barakar sandstones and grits furnish excellent material for mill-stones.

Classification—During recent years Dr. G. de P. Cotter, in an attempt to subdivide the Gondwana system on a palaeobotanical basis, has found it more appropriate, on the evidence of an interesting suite of plant fossils obtained from the Parsora beds of South Rewah, to include among the Lower Gondwanas the thick zone of strata which overlies the Damuda series and underlies the Rajmahal, embodying in fact the group that has been here treated as Middle Gondwana. Dr. Cotter named the strata in question the *Panchet series* (divided into three stages—*Panchet*, *Maleri* and *Parsora*), and grouped them along with the Talchir and Damuda series in the Lower Gondwana. In C. S. Fox's scheme the Maleri and Parsora series are included in the Upper Gondwanas while the Panchets are grouped with the Lower.

The flora of the beds placed in the Parsora stage by Cotter still needs a critical examination. Possibly the fossils belong to two distinct horizons, the older (containing a typical *Glossopteris* flora) definitely belonging to the Lower Gondwanas, the younger (with *Thinnfeldia* as the dominant genus) belonging to the Middle or Upper Gondwanas. According to Seward and Sahni the affinities of the latter flora are also distinctly with the Lower rather than with the Upper Gondwana (see page 184). Sahni held the view that on the palaeobotanical evidence the Parsora beds cannot possibly be classed as Jurassic. The classification that is here adopted was originally based on the views of Feistmantel and Vredenburg, but chiefly on lithological grounds and the life and physical conditions of the period embracing the Middle Gondwanas, which were strikingly different from those prevailing in the Damuda and Rajmahal areas.

The presence of red beds, indicating arid or semi-desert conditions supervening on the damp forest climate of the Damuda period, the Triassic affinities of the fossil reptiles and stegocephalian amphibia, and the coincidence of the Palaeo-Mesozoic boundary at the base of the Panchets with the unconformity at the top of the Raniganj series are features distinguishing the Middle Gondwana group. The total extinction of *Gangamopteris* and the *Sphenophyllales* after the Damuda epoch is widespread, and denotes a datum-line of some importance. The Middle Gondwana was also the epoch of most extensive land-conditions in India. During the Upper Gondwana, epeiric seas began to encroach on its borders from the north-west and south-east.

Lower Gondwana of the Himalayas—At several localities along the foot of the Himalayas, from Hazara to Assam, strips of Lower Gondwana rocks are found sandwiched in between the Tertiaries and Older Himalayan strata, sometimes with coal-seams of Barakar or other horizons. These outcrops are generally narrow and structurally much disturbed, appearing from under the eroded cover of overthrust sheets of pre-Cambrian strata, "window", in the Middle Himalayas. Only in the Kashmir area do these rocks attain any development and exhibit normal stratigraphic relations to the Permo-Carboniferous. (Chapter XI.)

Besides a strip of Gondwanas adpressed against the Siwalik foot-hills in the Darjeeling sector of W. Sikkim, there is an interesting occurrence of a Lr. Gondwana coal-field with thick coal seams, underlying the dolomites and phyllites of the Daling series. In the Ranjit Valley north of Darjeeling, the coal-field is exposed from underneath the eroded cover of overthrust sheets of the pre-Cambrian and now appearing as a "window". Intercalated in the tillites and boulder-beds of Talchir affinities and underlying the coarse sandstones and shales containing *Glossopteris*, *Vertebraria* and *Schizoneura* are marine bands with fossil *Spirifer*, *Pro-Productus*, *Uridesma*, *Conularia* and *Chonetes*.

The association of marine fossils with Talchir boulder-conglomerate suggests contiguity of the edge of the Gondwana foreland to the Himalayan geosynclinal sea and the existence of the coal-bearing fresh water strata so deep within the mountains suggests involvement of the Gondwana continental shelf in the complicated inversions of Himalayan orogeny.

The Gondwana coast-line in the N.-W. extends beyond the Central Himalaya to as far as the S.-W. foot of the Karakoram, where Norin (1946) has observed fossiliferous Talchir rocks. During the Mid. Carboniferous, West Kashmir was a land area—an archipelago of volcanic islands joining the Gondwana to Angaraland (Wadia, 1937).

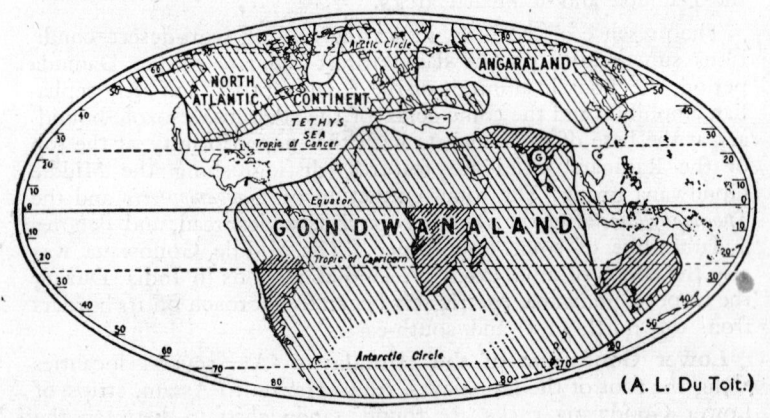

FIG. 17a.—A restoration of Gondwana Continent as it existed in early Mesozoic.

CHAPTER X

The Gondwana System (*continued*)
2. Middle Gondwanas

BETWEEN the upper beds of the Damuda series and the next overlying group of strata, distinguished as the Panchet, Mahadev, Maleri and Parsora series, there is an unconformable junction; in addition there exists a marked discordance in the lithological

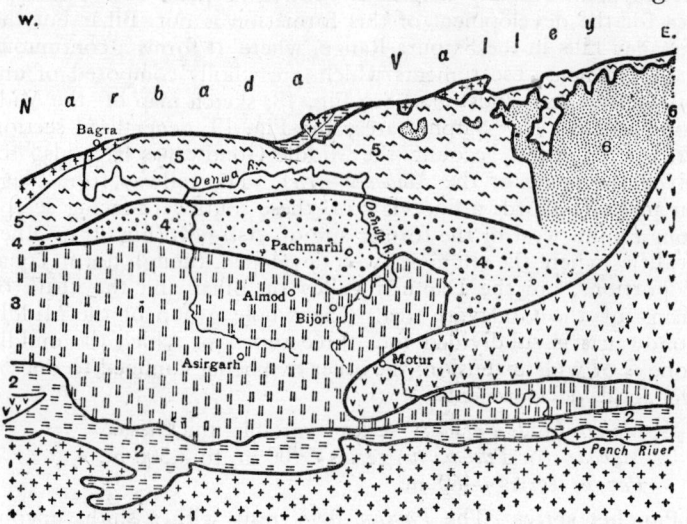

FIG. 18.—Sketch map of the Gondwana rocks of the Satpura area.
1. Archaean.   4. Pachmarhi (Kamthi).   6. Jabalpur.
2. Talchir.   5. Denwa (Maleri) and Bagra.   7. Deccan Trap.
3. Damuda.
Medlicott, *Mem. G.S.I.* vol. x. (1873).

composition and in the fossil contents of these groups. For these reasons the series overlying the coal-bearing Damudas have been separately grouped together under the name of Middle Gondwanas by E. W. Vredenburg.[1] Usually it is the practice to regard

[1] *Summary of the Geology of India*, Calcutta, 1910.

a portion of the latter group as forming the upper portion of the Lower Gondwanas, and the remaining part as belonging to the bottom part of the Upper Gondwanas, but, in view of the above dissimilarities, as well as of the very pronounced lithological and climatic resemblance of the Middle Gondwanas to the Triassic system of Europe, and more so on stratigraphic grounds, it is appropriate to regard the middle division as a separate section of the Gondwana system.

Rocks—The rocks which constitute the Middle Gondwanas are a great thickness of massive red and yellow coarse sandstones, conglomerates, grits and shales, altogether devoid of coal-seams or of carbonaceous matter in any shape. Vegetation, which flourished in such profusion in the Lower Gondwanas, became scanty, or entirely disappeared, for the basins in which coarse red sandstone were deposited must have furnished very inhospitable environments for any luxuriant growth of plant life. The type area for the development of this formation is not Bihar but the Mahadev hills in the Satpura Range, where it forms a continuous line of immense escarpments which are wholly composed of unfossiliferous red sandstone. (See Fig. 18, sketch map of the Mid-Gondwanas of the Satpura area, and Fig. 19, generalised section across it.) On this account the Middle Gondwanas have also received the name of the *Mahadev series*. The railway from Nagpur to Itarsi affords a fine view of these southward-facing scarps from a point west of the former site of Asirgarh (Fig. 18). Other localities where the strata are well developed, though not in equal proportions, are the Damuda valley of Bihar and the chain of basins of the Godavari area. The whole group of the Middle Gondwanas is subdivided into three series, of which the middle alone is of wide extension, the other two being confined to one or two local developments:

Maleri (and *Parsora*) series—variable thickness.
Mahadev (or *Pachmarhi*) series—900 m. to 2,500 m.
Panchet series—460 m.

Panchet series—The *Panchet series* rests with a slight unconformity on the denuded surface of the Raniganj stage but at some localities the Panchets overlap on to the Barakar stage. The beds consist of alternations of fine red clays and coarse, micaceous and felspathic sandstones, occasionally containing rolled fragments of Damuda rocks. The felspar in the sandstones is in undecomposed grains. Characteristic Panchet plant fossil are *Schizoneura gondwanensis*, *Glossopteris*, *Vertebraria indica*, *Pecopteris concinna*, *Cyclopteris*, *Thinnfeldia*. The group is of importance as containing many well-preserved remains of vertebrate animals, affording us a glimpse of the higher land-life that inhabited the Gondwana

MIDDLE GONDWANAS

continent. These vertebrate fossils consist of the teeth, scales, scutes, jaws, vertebrae and other bones of lacustrine and fresh-water fishes, amphibians and reptiles. Three or four genera of labyrinthodonts (belonging to the extinct order *Stegocephalia* of the amphibians) have been discovered, besides several genera of primitive and less differentiated reptiles.

Panchet fossils:
>(Amphibia) *Gonioglyptus*, *Glyptognathus*, and *Pachygonia*; (Fish) *Amblypterus*; (Reptiles) *Dicynodon* and *Ptychosiagum* and the dinosaur *Epicampodon*. The fresh-water crustacean *Estheria* is very abundant at places.

Mahadev series—The *Mahadev series*, locally also named *Pachmarhi*, is the most conspicuous and the best-developed member of the Middle Gondwana in Madhya Pradesh. Near Nagpur it consists of some 1,200 m. of variously coloured massive sandstones, with ferruginous and micaceous clays, grits and conglomerates.

The most typical development of the series is, however, in the Mahadev and Pachmarhi hills of the Satpura range, where it is exposed in the gigantic escarpments of these hills. It unconformably overlies the *Bijori stage* there (*Raniganj* stage of Damodar valley area). Here the series is composed essentially of thick-bedded massive sandstones, locally called Pachmarhi sandstones, variously coloured by ferruginous matter; in addition to sandstone there are a few shale beds which also contain a great deal of ferruginous matter, with sometimes such a concentration of the iron oxides in them locally that the deposits are fit to be worked as ores of the metal. The sandstones as well as shales are frequently micaceous. The shales contain beautifully preserved leaves of seed-

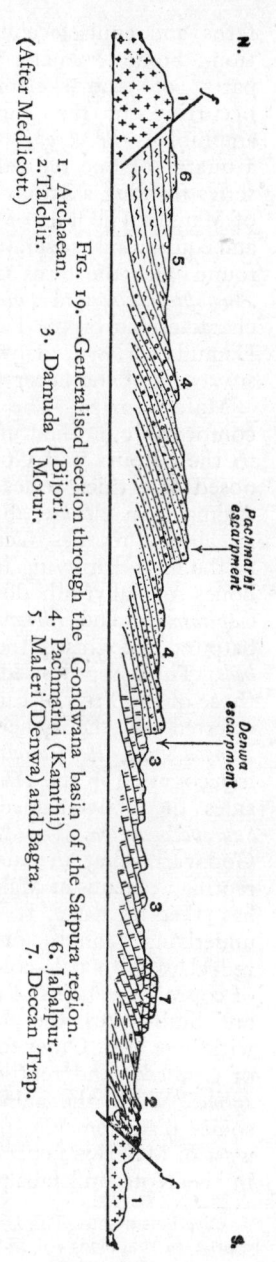

FIG. 19.—Generalised section through the Gondwana basin of the Satpura region. 1. Archaean. 2. Talchir. 3. Damuda {Bijori. Motur. 4. Pachmarhi (Kamthi) 5. Maleri (Denwa) and Bagra. 6. Jabalpur. 7. Deccan Trap. (After Medlicott.)

ferns and equisetaceous plants along their planes of lamination. Some animal remains are also obtained, including parts of the skeletons of vertebrates similar to those occurring in the Panchet beds. The most important is an amphibian—*Brachyops*. This labyrinthodont was obtained from a quarry of fine red sandstone which lies at the bottom of the series forming a group known as the *Mangli beds* near the village of Mangli. The flora of the Pachmarhi series consists of seed-ferns and equisetums, several species of *Vertebraria* and *Phyllotheca* being found with the ferns *Glossopteris*, *Gangamopteris*, and *Pecopteris*, *Angiopteridium* and *Thinnfeldia*, the species *T. hughesi* being very characteristic of the Pachmarhi. This flora resembles that of the Damuda series in many of its forms, being for the most part the survivors of the latter flora.

Maleri series—The *Maleri* (or *Denwa*) *series* comes generally conformably on the top of the last. Its development is restricted to the Satpura and Godavari regions. Lithologically it is composed of a thick series of clays with a few beds of sandstones. Animal remains are abundant. The shales are full of coprolitic remains of reptiles. Teeth of the Dipnoid fish *Ceratodus*, similar to the mud-fish living in the fresh waters of the present day, and bones of labyrinthodonts like *Mastodonsaurus*, *Gondwanosaurus*, *Capitosaurus* and *Metopias* are met with in the Maleri rocks of Satpura, recognised there under the name of *Denwa* and *Bagra beds*. Three reptiles, identical in their zoological relations with those of the Trias of Europe, are also found in the rocks; they are referred to the genera *Hyperodapedon* (order *Rhynchocephalia*), *Belodon*, and *Parasuchus* (order *Crocodilia*). The Maleri horizon is recognised in the *Tiki beds* in south Rewah, which contain, besides the above-named reptiles, *Coelurosauria*, *Brachysuchus* and *Sauropodomorpha*. The Maleri group is well represented in the Godavari valley in Andhra also, and it is from the discovery of reptilian remains at Maleri, a village near Sironcha, that the group has taken its name. It here rests with an unconformity on the underlying Mangli, or Panchet, beds and consists of bright red clays with pale-coloured sandstone beds. The shales are full of coprolite remains of reptiles together with their teeth, vertebrae and limb-bones, the above three fossil genera having been met with here also. Other fossils from the same locality include species of *Ceratodus* and reptiles of the genera *Hyperodapedon* and *Parasuchus*. While the animal fossils clearly indicate a Triassic age, some plant-remains recorded by Feistmantel[1] from Naogaon west of Maleri are characteristic Upper Gondwana fossils common in the Kota and Jabalpur stages, and would point definitely to a

---

[1] See Feistmantel, *Pal. Indica, Fossil Flora of the Gondwana System* (1877), vol. ii. pt. 2, p. 16; (1879) vol. i., pt. 4, pp. 198–208.

Jurassic horizon. These species are *Araucarites cutchensis* and *Elatocladus jabalpurensis*.[1]

The Maleri group is succeeded by the *Kota stage*. Its affinities, however, are with the Upper Gondwanas, and it will be described in connection with them. The combined groups were sometimes designated as the *Kota-Maleri stage*. Reptilian fossils have also been collected from the *Tiki beds* of South Rewah, representing approximately the Maleri horizon of other Gondwana centres. The Tiki sandstones and shales have yielded some fragmentary bones, among which are maxillae and vertebrae of *Hyperodapedon*, teeth and other relics of Dinosaurs, together with shells of the fresh-water lamellibranch *Unio*.

The *Parsora stage*: These beds in South Rewah, corresponding roughly with the Rhaetic stage of the Trias, form the typical Middle Gondwanas of Feistmantel. The Parsoras have yielded a flora of somewhat uncertain affinities containing elements of both Lower and Upper Gondwana type which still await a critical examination. Among the fossils collected from the villages of Parsora and Chicharia the dominant genus is *Thinnfeldia* (*Dicroidium*). This is represented by *T. (D.) hughesi* and several species allied to those known from other parts of Gondwanaland, where the introduction of the *Thinnfeldia* element marks the later (Permo-Triassic) phases of the *Glossopteris* flora. From localities further south a flora apparently somewhat older, with *Glossopteris* as the chief genus, had been collected.

A recent work in the Gondwanas of South Rewa basin by K. M. Lele has brought to light a large number of fossil plant remains from the Panchet, Mahadev and Maleri (Parsora) series to supplement the hitherto known scanty and unsatisfactory flora from this division of the Gondwanas. In addition to numerous *Thinnfeldia* species, this flora comprises the following genera: *Desmophyllum, Neocalamites, Cladophlebis, Triletes, Dadoxylon, Lycopodites, Sphenopteris, Taeniopteris, Samaropsis, Cordaicarpus, Pterophyllum, Pseudoctenis, Baiera, Noeggerathiopsis* and *Araucarite s*.

The above floral assemblage is quite distinctive and imparts to these three series, grouped here under the Middle Gondwana, an individuality separate from the Lr. Gondwanas, characterised by the preponderance of *Glossopteris*, as well as from the overlying group of Up. Gondwanas characterised by the genus *Ptilophyllum*.

Triassic age of the Middle Gondwanas—From the foregoing account of the Middle Gondwanas it must have been clear that they agree in their lithology with the continental facies of the Triassic (the New Red Sandstone) system of Europe. At the same time the terrestrial forms of life that are preserved in them indicate that they are as distinct biologically as they are physically

---

[1] Sahni, *Pal. Indica*, vol. xi. (1931), pp. 115–16.

from the underlying Lr. Gondwanas with their coal seams and luxuriant plant life as well as from the succeeding Up. Gondwana system with its more advanced plant life of Cycads, Conifers and Ferns. They bear distinct signs of arid conditions in the barren red sandstones and shales and the scanty desert fauna and flora. There are, however, no indications in these rocks of that wonderful differentiation of reptilian life which began in the Triassic epoch in Europe and America, and gave rise, in the succeeding Jurassic period, to the numerous highly specialised races of reptiles that adapted themselves to life in the sea and in the air as much as on the land, and performed in that geological age much the same office in the economy of nature as is now performed by the class of Mammals.

3. THE UPPER GONDWANA SYSTEM

Distribution—Upper Gondwana rocks are developed in a number of distant places in the Peninsula, from the Rajmahal hills in Bengal to the neighbourhood of Madras. The outcrops of the Upper Gondwanas, as developed in their several areas, *viz.*, Rajmahal hills, Damuda valley, the Satpura hills, the Mahanadi and Godavari valleys, Kutch and along the eastern coast, are designated by different names, because of the difficulty of precisely correlating these isolated outcrops with each other. It is probable that future work will reveal their mutual relations with one another more clearly, and will render possible their grouping under one common name. In Kutch and along the Coromandel coast, beds belonging to the upper horizon of the Gondwanas are found interstratified with marine fossiliferous sediments, a circumstance of great help to geologists in fixing the time-limit of the Upper Gondwanas, and determining the homotaxis of the system in the stratigraphical scale.

Lithology—Lithologically the Upper Gondwana group is composed of the usual massive sandstones and shales closely resembling those of the Middle Gondwanas, but is distinguished from the latter by the presence of some coal-seams and layers of lignitised vegetable matter, and a considerable development of limestones in some of its outcrops, while one outcrop of the Upper Gondwanas, *viz.* that in the Rajmahal hills, is quite distinct from the rest by reason of its being constituted principally of volcanic rocks. This volcanic formation is composed of horizontally bedded basalts contemporaneously erupted, which attain a great thickness.

Rajmahal series—Upper Gondwana rocks are found in Bengal and Bihar at two localities, the Damodar valley and the Rajmahal hills, some 48 km. N.E. of the Raniganj coal field, the latter being the more typical locality. The Upper Gondwanas in the

THE UPPER GONDWANA SYSTEM 187

Rajmahal hills rest unconformably on the underlying Barakar stage. The lowest beds above the break are known under the name of the *Dubrajpur sandstone*. The *Rajmahal series* consists of about 600 m. of bedded basalts or dolerites, with about 30 m. of interstratified sedimentary beds (*inter-Trappean beds*) of siliceous and carbonaceous clays and sandstones. Almost the whole mass of the Rajmahal hills is made up of the volcanic flows, together with these inter-trappean sedimentary beds. The shales have turned porcellanoid and lydite-like on account of the contact-effects of the basalts. The basalt is a dark-coloured, porphyritic and amygdaloidal rock, commonly fine-grained in texture. When somewhat more coarsely crystalline it resembles a dolerite. The amygdales are filled with beautiful chalcedonic varieties of silica, calcite, zeolites or other secondary minerals. A radiating columnar structure due to "prismatic" jointing is produced in the fine-grained traps at many places. It is probable that these superficial basalt-flows of the Rajmahal series are connected internally with the dykes and sills that have so copiously permeated the Raniganj and other coal-fields of the Damuda region, as their underground roots. The latter are hence the hypabyssal representatives of the subaerial Rajmahal eruptions. Among these dykes mica-peridotites, lamprophyre, minette and kersantite types have been found.

The andesitic trap of Sylhet, in the Khasi hills of Assam, unconformably underlying the Upper Cretaceous, is probably an eastward continuation of the Rajmahal trap.

Rajmahal flora—The silicified shales of the Rajmahal beds have yielded a very rich flora in which the fossil Cycads (Bennettitales) are the predominant group. Next in order of abundance are the Ferns and Conifers. The cycad genera comprise many types of leaves (*e.g. Ptilophyllum, Pterophyllum, Dictyozamites, Otozamites, Nilssonia, Taeniopteris*), also a few flowers (*Williamsonia*) and stems (*Bucklandia*). The stem known as *Bucklandia indica* bore leaves of the *Ptilophyllum* type and *Williamsonia* flowers; the connections of the other leaf genera are still unknown. The most important Fern genera are *Marattiopsis, Cladophlebis, Coniopteris, Gleichenites, Pecopteris* and *Sphenopteris*. The Coniferales include several kinds of vegetative shoots (*Elatocladus, Brachyphyllum, Retinosporites*), detached cones and scales (*Conites, Ontheodendron, Araucarites*) and wood (*Araucarioxylon*). The Equisetales are represented by *Equisetites* and the Lycopodiales by *Lycopodites*. Among the Incertae are some genera (*Rajmahalia, Homoxylon, Pentoxylon*, etc.) of much palaeobotanical interest. *Homoxylon rajmahalense* is a type of fossil wood which closely resembles the wood of some Jurassic Cycads as well as that of some primitive modern angiosperms. It therefore supports the well-known theory of the Bennettitalean origin of angiosperms. The Rajmahal flora was till recently known almost exclusively from impressions. Recent anatomical studies have considerably advanced our knowledge of this classical flora.

The Rajmahal stage can fitly be called an age of fossil cycads, from the predominance of the Bennettitales. The flora presents a sharp contrast with those of the Lower and Middle Gondwanas. It wears a distinctly more familiar aspect, the affinities of the great majority of the genera being known. The Pteridosperms and the Cordaitales have disappeared. The Equisetales have dwindled into insignificance. The Ferns now claim an important place, and most of them can be assigned to recent families. The conifers, formerly a small group, are now on the increase; in the collateral Kota and Jabalpur stages of the Upper Gondwana they are as important an element in the flora as the cycads, while in the succeeding Umia stage they actually dominate the flora.

Satpura and Madhya Pradesh

Jabalpur stage—Upper Gondwana rocks, of an altogether different facies of composition from that at Rajmahal, are developed on a very large scale in these areas. The base of the series rests unconformably on the underlying Maleri beds locally known under the name of *Denwa* and *Bagra beds*, and successively covers, by overlapping, all the older members of the Middle and Lower Gondwanas exposed in the neighbourhood. The rocks include two stages: the lower *Chaugan* and the upper *Jabalpur stage*. The Chaugan stage consists of limestones, clays and sandstones, with boulder conglomerates. It is succeeded unconformably by the next stage, named after the town of Jabalpur. The rock components of the Jabalpur stage are chiefly soft massive sandstones and white or yellow shales, with some lignite and coal seams, and in addition a few limestone bands. The Jabalpur stage is of palaeontological interest because of its having yielded a rich Jurassic flora, rather distinct from that of the preceding series and of somewhat newer age, *viz*. Lower Oolite. It differs from the Rajmahal flora mainly in its containing a greater proportion of conifers, *viz*. *Elatocladus* (several species), *Retinosporites*, *Brachyphyllum*, *Pagiophyllum*, *Desmiophyllum*, *Araucarites*, and *Strobilites*, and in the much reduced number of cycads.

At Jabalpur this stage is overlain by the Lameta group of Cretaceous strata, remarkable for their containing many fossil remains of dinosaurs.

Godavari Basin

Kota stage—A narrow triangular patch of Upper Gondwana rocks occurs in the Godavari valley south of Chanda. The rocks are of the same type as those of the Satpuras, with the exception of the top member, which is highly ferruginous in its constitution. At places the oxides of iron are present to such an extent as to be of economic value. Here also two stages are recognised : the lower *Kota stage*, some 610 m. in thickness, and the upper *Chikiala stage*, about 150 m. composed of highly ferruginous sandstones, poor-quality coal-seams and conglomerates. The Kota stage is fossiliferous, both plant and animal remains being present in its rocks in large numbers. The Kota stage, which overlies the Maleri stage described above, consists of loosely consolidated sandstone, with a few shale beds and with some limestones. From the last beds fossil fishes, *Lepidotus*, *Tetragonolepis*, *Dapedius*, and crustaceans have been obtained. A fauna of dinosaurs has recently been discovered in clays and sandstones lying immediately below the middle Kota limestone band. The plants include the conifers *Palissya*, *Araucarites* and *Cheirolepis*, and numerous species of cycads belonging to *Cycadites*, *Ptilophyllum*, *Taxites*, etc., resembling the Jabalpur forms. The Chikiala stage is unfossiliferous, being often strongly ferruginous (haematitic) and conglomeratic.

Gondwanas of the East Coast

The Coastal system—Along the Coromandel coast, between Vishakapatnam and Tanjore, there occur a few small isolated outcrops of the Upper Gondwanas along a narrow strip of country between the gneissic country and the coast-line. These patches are composed, for the most part, of marine deposits formed not very far from the coast, during temporary transgressions of the sea, containing a mingling of marine, littoral organisms with a few relics of the plants and animals that lived near the shore. Near the Peninsular mainland there are consequently to be seen in these outcrops both fossil plants of Gondwana facies and the marine or estuarine molluscs including *ammonites*. In geological horizon the different outliers correspond to all stages from the Rajmahal to the uppermost stage (Umia).

Rajahmundry outcrop—The principal of these outcrops is the one near the town of Rajahmundry on the Godavari delta. It includes three divisions:

Tripetty sandstone—45 metres.
Raghavapuram shales—45 metres.
Golapilli sandstones—90 metres.

This succession of beds rests unconformably over strata of Raniganj horizon, termed *Chintalpudi sandstones*. Lithologically it is composed of littoral sandstones, gravel and conglomerate rock, with a few shale-beds. The latter contain some marine lamellibranchs (*e.g.* species of *Trigonia*, including *T. ventricosa*) and a few species of ammonites. Intercalated with these are some beds containing impressions of the leaves of cycads and conifers.

Ongole outcrop—Another outcrop of the same series of beds is found near the town of Ongole, on the south of the Krishna. It also consists of three subdivisions, all named after the localities :

Pavalur beds—red sandstone.
Vemavaram beds—shales.
Budavada beds—yellow sandstone.

The Vemavaram shales contain a very rich assemblage of Gondwana plants, related in their botanical affinities to the Kota and Jabalpur plants.

Madras group—A third group of small exposures of the same rocks occurs near Madras, in which two stages are recognised. The lower beds form a group which is known as the *Sripermatur beds*, consisting of whitish shales with sandy micaceous beds containing a few cephalopod and lamellibranch shells in an imperfect state of preservation; the plant fossils obtained from beds associated in the same horizon correspond in facies to the Kota and Jabalpur flora. The Sripermatur beds are overlain by a series of coarser deposits, consisting of coarse conglomerates interbedded with sandstones and grits, which contain but few organic remains. This upper division is known as the *Sattavadu beds*. Solitary outcrops of these rocks containing fossil cycads and conifers extend to Tiruchirapalli, Madurai and Ramnad. At Utatur the Upper Gondwanas are found underlying the Cenomanian marine beds.

Cuttack—One more similar exposure, occurring far to the north on the Mahanadi delta, is seen at Cuttack. It is composed of grits, sandstones and conglomerates with white and red clays. The sandstone strata of this group are distinguished as the *Athgarh sandstones*. They possess excellent qualities as building stones, and have furnished large quantities of building material to numerous old edifices and temples, of which the temple of Jagan Nath Puri is the most famous.

Age—A middle Jurassic age was ascribed to these coastal Gondwanas, but the discovery of a suite of better preserved ammonites from Budavada and Raghavapuram proves a considerably newer horizon for these beds, Lower Cretaceous (Barremian). The ammonites are *Holcodiscus, Lytoceras, Gymnoplites* and *Hemihoplites*.

The identification of angiospermous fossil wood *Homoxylon*, a magnoliaceous dicotyledon and the flower of *Williamsonia sewardi* from the Rajmahal series (the flora of which is essentially identical with that of the coastal Gondwanas) by Sahni lends support to the inference that both the series are probably of Neocomian or still later age.

Sri Lanka Gondwanas—The coarse sandstones, grits and arkose, containing cycads, ferns and pteridosperms, which are found in two small isolated basins in Sri Lanka faulted into the Archaeans—the *Tabbowa* and *Andigamma beds*—belong probably to the Tamilnadu group of Upper Gondwanas. It is probable that some more occurrences of these rocks further north in the island are concealed under the alluvium and the Miocene limestones (*Jaffna beds*).

Gondwanas of the West Coast: Umia Series

Upper Gondwanas of Kutch—The highest beds of the Upper Gondwanas are found in Kutch, at a village named Umia. They rest on the top of a thick series of marine Jurassic beds (to be described with the Jurassic rocks of Kutch in a later chapter). The *Umia series*, as the whole formation is called, is a very thick series of marine conglomerates, sandstones and shales, in all about 900 m. in thickness. The special interest of this group lies in the fact that with the topmost beds of this series, containing the relics of various cephalopods and lamellibranchs, there occur interstratified a number of beds containing plants of Upper Gondwana facies, pointing unmistakably to the prevalence of Gondwana conditions at the period of deposition of this series of strata. The marine fossils are of uppermost Jurassic to lower Cretaceous affinities, and hence serve to define the upward stratigraphic limit of the great Gondwana system of India within very precise bounds. The Umia plant-remains are thought to be the newest fossil flora of the Gondwana system. The following is the list of the important forms :

(Conifers) *Elatocladus, Retinosporites, Brachyphyllum, Pagiophyllum, Araucarites.*
(Cycads) *Ptilophyllum, Williamsonia, Taeniopteris.*
(Ferns) *Cladophlebis.*

Some of the species of these genera are allied to the Jabalpur species, others are distinctly newer, more highly evolved types.

The Umia beds have also yielded the remains of a reptile, a species belonging to the famous long-necked *Plesiosaurus* of the European Jurassic. It is named *P. indica*.

In **Northern Saurashtra** there is a large patch of Jurassic rocks occupying the region near Dhrangadhra and Wadhwan, consisting of about 300 m. of horizontally bedded sandstones. The lower part, containing some carbonaceous beds and ferruginous slates, has fossils of Jabalpur affinities, while the upper part corresponds to the Umia group of Kutch in geological horizon. It has yielded conifers and cycads resembling the Umia plants.

Economics—The Upper Gondwana rocks include several coal-seams, but they are not workable. Some of the fine-grained sandstones, *e.g.*, those of Cuttack, Athgarh, Tirupati and Ahmednagar, are much used for building purposes, while the clays obtained from some localities are utilised for a variety of ceramic manufactures. The soil yielded by the weathering of the Upper Gondwanas, as of nearly all Gondwana rocks, is a sandy shallow soil of poor quality for agricultural uses. Hence outcrops of the Gondwana rocks are marked generally by barren landscapes or else they are covered with a thin jungle. The few limestone beds are of value for lime-burning, while the richly haematitic or limonitic shales of some places are quarried for smelting purposes or use as ochres. The coarser grits and sandstones are cut for millstones.

REFERENCES

W. T. Blanford, The Ancient Geography of Gondwanaland, *Rec. G.S.I.* vol. xxix. pt. 2, 1896.

Charles Schuchert and Bailey Willis, Gondwana Land Bridges and Isthmian Links, *Bulletin of the Geological Society of America*, vol. xliii., 1932.

T. H. Huxley, R. D. Lydekker, etc., Fossil Vertebrata of Gondwana System, *Pal. Indica*, Series IV. pts. 1–5 (1865–1885).

G. de P. Cotter, Revised Classification of the Gondwana System, *Rec. G.S.I.* vol. xlviii. pt. 1, 1917.

A. C. Seward and B. Sahni, Indian Gondwana Plants : A Revision, *Pal. Indica*, vol. vii. mem. 1, 1920.

B. Sahni, Revisions of Indian Fossil Plants, *Pal. Indica*, N.S. vol. ix., 1928–1931 and vol. xx. mems. 2 and 3, 1932.

C. S. Fox, The Gondwana System and Coal Deposits of India, *Mem. G.S.I.* vols. lvii.–lix., 1931–1934.

E. R. Gee, Geology and Coal of the Raniganj Coalfied, *Mem. G.S.I.* vol. lxi., 1932.

H. Crookshank, Gondwanas of North Satpura, *Mem. G.S.I.* vol.lxvi., pt. 2, 1936.

A. L. Du Toit, *our Wandering Continents* (Oliver & Boyd), London, 1937.

Wadia, D. N., Upper Palaeozoic Land-bridge between the Northern and Gondwana Continents, *Proc.* xvii. *Intn. Geol.* Congr., Moscow, 1937.

Norin, E., Geological Exploration in Western Tibet. Sino-Swedish Expedition, Stockholm, 1946.

CHAPTER XI

Upper Carboniferous and Permian Systems

The commencement of the Aryan era—In the last two chapters we have followed the geological history of the Peninsula upto the end of the Jurassic period. In the other provinces of the Indian region a different order of geological events was in progress during this long cycle of ages.

As referred to before, the era following the Middle Carboniferous was of great earth-movements, the *Hercynian*, both in Central Asia (whose chief mountain-ranges are of Hercynian orogeny) as well as in North India : sedimentation was interrupted in the various areas of deposition, the distribution of land and sea was readjusted, and numerous changes of physical geography profoundly altered the face of the continent. As a consequence there is, almost everywhere in India, a very marked break in the continuity of deposits, represented by an unconformity at the base of the Permo-Carboniferous. Before sedimentation was resumed, these earth-movements and crustal readjustments had resulted in the easterly extension over the whole of Northern India, Tibet and China of the great Mediterranean sea of Europe, which in fact at this epoch girdled almost the whole earth as a true *mediterranean* sea, separating the great Gondwana continent of the south from the Eurasian continent (Angaraland) of the northern hemisphere. The southern shores of this great sea, which has played such an important part in the Mesozoic geology of the whole Indian region—the *Tethys*—coincided with the central chain of snow-peaks of the Himalayas, beyond which it only rarely transgressed to any extent ; but, to the east and west of the Himalayan chain, bays of the sea spread over areas of Upper Burma and Baluchistan, a great distance to the south of this line, while an arm of the same sea extended towards the Salt-Range and occupied that region, with but slight interruptions, almost up to the end of the Eocene period. It is in the zone of deep-water deposits that began to be formed on the floor of this central sea at this time that the materials for the geological history of those regions are

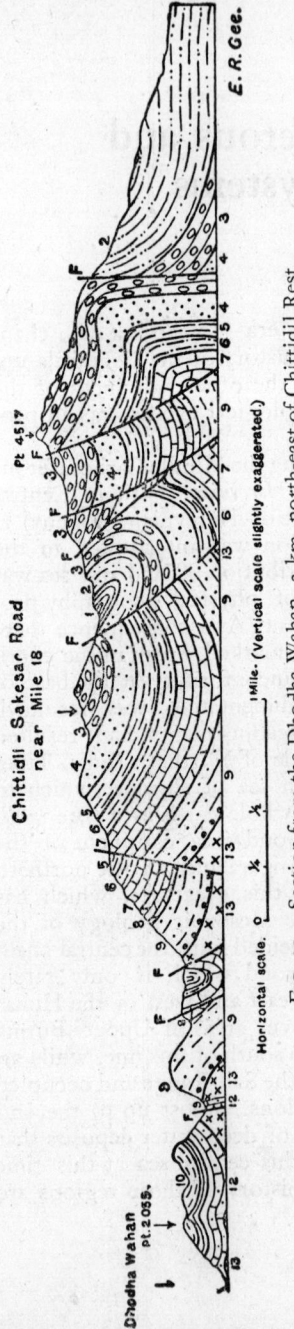

FIG. 20.—Section from the Dhodha Wahan, 2 km. north-east of Chittidil Rest House, running due north across the western part of Sakesar ridge. (The line of the section is 500 m. west of long. 71°55').

1. Lower Siwaliks (Chinji stage).
2. Lower Siwaliks (Kamlial stage).
3. Nummulitic (Laki and Ranikot).
4. Jurassic.
5. Ceratite beds (Trias).
6. Upper Productus beds.
7. Middle Productus Limestones.
8. Lower Productus beds.
9. Speckled Sandstone series (Talchir boulder-bed at base).
10. Salt Pseudomorph beds.
11. {Magnesian Sandstones. Neobolus Shales.
12. Purple Sandstones.
13. Saline series.

F—Fault usually reversed.

preserved for the long succession of ages, from the Permian to the Eocene, the Aryan era of Indian geology.

The nature of geosynclines—Portions of the sea-floor subsiding in the form of long narrow troughs concurrently with the deposition of sediments, and thus permitting an immense thickness of deep-water deposits to be laid down over them without any intermission, are called *Geosynclines*. It is the belief of some geologists that the slow continual submergence of the ocean bottom, which renders possible the deposition of enormously thick sediments in the geosynclinal tracts, arises in the first instance from a disturbance of the *isostatic* conditions of that part of the crust, further accentuated and enhanced by the constantly increasing load of sediments over localised tracts. The adjacent areas, on the other hand, which yield these sediments, have a tendency to rise above their former level, by reason of the constant unloading of their surface due to the continued exposure to the denuding agencies. They thus remain the feeding-grounds for the sedimentation-basins. This state of things will continue till the isostatic equilibrium of the region has been restored by internal readjustment in the sub-crust. At the end of this cycle of processes, after prolonged intervals of time, a reverse kind of movement will follow in this flexible and comparatively weak zone of the crust, rendered more plastic by the rise of the isogeotherms, compressing and elevating these vast piles of sediments into a mountain-chain, on the site of the former geosyncline.

Geosynclines are thus long narrow portions of the earth's outer shell which are relatively the weaker parts of the earth's circumference, and are liable to periodic alternate movements of depression and elevation. It is such areas of the earth which give rise to the mountain-chains when they are, by any reason, subjected to great lateral or tangential compression. Such compression occurs, for instance, when two large adjacent blocks of the earth's crust—horsts—sink during secular movements caused by some thermal readjustments in the sub-crustal *sima*, or mantle layers through unequal radio-active heating or deep convection currents upon its continual loss of internal heat. The bearing of these conceptions on the elevation of the Himalayas, subsequent to the great cycle of Permo-Eocene deposits on the northern border of India, is plausible enough. The Himalayan zone is, according to this view, a geosynclinal tract squeezed between the two large continental masses of Eurasia and Gondwanaland. This subject is, however, one of the unsettled problems of modern geology, and one which is yet *sub judice*.

The records of the Himalayan area which we have now to study reveal an altogether different geological history from what we have known of the Gondwana sequence. It is as essentially a

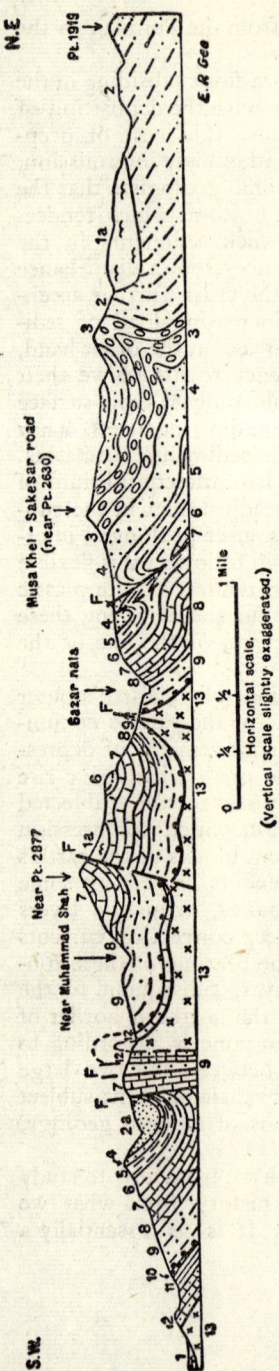

FIG. 21.—Section across the Salt-Range, taken N.E.–S.W., from the exit of the Khanzaman *nala*, 2·4 km. approx. E.S.E. of Childeru (lat. 32° 33′ long 71° 46′).

1. Recent alluvium.
1a. Pleistocene alluvium.
2. Lower Siwalik (Kamlial stage).
2a. Siwalik.
3. Nummulitic (Laki and Ranikot).
4. Jurassic.
5. Trias.
6. Upper Productus beds.
7. Middle Productus Limestones.
8. Lower Productus beds.
9. Speckled Sandstone series (Talchir boulder-bed at base).
10. Salt Pseudomorph beds.
11. { Magnesian Sandstones. Neobolus Shales. }
12. Purple Sandstones.
13. Saline series.
F.—Fault, usually reversed.

history of the oceanic area of the earth and of the evolution of the marine forms of life, as the latter is a history of the continental area of the earth and of the land plants and animals that inhabited it. This difference emphasises the distinction between the stable mass of the peninsula and the flexible, relatively much weaker extra-Peninsular area subject the periodic movements of the crust. In contrast to the Peninsular horst, the latter is called the geosynclinal area.

The Upper Carboniferous and Permian—The Upper Carboniferous and Permian systems are found perfectly developed in two localities of extra-Peninsular India, one in the western part of the Salt-Range and the other in Kashmir and the northern ranges of the Himalayas.

I. UPPER CARBONIFEROUS AND PERMIAN OF THE SALT-RANGE

After the Cambrian Salt-pseudomorph shales the next known series of deposits that was laid down in the Salt-Range area belongs to these systems. Sometime after the Cambrian, the Salt-Range, like the Peninsula, became a bare land area exposed to denudational agencies, but, unlike the Peninsula, it was brought again within the area of sedimentation by the late Carboniferous movements. From this period to the close of the Eocene, a branch of the great central sea to the north spread over this region and laid down the deposits of the succeeding geological periods, with a few slight interruptions. These deposits are confined to the western part of the Range, beyond longitude 72° E., where they are exposed in a series of more or less parallel and continuous outcrops running along the strike of the range. In the eastern part of these mountains, Permo-Carboniferous rocks are not met with at all, the Cambrian group being there abruptly terminated by a fault of great throw, which has thrust the Nummulitic limestone of Eocene age in contact with the Cambrian.

The Permo-Carboniferous rocks of the western Salt-Range are a thick series of highly fossiliferous strata. A two-fold division is discernible in them: a lower one composed of sandstones and an upper one mainly of limestones, characterised by an abundance of the brachiopod *Productus*, and hence known as the *Productus limestone*. The Productus limestone constitutes one of the best developed geological formations of India, and, on account of its perfect development, is a type of reference for the Permian system of the other parts of the world.

The table below shows the chief elements of the Permo-Carboniferous system of the Salt-Range:

GEOLOGY OF INDIA

Productus limestone 210 m.	Upper 60 m.	*Chideru Stage.* *Jabi* ,, *Kundghat* ,,	Marls and sandstones. Sandy limestones. Sandstones with *Bellerophon*.	Thuringian.
	Middle 90 m.	*Kalabagh* ,, *Virgal* ,,	Crinoidal limestones with marls and dolomites. Cherty limestones.	Punjabian.
	Lower 60 m.	*Katta* ,, *Amb* ,,	Brown sandy limestones. Calcareous sandstones, *Fusulina* limestone.	Artinskian.
Speckled Sandstones 210 m.		Speckled sandstones, 90 m. *Conularia beds*, 60 m.	Clays, grey and blue. Mottled sandstones. Olive shales and sandstones. *Conularia* and *Eurydesma*.	Upper Carboniferous. (Uralian.)
		Boulder-bed, 3–60 m. } Talchir Stage	Glaciated boulders in a fine matrix.	

Boulder-beds—The basement bed of the series is a boulder-conglomerate of undoubted glacial origin, which from its wide geographical occurrence in strata of the same horizon, in such widely separated parts of India as Hazara, Simla, the Salt-Range, Rajasthan, Bihar, Orissa and other localities wherever the Lower Gondwana rocks have been found, has been made the basis of an inference of a Glacial Age at the commencement of the Upper Carboniferous period throughout India. The evidence for this Ice Age in India lies in the existence of the characteristic marks of glacial action in all these areas, *viz*. beds of compacted "boulder-clay" or glacial drift, resting upon an under surface which is often sharply defined by being planed and striated by the glaciers. The most striking character of a boulder-clay is its heterogeneity, both in its component materials, which have been transported from distant sources, and in the absence of any assortment and stratification of these materials. Many of the boulders in the boulder-bed of the Salt-Range are striated and polished blocks of the Malani rhyolites, felsites and granites of Vindhyan age—an important formation of Rajasthan. These are intermixed with smaller pebbles from various other crystalline rocks of the same area, and embedded in a fine dense matrix of clay. Besides striations and polishing, a certain percentage of the pebbles and boulders shows distinct "faceting". The Aravalli region must have been the home of snow-fields nourishing powerful glaciers at this time, as the size of the boulders as well as the distances to which they have been transported from their source clearly testify to the magnitude of the glaciers radiating from it.

Boulder-beds similar to that of the Salt-Range, and also like it composed of ice-borne boulders of Malani rhyolites and other crystalline rocks, are found in Rajasthan in Marwar (Jodhpur region) and are known as the Bap and Pokaran beds, from places of those names. At the latter place there occur typical *roches moutonnées*. The Talchir boulder-bed is homotaxial with the glacial beds associated with the *Eurydesma beds* of south-east Australia.

The Speckled sandstones—The boulder-bed is overlain by a group of olive shales and sandstones forming the lower part of the Speckled sandstone series and designated as the Conularia beds, because of their containing the fossil *Conularia* enclosed in calcareous concretions. The genus *Conularia* is of doubtful systematic position and, like *Hyolithes*, is referred to the Pteropoda, or at times to some other sub-order of the Gastropoda, or even to some primitive order of the Cephalopoda. Associated fossils are *Pleurotomaria, Eurydesma, Bucania, Nucula, Pseudomonotis, Chonetes, Aviculopecten*, etc. These fossils are of interest because of their close similarity to the fauna of the Permo-Carboniferous of Australia, which also contains, intercalated at its base, a glacial formation in every respect identical with that of the Talchir series. The Conularia beds are succeeded by a series of mottled or speckled red sandstones, from 90 to 150 m. in thickness, inter-bedded with red shales. The whole group is current-bedded, and gives evidence of deposition in shallow water. From the mottled or speckled appearance of the sandstone, due to a variable distribution of the colouring peroxide of iron, the group is designated the *Speckled sandstones*.

The Productus limestone—This group is conformably overlain by the Productus limestone, one of the most important formations of India, and one which has received a great deal of attention from Indian geologists, being the first fossiliferous rock-system to be discovered in India. It is fully developed in the central and western part of the range, but thins out at its eastern end. About 210 m. of limestones are exposed in a series of fine cliffs near the Nilawan valley, and thence continue westwards along the Salt-Range right up to the Indus gorge, beyond which the group disappears gradually. The best and the most accessible outcrops of the rocks are in the Warcha valley[1] and Chideru hills in the neighbourhood of Musa Khel, west of the Son Sakesar plateau. The greater part of the Productus limestone is a compact, crinoidal magnesian limestone sometimes passing into pure crystalline dolomite,

---

[1] *Records*, G.S.I. vol. lxii., pt. 4, 1930.

associated with beds of marl and sandstones. It contains a rich and varied assemblage of fossil brachiopods, corals, crinoids, gastropods, lamellibranchs, cephalopods, fusulinae and plants, constituting the richest Upper Palaeozoic fauna anywhere discovered in India, to which the faunas of the other homotaxial deposits are referred. An added interest is the commingling of Lower Gondwana *Glossopteris* flora with the lower stage of the Productus limestone, crowded with a rich brachiopod fauna, and also almost immediately above the Talchir boulder-bed. The abundance and variety of the Productus fauna has thus led to the name *Punjabian* being given to the series of Middle Permian strata coming between the Artinskian and Thuringian. The stage name of Punjabian has also been used in the past to include the strata from the boulder-bed to the top of the Speckled sandstone (Uralian to Artinskian). On a palaeontological basis the Productus limestone is divided into three sections: the Lower, Middle and Upper.

With the lower beds of the Lower Productus limestone there comes a sudden change in the character of the sediments, accompanied by a more striking change in the facies of the fauna, almost all the species of the Speckled sandstone group disappearing from the overlying group. The lower 60 m. carry many beds of Fusulina limestone with *Parafusulina*. It is composed of soft calcareous sandstones, full of fossils, with coal-partings at the base. *Productus cora*, *P. semireticulatus* and *P. spiralis* are the characteristic species of this division. Associated with these, in the coal-partings, are the genera *Glossopteris* and *Gangamopteris*, of Damuda affinities suggesting the vicinity of the coast of the Gondwana mainland. Two stages are present: the lower, more arenaceous stage is well seen at Amb village, and is known as the Amb beds, and the upper calcareous stage is known as the Katta beds.

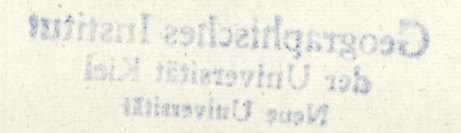

The Middle is the thickest and most characteristic part of the Productus limestone, consisting of from 60 to 90 m. of blue or grey limestone, which forms the high precipitous escarpments of the mountains near Musa Khel. Dolomite layers, which are frequent, are white or cream-coloured, and from the greater tendency of dolomite to occur in crystalline form they are much less fossiliferous owing to the obliteration of the fossils attending the recrystallisation process. Marly beds are common, and are the best repositories of fossils, yielding them readily to the hammer. The limestones are equally fossiliferous, but the fossils are very difficult to extract, being visible only in the weathered outcrops at the surfaces. Many of the fossils are silicified, especially the corals. There is also an intercalation of plant-bearing lower Gondwana shales and sandstones. *P. lineatus* is a common brachiopod species in the Middle Productus. Flint and chert concretions are abundantly distributed in the limestones. This division also includes two stages, Virgal and Kalabagh, the latter containing the ammonoids *Xenaspis* and *Foordoceras*.

The Upper Productus group is much less thick, hardly reaching 30-60 m. at places. The group is more arenaceous, being composed of sandstones with carbonaceous shales, with subordinate bands of limestone and dolomite. Silica is the chief petrifying agent here also. *P. indicus* is a common species. Fossils are numerous, but they reveal a striking change in the fauna, which separates this group from the preceding group. The most noteworthy feature of this change is the advent of cephalopods of the order *Ammonoidea*, represented by a number of its primitive genera. The topmost stage of the Upper Productus forms a separate stage by itself, known as the *Chideru beds*. They show a marked palaeontological departure from the underlying ones in the greatly diminished number of brachiopods and the increase of lamellibranchs and cephalopods. They are thus to be regarded, from these peculiarities, as a sort of transition, or "passage beds", between the Permian and the Triassic. The Chideru beds pass conformably and without any notable change into a series of *Ceratites*-bearing beds of Lower Triassic age.

Productus fauna—The following are lists of the more characteristic fossil genera, many of which are represented by numerous species, of the three divisions of the Productus limestone:

Upper Productus: (Ammonites) *Xenodiscus, Cyclolobus, Medlicottia, Arcestes, Sageceras, Popanoceras, Tainoceras*; (Brachiopods) *Productus, Oldhamina, Derbya, Chonetes, Martinia, Aulostegia*; (Gastropods) *Bellerophon, Euphemus,* etc.; (Lamellibranchs) *Schizodus, Lima, Gervillia*; (Polyzoa) *Entolis, Synocladia,* etc.

Middle Productus: (Brachiopods) *Productus, Spirifer, Spiriferina, Athyris, Lyttonia, Oldhamina, Richthofenia, Reticularia, Hemyptychina, Marginifera, Notothyris*; (Lamellibranchs) *Oxytoma, Pseudomonotis*; (Polyzoa) *Fenestella, Thamniscus, Acanthocladia*; (Worm) *Spirorbis*; (Corals) *Zaphrentis, Lonsdaleia, Stenopora*; (Gastropod) *Macrocheilus*; (Cephalopods) *Xenaspis, Nautilus, Orthoceras*.

Lower Productus: (Brachiopods) *Productus* (*P. cora, P. semireticulatus, P. spiralis*), *Spirifer, Spiriferina, Athyris royssii, Orthis, Reticularia, Richthofenia, Martinia, Dielasma, Streptorhynchus, Strophalosia*; (Foraminifers) *Fusulina, Parafusulina*.

[The following fossils may be considered characteristic of the Salt-Range Productus limestone :

Gastropods: *Euomphalus, Macrocheilus, Naticopsis, Phaseonella, Pleurotomaria, Murchisonia, Bellerophon* (*Bucania, Stachella, Euphemus,* and several other genera of the family *Bellerophontidae*), *Hyolithes* and *Entalis*.

Lamellibranchs: *Cardiomorpha, Lucina, Cardinia, Schizodus, Aviculopecten, Pecten* (two species).

Brachiopods: These are the most abundant, both as regards species and individuals. *Dielasma* is represented by ten species, *Notothyris* (eight species), *Lyttonia* (three species), *Camarophoria* (five species), *Spirigerilla* (ten species), *Athyris* (ten species), *Spirifer* (eight species), *Martiniopsis, Strophomena, Streptorhynchus, Derbya* (eight species), *Leptaena, Chonetes* (fourteen species), *Strophalosia, Productus* (fifteen species).

Polyzoa: *Polypora, Goniocladia*.

Crinoids: *Poteriocrinus, Philocrinus, Cyathocrinus*, etc.

Corals: *Pachypora, Michelinia, Amplexus, Clisiophyllum*.

Ganoid and other fishes, plants, etc.]

The Productus fauna shows several interesting peculiarities. While the fauna as a whole is decidedly Permian, the presence in it of several genera of true *Ammonites* and of a lamellibranch like *Oxytoma* and a *Nautilus* species, which in other parts of the world are not met with in rocks older than the Trias, gives to it a somewhat newer aspect. The most noteworthy peculiarity, however, is the association of such eminently Palaeozoic forms as *Productus, Spirifer, Athyris, Bellerophon*, etc. with cephalopods of the order *Ammonoidea*. All forms which can be regarded as transitional between the goniatites and the Triassic ceratites are found, including true ammonites like *Cyclolobus, Medlicottia, Popanoceras, Xenodiscus, Arcestes*, etc. Some of these possess a simple pattern of *sutures* resembling those of the *Goniatites* (sharply folded) or *Clymenia* (simple zig-zag *lobes* and *saddles*), while others show an advance in the complexity of the sutures approaching those of some Mesozoic genera.

The Anthracolithic systems of India—The lower part of the Salt-Range Productus limestone group is, from fossil evidence, the homotaxial equivalent of the Permo-Carboniferous of Kashmir, Spiti and the Northern Himalayas generally. The term "anthracolithic" is used by some authors as a convenient term to express the closely connected Carboniferous and Permian systems of rocks and fossils in those areas, *e.g.* the Shan States of Burma, which exhibit an intimate stratigraphic as well as palaeontological connection with one another, and where it is difficult to separate the Carboniferous from the Permian.

II. THE UPPER CARBONIFEROUS AND PERMIAN SYSTEMS OF THE HIMALAYAS

The Himalayan representatives of the Productus limestone are developed in the northern or Tibetan zone of the Himalayas along

CARBONIFEROUS AND PERMIAN SYSTEMS

their whole length from Kashmir to Kumaon and beyond to the Everest region. They are displayed typically at two localities, Spiti and Kashmir, where they have been studied in great detail by the Geological Survey of India.

Spiti

In Chapter VIII we have followed the Palaeozoic sequence of the area up to the Fenestella shales of the Po series. Resting on the top of the Fenestella shales in our type section, but at other places lying over beds of varying horizons from the Silurian to the Carboniferous, is a conglomerate layer of variable thickness, belonging in age to the Upper Carboniferous or Permian. This conglomerate, as has been stated before, is an important *datum-line* in India, for it is made the basis of the division of the fossiliferous rock-systems of India into two major divisions, the Dravidian and Aryan. The Aryan era, therefore, commences in the Himalayas with a basement conglomerate, as it commenced in the Salt-Range and in the Peninsula with the glacial boulder-bed.

The Productus Shales—

The conglomerate is succeeded by a group of calcareous sandstones, containing fossil brachiopods of the genera *Spirifer*, *Productus*, *Spiriferina*, *Dielasma* and *Streptorhynchus*, representing the

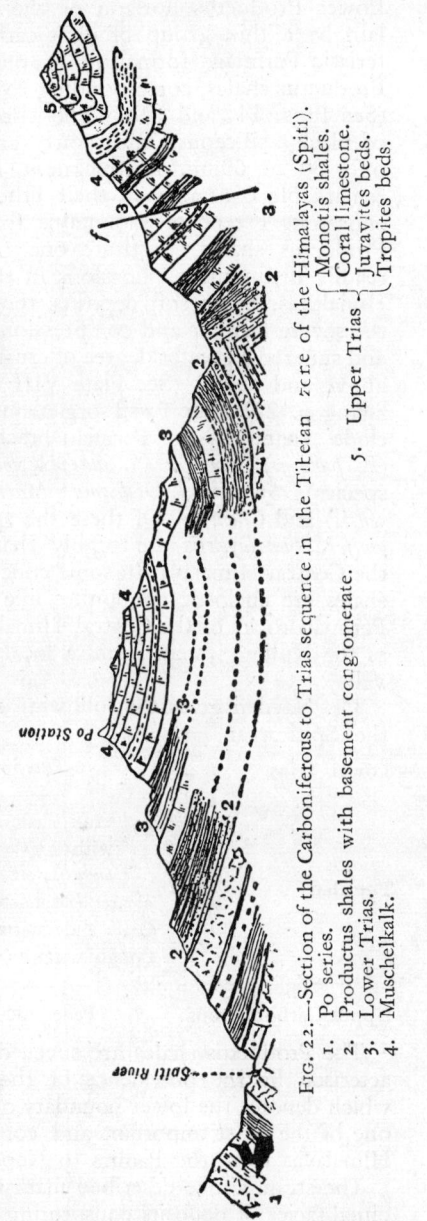

FIG. 22.—Section of the Carboniferous to Trias sequence in the Tibetan zone of the Himalayas (Spiti).
1. Po series.
2. Productus shales with basement conglomerate.
3. Lower Trias.
4. Muschelkalk.
5. Upper Trias { Monotis shales. Coral limestone. Juvavites beds. Tropites beds.

Lower Productus horizon of the Salt-Range. These are overlain by a thin group of dark carbonaceous shales, the characteristic Permian formation of the Himalayas, known as the Productus shales, corresponding to the Upper Productus horizon. (See Figs. 14 and 22). The Productus shales are a group of black, siliceous, micaceous and friable shales. They are only 30 to 60 m. in thickness, but are distinguished by a remarkable constancy in their lithological composition over the enormous extent of mountains from Kashmir to Nepal. The Productus shales constitute one of the most conspicuous and readily distinguished horizons in the Palaeozoic geology of the Himalayas. Being soft deposits, they have yielded more freely to the severe flexures and compression of this part of the mountains and suffered a greater degree of crushing than the more rigid strata above and below. (See Plate VIII facing p. 151, also Plate XII facing p. 222) The fossil organisms entombed in the shales include characteristic Permian brachiopod species of *Productus* (*P. purdoni*), *Spirifer* (*S. musakheylensis*, *S. rajah*, and five other species), *Spirigera*, *Dielasma*, *Martinia*, *Marginifera* (*M. himalayensis*) and *Chonetes*. Of these the species *Spirifer rajah* and *Marginifera himalayensis* are highly characteristic of the Permian of the Central Himalaya. In some concretions contained in the black shales are enclosed ammonites like *Xenaspis* and *Cyclolobus*. The Permian rocks of the Central Himalaya have been also designated as the Kuling system from a locality of that name in the Spiti valley.

Dr. Hayden gives the following sequence of Permian strata in the Spiti area:

Lower Trias.	*Otoceras* zone of Lower Trias.
Permian.	Productus shales: black or brown siliceous shale with *Xenaspis*, *Cyclolobus*, *Marginifera himalayensis*, etc.
	Calcareous sandstone with *Spirifer*.
	Grits and quartzites.
	Conglomerates (varying in thickness).
Slight unconformity	
Upper Carboniferous.	Fenestella shales of Po series.

The Productus shales are succeeded by a group of beds characterised by the prevalence of the Triassic ammonite *Otoceras*, which denotes the lower boundary of the Trias of the Himalayas, one of the most important and conspicuous rock-systems of the Himalayas from the Pamirs to Nepal.

The strata above described mark the beginning of the geosynclinal facies of deposits constituting the northern or Tibetan zone

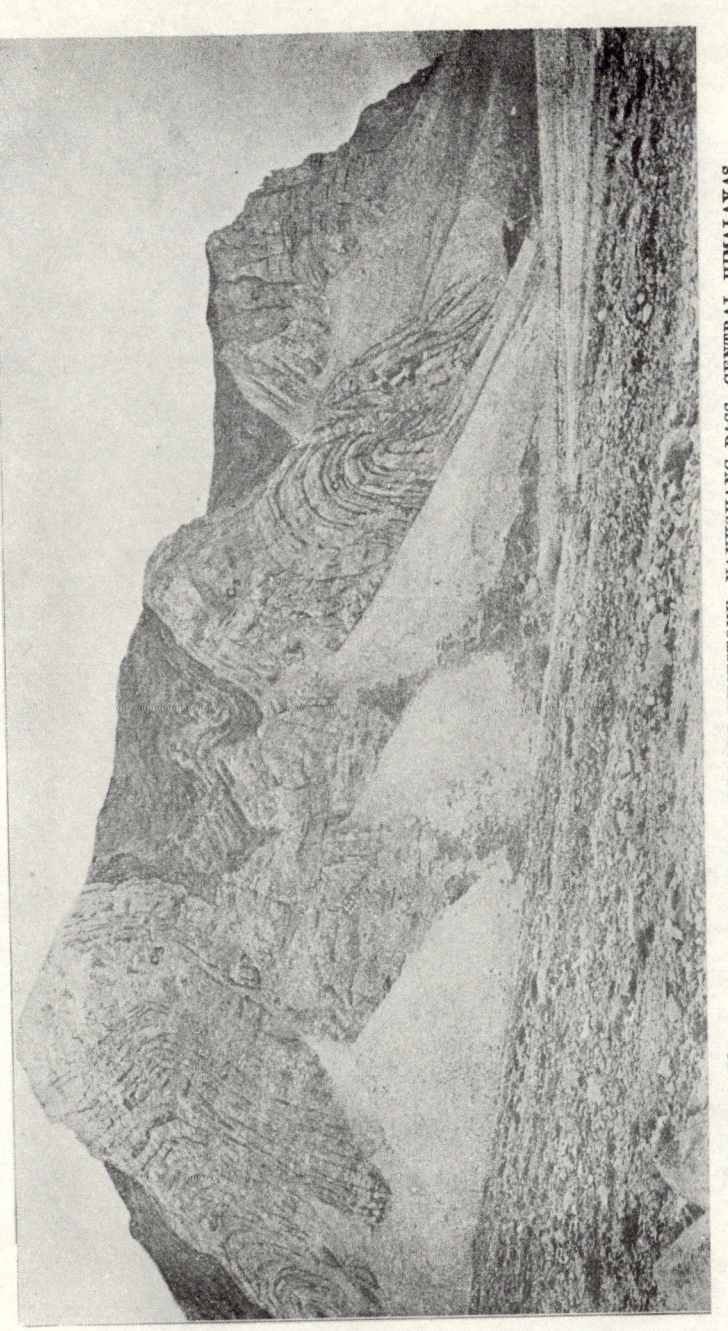

PLATE XI. CONTORTED CARBONIFEROUS LIMESTONE, NANKSHANG PASS, CENTRAL HIMALAYAS.

(C) Overlain by the black Productus shales. Notice the unconformable junction between the Carboniferous and Permian; also the fan-taluses at the base of the cliff. (*Geol. Survey of India, Mem.* vol. xxiii.)

of the Himalayas. As yet the strata are composed of shales and sandstones, indicating proximity of the coast and comparatively shallow waters, but the overlying thick series of the Triassic and Jurassic systems are wholly constituted of limestones, dolomites and calcareous shales of great thickness, giving evidence of the gradual deepening of the ocean bottom.

Carboniferous and Permian of Mt. Everest (Sikkim)

Carboniferous and Permian rocks cover considerable areas of Nepal, Sikkim and Bhutan, the crestal portion of Mt. Everest being composed of the *Mt. Everest limestone* 300–600 m. thick. This is a massive sandy limestone, believed to be of Carboniferous age, dipping northwards and overlain conformably by about 600 m. of fossiliferous Permian limestone and sandstone series (*Lachi series*). The Lachi series is in turn overlain by a continuous succession of Triassic, Jurassic and Cretaceous strata which cover a large extent of southern Tibet.

The Lachi series has yielded some fairly well preserved Permian Brachiopods of *Productus* and *Spirifer* affinities.

Kashmir

In keeping with the rest of the Palaeozoic systems, the Carboniferous and Permian are developed on a large scale in Kashmir. The Upper Carboniferous consists of a thick (over 2,400 m.) volcanic series—*Panjal Volcanic series*—of bedded tuffs, slates, ash-beds and andesitic to basaltic lava-flows (*Panjal Trap*). The slaty tuffs contain at places marine fossils allied to the fauna of the Productus limestone. A most interesting circumstance in connection with the Permian of Kashmir is the association of both the Gondwana facies of fluviatile deposits, containing seed-ferns like *Gangamopteris* and *Glossopteris*, and the marine deposits containing the characteristic fossils of the age. The Gondwana beds (known as the *Gangamopteris beds*), which are the local representatives of the Talchir-Damuda series of the Peninsula, are overlain by the marine Permian beds (*Zewan series*), containing a brachiopod fauna identical in many respects with that of the Productus limestone.

The Mid-Palaeozoic Unconformity of North-West Kashmir

—While the records of the Palaeozoic from the Silurian to the Permian are continuous in the Spiti Himalayas as well as in eastern Kashmir, the geological record of north-western Kashmir and Hazara during the greater part of this interval is a total blank. With the exception of small patches of Muth Quartzites, the Silurian system of Kashmir, west of the Wular lake, is succeeded by the Panjal Volcanic series which is not older than the Uralian at the earliest. This is the most widespread regional unconformity in the geological records of North-West India, equally well seen in Hazara, the western Pir Panjal and the Punjab Salt-Range. The Hazara unconformity is proved by the Hazara (Dogra) slates underlying with an angular unconformity a glacial boulder-conglomerate which is now accepted as of Talchir age. In the Salt-Range, Cambrian beds with a *Neobolus* fauna are overlain by a boulder-bed at the base of the Productus limestone with an intervening group of Damuda plant-bearing sandstones. This widespread unconformity is proof of the prevalence of continental conditions during the Devonian and the greater part of the Carboniferous. The existence of a Punjab-Kashmir-Hazara land-mass during the Dravidian era is a well-established fact in the palaeogeography of North-West India.

This mid-Palaeozoic land-mass of Kashmir performed one important function: it must have served as a land-bridge between Gondwanaland and the great northern Eurasian continent (Angaraland). It was through this land-bridge the terrestrial vegetation of the Indian portion of Gondwanaland established some links with Angaraland.

When, at the end of the Dravidian era, the earth movements which supervened ushered in a new sedimentary period on the surface of the great continent of Gondwanaland to the south of the Himalayan sea, this part of Kashmir, for a brief interval, formed the northernmost frontier of Gondwanaland and was occupied by a characteristic land vegetation—the *Glossopteris* flora, some typical members of which are found entombed at six or seven widely scattered sites extending as far north as the south flank of the Zanskar.

In all parts of Kashmir west of the Sind valley, this unconformity is clearly revealed, its effect being in some places exaggerated by a progressive overlap of the Panjal Volcanic series.

It was with the commencement of the Uralian that the Productus sea of Spiti extended westward and overspread Kashmir, Hazara and the Salt-Range, ushering in the long period of Tethyan marine sediments that ceased only with the Middle Eocene.

Tanawal series—In the Purana and metamorphic belt of the N.W. Himalayas, extending from Kaghan to Jammu, a voluminous series, upto thousands of metres thick, of metamorphosed rocks of markedly arenaceous composition—banded argillaceous quartzites, grits, phyllites and quartz-schists, with clastic as well as crush-conglomerates—occurs in a number of fold-faulted, disturbed longitudinal basins, one to four miles across the strike. These have been named from the Tanawal country in Hazara, in which similar rocks were first recognised by Wynne. Their field relations with the Purana rocks, among which they lie, are so distorted that it is often difficult to decide whether they are older or newer than these. Their grade of stress metamorphism is sometimes higher than that of the Puranas. However, from some evidence that the upper quartzite masses are, in a few cases, really silicified limestones of the Sirban type (the "Infra-Trias" series) it is possible to infer that the whole group is newer than the slate series; but beyond suggesting that the Tanawals bridge the gap between these slate series and the Permo-Carboniferous, no definite age can at present be ascribed to this group. It is possible that the lower part of the Tanawals may be coeval with so old a formation as the Muth series. In the Poonch Pir Panjal these rocks show a clear lateral passage into the Agglomeratic Slate series of Upper Carboniferous age. The whole group is entirely devoid of fossils.

In the Simla and Garhwal area the formation which succeeds the Simla slates is the *Jaunsar series* or the *Nagthat series*, both unfossiliferous and of uncertain stratigraphic position, similar in this respect to the equally obscure Tanawals. At many localities, however, the Simla slates are overlain unconformably by the Blaini series, the Upper Carboniferous age of which is now regarded as proved beyond serious doubt.

From the nature of their occurrence in disconnected isolated basins, away from the wide sedimentary terrains, and their barren nature it is conjectured that the Jaunsars and Tanawals are a continental system of mid-Palaeozoic deposits, laid down in depressions of the Hazara-Kashmir land mass.

Upper Carboniferous. The Panjal Volcanic Series. Middle Carboniferous earth movements—During the last of the deposition of the Fenestella shale-beds, the physical geography of the Kashmir area underwent a violent change, and what was before a region of quiet marine sedimentation was converted into a great theatre of vulcanicity, whereby an enormous superficial extent of the country was converted into a volcanic region, such as Java and Sumatra in the Malay Archipelago of the present day. The clastic and liquid products of these volcanoes buried large areas of Kashmir under 2,000–2,500 metres of lavas and tuffs. The volcanic activity was most intense during the Permian when it reached its climax, after which it diminished greatly; though at isolated centres, as in Gurais, it persisted up to the Upper Triassic period.

Physical history at the end of the Dravidian era—The earth-movements and physiographic revolutions, with which this igneous outburst was associated in the Kashmir area, were connected and contemporaneous with the crust-movements in other parts of India at the end of the Dravidian era. This was the epoch of many far-reaching changes on the face of India, as we have seen in Chapter VIII. These changes put an end to the continental phase in Kashmir and to the epoch of Gondwana conditions which had invaded Kashmir, converting it in fact into a north-western province of that continent.

This Gondwana epoch in the history of Kashmir was thus of but short duration. For the sea soon resumed its hold over this area in the Permian times and commenced to throw down its characteristic deposits in the geosynclinal of the Tethys, which once more brought Kashmir within the "Tibetan" zone of the Himalayas. The marine Permian of Kashmir, as we shall see, is both in its physical and biological characters on a par with the Productus limestone of the Salt-Range and the Productus shales of Spiti and other Himalayan areas.

Agglomeratic Slates and Trap—Rocks of this series are divisible into two broad sections: the lower—a thick series of pyroclastic slates, conglomerates and agglomeratic products, upto thousands of metres in thickness, and called by Middlemiss the "Panjal agglomeratic slates"; and the upper—the "Panjal traps", an equally thick series of bedded andesitic and basaltic traps generally overlying the agglomerates. The series covers an enormous superficial area of the country, being only next in areal distribution to the gneissic rocks. It builds the majority of the high peaks surrounding the Jhelum valley from the Shamsh Abari to the Kolahoi (5,428 m.)

Distribution—It is specially well developed in the Panjal range, of which it forms the principal substratum, being visible as prominently on its sides and summit as in its centre for the entire length of the range from the Kishenganga valley in Karnah to its termination at the Beasavi (see Pl. XV). This circumstance gives the name Panjal to the series. These rocks also form the black hill-masses on the north-west continuation of the mountain range, beyond Nun Kun to as far as Hazara. With this exception, the Punjal volcanics are largely confined to the geographical limits of Kashmir. They were once believed to extend northwards to Baltistan and Skardu and north-east to Ladakh. Lydekker has mapped extensive outliers of Punjal Trap in Ladakh as far as the Changchenmo valley. These have been found, by later detailed surveys, to belong to another of vulcanicity—Up. Cretaceous (Dras Volcanic Series).

The stratigraphical position of these deposits is noteworthy. The Panjal volcanic series commences from varying horizons, from the Moscovian, Uralian, or even Permian, in different localities and extends in its upper limit, likewise, to the Lower Permian in some places and the Upper Trias in others. Both the lower and upper limits are generally precisely dated by intercalation with known fossiliferous horizons. In the Vihi district the volcanic eruptions die out with the Lower Permian; in the Lidar with the end of the Permian; while in Gurais the vulcanicity did not end till well into the Upper Trias. The erratic nature of the traps as a stratigraphic unit is thus evident.

Nature of the Panjal slate-agglomerate—The mode of origin of the lower part of the Panjal volcanic series, or what has been called the "agglomeratic" slates, is not easy to understand. Much of it is composed of a fine greywacke-like matrix with embedded angular grains of quartz. But the rock does not appear to be an ordinary sedimentary deposit, inasmuch as the embedded fragments are quite angular and often become very large in size at random. They are pieces of quartzite, slate, porphyry, granite, etc., irregularly dispersed in a fine-grained matrix. The rock is generally unfossiliferous throughout, though at a few localities several interesting suites of fossils have been discovered[1] which are identical with forms entombed in the underlying Fenestella series. The most common forms are *Productus, Spirifer, Chonites, Dielasma, Camarophoria, Strophalosia, Leptaena, Streptorhynchus, Spiriferina, Eurydesma, Aviculopecten, Sanguinolites, Conocardium, Fenestella, Euphemus* and *Pleurotomaria*. That such a rock could not have been the product of any simple process of sedimentation, whether subaerial or submarine, is quite clear, and the origin of the deposit so widespread and of such uniform character is a problem.

One view is that the rock is a joint product of explosive volcanic action combined with ordinary subaerial deposition; the other, a diametrically opposite view, is that it is due to frost-action under glacial or arctic conditions, the frost-weathered débris being subsequently transported by floating ice-masses to lakes. Middlemiss favours the former view, as being more in keeping with the actual circumstances of the case and as congruent with the lava-eruptions that succeeded it, though he points out that the absence of glass particles, pumice fragments and other products usually associated with tuffs is irreconcilable with this view. Later work in the Pir Panjal has established the pyroclastic nature of large parts of this formation beyond any doubt. The matrix of the slate often is full of devitrified and altered glass with phenocrysts of felspars.[2] The presence of Lower Gondwana plants in beds immediately overlying the volcanics favours the inference that the slate-conglomerate is a glacial deposit corresponding to the Talchir boulder-beds. No faceted or striated pebbles[3] are, however, seen in the slates; on the contrary the pebbles are frequently

---

[1] H. S. Bion, *Pal. Indica* N. S. vol. xii, 1928 ; F. C. Reed, *Pal Indica*, N. S. vol. xx. mem 1, 1932.

[2] Wadia, *Mem. G.S.I.* vol. li, pt. 2, 1928.

[3] At a few local spots numerous faceted glacical pebbles are found embedded in the slates.

quite angular. The following section gives a general idea of the rocks of the Panjal series.

Aggregate thickness some thousands of metres.
- 5. Bedded green and purple traps, several thousand metres thick.
- 4. Greenish ash-beds, slates and agglomeratic quartzites with amygdaloidal traps.
- 3. Black and grey agglomeratic slates (tuffs) with thick beds of conglomerate containing sub-angular pebbles of quartzite and slate.
- 2. Whitish quartzite and sandstones.
- 1. Black agglomeratic slates (tuffs) with angular or sub-angular pebbles of quartz, slate and gneiss.

The Agglomeratic slates of Nagmarg and Bren contain Lower Gondwana plants, associated with a series of sandstones and shales containing a marine brachiopod fauna and *Eurydesma*. This horizon corresponds with the *Eurydesma* horizon of the Salt-Range Productus series.

Panjal lavas. Petrology—Over the agglomeratic slates there comes a great thickness of distinctly bedded massive lava-flows. In composition the lava is a basic variety of augite-andesite or basalt of acidity varying from 49 to 60 per cent, of a prevailing dark or greenish colour, the green colour being due to the alteration of augite and other constituents into epidote. Acid and intermediate differentiation-products also occur locally and in small masses, *e.g.* trachyte, ceratophyre, rhyolite, acid tuffs, etc. The rock is usually non-porphyritic and very compact in texture, but porphyritic varieties are sometimes, and amygdaloidal varieties are often, met with. In microscopic structure the lavas are a microcrystalline aggregate of plagioclase felspar and finely granular augite, with traces of yet undevitrified glassy matrix. Magnetite is very common in irregular grains and crystals. No olivine is present, nor any well-formed crystals of augite. The structure is hemicrystalline throughout, only minute prisms of white turbid felspar being detected in a finely granular aggregate, but in some varieties there are large prismatic phenocrysts of felspar arranged in star-shaped or radiating aggregates giving rise to what is called *glomero-porphyritic* structure. Some varieties are amygdaloidal, the amygdules being composed of silica or epidote or rarely of some zeolites. The lavas often show wide-spread alteration of the nature of epidotisation, chloritisation, and silicification. Devitrification is most common. Green chlorite is commonly present in the felspars, and epidote is a universal secondary product resulting from the interaction between augite and plagioclase.

When the lavas are interbedded with the slates, the contact metamorphism induced in both the rocks is of very marked degree,

CARBONIFEROUS AND PERMIAN SYSTEMS

the two becoming quite indistinct from each other. At Gagribal, near Srinagar, such an intimate association of the two kinds of rocks is seen. Sills and dykes of coarse-textured dolerite are frequent in the bedded trap-flows.

The individual flows vary in thickness from a few centimetres to six metres or more, and are markedly lenticular. There are no fresh-water sedimentary intercalations of the nature of "inter-trappean" beds, but in the body of the traps there are found considerable thicknesses of inter-trappean marine fossiliferous limestones of Permian (Sirban), and Lower and Middle Trias age. These limestones are obviously fossiliferous and show a gradual passage into ash-beds and traps above and below. Such inter-trappean limestones of thicknesses varying from 15–300 m. are observed in the mountains north of the Wular, in the Uri district and in the Kaghan valley, Hazara. The total aggregate thickness of the lava-flows measures thousands of metres, 2,000–2,500 m. being seen in the cliffs above the Wular. But this development is often purely local; over large areas the trap is missing, its place being occupied by agglomerate slate.

Age and vertical extension of Panjal lavas—The upper limit of the Panjal lava-flows in Vihi is clearly defined by the directly overlying plant-bearing beds of Lower Gondwana facies, which in turn are immediately succeeded by marine Permian rocks. In other cases, however, the flows have been found to extend to a much higher horizon, as far as the Upper Triassic, a few flows being found locally interbedded with limestone of that age. In general the Panjal volcanoes ceased their eruptive activity in the Permian. These subaerial volcanic eruptions therefore bridge over the gap which is usually perceived at the base of the Permian in all other parts of India.

In addition to lava-flows there are seen dykes and laccolithic masses of a gabbroid and doleritic magma, cutting through both the Panjal slates and traps or earlier rocks in several parts of Kashmir.

Lower Gondwana of Kashmir

Gangamopteris Beds

Distribution—The Panjal traps are directly and conformably overlain in several parts of Kashmir by a series of beds containing *Gangamopteris* and *Glossopteris*, so eminently characteristic of the Talchir and Damuda series of the Peninsular Gondwanas. The Gondwana plant-bearing beds have been met with at seven localities, *viz*. on the north-east slopes of the Pir Panjal, at Banihal pass, Golabgarh pass and near Gulmarg; on the opposite side of the Jhelum valley, in Vihi; near Srinagar; at Marahom near Bijbiara; and at Nagmarg on the Wular lake. Of these, the exposures at Risin and Zewan in the Vihi district are the most noteworthy because of their directly underlying fossiliferous Permian limestones, a circumstance which clearly establishes their exact stratigraphic horizon. This is illustrated in the section in Fig. 23, p. 214. This series of beds is known as the *Gangamopteris beds* from the most prevalent seed-fern, impressions of whose leaves are well preserved in the black or grey "shales", which in their composition are black glassy tuffs, almost entirely composed of isotropic obsidian-like glass. A fossiliferous outcrop of these beds is visible at the Golabgarh pass of the Pir Panjal, one of the passes on the range leading from the province of Jammu to Kashmir.

Lithology—The Gangamopteris beds are composed of a variable thickness of cherts, siliceous shales, carbonaceous shales and flaggy beds of quartzite, which in their constitution are largely pyroclastic. The thickness varies from a few metres at some of the Vihi outcrops to some hundreds of metres in the outcrop at the Panjal range. A peculiar rock of this series is a "novaculite", well seen at Barus and at Khunmu. It is a compact chert-like rock of white or cream colour, which has replaced an original limestone by silicification, forming the base of the series and directly overlying the traps. The black shales of many of the outcrops of the Gangamopteris beds are likewise frequently silicified. On the southwest flank of the Pir Panjal, Gondwana beds (? Upper Tanawals) constitute a thick series of deposits upto thousands of metres in thickness consisting of partly metamorphosed shales, phyllites, quartzose grits and sandstones, the latter showing extensive ripple-marking, cross-bedding and colour-banding. The series is generally barren of recognisable fossils, but from its position above the

Zewan series.	*Protoretepora* limestone.		Permian.
Ganga-mopteris beds.	Earthy sandstones, calcareous above, passing into Zewan limestones.	70 m.	
	Hard, compact black shales with *Glossopteris*; hard grey sandstones and interbedded shales with *Psygmophyllum, Gangamopteris* and *Vertebraria*.	122 m.	Lower Gondwana (Artinskian).
	Thin-bedded, buff-coloured compact siliceous and carbonaceous shales.	55 m.	
	Basal conglomerate.	2 m. (approx.)	Upper Carboniferous.
	Panjal traps and ash-beds.		

Dogra slates in wide synclinal basins, with a basal boulder-conglomerate, and its conformable relations to the Agglomeratic Slate series, it is tentatively referred to the Lower Gondwanas.[1]

The Golabgarh section—The section below gives the chief components of the series viewed at the Golabgarh Pass.[2]

Fossils—The Gondwana fossils include plant impressions together with parts of the skeletons of labyrinthodonts and fishes. The plants are chiefly obtained from the Golabgarh outcrop, while the vertebrate remains were obtained from Risin and Khunmu. The plants include a species of *Gangamopteris* sufficiently distinct from those of the Peninsula to be named *G. kashmirensis*. Other fossils are *Glossopteris indica*, *Vertebraria indica*, *Callipteridium*, *Cordaites* (*Naeggerathiopsis*) and leaves of *Psygmophyllum*, a genus related to *Ginkgo*. The vertebrate fossils consist of the scales, fins, portions of skulls, a mandible, and fragments of the hind-limbs of *Amblypterus* (a cartilaginous *ganoid* fish), together with fragmentary remains of a species of labyrinthodont *Archegosaurus*, and a cranium of an *Actinodon* species, *A. risinensis*.

Age—The exact horizon represented by the Gangamopteris beds, in terms of the typical Gondwana sequence, cannot be determined with the help of the plant-remains alone, although the occurrence of *Gangamopteris* suggests a relatively low horizon in the Gondwana series. But the association of this meagrely known flora with marine strata below and above (*viz*. the Middle Carboniferous Fenestella shales and the Permian Zewan beds) is an event of the greatest importance in the stratigraphic records of India. It has helped to solve one of the most difficult problems of Indian geology—the settlement of the precise horizon of the Lower Gondwana system of India.

The plants resemble the characteristic Lower Gondwana types of South Africa, Australia and other countries of the southern hemisphere, and are thus very interesting as affording us a glimpse into the geography of the northernmost limit of the Gondwana continent which not included within its borders all these countries.

The Permian. The Zewan Series. The Zewan beds—The Permian deposits, the local representatives of the Productus limestone of the Salt-Range and of the Productus shales of Spiti, make a very well-marked horizon in the geology of Kashmir. These deposits have been known since an early date as the *Zewan beds*, from their exposure at the village of Zewan in the Vihi district. At this particular locality the Gangamopteris beds are overlain by a series of fossiliferous shales and limestones containing crowds

---

[1] *Mem. G.S.I.* vol. li. pt. 2, 1928.

[2] Middlemiss, *Rec. G.S.I.* vol. xxxvii. pt. 4, 1909. For another section at Zewan see Hayden, *Rec. G.S.I.* vol. xxxv. pt. 1, 1908.

of fossil brachiopods and polyzoa. In other parts of Vihi this series is more fully formed, the portion representative of the typical Zewan section being succeeded by another thick group of limestones and shales underlying the Lower Triassic beds. The term "Zewan series" has consequently been amplified to receive the entire succession of beds between the Gangamopteris and the Lower Triassic beds. The base of the Zewan series is argillaceous in composition, the shales being crowded with the remains of *Protoretepora*, a polyzoon resembling *Fenestella*. The upper part ia calcareous, the limestone strata preponderating. In a few shales, intercalated among the latter, is contained a fauna resembling that of the Productus shales of Spiti and other parts of the central Himalayas. Over the top of the series there lie thin bands of hard limestone and shales bearing *Pseudomonotis*, *Danubites* and other ammonites, marking a Lower Trias limit.

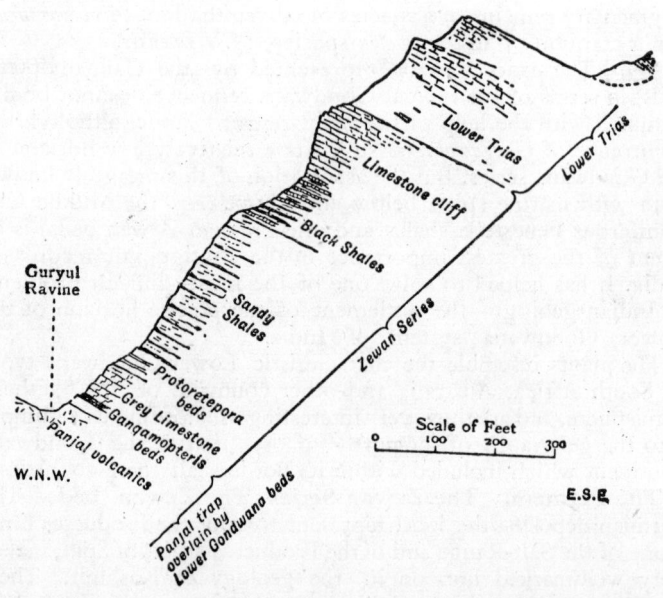

FIG. 23.—Section of the Zewan series, Guryul Ravine.
(Middlemiss, *Rec. Geological Survey of India*, vol. xxxvii. pl. 4.)

A thin but continuous band of Zewan rocks is seen along the south-west hills of Vihi, and is co-extensive with the much more prominent Triassic outcrop. A few thin isolated outcrops of the series are noticed in the Pir Panjal on either side of the central axis, overlying the trap. A more voluminous development of the

Permian is witnessed in the watershed area of the Upper Sind and Lidar valleys, normally underlying the Lower Trias.

The following section, very well exposed in a ravine near Khunmu (Guryul ravine), is reproduced from Middlemiss and Hayden:

Meekoceras zone of the Lower Trias.

Shales and limestone, thin-bedded. Fossils: *Pseudomonotis, Bellerophon, Danubites, Flemingites*. } 30 m.

Dark arenaceous shales, micaceous and carbonaceous, with limestone intercalations at base. Fossils: *Marginifera himalayensis, Pseudomonotis*, etc. } 90 m.

Shales and limestone, crowded with *Protoretepora, Athyris royssii, Productus, Dielasma*, etc. } 10 m.

Dark grey limestone with shale partings. Fossils: *Athyris, Notothyris*, etc. } 20 m.

Novaculites and tuffaceous strata of the *Gangamopteris beds*.

Fossils—Fossils are present in large numbers in the Zewan beds. They include one *Nautilus* and two genera of ammonites, *Xenaspis* and *Popanoceras*. The lamellibranchs are *Pseudomonotis, Aviculopecten* and *Schizodus*; but the most predominant groups are the brachiopods and polyzoa. The former are represented by *Productus cora, P. spiralis, P. purdoni, P. gangeticus, P. indicus, Spirifer rajah* (the most numerous), *Dielasma, Martinia, Spirigera, Spiriferina, Marginifera vihiano, M. himalayensis, Lyttonia, Camarophoria, Chonetes, Derbya*, etc. Among polyzoa the species *Protoretepora ampla* is present in overwhelming numbers at some horizons. Its fan-shaped reticulate-structured zoaria resemble those of *Fenestella*, but actually it belongs to a slightly different zoological family. *Acanthocladia* also is a frequent form. *Amplexus* and *Zaphrentis* are the more common corals.

Age of the Zewan series—From the palaeontological standpoint the Zewan series is correlated with the Middle Permian system of Europe, a conclusion amply corroborated by the stratigraphic relations of the series to the Lower Trias. An interesting fact revealed by the Zewan fauna is the exact parallelism of these deposits with the middle and upper part of the Productus limestone of the Salt-Range, most of the genera and many of the species being common to the two regions. A comparison of the faunas with the Productus (Kuling) shales of the central Himalayas also brings out the closest zoological affinities between these three homotaxial members of the Indian Permian and Permo-Carboniferous systems.[1]

---

[1] Dr. Diener, *Pal. Indica*, N.S. vol. v. mem. 2, 1915.

Permian of Jammu[1]

Within the sub-Himalayan zone of Jammu, representatives of the unfossiliferous limestone, Sirban limestone of Hazara (*Infra-Trias series*), of presumably Permian or Permo-Carboniferous age, crop out in a chain of large and small inliers extending from Riasi to the Poonch valley. This is a very unusual circumstance, which finds only one parallel in the Tal series of the Nepal Himalayas. In Jammu, mountainous masses of white or blue-grey dolomitic limestone are laid bare by the removal of the overlying Eocene and Murree series from anticlinal tops. The most notable of the inliers thus exposed forms a conspicuous landmark near Riasi (the Trikuta hill). To the west of this is a series of hog-backed masses of the same limestone laid bare in denuded anticlines, generally faulted in their steep south limbs against the younger Tertiaries of Jammu. The limestone, over 450 m. thick, is entirely barren of organic remains and, its stratigraphic relations being nowhere exposed, it was doubtfully referred to the Kioto limestone of Spiti and named the "Great limestone". During later Survey work, however, some clue to the identity of the rock has been discovered in the intercalation of the base of the limestone with Agglomeratic slate—an association often noticed in the Sirban limestone of the Kaghan valley. There is also a close lithological similarity between these outcrops.

In its petrological characters this limestone shows analogy also with the unfossiliferous Krol limestone of the Simla-Chakrata area, constituting a wide and long belt of post-Blaini limestone and associated rocks.

The Riasi limestone possesses considerable economic importance and forms one of the few noticeably mineralised rock formations of the North-West Himalayas. Important lodes of zinc and copper are found in the limestone, with veins of nickeliferous pyrites and galena. The sulphidic ores of zinc and copper are probably metasomatic replacements, while galena and pyrites are vein-fillings. (See Fig. 34, p. 316.)

Krol Series of Simla and Kumaon

Simla Hills—With the exception of the Jaunsars and some intervening limestones and slates of uncertain position (*Shali limestone*), the system of deposits which comes next above the Simla slates is referred to the Upper Carboniferous and Permian with a high degree of probability. As in Hazara, the bottom bed is a glacial boulder-bed—the *Blaini conglomerate*—unconformably reposing on the Simla slates or the Jaunsars, succeeded by pink-coloured dolomitic limestones. Over these comes a thick series

---

[1] D. N. Wadia, *Rec. G.S.I.* vol. lxxii. pt. 2, 1937.

of carbonaceous shaly slates, with brown quartzite partings—
—the *Infra-Krol series*—which have been provisionally correlated
with the Lower Gondwanas of the Peninsula. A few micro-fossils
believed to be of Lr. Gondwana affinities have been found in
black shales associated with these beds. The succeeding series
consists of a thick group of massive blue limestones and shales,
underlain by partly consolidated, coarse sandstones, referred to
as the *Krol series*, from their building the conspicuous mountain
of that name near Solon. As with the rest of the formations of the
Simla area, the Krol limestones are entirely barren of fossils. The
inference that they are homotaxial with the Sirban limestone of
Hazara and the richly fossiliferous Productus group of the Salt-
Range and the Zeewans of Kashmir is based on the probable parallelism of the sequence commencing with the Blaini glacial boulder-
bed (? Talchir) in these areas.

The most prominent development of the Krol series is in the
Outer Himalaya of Simla, extending from near Subathu to Naini
Tal, a distance of 290 km. A very perfect stratigraphic sequence
has been worked out in this area by J. B. Auden, which has re-
vealed the presence of a number of thrusts causing overriding of
Tertiary rocks by the much older rocks we are considering here.
In the neighbourhood of Solon and Subathu, Eocene and Oli-
gocene rocks are exposed as inliers ("windows") by the erosion
of the superjacent overthrust masses of these presumed Permo-
Carboniferous rocks.[1]

The probable equivalent of the fossiliferous Upper Carboniferous and Permian of Kashmir is this thick pile of sediments, for
to greater part obviously marine, but showing oscillation to fresh-
water and terrestrial conditions, coming over the *Blaini boulder-bed*—
a glacial till consisting of ice-scratched pebbles in a fine matrix.
Though quite barren of fossils, the *Krol series* is of high interest
because of its tectonic complexity and the greatly involved thrust-
sheets (*nappes*), brought to light in Garhwal Kumaon mountains.
These nappes have their roots in the crystalline zone of the Central Himalaya ranges and they have moved bodily southwards
in thick, deeply flexed and eroded recumbent folds covering wide
stretches of the Lesser Himalayas, reaching up to the Tertiary
foot-hills and at times encroaching over them. The Krol belt of
the Kumaon Himalayas builds an important section of the middle
Himalayas as shown during late years.[2]

Karakoram and Chitral

Fossiliferous Permian or Permo-Carboniferous strata, mainly

---

[1] *Rec. G.S.I.* vol. lxvii. pt. 4, 1934.

[2] W. D. West, *Mem. G.S.I.* vol. liii, 1928; J. B. Auden, *Rec. G.S.I.* vol. lxvii. pt. 4, 1934.

limestones, are observed extensively formed in the Karakoram[1]. According to the findings of the Italian Expedition of 1913-14, the mountains of Gasherbrum, the Golden Throne, and the Crystal and Bride Peaks are built of these limestones. Permian limestones have also been observed in the Shaksgam valley of the range.

A great thickness of *Fusulina* limestone of Permian or Upper Carboniferous age occurs among the crystalline limestones of the Tirich valley in Chitral. Outcrops of the Fusulina limestone extend from Chitral into Russian Turkestan.

Hazara

As in the western parts of Kashmir, the Palaeozoic record of Hazara is confined to representatives of the Upper Carboniferous and the Permian. On the upturned truncated edges of Purana slates, the contemporaries of Attock and Dogra slates, there comes a *boulder-conglomerate*, the *Tanakki* boulder-bed, composed of faceted and striated boulders set in a fine silty matrix. This boulder-bed (tillite), regarded as the contemporary of the Talchir and Salt-Range glacial conglomerate, is followed by a series of purple and speckled sandstones and shales, the whole overlain by dolomitic limestones, over 600 metres in thickness. The limestone is compact and well bedded, of purple, grey and cream colours; its weathering is very peculiar, giving rise to blocks with deeply incised cuts and grooves. The rock is wholly unfossiliferous, but from its intimate association in Kaghan with the Panjal Volcanic series and the occurrence of the glacial boulder-bed at its base there is now little room for doubting its Upper Carboniferous or Permo-Carboniferous age. The above Hazara sequence was formerly regarded as probably Devonian and named "Infra-Trias" from its immediately underlying the more conspicuous Trias limestone of the Sirban mountain, a prominent mountain near Abbottabad. (Fig. 26).

Burma

We have seen in Chapter VIII that there is in Upper Burma (Northern Shan States) a conformable passage of the Devonian and Carboniferous to strata of the Permian age in the great limestone formation constituting the upper part of what is known there as the Plateau limestone. (See also Fig. 15, p. 157) In the upper beds of these limestones there is present a fauna[2] of brachiopods, corals, polyzoa, etc. which shows on the whole fairly

---

[1] De Terra, *Forschungen im westlichen Kun Lun und Karakoram-Himalaya*, Berlin, 1932.

[2] Anthracolithic Faunas of the Southern Shan States, *Rec. G.S.I* vol. lxvii. pt. 1, 1933.

close relations to the Productus limestone of the Salt-Range and the Productus shales of the Spiti Himalayas and the Zewan series of Kashmir. From these affinities between the homotaxial faunas of the Indo-Burma region, Dr. Diener, the author of many memoirs on the faunas, considers all these regions as belonging to the same zoogeographical province, their differences being ascribed to the accidents of environment, isolation through temporary barriers, and differences in the depth and the salinity of waters, etc.

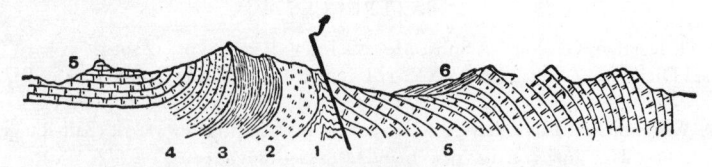

FIG. 24.—Palaeozoic rocks of the N. Shan States.
1. Chaung Magyi series (Cambrian ?).
2. and 3. Naungkangyi series (Ordovician).
4. Namshim beds (Silurian).
5. Plateau limestone (Devonian and Permo-Carboniferous).
6. Napeng beds (Upper Triassic).
La Touche, *Mem. G.S.I.* xxxix. pt. 2, 1913.

The Permo-Carboniferous rocks of Burma contain two foraminiferal limestones : the *Fusulina* limestone and the *Schwagerina* limestone, from the preponderance of these two genera of Carboniferous and Permian foraminifers.

III. MARINE PERMO-CARBONIFEROUS OF THE PENINSULA

An extraordinary occurrence has been recorded[1] at Umaria, in Vindhya Pradesh, of a thin and solitary band of marine Productus limestone in the midst of fresh-water coal-bearing beds belonging to the Barakar stage of the Damuda series (Lower Gondwana system).

The marine intercalation is only three metres thick and conformably underlies the sandstone and grit strata of normal Barakar facies, exposed in a cutting in the Umaria coal-field. It unconformably overlies the Talchir boulder-bed. The limestone bed in made up entirely of the fossil shells of *Productus*, the only other fossils present being *Spiriferina* and *Reticularia*.

Cowper Reed considers the Umaria fauna to be quite local and unique, showing no clear affinities with the near-by Salt-Range

---

[1] K. P. Sinor, *Mineral Resources of Rewa State*, p, 21, 1923.

province, but rather with the Himalayan and Russian Permo-Carboniferous province.

This bed must be regarded as a solitary record of an evanescent transgression of the sea-waters into the heart of the Peninsula, either from the north through Rajasthan or from the west coast, induced by some diastrophic modification of the surface of the land, which, however, must have been of a transient nature and must have soon ceased to operate.

REFERENCES

H. H. Hayden, Geology of Spiti, *Mem. G.S.I.* vol. xxxvi. pt. 1, 1904.

Karl Diener, *Pal. Indica*, Series XV. vol. i. pts. 2, 3, 4 and 5 (1897-1903); *Pal. Indica*, New Series, vol. iii. mem. iv. (1911); vol. v. mem. ii. (1915).

W. Waagen, *Pal. Indica*, Series XIII. vol. i. pts. 1-7 (1879-1887), (Salt-Range Fossils); and vol. iv. pts. 1 and 2 (1889-1891).

D. N. Wadia, Permo-Carboniferous in the Tertiary zone of Kashmir Himalayas, *Rec. G.S.I.* vol. lxxii, pt. 2, 1937.

J. B. Auden, Geology of the Krol Belt, Simla Himalaya; *Rec. G.S.I.* vol. lxvii. pt. 4, 1934.

C. L. Griesbach, Notes on Central Himalayas, *Rec. G.S.I.* vol. xxvi. pt. 1, 1893.

A. B. Wynne, Geology of the Salt-Range, *Mem. G.S.I.* vol. xiv., 1878.

F. R. C. Reed, Marine Fauna from the Umaria Coalfield; *Rec. G.S.I.* vol. lx. pt. 3, 1928; Permo-Carboniferous faunas, *Pal. Ind.*, N.S. vols. xvii., xx., xxiii. 1931-1944.

L. R. Wager, The Lachi Series of N. Sikkim, *Rec. G.S.I.* vol. lxxiv. pt. 2, 1939.

close relations to the Productus limestone of the Salt-Range and the Productus shales of the Spiti Himalayas and the Zewan series of Kashmir. From these affinities between the homotaxial faunas of the Indo-Burma region, Dr. Diener, the author of many memoirs on the faunas, considers all these regions as belonging to the same zoogeographical province, their differences being ascribed to the accidents of environment, isolation through temporary barriers, and differences in the depth and the salinity of waters, etc.

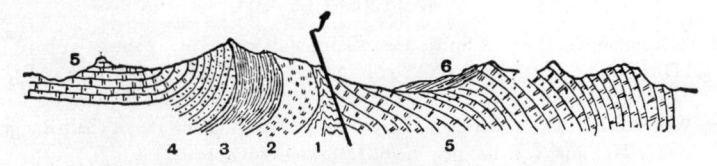

FIG. 24.—Palaeozoic rocks of the N. Shan States.
1. Chaung Magyi series (Cambrian ?).
2. and 3. Naungkangyi series (Ordovician).
4. Namshim beds (Silurian).
5. Plateau limestone (Devonian and Permo-Carboniferous).
6. Napeng beds (Upper Triassic).

La Touche, *Mem. G.S.I.* xxxix. pt. 2, 1913.

The Permo-Carboniferous rocks of Burma contain two foraminiferal limestones : the *Fusulina* limestone and the *Schwagerina* limestone, from the preponderance of these two genera of Carboniferous and Permian foraminifers.

III. MARINE PERMO-CARBONIFEROUS OF THE PENINSULA

An extraordinary occurrence has been recorded[1] at Umaria, in Vindhya Pradesh, of a thin and solitary band of marine Productus limestone in the midst of fresh-water coal-bearing beds belonging to the Barakar stage of the Damuda series (Lower Gondwana system).

The marine intercalation is only three metres thick and conformably underlies the sandstone and grit strata of normal Barakar facies, exposed in a cutting in the Umaria coal-field. It unconformably overlies the Talchir boulder-bed. The limestone bed in made up entirely of the fossil shells of *Productus*, the only other fossils present being *Spiriferina* and *Reticularia*.

Cowper Reed considers the Umaria fauna to be quite local and unique, showing no clear affinities with the near-by Salt-Range

---

[1] K. P. Sinor, *Mineral Resources of Rewa State*, p, 21, 1923.

province, but rather with the Himalayan and Russian Permo-Carboniferous province.

This bed must be regarded as a solitary record of an evanescent transgression of the sea-waters into the heart of the Peninsula, either from the north through Rajasthan or from the west coast, induced by some diastrophic modification of the surface of the land, which, however, must have been of a transient nature and must have soon ceased to operate.

REFERENCES

H. H. Hayden, Geology of Spiti, *Mem. G.S.I.* vol. xxxvi. pt. 1, 1904.

Karl Diener, *Pal. Indica,* Series XV. vol. i. pts. 2, 3, 4 and 5 (1897–1903); *Pal. Indica,* New Series, vol. iii. mem. iv. (1911); vol. v. mem. ii. (1915).

W. Waagen, *Pal. Indica,* Series XIII. vol. i. pts. 1–7 (1879–1887), (Salt-Range Fossils); and vol. iv. pts. 1 and 2 (1889–1891).

D. N. Wadia, Permo-Carboniferous in the Tertiary zone of Kashmir Himalayas, *Rec. G.S.I.* vol. lxxii, pt. 2, 1937.

J. B. Auden, Geology of the Krol Belt, Simla Himalaya; *Rec. G.S.I.* vol. lxvii. pt. 4, 1934.

C. L. Griesbach, Notes on Central Himalayas, *Rec. G.S.I.* vol. xxvi. pt. 1, 1893.

A. B. Wynne, Geology of the Salt-Range, *Mem. G.S.I.* vol. xiv., 1878.

F. R. C. Reed, Marine Fauna from the Umaria Coalfield; *Rec. G.S.I.* vol. lx. pt. 3, 1928; Permo-Carboniferous faunas, *Pal. Ind.,* N.S. vols. xvii., xx., xxiii. 1931–1944.

L. R. Wager, The Lachi Series of N. Sikkim, *Rec. G.S.I.* vol. lxxiv. pt. 2, 1939.

CHAPTER XII

The Triassic System

Introduction—The Productus shales (Kuling system) of the Himalayas and the Chideru stage of the Productus limestone of the Salt-Range are succeeded by a more or less complete development of the Triassic system. The passage in both cases is quite conformable and even transitional, no physical break in the continuity of deposits being observable in the sequence. The Triassic system of the Himalayas, both by reason of its enormous development in the northern geosynclinal zone as well as the wealth of its contained faunas, makes a conspicuous landmark in the history of the Himalayas. The abundance of its cephalopod fauna is such that it has been the means of a *zonal* classification of the system (*zones* are groups of strata of variable thickness, but distinguished by the exclusive occurrence, or predominance, of a particular species, the zone being designated by the name of the species). In north-west Nepal, Spiti, Garhwal and Kumaon, and extension of the same axis in Kashmir, the Trias attains a development of more than 1,000 m., containing three well-marked subdivisions, corresponding respectively to the Bunter, Muschelkalk and Keuper of Europe. In fauna as well as in lithology there is a remarkable similarity of facies between the Himalayan Trias and the Trias of the Eastern Alps.

Other regions where the Trias occurs, either completely developed or in some of its divisions, are the Salt-Range, Baluchistan and Burma. In the Salt-Range the Triassic system is confined to the Lower Trias and the lower part of the Middle Trias, while in Baluchistan and Burma it is confined to the Upper Triassic stages only. In the two latter areas it assumes an argillaceous facies of shales and slates, whereas in the Himalayan region the system is entirely composed of limestone, dolomites and calcareous shales.

Principles of classification of the geological record—With the Trias we enter the Mesozoic era of geology; and before we proceed further we might at this stage enquire into the basis for the classification of the geological record into systems and series, and consider whether the interruptions or "blanks" in the course of the earth's history, which have led to the creation of the chief divisions, in the first instance in some parts of the world, were necessarily world-wide in their effects and applicable to all parts of the world.

In Europe the geological record is divided into three broad sections or groups: the Palaeozoic, Mesozoic and Cainozoic, representing three great eras in the history of the development of life on the earth, each of which is separated from the one overlying it by an easily perceptible and comparatively widespread physical break or "unconformity". Whether these divisions, so well marked and natural in Europe, where they were first recognised, are as well marked and natural in the other parts of the world, and whether these three, with their subdivisions, should be the fundamental periods of earth-history for the whole world are subjects over which the opinion of geologists is sharply divided. In the geological systems of India, as in the other regions of the earth, although the distinctive features of the organic history of the Palaeozoic, Mesozoic and Cainozoic are clearly evident as we ascend in the stratigraphic scale, we cannot detect the sharp breaks in the continuity of that history at which one great time-interval ends and the next begins. Just at these parts the geological record appears to be quite continuous in India, and any attempt at setting a limit would be as arbitrary as it would be unnatural. On the other hand, there are great interruptions or "lost intervals" in the Indian record at other stages (where the European record is quite continuous) at which it is much more natural to draw the dividing lines of its principal divisions—the *groups*. As we have already seen, Sir T. H. Holland has accomplished this in his scheme of the classification of the Indian formations. Though generally adopted in India, and best suited to the rather imperfect character of the geological record as preserved in India, such a classification and nomenclature may not be acceptable to those geologists who hold that the grand divisions of geology are universal and applicable to the whole world. The subject is difficult to decide one way or the other, but for the information of the student the following view, which summarises the arguments of the latter class of geologists with admirable lucidity, is given *verbatim* from the work of Professors T. C. Chamberlin and R. D. Salisbury:[1]

"We believe that there is a natural basis of time-division, that it is recorded dynamically in the profounder changes of the earth's history, and that its basis is world-wide in its applicability. It is expressed in interruptions of the course of the earth's history. It can hardly take account of all local details, and cannot be applied with minuteness to all localities, since geological history is necessarily continuous. But even a countinuous history has its times

---

[1] *Advanced Geology*, vol. iii., Early History.

PLATE XII. FOLDED TRIAS BEDS, DHAULI GANGA VALLEY, CENTRAL HIMALAYAS.

C. Upper Carboniferous white quartzite. P. Permian black Productus shales. T. Lower Trias limestone, *Otoceras* stage.
(*Geological Survey of India, Mem.* vol. xxiii.)

and seasons, and the pulsations of history are the natural basis for its divisions.

"In our view, the fundamental basis for geologic time-divisions has its seat in the heart of the earth. Whenever the accumulated stresses within the body of the earth overmatch its effective rigidity, a readjustment takes place. The deformative movements begin, for reasons previously set forth, with a depression of the bottoms of the oceanic basins, by which their capacity is increased. The epicontinental waters are correspondingly withdrawn into them. The effect of this is practically universal, and all continents are affected in a similar way and simultaneously. This is the reason why the classification of one continent is also applicable, in its larger features, to another, though the configuration of each individual continent modifies the result of the change, so far as that continent is concerned. The far-reaching effects of such a withdrawal of the sea have been indicated repeatedly in preceding pages. Foremost among these effects is the profound influence exerted on the evolution of the shallow-water marine life, the most constant and reliable of the means of intercontinental correlation. Second only to this in importance is the influence on terrestrial life through the connections and disconnections that control migration. Springing from the same deformative movements are geographic and topographic changes, affecting not only the land, but also the sea currents. These changes affect the climate directly, and by accelerating or retarding the chemical reactions between the atmosphere, hydrosphere, and lithosphere, affect the constitution of both air and sea, and thus indirectly influence the environment of life, and through it, its evolution. In these deformative movements, therefore, there seems to us to be a universal, simultaneous, and fundamental basis for the subdivision of the earth's history. It is all the more effective and applicable, because it controls the progress of life, which furnishes the most available criteria for its application in detail to the varied rock formations in all quarters of the globe.

"The main outstanding question relative to this classification is whether the great deformative movements are periodic rather than continuous, and co-operative rather than compensatory. This can only be settled by comprehensive investigation the world over; but the rapidly accumulating evidence of great base-levelling periods, which require essential freedom from serious body deformation as a necessary condition, has a trenchant bearing on the question. So do the more familiar evidences of great sea transgressions, which may best be interpreted as a consequence of general base-levelling and concurrent sea-filling, abetted by continental creep during a long stage of body quiescence. It is too early to affirm, dogmatically, the dominance in the history of the earth of great deformative movements, separated by long intervals of essential quiet, attended by (1) base-levelling, (2) sea-filling, (3) continental creep, and (4) sea-transgression; but it requires little prophetic vision to see a probable demonstration of it in the near future. Subordinate to these grander features of historical progress, there are innumerable minor ones, some of which appear to be rhythmical and systematic, and some irregular and irreducible to order. These give rise to the local epochs and episodes of earth-history, for which strict intercontinental correlation cannot be hoped, and which must be neglected in the general history as but the individualities of the various provinces."

Triassic System

The Trias is one of the best developed sedimentary formations of the Inner Himalayas from Hazara to Nepal. In Sikkim its development is on a lesser scale. Impressive sections of the system are exposed on the south flank of the Great Himalaya range from Kashmir to Byans in eastern Kumaon. There is a close faunistic relation between these distant outcrops, but there exists considerable variation in the facies of the lithological assemblages and in the relative thickness of the Lower, Middle and Upper divisions. The Lower Trias is over 100 m. thick in Kashmir, 50 m. in Byans on the western border of Nepal, and only 12 m. in Spiti. The Middle Trias is 275 m. thick in Kashmir; the entire thickness is restricted to the Muschelkalk, the Ladinic stage being absent. In the Spiti, Kumaon and Byans regions it is less than one-third of this, though they contain a rich fauna representative of both the Muschelkalk and the Ladinic. The Carnic and Noric stages of the Upper Trias in Kashmir are seen in the Gurais region in a 1,200-1,800 m. pile of massive well-bedded limestones, though the Upper stage here is barren of fossils. In Spiti both these stages are represented by about 1,000 m. of dominantly calcareous strata, while in the Kumaon and Garhwals mountain the Upper Trias is not as calcareous. Its thickness has diminished to 600 to 450 m. but it has several rich fossiliferous horizons. The topmost beds of the Upper Trias in all the above areas are the *Megalodon limestone*, which forms the basal portion of the great *Kioto limestone* of Spiti and Kumaon, a vast limestone and dolomite formation of Lias age.

In the Sikkim region of Eastern Himalayas, the Everest region, a series of quartzites and shales several hundred metres thick overlie the *Lachi series* (Permian) and are believed to represent some Trias horizon, on the ground of their containing ammonoids and molluscs of Triassic affinities (*Tso Lhamo series*).

The Triassic system of Spiti—Triassic rocks are developed along the whole northern boundary of the Himalayas, constituting the great scarps of the plateau of Tibet. (See Figs. 14, 22 and 25.) A perfect section of these rocks, showing the relations of the Trias to the systems below and above it, is exposed at Lilang in Spiti. The following section from Dr. Hayden's *Memoir* gives a clear idea of the classification of the system:[1]

Jurassic : (Rhaetic ?)   Massive *Megalodon* limestone.

Keuper 850 m.	Quartzites with shales and limestones: *Lima*, *Spirigera*. "*Monotis* shale": sandy and shaly limestone. Coral limestone. *Juvavites* beds: sandstones, shales and limestones. *Tropites* beds: dolomitic limestone and shales. Grey shales: shaly limestone and shales with *Spiriferina*, *Rhynchonella*, *Trachyceras*, etc. *Halobia* beds: hard dark limestone with *Halobia*, *Arcestes*, etc.
Muschelkalk 120 m.	*Daonella* limestone: thin black limestone with shales, *Daonella*, *Ptychites*. Limestone with concretions. Grey limestone with *Ceratites*, *Sibirites*, etc. Nodular limestone (Niti limestone).
Bunter 15 m.	Nodular limestone. Limestone and shale with *Aviculopecten*. *Hedenstroemia* zone. *Meekoceras* zone, *M. varahaa*. *Ophiceras* zone, *O. sakuntala*. *Otoceras* zone, *O. woodwardi*.

Permian :   Productus shales.

Triassic fauna—The Lower Trias is thin in comparison with the other two divisions of the system, and rests conformably on the top of the Productus shales. The rocks are composed of dark-coloured shales and limestones, with an abundant ammonite fauna. Besides those mentioned in the section above, the following genera are important: *Tirolites*, *Ceratites*, *Danubites*, *Flemingites*, *Stephanites*, with *Pseudomonotis*, *Rhynchonella*, *Spiriferina*, and *Retzia*.

The middle division is thicker and largely made up of concretionary limestones. This division is also widespread and capable of detailed subdivision into stages and zones, which preserve a uniform character, both faunistic and lithological, over Spiti,

---

[1] Geology of Spiti, *Mem*, *G.S.I.* vol. xxxvi. pt. 1, p. 90.

Painkhanda, Byans and Johar. This division possesses a great palaeontological interest because of the rich Muschelkalk fauna it contains, resembling in many respects the Muschelkalk of the Alps.

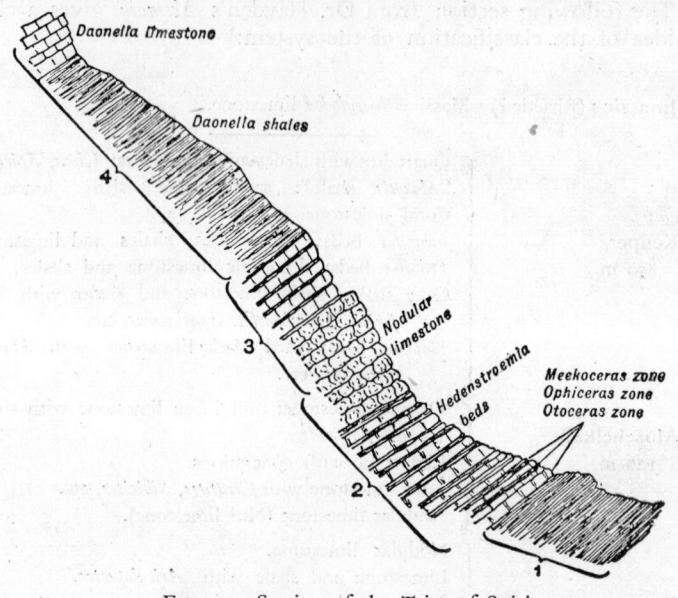

FIG. 25.—Section of the Trias of Spiti.

1. Productus shales (Permian).
2. Lower Trias.
3. Muschelkalk (lower part).
4. Muschelkalk (upper part).

after Hayden, *Mem. G.S.I.*, vol. xxxvi. pl. 1.

The upper Muschelkalk is especially noted for the number and variety of its cephalopod fossils; it forms indeed the richest and most widely spread fossil horizon in the central and N.W. Himalaya. It is capped by the Ladinic stage, composed of *Daonella* limestones and slates. The most typical fossil belongs to the genus *Ceratites*; besides it are the other cephalopods *Ptychites*, *Trachyceras*, *Xenaspis*, *Monophyllites*, *Gymnites*, *Sturia*, *Proarcestes*, *Isculites*, *Hollandites*, *Dalmanites*, *Haydenites*, *Pinacoceras Buddhaites*, *Nautilus* (sp. *spitiensis*), *Pleuronautilus*, *Syringonautilus* and *Orthoceras*. The brachiopods are *Spiriferina* and *Spirigera*; *Daonella* and *Halobia* are the leading lamellibranchs.

The uppermost division of the Trias is by far the thickest, and is composed of two well-marked divisions—dark shales and marl beds in the lower part, and thick grey-coloured limestone and dolomite in the upper, with an abundant cephalopod fauna, whose distribution often characterises well-marked zones. The lower of the two divisions corresponds to the Carnic and Noric stages of the Alpine Trias, while the uniform mass of limestones overlying it probably represents the Rhaetic of the Alps (cf. Kioto limestone, p. 238).

The faunistic resemblance between the Triassic rocks of the Himalayas and Alps suggests open sea communication maintained by the Tethys between these two areas since the beginning of the Permian. This sea provided a free channel of migration and intercommunication between the marine inhabitants of the central zone of the earth from the Mediterranean shores of France to the eastern borders of China, and maintained this waterway up to the beginning of the Eocene period. The commonest fossils are again: (Ammonites) *Joannites, Halorites, Trachyceras, Tropites, Juvavites, Sagenites, Sirinites, Hungarites, Gymnites, Ptychites, Griesbachites.* Lamellibranchs are also numerous; the most commonly occurring forms are *Lima, Daonella, Halobia, Megalodon, Monotis, Pecten, Avicula, Corbis, Modiola, Mytilus, Homomya, Pleuromya,* with the addition of the aberrant genera *Radiolites* and *Sphaerulites* of the *Rudistae* family of the lamellibranchs. The brachiopods are very few, both as regards number and their generic distribution, being confined to *Spirigera, Spiriferina, Rhynchonella,* and their allied forms.

The Triassic fauna shows a marked advance on the fauna of the Productus limestone. The most predominant element of the former is cephalopods, while that of the latter was brachiopods. This is the most noteworthy difference, and signalises the extinction of large numbers of brachiopod families during the interval. The brachiopods can be said to enter on their decline after the end of the Palaeozoic era, a decline which has steadily persisted up to the present. During the Mesozoic era the brachiopods were represented by three or four genera like *Terebratula, Rhynchonella, Spirigerina,* etc. The place of the brachiopods is taken by the lamellibranchs, which have greatly increased in genera and species. The cephalopods, the most highly organised members of the Invertebrata, will henceforth occupy a place of leading importance among the fauna of the succeeding Mesozoic systems.

"**Exotic" Trias of Malla Johar and Chitichun**—Large blocks and masses of Trias limestone of all sizes up to mountainous masses are found lying in confused stratigraphic disorder, over various Mesozoic formations, at the above localities on the Tibetan border of Kumaon. The lithological as well as fossil facies of these blocks is quite different from any known in the Himalaya and is evidently "foreign", being allied to the Trias of the Eastern Alps. These Triassic limestone blocks are mixed up with blocks of foreign Permian, Jurassic and Cretaceous limestones. It is quite evident that these Mesozoic rock-masses are not in their original site of deposition, but are truly exotic and have been transported from a distant locality by an agency that is not yet certainly established (shattering by explosive volcanic action and transport by lava-flows, or an overthrust sheet from a northern region severed by denudation into detached masses).[1] This subject is discussed again on p. 258 in connection with the exotic Cretaceous of the same region in Kumaon.

Hazara (the Sirban Mountain)

The Trias is found in Hazara occupying a fairly large area in the south and south-east districts of this province, resting on the presumed Permo-Carboniferous series of sediments, underlain by a glacial conglomerate, which was formerly referred to as the Infra-

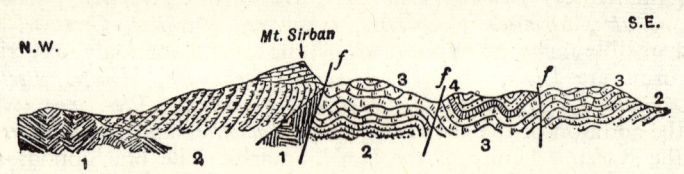

FIG. 26.—Diagrammatic section of Mt. Sirban, Hazara.

1. Slate series.   3. Trias.
2. Permo-Carboniferous.   4. Jurassic, Cretaceous.

After Middlemiss, *Memoir, Geological Survey of India*, vol. xxvi.

Trias. The Triassic system of Hazara consists, at the base, of about 30 m. of felsitic or devitrified acid lavas of rhyolitic composition, succeeded by a thick formation of rather poorly fossiliferous limestone, in which the characteristic Upper Triassic fossils of the other Himalayan areas are present. The Lower and Middle Trias are absent from Hazara. The limestone is thickly bedded, of a grey colour, sometimes with an oolitic structure. Its thickness varies from 150 to 370 m. These rocks form the base of a nearly complete Mesozoic sequence in Hazara, which though consideraby thinner, is similar in most respects to that of the geosynclinal zone of the Northern Himalayas, so typically displayed in the sections in the Spiti Valley and in Hundes.

Mt. Sirban—A locality famous for geological sections in Hazara is Mt. Sirban, a lofty hill lying to the south of Abbottabad. Most of the formations of Hazara are exposed with wonderful clearness of detail in a number of sections along its sides (see Figs. 26 and 27), in which one can trace the whole stratigraphic sequence from the base of the Permo-Carboniferous to the Nummulitic limestone. The sections revealed in this hill epitomise in fact the geology of a large part of the North-West Himalaya.[2]

---

[1] A. Heim and A. Gansser, Central Himalaya, Zurich, 1937.
[2] *Mem. G.S.I.* vol. ix, pt. 2, 1872, and *Mem. G.S.I.* vol. xxvi., 1896.

THE TRIASSIC SYSTEM

On the south border of Hazara, in the Kala Chitta hills of Attock district, some strips of Trias limestone (Kioto or *Megalodon* limestone) are laid bare in a series of denuded isoclinal folds of the Nummulitic limestones which constitute these hill-masses.

The Trias of the Salt-Range

The Ceratite beds—The Trias is developed, though greatly reduced in its proportions, in the western part of the Salt-Range. The outcrop of the system commences from the neighbourhood of the Chideru hills, and thence continues westward up to a great distance beyond the Indus. It caps the underlying Productus limestone (Chideru beds), and accompanies it along a great length of the Range until the disappearance of the latter beyond Kalabagh and the Shekh Budin hills. The Triassic rocks of the Salt-Range proper comprise only the Lower Trias and a small part of the Middle Trias in actual stratigraphic range but these horizons are completely developed, and they include all the cephalopod zones worked out in the corresponding divisions of the Spiti section. In the trans-Indus Salt-Range the complete Triassic sequence is developed, including the Upper Trias, the whole system being about 150 m. in vertical extent. On account of the abundance of the fossil ammonite genus *Ceratites* the Lower Trias of the Salt-Range is known as the *Ceratite beds*. The rocks comprising them are about thirty metres of thin flaggy limestone, which overlie the Chideru stage quite conformably, from

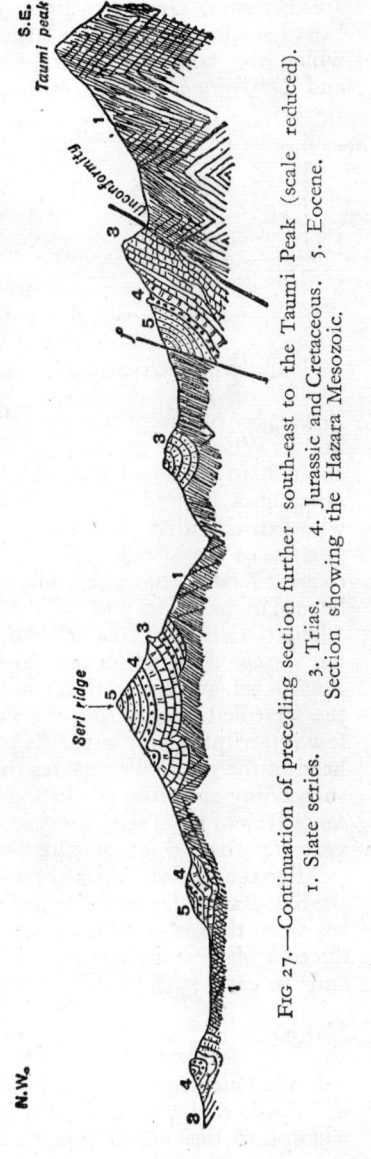

FIG 27.—Continuation of preceding section further south-east to the Taumi Peak (scale reduced). Section showing the Hazara Mesozoic. 1. Slate series. 2. 3. Trias. 4. Jurassic and Cretaceous. 5. Eocene.

which also they are indistinguishable lithologically. Overlying beds are grey limestones and marls, nodular at places. Besides *Ceratites*, which is the leading fossil, the other ammonites are *Ptychites*, *Gyronites*, *Flemingites*, *Koninckites*, *Prionolobus*, etc. Fossil shells are found in large numbers in the marly strata, of which the common genera are *Cardinia*, *Gervillia*, *Rhynchonella* and *Terebratula*. A very curious fossil in the Ceratite beds is a

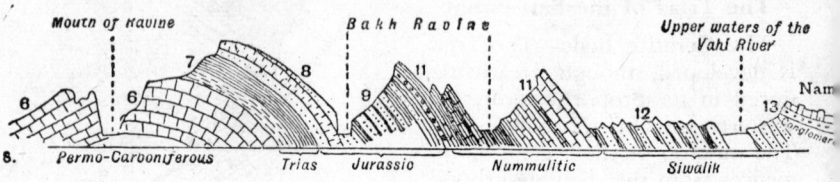

FIG. 28.—Section through the Bakh Ravine from Musa Khel to Nammal. About natural scale, 7.5 cm.=1.609 m.
Wynne, Salt-Range, *Memoir*, G.S.I. vol. xiv.

Bellerophon of the genus *Stachella*, the last survivor of the well-known Palaeozoic gastropod. The Ceratite beds are succeeded by about 30 to 60 metres of Middle Trias (Muschelkalk), composed of sandstones, crinoidal limestone and dolomites full of cephalopods, whose distribution characterises zones corresponding to the lower portion of the Middle Trias of Spiti and Kashmir. Some of the clearest sections of these and younger Mesozoic formations are to be seen in the gullies and nullahs of the Chideru hills of the range.

There is a deep ravine near Musa Khel, the Nammal gorge, which has dissected the whole breadth of the mountain from Nammal to Musa Khel, and the section laid bare in its precipices comprehends the stratified record from the Permian to the Pliocene, with but few interruptions or gaps. As one walks along the section to the head of the gorge, one passes in review the rock-records of every succeeding age from the Productus limestone, through the representatives of the Trias, Jurassic, Eocene and Miocene, with at the very top the Upper Siwalik boulder-conglomerates.

After the Middle Trias there comes a gap in the continuity of the Salt-Range deposits, indicating a temporary withdrawal of the sea from this area. This cessation of marine conditions has produced a blank in its geological history covering the Upper Trias and the early part of the Jurassic period.

Baluchistan

In the Quetta and Zhob districts of North Baluchistan, outcrops of Triassic rocks, appearing as inliers in the anticlines of the more widespread Lias development, are marked by the exclusive prevalence of the uppermost Triassic or Rhaetic stage, no strata referable to the Lower and Middle Trias being found in this province. The rocks are several thousand metres of shales and slates, with a few intercalations of limestone. They contain the Upper Trias species of *Monotis* and a few ammonites like *Didymites*, *Halorites*, *Rhacophyllites*.

The Trias of Baluchistan rests unconformably on an older *Productus*-bearing limestone, enclosing a foraminiferal limestone, *Fusulina* limestone, of Permo-Carboniferous age.

Kashmir

As is generally the case with the other rock systems, the development of the Trias in Kashmir is on much the same scale as in Spiti, if indeed not on a larger scale. A thick series of compact blue limestone, slates and dolomites is conspicuously displayed in many of the hills bordering the valley to the north, while they have entered largely into the structure of the higher parts of the Sind, Lidar, Gurais and Tilel valleys and of the north-east flanks of the Pir Panjal (p. 359). The Trias of Kashmir, in common with the whole length of the North Himalayas from the Pamirs to Nepal, is on a scale of great magnitude, although because of its lack of richness in fossils, compared with the Trias of Spiti, the system has not been subdivided zonally to the same extent as the latter. A superb development of limestones and dolomites of this system is exhibited in a series of picturesque escarpments and cliffs forming the best part of the scenery north of the Jhelum. The Trias attains great dimensions farther north in the upper Sind, Lidar and Wardwan valleys, and again in Gurais, Tilel and Central Ladakh, thence extending as far as the Karakoram and Lingzhithang plains. Another locality for the development of the Trias, principally belonging to its upper division, is the Pir Panjal, of which it is the youngest constituent rock-group, capping the volcanic beds over the whole stretch of the range from beyond the Jhelum to Kishtwar. A great part of the Triassic on the north-east flanks, however, is obscured under later formations such as the Karewas and moraine débris.

Lithology—Limestones are the principal components of this system. The rocks are of a light blue or grey tint, compact and homogeneous, and sometimes dolomitic in composition. They are thin-bedded in the lower part of the system, with frequent interstratifications of black sandy and calcareous shales, but towards the top they become one monotonously uniform group of thickly bedded limestones. They compose a very picturesque feature of the landscapes, noticeable from all parts of the country by the light coloration of their outcrops and their graceful long and undulating folds, interspersed with areas of close plication

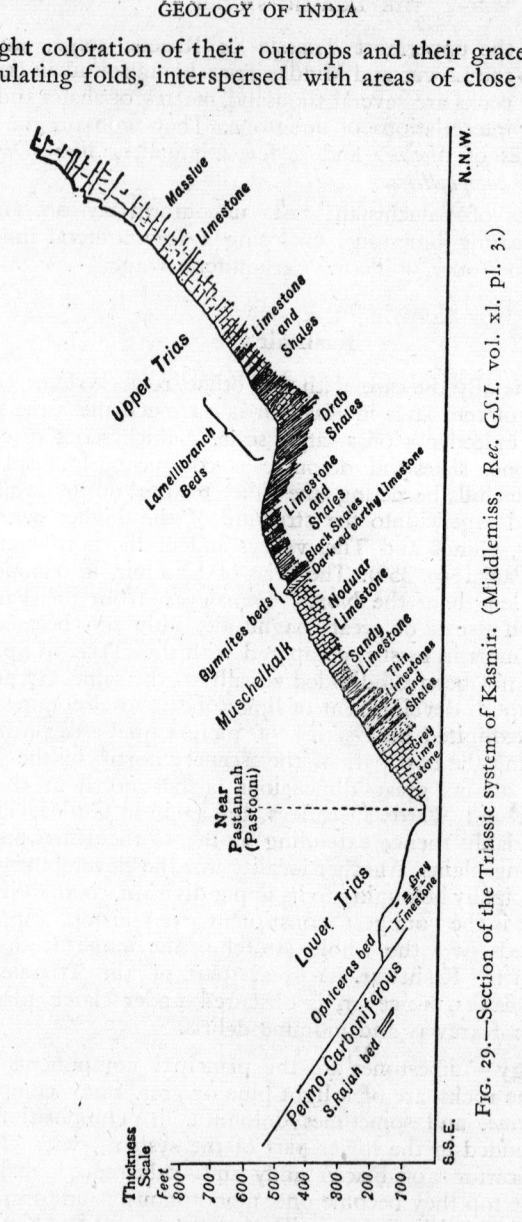

FIG. 29.—Section of the Triassic system of Kasmir. (Middlemiss, *Rec. G.S.I.* vol. xl. pl. 3.)

and inversions, both of which characteristics bring them out in strong relief against the dark-coloured, craggy lavas and slates of the underlying Panjals. Numerous springs of fresh water issue from the cliffs and prominences of these limestones at the southeast end of the valley, and form the sources of the Jhelum; the best known of these are the river-like fountains of Achabal and Vernag and the multitudinous springs of Anantnag and Bhawan. The lower and middle sections of the system are rich in fossils, the abundance of the Cephalopoda and the peculiarities of their vertical range in the strata being the means of a fairly detailed zonal classification of the system, all the zones of which are related to the corresponding ones of Spiti. The upper division of the Trias is largely barren of fossils. The following succession of the Triassic strata may be taken as typical:

Upper Trias.
(Many thousand metres thick.)
- Unfossiliferous massive limestone with occasional corals and crinoids, *Calamophyllia*.
- *Spiriferina stracheyi* and *S. haueri* zones.
- Lamellibranch beds.

Middle Trias.
(About 300 m.)
- *Ptychites* horizon: sandy shales with calcareous layers.
- Ceratite beds: „ „ „ „
- *Rhynchonella trinodosi* beds: „ „ „
- *Gymnites* and *Ceratite* beds: „ „ „
- Lower nodular limestone and shales.
- Interbedded thin limestones, thick black shales and sandy limestones.

Lower Trias.
(Over 100 m.)
- *Hungarites* shales (position uncertain).
- *Meekoceras* limestones and shales.
- *Ophiceras* limestones.
- *Otoceras* beds (seen at a few localities only).

Lower Trias—At all the Permian localities referred to on pages 213–215 the Zewan series shows a conformable passage upwards into a series of limestone strata, which in their fossil ammonites are the exact parallels of the *Ophiceras* and *Meekoceras* zones of Spiti. The *Otoceras* zone is recognised in the Sind valley, at the base of the Lower Trias, curiously containing some *Productus*, a survival from the Palaeozoic. These in turn pass upwards, after the intervention of a shaly zone (the *Hungarites* zone), into the great succession of Middle Triassic limestones and shales. The best sections of the Lower Trias are those laid bare at Pastanah and at Lam, two places on the eastern border of the Vihi district, though the sections are somewhat obscured by jungle-growth. Fossil ammonites are *Xenodiscus* (seven species), *Otoceras*, *Ophiceras* (*O. sakuntala* and five other species), *Flemingites*, *Vishnuites*, *Hungarites*, *Meekoceras*, *Sibirites*, and a new genus of ammonite, *Kashmirites*. Other cephalopods are *Orthoceras* and *Gryphoceras*; the lamellibranch, *Pseudomonotis*, is a type form.

Middle Trias—Sections of the Middle Trias, or Muschelkalk, are visible at many points in Vihi, *e.g.* at Pastanah, Khrew and Khunmu above Pailgam in the Lidar, and in some of the tributary valleys of the Upper Sind. The limestones of this part of the Trias are more frequently interbedded with shales, the latter being often black and arenaceous. The Muschelkalk has yielded a very diversified fauna of cephalopods indicating the very high degree of specialisation reached by this class of animals, particularly the order of the ammonites. The specific relations of the types are in all respects like those of the other parts of the Himalayas.

The Muschelkalk fauna—The principal forms of the Muschelkalk fauna of Kashmir are *Ceratites* (sixteen species), *Hungarites, Sibirites, Isculites, Pinacoceras, Ptychites, Gymnites* (spp. *sankara, vasantsena* and other species), *Buddhaites*. The nautiloidea are *Syringonautilus, Gryphoceras, Paranautilus, Orthoceras*. The lamellibranch genera are *Myophoria, Modiola, Anomia, Anodontophora*; the brachiopods are *Spiriferina stracheyi, Dielasma* and *Rhynchonella*; the gastropods are represented by a species of *Euomphalus* and the aberrant genus *Conularia*.

Upper Trias—The Muschelkalk is succeeded, in all the above-noted localities, by an enormous development of the Upper Triassic strata, which are mostly unfossiliferous but for a zone of coral-, lamellibranch- and brachiopod-bearing beds included in the lower part. An Upper Triassic crinoidal limestone is widely distributed in moraine heaps clothing the N.E. slopes of the Pir Panjal, but for the greater part the formation is an unvarying succession of thick massive unfossiliferous limestone. It is this limestone which builds the range of high hills and precipices so conspicuous by their colouring in the Vihi and the Islamabad districts.

A broad and continuous belt of barren, light and dark grey, Upper Trias dolomites and limestones stretches from north of Pailgam, through the head-waters of the Sind, to beyond Gurais. At the latter locality the Kishenganga river has excavated through this limestone a broad U-shaped valley bounded on both sides by an imposing line of precipices, towerin 1,200 to 1,800 metres above the flat scree-strewn bottom. The Lower and Middle Trias are missing in Gurais, the upper flows of the Panjal trap showing a conformable passage into the Upper Trias. In Tilel the lower part of the Trias is scantily developed in the south slopes of the valley. The Trias is developed in great force along the margin of the Tethyan geosyncline north of the Great Himalaya Range, from Kargil to beyond Spiti, associated with less conspicuously formed Permo-Carboniferous.

Near Baltal the Upper Trias forms the mountains surrounding Kolahoi (5,426 m.) and exhibits a great deal of complex folding.

In some of the major synclinal flexures of this series, between Baltal and Zoji La, it is probable that Jurassic strata of Lias or Lower Oolite age are exposed, containing a few badly preserved ammonites and belemnites. The group of Amarnath peaks (5,270 m.) with the sacred cave on its south flank is composed of Upper Trias limestone and dolomite, at some places altered to gypsum.

The Triassic limestone has furnished an abundant building material to the architects of ancient Kashmir in the building of their great temples and edifices, including the famous shrine of Martand.

Relation of the Kashmir and Spiti provinces during the Upper Trias—A broad belt of Triassic rocks, mostly poorly fossiliferous Upper Trias limestones, stretches from southern Ladakh, through 480 km. of intervening mountains, connecting up with the Trias capping the classical Palaeozoic basin exposed in the Spiti Valley. It is part of the Permo-Trias sequence so well developed from western Kashmir to the Everest region in Nepal. The fauna of the Upper Trias is quite poor in comparison to that of the Lower and Middle divisions. Cephalopods are almost absent. The few lamellibranchs include *Myophoria, Gervillia, Pseudomonotis, Lima, Pecten, Pleurophora, Trigonodus*. The brachiopods are *Spiriferina haueri, Dielasma, Rhynchonella; Calamophyllia* is a common coral; crinoids; *Marmolatella*, etc. The rarity of the zone fossils *Halobia* and *Daonella*, and the almost complete absence in Kashmir of the cephalopods that are so numerous and highly diversified in the Spiti Upper Trias, suggest some sudden and effective interruption in the free intercourse and migrations of species that had existed between the seas of the two areas for such long ages. This intercourse appears to have been partly re-established during the Jurassic, though not on the former scale, for the fauna of the later ages, that has been discovered in Kashmir up to now, is quite scanty and impoverished in comparison with the Spiti fauna.

Burma

A very similar development of the Triassic system, also restricted to the uppermost (Rhaetic or Noric) horizon, occurs in the Arakan Yoma of Burma. The fossils are a few ammonites and lamellibranchs, of which *Halobia* and *Monotis* are the most common.

The only known occurrence of Lower Triassic rocks and fossils in Burma was recorded by M. R. Sahni at Na-hkam in the Northern Shan States. Among the genera represented are *Ophiceras, Juvenites, Hemiprionites, Naticopsis, Platyceras, Lingula*, etc.

The fauna is of shallow-water facies, ammonites forming the dominant element, while the calcareous brachiopods are entirely absent. It therefore presents a striking contrast to the Burmese anthracolithic faunas in which calcareous brachiopods predominate and ammonites are absent. With the exceptions of a few fragments of Upper Trias ammonites from the Burmo-Siamese frontier and a Turonian species from Ramri Island, the Na-hkam fauna contains the only ammonites so far known from Burma.

Napeng series—Also, what are known as the Napeng beds occur in a number of scattered small outcrops in the Northern Shan States. (See Fig. 24, p. 219). The beds are composed of highly argillaceous, yellow-coloured shales and marls, with a few nodular limestone strata. The fossils are *Avicula contorta, Myophoria, Gervillia praecursor, Pecten, Modiolopsis, Conocardium*, etc. Although some of these are survivals of Palaeozoic genera, the other fossils leave no doubt of the Triassic age of the strata, while the specific relations of the latter genera suggest a Rhaetic age.[1]

REFERENCES

Karl Diener, Trias of the Himalayas, *Mem. G.S.I.* vol. xxxvi. pt. 3, 1912; Triassic Fauna of the Himalayas, *Pal. Indica*, Series XV. vol. ii, pts. 1 and 2, 1897.

E. Mojsisovics, Cephalopoda of the Upper Trias, *Pal. Indica*, Series XV. vol. iii. pt. 1, 1899.

W. Waagen, Fossils of the Ceratite Beds, *Pal. Indica*, Series XIII. vol. ii., 1895.

D. N. Wadia, Cambrian-Trias Sequence of N.W. Kashmir, *Rec. G.S.I.* vol. lxviii., pt. 2, 1934.

H. H. Hayden, Geology of Spiti, *Mem. G.S.I.* vol. xxxvi., pt. 1, 1904.

---

[1] *Mem. G.S.I.* vol. xxxix. pt. 2, 1913.

CHAPTER XIII

The Jurassic System

Instances of Jurassic development in India—In the geosynclinal zone of the Northern Himalayas, Jurassic strata conformably overlie the Triassic covering wide areas in south Tibet, south Ladakh, Spiti, Nepal and Bhutan. The succession is quite normal and transitional, the junction-plane between the two systems of deposits being not clearly determinable in the type section at Lilang in Spiti. Marine Jurassic strata are also found in the Salt-Range, representing the middle and upper divisions of the system (Oolite). The system is developed on a much more extensive scale in Baluchistan, both as regards its vertical range and its geographical extent. A temporary invasion of the sea (marine transgression), in the latter part of the Jurassic gave rise to a thick series of shallow-water deposits in Rajasthan and in Kutch. A fifth instance of Jurassic development, also the result of a marine transgression, is witnessed on the East Coast of the Peninsula, where an oscillation between marine and terrestrial conditions has given rise to the interesting development of marine Upper Jurassic strata intercalated with the Upper Gondwana formation.

Life during the Jurassic—Cephalopods, especially the ammonites, were the dominant members of the life of the Jurassic in all the above noted areas. Although perhaps they reached the climax of their development at the end of the Trias in the Himalayan province, they yet occupied a place of prominent importance among the marine forms of life of this period, and are represented by many large and diversified forms with highly complex sutured shells. Nearly 1,000 species and over 150 genera have been found in the Jurassic rocks of Kutch; the majority of the species are new and restricted to the west Indian province. Lamellibranchs were also very numerous in the Jurassic seas, and held an important position among the invertebrate fauna of the period. A rich Jurassic flora of cycads and conifers peopled the land regions of India. The lower classes of phanerogams had already appeared and taken the place of the seed-fern (pteridosperm) and the horse-tail of the Permo-Carboniferous period. The land was also inhabited by a varied population of fish, amphibia and several orders of reptiles, besides the terrestrial invertebrates. We have already dealt with the relics of the latter class of organisms in the description of the Gondwana system.

JURASSIC OF THE CENTRAL HIMALAYAS
Spiti

Kioto limestone—In the Zanskar range of Spiti, Garhwal and Kumaon, as far as the west frontier of Nepal, the Upper Trias (Noric stage) is succeeded by a series of limestones and dolomites of great thickness, the lower part of which recalls the Rhaetic of the Alps, while the upper is the equal of the Lias and part of the Oolite. The bottom beds of the series, containing shells of *Megalodon*, pass up into a massive limestone, some 600 to 900 m. thick, called the Great limestone, from its forming lofty precipi-

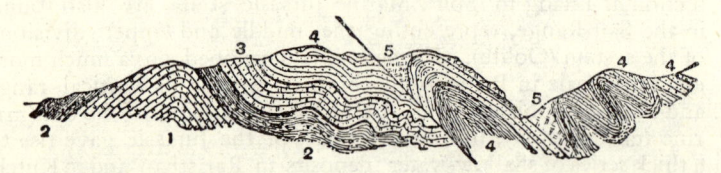

FIG. 30.—Section of the Jurassic and Cretaceous rocks of Hundes.

1. Kioto limestone.
2. Spiti shales.
3. Giumal sandstone.
4. Cretaceous flysch.
5. Basic igneous rocks.

After Von Krafft, *Mem, G.S.I.* xxxii. pt. 3.

tous cliffs facing the Punjab Himalayas. It is better known under the name of the Kioto limestone. The lithological characters of this limestone indicate the existence of a constant depth of clear water of the sea during its formation. The passage of time represented by this limestone is from Rhaetic to Middle Oolite, as evidenced by the changes in its fauna. There is an insignificant break in the sequence succeding the Callovian. The highest beds of the Kioto limestone are fossiliferous, containing a rich assemblage of belemnites and lamellibranchs, and are known as the *Sulcacutus beds* from the preponderance of the species *Belemnites sulcacutus*. The greater part of the Kioto limestone—the middle—is unfossiliferous. A fossiliferous horizon—the *Megalodon limestone*—occurs again at the base, containing numerous fossil shells of *Megalodon* and *Dicerocardium*. Other fossils are *Spirigera*, *Lima*, *Ammonites*, *Belemnites*, with gastropods of Triassic affinities. This lower part of the Kioto limestone is also sometimes designated as the *Para stage*, while the part above the Megalodon limestone is known as the *Tagling stage*.

Spiti shales—The Kioto limestone is overlain conformably by the most characteristic Jurassic formation of the inner Himalayas, known as the *Spiti shales*. (See Fig. 30.) These are a group of splintery black, almost sooty, micaceous shales, about 100 to 150 m. thick, containing numerous calcareous concretions, many of which enclose a well-preserved ammonite shell or some other fossil as a nucleus (*saligram*). These shales enclose pyritous nodules and ferruginous partings and, towards the top, impure limestone intercalations. The whole group is very soft and friable, and has received a great amount of crushing and compression. These black or grey shales show a singular lithological persistence from one end of the Himalayas to the other, and can be traced without any variation in composition from Hazara and the northern confines of the Karakoram range on the west to as far as Sikkim on the east. They also cover vast tracts of Southern Tibet overlain by Eocene and Cretaceous. These Upper Jurassic shales, therefore, are a valuable stratigraphic "reference horizon", of great help in unravelling a confused mass of strata, in mountainous regions, where the order of superposition is obscured by repeated folding and faulting.

Fauna of the Spiti Shales—The Spiti shales are famous for their great faunal wealth, which has made great contributions to the Jurassic geology of the world. The ammonites are the preponderant forms of life preserved in the shales. The enumeration of the following genera gives but an imperfect idea of the great diversity of cephalopod life : *Phylloceras*, *Lytoceras*, *Hoploceras*, *Hecticoceras*, *Oppelia*, *Aspidoceras*, *Holcostephanus* (*Spiticeras*)—the most common fossil, *Hoplites*, *Perisphinctes* and *Macrocephalites*, each of the genera being represented by a large number of species. *Belemnites* are very numerous as individuals, but they belong to only two genera, *Belemnites* (*B. gerardi*) and *Belemnopsis*. The principal lamellibranch genera are *Avicula* (*A. spitiensis*), *Pseudomonotis*, *Aucella*, *Inoceramus* (*I. gracilis*), *Lima*, *Pecten*, *Ostrea*, *Nucula*, *Leda*, *Arca* (*Cucullaea*), *Trigonia*, *Astarte*, *Pleuromya*, *Cosmomya*, *Homomya*, *Pholadomya*. Gastropod species belong to *Pleurotomaria* and *Cerithium*.

The fauna of the Spiti shales indicates an uppermost Jurassic age—Portlandian and Purbeckian. They pass conformably into the overlying Cretaceous sandstone of Neocomian horizon (Giumal sandstone).

Upper Jurassic deposits of *Spiti shales* facies cover large areas of central and southern Tibet, according to the accounts of Sven Hedin, and are overlain by an enormous spread of Cretaceous.

The Jurassic is folded into long isoclinal belts, carrying outlier strips of the Cretaceous and Eocene.

The following table shows in a generalised manner the Jurassic succession of the Central Himalaya:

	Giumal sandstone.	Lower Cretaceous.
Spiti shales (150 m.)	⎧ Lochambel beds. ⎨ Chidamu beds. ⎩ Belemnites beds.	Portlandian.
Kioto limestone (900 m.)	⎧ Sulcacutus beds. ⎪ Great thickness of massive limestones, unfossiliferous (Tagling stage). ⎩ Megalodon limestone (Para stage).	Callovian. Lias (Rhaetic in part).
	Monotis shale.	Noric.

A very similar Jurassic succession is met with in the Niti pass 240 km. east of Spiti, in the Kumaon mountains (Shalshal), and further south-east in the Byans area near Nepal. An "exotic" facies of Jurassic occurs in a volcanic breccia spread over Johar on the Kumaon border of Tibet (see p. 258) containing blocks of limestone. The fossils present in this limestone indicate affinities with the Alpine Jurassic.

Eastern Himalayas—Mt. Everest Region

Vast tracts of the Himalayas east of the Ganges—the Nepal and Assam Himalayas—are yet geologically unknown, but the successive Mt. Everest expeditions have elucidated the geology of the neighbouring tracts north of Darjeeling and Sikkim. Hayden, Heron, Odell, and Wager have geologically surveyed large areas of Sikkim and Southern Tibet.

To the north of the Central Axial Range, culminating in the peak of Mt. Everest, there lies a broad zone of much folded Jurassic strata composed of black shales and argillaceous sandstones, probably the eastern continuation of Spiti shales. The Jurassic shales contain a few obscure ammonites, belemnites and crinoids. In the tightly compressed and inverted folds of Jurassic are outliers of Cretaceous and Eocene (*Kampa system*), the latter containing *Alveolina* limestone; while underlying the Jurassic shales are dark limestones and shales of Triassic affinities (*Tso Lhamo series*). Below these, and pressed against the crystalline rocks at the foot of Mt. Everest on its north slopes, is a thick series of metamorphosed fossiliferous limestone, quartzites and shales (*Lachi series*), 460 m. thick, which has yielded some well preserved *Productus* and *Spirifer*, probably of Permian age. Immediately underlying these, is a dark grey limestone, dipping northwards, about 300 m. thick, followed by a yellow slabby schistose limestone, (300 m.) which together form the actual summit of Mt. Everest—the *Mt. Everest limestone* series (Upper Carboniferous). The thick zone of rocks which come below the Mt. Everest limestone consists of slaty rocks with limestone bands (*Everest Pelitic series*) ? Carboniferous, 1,220 m. thick, underlain by metamorphosed foliated slates and schists referable to the *Daling* series. This zone is extensively plicated and injected by granite of Tertiary age.

The prevalent tectonic strike of the Nepal mountains is east to west, the regional dip being to the north. To the east of the river Arun the strike undergoes a sharp bend to the north-east.

South of the Everest-Kanchenjunga group, to as far as the Darjeeling Duars, the geology is more complicated, the rocks being a complex of crystalline metamorphic schists, gneisses, and Tertiary injection-granites, with here and there patches of the Dalings. The latter series ends abruptly to the south of Kurseong by a mechanical thrust contact in a narrow belt of coal-bearing Gondwana rocks of Damuda age. These in turn are inverted and thrust over the Upper Tertiary Siwalik belt of the foot-hills, the structural relations being here similar to those observed all along the south foot of the Himalayas west of the Ganges (pp. 179–180).

Jurassic of Garhwal Sub-Himalaya. Tal series—In the Lesser Himalayas east of the Ganges, in the zone lying between the outer Tertiary zone and the inner crystalline zone of the main snow-covered range, and distinguished by the exclusive occurrence in it of highly metamorphosed old (Purana) sediments (allied to the Tanawals of Kashmir and the Jaunsars of Simla-Garhwal), there is noted an exceptional development of patches of fossiliferous Jurassic ?) beds underlying the Eocene Nummulitic limestone and overlying the Krol beds. This is one of the rare instances of fossiliferous pre-Tertiary rocks being met with south of the central axis of the Himalayas, and is therefore interesting as indicating a trespass of the shores of the Tethys beyond its usual south border (see p. 193). The fossils are few and indeterminable specifically; they are fragments of belemnites, corals and gastropods. These beds, known as the *Tal series*, overlie older limestones belonging to the Deoban or Krol series.

Lithologically the Tal series consists of about 1,500 m. thick dark greywackes, sandstones and black shales with fossil plant-impressions in the lower part of the series and arenaceous blue limestones containing comminuted shells in the upper. The principal outcrops of the series are in the Tal tributary of the Ganges, covering large areas in western Garhwal. The fossils are not very helpful in indicating the exact age of the rocks, but they approximately indicate Jurassic affinities.

Hazara

Spiti shales of Hazara—The Jurassic system is developed in Hazara, both in the north and south of the province. The two developments, however, are quite distinct from one another, and exhibit different facies of deposits. The northern exposure is similar both in its lithological and palaeontological characters to the Jura of Spiti, and conforms in general to the geosynclinal facies of the Northern Himalaya. The Spiti shales are conspicuous at the top, containing some of the characteristic fauna; they overlie a dark-grey massive limestone, 300 m. thick, containing *Megalodon* and *Diceracardium*. But the Jurassics of south Hazara differ abruptly from the above, both in their composition and their fossils; they show greater affinity to the Jurassic outcrops of the Salt-Range, which are characterised by a coastal, more arenaceous facies of deposits. The Spiti shales of the Northern zone have yielded these fossils ; *Oppelia, Perisphinctes, Belemnites, Inoceramus, Cucullaea, Pecten, Corbula, Gryphaea, Trigonia*.

To the south of Hazara the Jurassics outcrop in a few inliers in the tightly squeezed isoclinal folds of the Kala Chitta range of hills of the Attock district. The horizons present are the top beds of the Kioto (Lias) limestone, Middle Jurassic and the Purbeckian (spiti shales), closely associated with the Giumal series of limestones and sandstones (Albian or Gault). These horizons are squeezed together in a compressed band and are not easy to separate. The fossils present in these rocks are *Velata, Lima, Plicatula, Gryphaea, Chlamys, Mayaites* and *Perisphinctes*.

Kashmir

The Jurassic of Ladakh—In the Spiti area, which in reality is the direct south-east extension of the Zanskar area of the Kashmir basin, it will be remembered that the following sequence of Jurassic deposits is known:

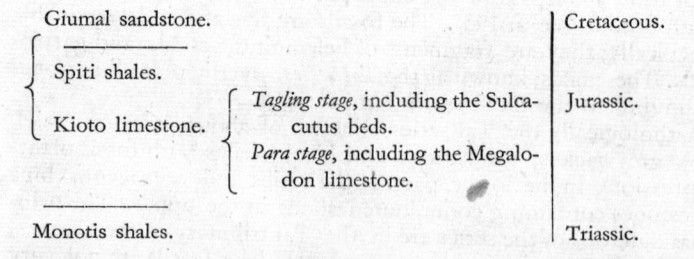

Giumal sandstone.		Cretaceous.
{ Spiti shales.		Jurassic.
{ Kioto limestone.	{ *Tagling stage*, including the Sulcacutus beds.	
	Para stage, including the Megalodon limestone.	
Monotis shales.		Triassic.

A sequence, roughly similar in many respects to this, is traceable in some outcrops in the central and southern parts of Ladakh, resting conformably upon the Upper Triassic limestone. These outcrops form part of a broad basin of marine Mesozoic rocks situated upon the inner flank of the Zanskar range, and are connected with the Jurassic formation of Spiti by lying on the same strike. The lower parts of a number of these outcrops, which include about 150 m. of dolomitic limestone, recall the Megalodon limestone, both in their constitution and in their fossil contents. At another locality this group is succeeded by light-blue limestone which from its contained fossils is referable to the Tagling stage.[1] The Tagling stage passes conformably up at several localities into the Spiti shales, that eminently characteristic Jurassic horizon of Himalayan stratigraphy. It is readily recognised by its peculiar lithology, its black, thin-bedded, carbonaceous and micaceous shales containing a few fossil-bearing concretions. The following fossils have been hitherto obtained from the Jurassic of Ladakh: *Megalodon*, *Avicula*, *Pecten*, *Cerithium*, *Nerinea*, *Phasianella*, *Pleurotomaria*; some *Ammonites*, including *Macrocephalus*; numerous fragments of *Belemnites*; and a few species of *Rhynchonella* and *Terebratula*.

With the exceptions described below, the Jurassic system has not been recognised in the Kashmir province proper. It is probable that the more detailed survey of the province will bring to light further outcrops of this system from remoter districts.

Jurassic of Banihal—An outcrop of the Jurassic system is found on the north side of the Banihal pass of the Pir Panjal in a tightly compressed syncline in the Upper Trias. A series of limestones, shales and sandstones therein, resting on the topmost beds of the Upper Trias, has yielded a few Jurassic cephalopods and lamellibranchs.

It is probable that similar outliers of the Jurassic exist in association with the extensive Triassic formation of the northern flank of the Pir Panjal between Banihal and Gulmarg, under cover of the Pleistocene glacial and Karewa deposits, which have sheeted the long gentle northern slopes of this mountain range.

As mentioned on pp. 234–235 the Upper Trias of Baltal appears to pass upwards into dark carbonaceous and pyritous shales, calcareous shales and limestones, containing some badly crushed and distorted ammonites and belemnites. These are probably of basal Jurassic, Lower Lias, age. These rocks are well displayed in the synclinal folds of the Upper Trias in the magnificent series of bare cliffs on the north side of the Sind above Sonamarg and in the

---

[1] The accuracy of this correlation, it must be realised, has never been sufficiently ascertained. Revision of the Kashmir sequence is needed. It is only when this work is completed that a full account of the Jurassic of Kashmir can be given in any detail such as we have given above of the Palaeozoic formations. The same is to be said about the Cretaceous.

Amarnath valley. They are well exposed in the cuttings of the Zoji La road above Baltal.

Baluchistan

Jurassic of Baluchistan—Marine Jurassic rocks, of the geosynclinal facies, and corresponding homotaxially to the Lias and Oolite of Europe, are developed on a vast scale in Baluchistan, and play a prominent part in its geology. The Liassic beds are composed of massive blue or black, crinoidal Oolite or flaggy limestone, interbedded with richly fossiliferous shales, attaining a thickness of more than 3,000 feet, in which the principal stages of the European Lias can be recognised by means of the cephalopods and other molluscs entombed in them. The Liassic limestones are overlain by an equally thick series of massive grey, thick-bedded limestone of Oolitic age, which is seen in the mountains near Quetta and the ranges running to the south. With the Callovian stage, however, there occurs a gap in the sequence which extends to the Neocomian. The top beds of the last-described limestone contain numerous ammonites, among which the genus *Macrocephalites*, represented by gigantic specimens of the species *M. polyphemus*, attains very large dimensions.

[The rock systems of Baluchistan are capable of classification into two broad divisions, comprising two entirely different types of deposits. One of these, the Eastern, is mainly characterised by a calcareous constitution and comprises a varied geological sequence, ranging in age from the Permo-Carboniferous upwards. This facies is prominently displayed in the mountain ranges of E. Baluchistan, constituting the Sind frontier. The other facies is almost entirely argillaceous or arenaceous, comprising a great thickness of shallow-water sandstones and shales, chiefly of Oligocene-Miocene age. The latter type prevails in the broad upland regions of W. Baluchistan, stretching from the Mekran coast northwards up to the southward confines of the Helmand desert. These differences of geological structure and composition in the two divisions of Baluchistan have determined in a great measure its principal physical features.[1]]

All the Mesozoic systems are well represented in East Baluchistan, and are very prominently displayed in the high ground extending meridionally from the Takht-i-Sulaiman mountain to the Mekran coast. In the broad arm or gulf of the Tethys which, as we have already stated, occupied Baluchistan almost since the commencement of its existence, a series of deposits was formed, representative of the ages that followed this occupation. Hence the main Mesozoic formations of the Northern Himalayas find

---

[1] E. W. Vredenburg. *Rec. G.S.I.* vol. xxxviii. pt. 3, 1910.

THE JURASSIC SYSTEM

their parallels in Baluchistan along a tract of country folded in a series of parallel anticlines and synclines of the Jura type, stretching in a north to south direction.[1]

Salt-Range

Jurassic of the Salt-Range—The middle and upper divisions of the Jurassic are represented in the Salt-Range, composed of sandstones of variegated colours, yellow limestones and gypseous and pyritous shales. They contain bands of yellow oolitic limestone resembling the "golden oolite" of Kutch. To the east of about longitude 72° the Jurassic is missing owing to the pre-Tertiary denudation, which was progressively more pronounced from west to east in the Salt-Range. In the north-western part of the Salt-Range, near Nammal and Khairabad, the Jurassic beds thicken to at least 350 m. and form prominent strike ridges. Still further west in the trans-Indus Salt-Range the gap in the sequence below the Tertiary becomes much less marked and the Jurassic system attains a thickness of 460 m. near Kalabagh, and nearly 600 m. in the Shekh Budin hills and the Surghar range. The Surghar range, north-west of Isa Khel, is composed almost wholly of Jurassic and Eocene (Nummulitic) rocks.

The Jurassic strata of the trans-Indus ranges may be summarised in the following table:

	White sandstones with dark shales, 18 m. Neocomian fossils.	Cretaceous.
Jurassic strata (150–460 m.).	Upper Jurassic—light-coloured, thin-bedded highly fossiliferous limestones, and blackish arenaceous shales. Fossils: *Pecten, Lima, Ostrea, Homomya, Pholadomya*, with several *ammonites, belemnites* and gastropods.	Middle and Upper Jurassic.
	Middle Jurassic—white and variously tinted soft sandstones and clays with lignite and coal-partings; pyritous (alum) shales with subordinate bands of limestone and haematite. Fossils: obscure plant remains—*Podozamites, Ptilophyllum, Belemnites, Pleurotomaria, Natica, Mytilus, Ancillaria, Pecten, Myacites, Nerinea, Cerithium, Rhynchonella*, etc.	
	Ceratite beds, 113 m.	Triassic.

---

[1] See Vredenburg's Map of Baluchistan, *Rec. G.S.I.* vol. xxxviii. pt. 3, 1910, and *Pal. Ind.* New Series, vol. iii. mem. 1, 1909—Introduction.

A section of the above type is seen near Kalabagh on the Indus; a fuller section is visible in the Shekh Budin hills and in the Surghar range.

A few coal or lignite seams occur irregularly distributed in the lower part, and are worked near Kalabagh, and provide on an average about 1,000 tonnes of coal per year; some haematite layers also occur. Fossil plants of Jabalpur affinities, enclosed in these beds, point to the vicinity of land. A few beds of a peculiar oolitic limestone, known as the "golden oolite", are found among these rocks. The rock is a coarse-grained limestone, the grains of which are coated with a thin ferruginous layer. Fossil organisms preserved within these rocks consist of an assemblage in which there are *Ostrea*, *Exogyra*, species of *Terebratula*, numerous gastropods, many *ammonites* and *belemnites*. The spines of numerous large species of echinoids, like *Cidaris*, and fragments of the tests of *irregular* echinoids, are frequent in the limestones.

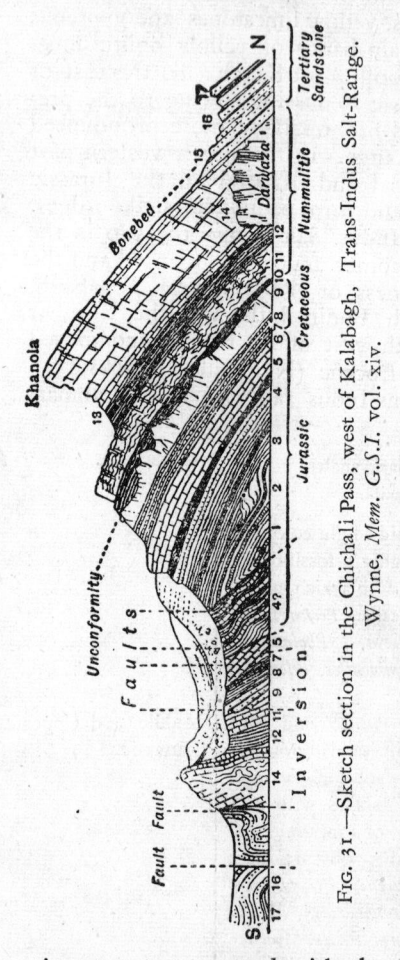

A rapidly varying lithological composition of a series of strata, such as that of the Jurassic of the Salt-Range, is suggestive of many minor changes during the course of sedimentation in that area; such, for instance, as changes in the depth of the sea, or of the height of the lands which contributed the sediments, or alterations in the courses of rivers and of the currents in the sea. The Salt-Range Jurassic sea was connected with the coastal sea of Kutch through Jaisalmer and West Rajasthan.

[It is in the west and trans-Indus part of the Salt-Range that the Mesozoic group is developed in some degree of completeness.

FIG. 31.—Sketch section in the Chichali Pass, west of Kalabagh, Trans-Indus Salt-Range. Wynne, *Mem. G.S.I.* vol. xiv.

In the eastern, cis-Indus part the Mesozoic group is, on the whole, rather incompletely developed.

The *structure* of this part of the Salt-Range is one of severe disturbance recalling the tectonic features of some parts of the Middle Himalayas. The strata by repeated folding and faulting have acquired a confused disposition.]

Marine Transgressions during the Jurassic period

After the emergence of the Peninsula at the end of the Vindhyan system of deposits, this part of India has generally remained a land area, a continental tableland exposed to the denuding agencies. No extensive marine deposits of any subsequent age have been formed on the surface of the Peninsula since that early date.

Nature of marine transgressions—In the Jurassic period, however, several parts of the Peninsula, *viz*. the coasts and the low-lying flat regions of the interior, like Saurashtra and a large part of Rajasthan, extending northwards to the Salt-Range, were temporarily covered by the seas which invaded the lands. These temporary encroachments of the sea over what was previously dry land are not uncommon in the records of several geological periods, and were caused by the sudden decrease in the capacity of the ocean basin by some deformation of the crust, such as the sinking of a large land-mass, or the elevation of a submarine tract. Such invasions of the sea on land, known as "marine transgressions", are of comparatively short duration and invade only low-level areas, converting them for the time into epicontinental seas. These temporary epicontinental seas should be distinguished from the geosynclinal or mediterranean seas. The series of deposits which result from these transgressions are clays, sands or limestones of a littoral type, and constitute a well-marked group of deposits, sometimes designated by a special name—the Coastal system. One example of the Coastal system we have already seen in connection with the Upper Gondwana deposits of the east coast. The remaining instances of marine transgressional deposits in the geology of India are the Upper Jurassic of Kutch and Rajasthan; the Upper Cretaceous of Tiruchinapalli, the Narmada valley and the Assam hills; the Eocene and Oligocene of Gujarat and Saurashtra; and the somewhat newer deposits of a number of places on the Coromandel coast.

Deposits which have originated in this manner possess a well-defined set of characters, by which they are distinguishable from the other normal marine shallow-water deposits. (1) Their thickness is moderate compared to the thickness of the ordinary marine deposits, or of the enormous thickness of the geosynclinal formations; (2) they, as a rule, cover a narrow strip of the coast only, unless lowlands extend farther inland, admitting the sea to the interior; (3) the dip of the strata is irregular and sometimes deceptive, owing to current-bedding and deposition on shelving banks. Generally the dip is seaward, away from the mainland; consequently the oldest beds are farthest inland while the newest are near the sea. In some cases, however, a great depth of deposition is possible during marine transgressions, as when tracts of the coast, or the continental shelf, undergo sinking, of the nature of trough-faulting, concurrently with deposition. Such was the case, for instance, with the basins in which the Jurassics of Kutch were laid down, in which the sinking of the basins admitted of a continuous deposition of thousands of feet of coastal detritus. Such block-faulting is quite in keeping with the horst-like nature of the Indian Peninsula, and belongs to the same system of earth movements as that which characterised the Gondwana period.

The marine Jurassic deposits of the coasts show an interesting feature: the marine transgression of these areas commences with the Callovian stage of the Oolite, the stage which marks a definite withdrawal of the sea from Baluchistan and a temporary pause in deposition in the Central Himalayas.

Kutch

The Jurassics of Kutch—Jurassic rocks occupy a large area of the Kutch State. It is the important formation of Kutch both in respect of the lateral extent it covers and in thickness. With the exception of a few small patches of ancient crystalline rocks, no older system of deposits is met with in this area. It is quite probable, however, that large parts of the country which at the present day are long, dreary wastes of black saline mud and silt (which form the Rann of Kutch) are underlain by a substratum of the Peninsualr gneisses together with the *Puranas*. A broad band of Jurassic rocks extends in an eastwest direction along the whole length of Kutch, and they also appear farther north in the islands in the Rann of Kutch. Structurally the Jurassic is thrown into three wide anticlinal folds, separated by synclinal depressions, with a longitudinal strike-fault at the foot of the southernmost anticline. The main outcrop attenuates in the middle, owing to the overlap of the younger deposits. The aggregate thickness of the formation is over 1,800 m., a depth quite incompatible with deposits of this nature, but for the explanation given above.

The large patch of Jurassic rocks in East Saurashtra around Dhrangadhra belongs to the same formation, and is an outlier of the latter on the eastern continuation of the same strike.

The Mesozoic of Kutch includes four series—*Patcham*, *Chari*, *Katrol* and *Umia*, in ascending order, ranging in age from Lower

THE JURASSIC SYSTEM 249

Oolite to Wealden. The base of the system is not exposed and the top is unconformably covered either by the basalts of the Deccan Trap formation or by Nummulitic beds (Eocene).

The following table, adapted from Dr. Oldham, gives an idea of the stratigraphic succession:

Umia (900 m).	Marine sandstones with *Crioceras*, etc., sandstone and shale with cycads, conifers, and ferns (Gondwana facies). Marine sandstone and conglomerate with *Perisphinctes* and *Trigonia*.	Wealden and uppermost Jurassic.
Katrol (300 m).	Sandstone and shale with *Perisphinctes* and *Oppelia*. Ferruginous red and yellow sandstone (*Kantkote* sandstones) with *Stephanoceras, Aspidoceras*.	Upper Oolite.
Chari (335 m).	Dhosa oolite, oolite limestone; *Peltoceras, Aspidoceras, Perisphinctes*. White limestones; *Peltoceras, Oppelia*. Shales with ferruginous nodules; *Perisphinctes, Harpoceras*. Shales with "golden oolite"; *Macrocephalites, Oppelia*.	Middle Oolite.
Patcham (300 m).	Grey limestones and marls with *Oppelia*, corals, brachiopods, etc. Yellow sandstones and limestones with *Trigonia, Corbula, Cucullaea*, etc. Base not seen.	Lower Oolite.

Patcham series—The lowest member, the Patcham series, occurs in the Patcham island of the Rann as well as in the main outcrop in Kutch proper. The lower beds are exposed towards the north, and are visible in many of the islands. The strata show a low dip to the south, *i.e.* seawards. The constituent rocks are yellow-coloured sandstones, and limestones, overlain by limestones and marls. The fossils are principally lamellibranchs and ammonites, but not so numerous as in the two upper groups, the leading genera being *Trigonia, Lima, Corbula, Gervillia, Exogyra*, and *Oppelia, Perisphinctes, Macrocephalites* (*M. triangularis*), *Sivajiceras, Stephanoceras*; some species of *Nautilus*.

Chari Series—The Chari series takes its name from a village near Bhuj, from where an abundant fauna corresponding with that of the Callovian stage of the European Jurassic has been obtained. It is composed of shales and limestones, with a peculiar red or brown, ferruginous, oolite limestone, known as the *Dhosa oolite*, at the top. There also occur, at the base, a few bands of what is known as the golden oolite, a limestone composed of rounded calcareous grains coated with iron and set in a matrix. The chief element of the fauna is cephalopods, some hundred species of ammonites being recognisable in them. The principal genera are *Perisphinctes, Phylloceras, Oppelia, Macrocephalites* (many species), *Harpoceras, Peltoceras, Aspidoceras, Reineckia, Mayaites, Choffatia, Indosphinctes, Aptychus, Grossouvria, Stephanoceras.* In addition there are three or four species of *Belemnites*, several of *Nautilus*, and a large number of lamellibranchs. The Chari group is palaeontologically the most important group of the Jurassic of Kutch, because it has furnished the greatest number of fossil species identical with known European types; it is divided into the following zones; *macrocephalus* beds, *rehmanni* beds, *anceps* beds, and *athleta* beds, underlying the Dhosa oolite; the Dhosa oolite, coming at the top of the Chari series, is the richest in ammonites, being divided into three well-defined zones.

The Katrol series—The Chari series is overlain by the Katrol group of shales and sandstones. The shales are the preponderant rocks of this series, forming more than half its thickness. The sandstones are more prevalent towards the top. The shales are variously tinted by iron oxides, which at places prevail to such an extent as to build small concretions of haematite or limonite. The Katrol series forms two long wide bands in the main outcrop in Kutch; the exposure where broadest is ten miles wide. Besides forms which are common to the whole system the Katrol series has, as its special fossils, *Harpoceras, Phylloceras, Lytoceras,* and *Aptychus.* The other Katrol cephalopods are *Hibolites, Aspidoceras, Waagenia, Streblites, Pachysphinctes, Katroliceras* (many species). The group is divided into lower, middle and upper, capped by the *Zamia shales,* containing fossil cycads and other plants. A few plants are preserved in the sandstone and shale beds belonging to the Zamia stage, but in such an imperfect state of fossilisation that they cannot be identified and named.

The Umia series—Over the Katrol group comes the uppermost division of the Kutch Mesozoic, the Umia series, comprising a thickness of 900 m. of soft and variously coloured sandstones and sandy shales. The lower part of the group is conglomeratic, followed by a series of marine sandstone strata in which fossils are rare except two species of *Trigonia, T. ventricosa* and *T. smeei,* which are, however, very typical. Over this there comes an intervening series of strata of sandstones and shales, which, both in their lithological as well as palaeontological relations are akin to the Upper Gondwana rocks of the more easterly parts of the Peninsula. The interstratification of these beds with the marine Jurassic should be ascribed to the same circumstances as those which gave rise to the marine intercalations in the Upper Gondwana of the east coast. After this slight interruption the marine conditions once more established themselves, since the higher beds of the Umia series contain many remains of ammonites and belemnites. The Umia group has wide lateral extent in Kutch, its outcrop being much the broadest of all the series. Its breadth, nevertheless, is considerably reduced by the overlapping of a large part of its surface by the Deccan Traps and still younger beds. The fossils yielded by the Umia series are species of *Williamsonia, Ptilophyllum, Elatocladus, Araucarites, Brachyphyllum,* Cycads and Conifers, which have been enumerated in the chapter relating to the Upper Gondwana. The marine fossils include the genera *Crioceras, Acanthoceras, Haploceras, Umiaites, Virgatosphinctes, Aulacosphinctes, Belemnites,* with *Trigonia smeei* and *Trigonia ventricosa.*

The Kutch Jurassic rocks are very rich in fossil cephalopoda. Out of the material lately collected, L. F. Spath has distinguished 114 genera, fifty-one of which are new to India, and nearly 600 species, a large percentage of which are of local or provincial type, unknown elsewhere. No resemblances are detected with the Mediterranean or with the North-West European province, nor is there seen any affinity with the Boreal province, but there exists a close faunal relationship between the Jurassic of Kutch and of Madagascar.

The rocks above described are traversed by an extensive system of trap dykes and sills and other irregular intrusive masses of large dimensions. In the north they become very complex, surrounding and ramifying through the sedimentary beds in an intricate network. The intrusions form part of the Deccan Trap series and are its hypogene roots and branches.

Saurashtra (Kathiawar)—The Jurassic outcrop of North-East Saurashtra, already referred to, is composed of soft white or ferruginous sandstones and pebble-beds or conglomerates. In this respect, as well as in its containing a few plant fossils, it is regarded as of Umia horizon. The sandstone is a light-coloured freestone, largely quarried at Dhrangadhra for supplying various parts of Gujarat with a much-needed building material.

Jurassic of Tamilnadu Coast—Strata containing marine Jurassic fossils are interbedded with Upper Gondwana beds in some outcrops between Guntur and Rajahmundry on the Tamilnadu coast (p. 189). The fossils, mostly ammonites, are badly preserved, but L. F. Spath regards them as showing Lower Cretaceous affinities.

Rajasthan

Jurassic of Rajasthan—The inroads of the Jurassic sea penetrated much farther than Kutch in a north-east direction, and overspread a great extent of what is now Rajasthan. Large areas of Rajasthan received the deposits of this sea, only a few patches of which are exposed to-day from underneath the sands of the Thar desert. It is quite probable that a large extent of fossiliferous rocks, connecting these isolated inliers, is buried under the desert sands.

Fairly large outcrops of Jurassic rocks occur in Jaisalmer and Bikaner. They have received much attention on account of their fossiliferous nature. A number of divisions have been recognised in them, of which the lowest is known as the *Balmir sandstone*; it is composed of coarse sediments—grits, sandstones and conglomerates, with a few badly preserved remains of dicotyledonous wood and leaves. The next group is distinguished as the *Jaisalmer limestone*, composed of highly fossiliferous limestones with dark-coloured sandstones. The limestones have yielded a number of fossils, among which the more typical are *Pholadomya, Corbula, Trigonia costata, Nucula, Pecten, Nautilus* and some *Ammonites*. This stage is regarded as homotaxial in position with the Chari series of Kutch.

The Jaisalmer limestone is overlain by a series of rocks which are referred to three distinct stages in succession: *Abur beds, Parihar sandstones* and *Badasar beds*. The rocks are red ferruginous sandstones, succeeded by a soft felspathic sandstone, which in turn is succeeded by a group of shales and limestones, some of which are fossiliferous.

Dr. La Touche, of the Geological Survey of India, has assigned a younger (Cretaceous) age to the Balmir beds, mainly from the evidence of the dicotyledonous plant fossils which they contain.

Jurassic rocks are also exposed in the southern part of Rajasthan, where a series of strata bearing resemblance to the above directly underlie Nummulitic shale beds of Eocene age.

Burma

Namyau beds—Jurassic strata are met with in the Northern Shan States, and are referred to as the *Namyau beds*, also sometimes designated the *Hsipau series*. (See Fig. 15, p.157.) The rocks are red or purple sandstones and shales, unfossiliferous in the main, but the few limestone bands have yielded a rich brachiopod fauna with some lamellibranchs of Upper Jurassic age. There are no ammonites present. This group of strata is underlain by shales and concretionary limestones, which have already been referred to as the equivalents of the Rhaetic or Napeng series of Burma.

Rocks belonging to the Lias horizon occur at Loi-an, near Kalaw, in the Southern Shan States, enclosing a few coal-measures, the plant-bearing beds of which have yielded, among fossil conifers and cycads, the characteristic species *Ginkgoites digitata*. The Loi-an coal-measures support a small coal-field.

REFERENCES

F. Stoliczka, *Mem. G.S.I.* vol. v. pt. 1, 1865.
T. H. D. La Touche, Geology of Western Rajputana, *Mem. G.S.I.* vol. xxxv. pt. 1, 1902.
A. B. Wynne, Geology of the Salt-Range, *Mem. G.S.I.* vol. xiv., 1878.
C. S. Middlemiss, Geology of Hazara, *Mem. G.S.I.* vol. xxvi., 1896.
V. Uhlig. *et al.*, Fauna of the Spiti Shales, *Pal. Indica*, Sers. XV. vol. iv. pts. 1 and 2, 1903-1914.
J. W. Gregory, *et al.*, Jurassic Fauna of Cutch, *Pal. Indica*, Sers. IX. vols. i to iii., 1873-1903.
A. B. Wynne, Geology of Cutch, *Mem. G.S.I.* vol. ix. pt. 1., 1872.
A. M. Heron, Geology of Mt. Everest, *Rec. G.S.I.* vol. liv. pt. 2, 1922.
L. F. Spath, Revision of the Jurassic Cephalopod Fauna of Cutch, *Pal. Ind.*, N.S. vol. ix., 1927-1933.
L. R. Wager in *Everest* 1933 (Hodder & Stoughton), 1934.
Raj Nath, Stratigraphy of Cutch, *Q.J.G.M.M.S.* vol. iv., 1933.
Heim and Gansser, Geological Structure of the Central Himalaya, Swiss Natural Science Society, *Mem.* vol. lxxiii. no. 1, Zurich, 1939.
A. von Krafft, Exotic Blocks of Malla Johar, *Mem. G.S.I.* vol. xxxii. pt. 3, 1903.

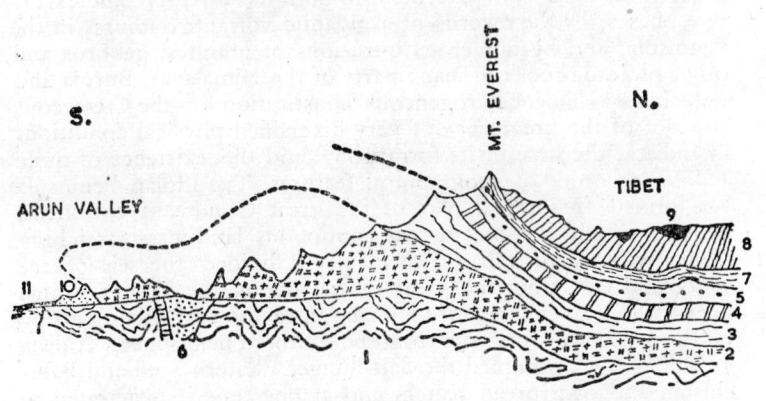

FIG. 31a. Sketch section across Mt. Everest region (diagrammatic).

11. Bengal Plain.
10. Tertiaries.
9. Cretaceous.
8. Jurassic.
7. Tso Lhamo Series.
6. Gondwana.
5. Lachi Series.
4. Everest Lime Stone.
3. Everest Pelitic Series.
2. Daling Series.
1. Archaean Peninsular Basement.

CHAPTER XIV

The Cretaceous System

Varied facies of the Cretaceous. The geography of India in the Cretaceous period—No other geological system shows a more widely divergent facies of deposits in the different areas of India than the Cretaceous, and there are few which cover so extensive an area of the country as the present system does in its varied forms. The marine geosynclinal type prevails in the Northern Himalayas and in Baluchistan; parts of the Coromandel coast bear the records of a great marine transgression during the Cenomanian Age, while right in the heart of the Peninsula there exists a chain of outcrops of marine Cretaceous strata along the valley of the Narmada. An estuarine or fluviatile facies is exhibited in a series of wide distribution in Madhya Pradesh and the Deccan. An igneous facies is represented, in both its intrusive and extrusive phases, by the records of a gigantic volcanic outburst in the Peninsula, and by numerous intrusions of granites, gabbros and other plutonic rocks in many parts of the Himalayas, Burma and Baluchistan. This heterogeneous constitution of the Cretaceous is proof of the prevalence of very diversified physical conditions in India at the time of its formation, and the existence of quite a different order of geographical features. The Indian Peninsula yet formed an integral part of the great Gondwana continent, which was still a more or less continuous land-mass stretching from Africa to Australia. This mainland divided the seas of the south and east from the great central ocean, the Tethys, which kept its hold over the entire Himalayan region and Tibet, cutting off the northern continents from the southern hemisphere. A deep gulf of this sea occupied the Salt-Range, Western Sind and Baluchistan and overspread Kutch, and at one time it penetrated to the very centre of the Peninsula by a narrow inlet through the present valley of the Narmada. The southern sea at the same time encroached on the Coromandel coast, and extended much further north, overspreading Assam and probably flooding a part of west Bengal. It is a noteworthy fact that no communication existed between these two seas—of Assam and the Narmada valley—although separated by only a small distance of intervening land.

THE CRETACEOUS SYSTEM

While such was the geography of the rest of India, towards the end of the Cretaceous the north-west part of the Peninsula was converted into a great centre of vulcanicity of a type which has no parallel among the volcanic phenomena of the modern world. Hundreds of thousands of square miles of the country between Southern Rajasthan and Dharwar, and in breadth almost from coast to coast, were inundated by basic lavas which covered, under thousands of feet of basalts, all the previous topography of the country, and converted it into an immense volcanic plateau.

We shall consider the Cretaceous system of India in the following order:

(i) *Cretaceous of the Extra-Peninsula*:
   Himalayan Regions.
   Sind and Baluchistan.
   Salt-Range.
   Assam.
   Burma.

(ii) *Cretaceous of the Peninsula*:
   Coromandel Coast.
   Narmada Valley.
   Lameta series: Infra-Trappean beds.
   Western India.

(iii) *Deccan Trap*.

CRETACEOUS OF THE EXTRA-PENINSULA

Northern Himalayas

Spiti: Giumal sandstone—That prominent Upper Jurassic formation, the *Spiti shales*, of the northern ranges of the Himalayas constituting the Tibetan zone of Himalayan stratigraphy, is overlain at a number of places by yellow-coloured siliceous sandstones and quartzites known as the Giumal sandstone. (See Fig. 30.) In the Spiti area the *Giumal series* has a thickness of about 100 m. The deep and clear waters of the Jurassic sea, in which the great thickness of the Kioto limestone was formed, had shallowed perceptibly during the deposition of the Spiti shales. The shallowing became more marked with the deposits of the next group containing fossils of Neocomian age. These changes in the depth of the sea are discernible as much by a change in the characters of the sediments as by changes in the faunas that are preserved in them. The deeper-water organisms have disappeared from the Giumal faunas, except for a few colonies where deep local basins persisted. The fossil organisms entombed in the Giumal sandstone include: (Lamellibranchs) *Cardium, Ostrea, Gryphaea, Pecten, Tellina, Pseu-*

domonotis, Arca, Opis, Corbis, Cucullaea, Tapes; (Ammonites) *Holcostephanus, Acanthodiscus, Perisphinctes, Hoplites*.

Chikkim series: Flysch—The Giumal series is succeeded in the area we are considering at present by a group of about 75 m. of white limestones and shales. Fossils are found only in the limestones, which underlie the shales. This group is known as the *Chikkim series*, from a hill of that name in Spiti. The Chikkim series is also one of wide horizontal prevalence, like the Spiti shales, outcrops of it being found in Kashmir, Hazara, Kumaon, Tibet, Afghanistan and Iran. The fossils that are preserved in the limestone are fragments of the guards of *Belemnites*, shells of the peculiar lamellibranch genus *Hippurites* (belonging to the family *Rudistae*) and a number of foraminifers, *e.g. Nodosaria, Cristellaria, Textularia, Dentallina*, etc., congeners of the foraminifers whose tiny shells have contributed to the chalk of Europe. In the areas adjacent to Spiti the Chikkim series is overlain by a younger series of unfossiliferous sandstones and sandy shales of the type to which the name *Flysch*[1] is applied (Fig. 30.) The Cretaceous flysch contains dense radiolarian cherts, indicating a deepening of the sea after the deposition of the Giumals, followed by ophiolite intrusions and its rapid filling up by the coarser littoral detritus. With the flysch deposits the long and uninterrupted geosynclinal conditions approached their end, and the Chikkim series may be regarded as the last legible chapter in the long history of the Himalayan marine period. The flysch deposits that followed mark the gradual emergence of land, and the receding of the shore-line further and further north. The Himalayan continental period had already begun and the first phase of its uplift into the loftiest mountain-chain of the world commenced, or was about to commence.

In the general retreat of the Tethys from the Himalayan province at this period, scattered basins were left at a few localities, *e.g.* in Central Tibet, Hundes and Ladakh. In these areas the sea retained its hold for a time, and laid down its characteristic deposits till about the middle of the Eocene, when further crustal deformations drove back the last traces of the sea from this part of the earth.

The Kampa system of Tibet—The geological composition of a large area of Central Tibet, lying between Ladakh and Shigatse, is now known from the rock and fossil collections brought by Sven

---

[1] The typical Flysch is a Tertiary formation of Switzerland, and is composed mainly of soft sandstones, marls, and sandy shales covering a wide extent of the country. Its age is Eocene or Oligocene. Fossils are rare or absent altogether. The term is, however, also applicable to similar deposits in other countries and of other ages than Eocene or Oligocene.

Hedin. An extensive spread of Cenomanian limestones covers thousands of square miles of the surface, underlain by Giumal sandstones (of Neocomian and Gault age) and the shales and sandstones of Spiti shales facies. The important Cretaceous fossils occurring in these limestones are *Praeradiolites*, several species of *Orbitolina* and *Choffatella*. This vast cover of Cretaceous rocks supports in the south a wide extent of Eocene rocks (*Kampa system*) and post-Eocene sediments, extending from Gartok, to the north-west of the Manasarowar lake, to the vicinity of Gyantse. In the north of this area the Cretaceous cover supports patches of newer Tertiary and Pleistocene sediments containing mammalian bones and other remains.

Chitral—In the Chitral area the Middle Cretaceous is represented by *Hippurites* limestone and *Orbitolites* limestone in narrow faulted bands which run along the general strike of the country (N. E.-S.W.). These pass upwards into the *Reshun conglomerate* of Upper Cretaceous or Lower Tertiary age[1].

Further evidence of the distinctly intrusive nature of extensive belts of granitoid gneiss of the Central Himalayan Gneiss facies has been recorded by Tipper in Chitral. Numerous bosses of granite are found to have invaded Mesozoic strata in some cases inferred, on fossil evidence, to be of Jurassic age.

Cretaceous of Malla Johar, N.W. Kumaon

150 m. of Giumal sandstone of the usual composition are overlain by *Belemnite* shale and these by the flysch type of sediments. The more interesting Cretaceous development of this area, however, is a mass of basic igneous rocks, overspreading the Flysch, which contain, embedded in them, blocks of sedimentary rocks of various ages ranging from Permian to Cretaceous. (See below, Exotic Blocks of Johar, p. 258.)

Plutonic and volcanic action during Cretaceous—The history of the latter part of the Cretaceous age, and the ages that followed it immediately, is full of the proofs of widespread igneous action on a large scale, both in its plutonic as well as in its volcanic phase. An immense quantity of magma was intruded in the pre-existing strata, as well as ejected at the surface over wide areas in Baluchistan, the North-West Himalayas, Ladakh, Kumaon and Burma. Masses of granites, gabbros and peridotites cut through the older rocks in bosses and veins, laccolites and sills, while the products of volcanic action (lava-flows and ash-beds) are found interstratified in the form of rhyolitic, andesitic and basaltic lava sheets, breccias and tuffs. The ultrabasic, peridotitic intrusions of these and slightly subsequent ages are at the present day found

---
[1]G. H. Tipper, *Rec. G.S.I.* vol. lv. p. 38, and vol. lvi. pp. 44-48, 1924.

altered into serpentine-masses bearing some useful accessory products that have been separated from them by the process of magmatic segregation. Of these the most important are the chromite masses in Baluchistan, the semi-precious mineral jadeite in Burma, and serpentine in Ladakh.

A great proportion of the granite which forms such a prominent part of the crystalline core of the Himalayas, forming the broad central belt between the outer Tertiary zone and the inner Tibetan zone, is also tentatively referred, in a great measure, to the igneous activity of this age. Three kinds of granites, as stated before, are recognised in the Himalayan central ranges, *viz*. biotite-granite, which is the most widely prevalent, hornblende-granite and tourmaline-granite, but it is quite probable that all the three have been derived by the differentiation of one originally homogeneous magma.

As will be alluded to later, this outburst of igneous forces is connected with the great physico-geographical revolutions of the early Tertiary period, revolutions which culminated in obliterating the Tethys from the Indian region and the severing of the Indian Peninsula from the Indo-African Gondwana continent.

Alpine Mesozoics in N. Kumaon

"**Exotic**" **Blocks of Johar**—According to von Krafft, the records of an extraordinary volcanic phenomenon are witnessed in connection with the Cretaceous rocks of the Kumaon Himalayas. Lying over the Spiti shales and Cretaceous Flysch of Johar, on the Tibetan frontier of Kumaon, are a number of detached blocks of sedimentary rocks of all sizes from ordinary boulders to blocks of the dimensions of an entire hill-mass. These lie in a confused pell-mell manner, in all sorts of stratigraphic discordance, on the underlying beds. From the evidence of their contained fossils these blocks are found to belong to almost every age from early Permian to the newest Cretaceous. But the fossils reveal another, more curious, fact that these rock-masses do not belong to the Spiti or Himalayan facies of deposits, but are of an entirely foreign facies of Permian, Triassic and Jurassic (more allied to the Alpine faunistic province) prevailing in a distant northern locality in Upper Tibet. Such a group of "exotic" or foreign blocks of rocks, out of all harmony with their present environments, were at first believed to be the remnants of denuded recumbent folds, or were ascribed to faulting, and were considered as analogous to the "Klippen" of the Alps. But from the circumstance of the close association, and sometimes even intermixing, of these blocks with great masses of early Tertiary volcanic products like basalts and andesites, an altogether novel method of origin has been suggested, *viz*. that these blocks were

torn by a gigantic volcanic explosion in North Tibet (such as is
connected with the production of volcanic agglomerates and
breccias), and subsequently transported in the lava inundation
to the positions in which they are now found. The mode in which
these blocks are scattered in confused disorder is not in disagree-
ment with the above view of their origin. These foreign, transported blocks on the Kumaon frontier are known in Himalayan
geology as the exotic blocks of Johar. Similar phenomena are
recorded in some other parts of the Himalayas as well.

In view of the *nappe* structures clearly observed and mapped in
the Kashmir, Simla and Garhwal Himalayas, it appears highly
probable that the Malla Johar blocks, some of which are found
building the tops of prominent mountains, may after all prove to
be tectonic phenomena and have to be regarded as the "klippen"
they were once conjectured to be, the severed frontal ends of
nappes or horizontally lying folds, whose main bodies have been
denuded away. According to Arnold Heim, these exotic blocks
of Johar not only occur as isolated masses but also form sheet-like
expanses covering several square miles of mountainous country.[1]

Kashmir

If the account of the Jurassic system of Kashmir is meagre, that
of the Cretaceous rocks is still more so. It is only at a few localities that rocks belonging to this system have been discovered; all
of these lie in distant unfrequented parts of Kashmir, either on
the Great Himalayan range between the Burzil and Deosai or in
the Zanskar range in the Rupshu province. The great developments of the Cretaceous rocks of Spiti and its surrounding places,
the Giumal sandstone, the Chikkim limestone and the enormous
flysch-like series, have not yet been recorded in Kashmir, though
from the fact of their occurrence in the western province of Hazara,
it is probable that these series may have their parallels in the
Skardo and Ladakh provinces of Kashmir in a few attenuated
outcrops at least.

The Chikkim series of Rupshu-Zanskar—Two or three small
patches of Cretaceous rocks occur in Rupshu which correspond
to the Chikkim series in their geological relations. They are composed of a white limestone, as in the type area, forming some of
the highest peaks of the range in Ladakh. No fossils, however,
have been obtained from them hitherto.

Cretaceous Volcanic Series of Astor, Burzil and Dras
A highly interesting group of volcanic rocks—laminated ash-

---

[1]Arnold Heim, *Himalayan Journal*, ix. p. 41, 1937; A. Heim & A. Gansser,
Central Himalaya, *Swiss Academy of Science*, Zurich, 1939 ; see also E. B. Bailey,
Nature, vol. 154. p. 752, 1944.

beds, tuffs, agglomerates, coarse agglomeratic conglomerates and bedded basaltic lava-flows, associated with marine Cretaceous limestones on the one hand and with a varied group of acid and basic plutonic intrusives—granites, porphyries, gabbro and peridotite—on the other, has lately been discovered during the geological survey of North Kashmir.[1] These rocks are folded into a synclinal trough lying among the Salkhalas, extending from south-east of Astor to beyond Dras, traversing the Great Himalayan range at the head of the Burzil valley in a 20 km. wide outcrop. The stratified volcanic series, several thousand feet in thickness, contains numerous sedimentary layers and lenticular intercalations of fossiliferous limestones and shales, with foraminifers, lamellibranchs, gastropods, ammonites and corals, among which the best preserved fossils are the Cretaceous foraminifer *Orbitolina* (cf. *O. bulgarica*).

Fully half the bulk of the Burzil outcrop is occupied by intrusive hornblende-granite, which has penetrated the basic volcanics in bathyliths and in anastomosing sills and veins, while massive stocks and bosses of pyroxenite (converted to serpentine) and gabbro are of local prevalence at various points in the outcrop.

It is clear that the belt of the Burzil-Dras Cretaceous volcanic series is in structural continuity with the wider development of the Lower Tertiary volcanics of the Upper Indus valley of Ladakh and Kargil, and constitutes its north-west prolongation along the strike.

The most interesting feature of the rocks we are now considering is the injection of fossiliferous Cretaceous sediments by a granite, one of the three varieties of common Himalayan granite, whose post-Cretaceous age is thus settled beyond doubt. This granite overspreads a large extent of the country from Astor to the Deosai plateau.

Ladakh—South of Ladakh, running along the Indus valley is a band of Cretaceous with ophiolites comprising the flysch facies as well as mafic and ultra-mafic volcanic rocks of the Dars Volcanic type just described. The former include the characteristic *Hippuritic* limestone and beds containing *Gryphaea vesiculosa*. The Cretaceous band runs all the way from Kargil, and beyond from as far as the Burzil Pass, along the main Himalayan strike to Rupshu. The Ladakh Cretaceous is compactly sandwiched in between the wide Jurassic-Triassic area to the south-west and the Indus Eocene belt running closely along the river valley to the north-east through Leh. The south border of the belt is one of tectonic contact, with north-directed thrusts.

Middle and Upper Cretaceous sediments containing *Orbitolina*

---

[1] D. N. Wadia, *Rec. G.S.I.* vol. lxxii. pt. 2, 1937.

and *Hippurites* are met with in Chitral underlying the Tertiary Reshun conglomerate.

The thick pile of volcanic ejectamenta, described above, with intercalated sedimentary layers and lenses of limestone containing *Orbitolina* and other foraminifers, corals, and echinoids, runs from south-east of Astor to beyond Dras in Ladakh. The mineral chromite is associated with the gabbro and serpentine intrusive into the series. There is local concentration of chromite into workable ore-masses which would be of economic importance in a more accessible locality. This is the north-west extension of the basal part of the much more extensive zone of Eocene volcanic and marine sediments of the upper Indus valley from Kargil to Hanle in S. E. Ladakh (p. 301). The vertical extent of this clastic volcanic series reaches several thousand metres and in its width the belt is over 20 km. across the strike where it is traversed by the Burzil valley. Dolerite, gabbro and pyroxenite masses and stocks, together with bathyliths of hornblende-granite, are injected into these rocks and have given rise to a varied suite of alteration products.

Hazara

The Hazara Cretaceous is the westernmost extension of marine geosynclinal formations constituting the Tibetan zone of Himalayas, which is found developed with varying degrees of completeness from Bhutan, through Nepal, N. Kumaon, Spiti and Kashmir. From Hazara, the geosynclinal belt turns southwards and passes through Sulaiman range westwards to the Zagros Range of Iranian Arc.

Representatives of the Giumal sandstone are found in north Hazara capping the Spiti Shales, in a group of dark-coloured, close-grained, massive sandstones, calcareous shales and shelly limestones containing *Ostrea* and *Trigonia*. The Giumal sandstone passes up into a very thin arenaceous limestone only some 3 to 6 m. thick, but containing a suite of fossils possessing affinities with the English *Gault*. The leading fossils are ammonites, of typically Cretaceous genera, like *Acanthoceras* (in great numbers), *Ancyloceras*, *Anisoceras*, *Lyelliceras*, *Hamites*, *Baculites*, etc., the latter forms being characterised by possessing an uncoiled shell. There are also many *Belemnite* remains. The Cretaceous limestone is overlain by a great development, about 120 m. of well-bedded grey limestone, succeeded by limestones of the Eocene system—the Nummulitic limestone—much the most conspicuous rock-group in all parts of the Hazara province. In south Hazara, the Giumal sandstones, with *Trigonia*, pass upwards into sand-stones with shelly limestones of Albian age. These are overlain by Nummulitic limestone. In both north and south Hazara, the junction with the Eocene is unconformable, marked by a layer of laterite.

The Giumal series also occurs in the Attock district; the Kala Chitta and Margala hills, near Rawalpindi, overlying the Spiti shales. The Gault, or Giumal series, has also been observed with identical fossil ammonites in the Kohat district and, in a somewhat more fully developed sequence, in the Samana range. It is there underlain by a lower Cretaceous stage with *Holcostephanus*.

Sind and Baluchistan

Cretaceous of Sind. Cardita beaumonti beds—Upper Cretaceous rocks indicating the Campanian and Maestrichtian horizons (Upper Chalk) are developed in Sind in one locality only, the Laki range. The bottom beds are about 100 m. of whitish limestones, containing echinoids like *Hemipneustes*, *Pyrina*, *Clypeolampas*, and a number of molluscs. Among the latter is the genus *Hippurites* so characteristic of the Cretaceous period in all parts of the world. This hippurite limestone is a local representative of the much more widely developed hippurite limestone of Iran, which is prolonged into south-eastern Europe through Asia Minor. It is succeeded by a group of sandstones and shales, often highly ferruginous, some beds of which contain ammonites like *Indoceras*, *Pachydiscus*, *Baculites*, *Sphenodiscus*, etc. These are in turn overlain by fine green arenaceous shales and sandstones, unfossiliferous and of a flysch type, attaining a great thickness. An overlying group of sandstone is known as the *Pab sandstone*. The top beds of this sandstone consist of olive-coloured shales and soft sandstones, the former of which are highly fossiliferous, the commonest fossil being *Cardita beaumonti*, a lamellibranch with a highly globose shell. This group is designated the *Cardita beaumonti beds*. Other fossils include *Ostrea*, *Corbula*, *Cytherea*; *Turritella*, *Natica*; *Caryophyllia*, *Smilotrochus*, and other corals; echinoderms; and some remains of vertebrae belonging to a species of crocodiles. The Cardita beds are both interstratified with as well as overlain by sheet of Deccan Trap basalts, one band of which is nearly 30 m. thick, of amygdaloidal basalt. The age of the Cardita beds, from the affinities of their contained fossils, is regarded as uppermost Cretaceous (Danian).

Cretaceous of Baluchistan—The Cretaceous system as found developed in Baluchistan is on a much more perfect scale than in Sind, covering a far wider extent of the country and attaining a greater thickness. In this area, moreover, the Lower Cretaceous horizons of Wealden and Greensand ages are also represented, having been recognised in a series of shales and limestones, resting upon the Jurassic rocks of Baluchistan, known respectively as the *Belemnite shales* and the *Parh limestone*. The lower, belemnite beds are a series of black shales crowded with the *guards* of belemnites.

THE CRETACEOUS SYSTEM   263

They are overlain by a conspicuous thick mass of variously coloured siliceous limestones, 450 m. in thickness, extending from the neighbourhood of Karachi to beyond Quetta in one almost continuous outcrop. The Parh limestone is in the main unfossiliferous except for a few shells, e.g. *Inoceramus*, *Hippurites*, and some corals.

The Upper Cretaceous sequence of Baluchistan rests with a slight unconformability on the eroded surface of the Parh limestone. This sequence is broadly alike in Sind and Baluchistan, and the account given above applies to both. In Baluchistan, however, the flysch deposits are found developed on a larger scale than in Sind, and form a wider expanse of the country. They are distinguished as the *Pab sandstone*, from the Pab range in Baluchistan. The upper beds of the Pab sandstone are the equivalents of the *Cardita beaumonti* beds of Sind.

The Upper Cretaceous of both Sind and Baluchistan, especially the *Cardita beaumonti* beds, is largely associated with volcanic tuffs and basalts, the local representatives of the Deccan Traps of the Peninsula. In Baluchistan there are also large bosses and dykes of gabbros and other basic plutonic rocks piercing through strata of this age.

It should be noticed that the upper parts of the Umia beds described with the Jurassic rocks of Kutch are of Lower Cretaceous age—Wealden.

Salt-Range

The Cretaceous system is rather inconspicuously developed in the cis-Indus Salt-Range, the principal fossiliferous outcrops being beyond the Indus in the Chichali hills, Makerwal and around Kalabagh. Only the lower Cretaceous is present; the rocks, consisting of white and yellow sandstones and *shales* with a basal stage of black shales and glauconitic sandy marls, *Belemnite shale*, rest upon the Jurassic and are overlain by the Eocene. A rich Neocomian fauna of cephalopods characterises the belemnite beds. The principal genera are *Holcostephanus* (very common), *Spiticeras*, *Neocomites*, *Blanfordiceras*, *Belemnopsis*, and *Hibolites*. Some reptilian and fish remains together with *Exogyra*, *Pecten*, *Pholadomya* and a few other molluscs are associated with these. Only the Lower Cretaceous up to the Albian is present: the Middle and Upper Cretaceous is missing from the Salt-Range area, the junction with the next succeeding Ranikot stage of the Eocene (Thanetian) being a marked hiatus, denoted by a bed of laterite.

Assam

With the exception of a narrow belt of interrupted Lower Gondwana outcrops stretching from Darjeeling to the Abor country, the oldest fossiliferous sediments of the Assam region belong to the Cretaceous system prominently seen in the Shillong plateau region. These are deposits of the Southern Sea which had no connection with the Himalayan Cretaceous Sea. In this area, Cretaceous sandstones lie on an irregular surface of *Shillong quartzites* and other metamorphic rocks. The basal bed is conglomerate, interbedded with sandstone, followed by glauconitic sands and carbonaceous sandstone which contains plant remains. There is much lateral variation and most of the sandstones are unfossiliferous, but below Cherrapunji there has been found in a series of massive sandstones (*Mahadek stage*) a large fauna indicating a Cenomanian horizon. The leading genera are *Hemiaster*, *Anisoceras*, *Baculites*, *Gryphaea*, *Pecten*, *Nerita*, *Spondylus*, *Inoceramus*, *Vola*, *Chlamys*, *Lyria*, *Rostellaria*, *Turritella*, etc., together with many plant remains. The organic remains of this group of beds prove their identity with the much better known and more perfectly studied Cretaceous of the south-east coast of Tiruchirapalli.

The *Cherra sandstone*, formerly regarded as the upper part of the Cretaceous, is now thought to be the lowest member of the Eocene (p. 318).

On the Shillong plateau (which includes a large part of the Garo hills and of the Khasi and Jaintia hills) the Cretaceous and overlying beds are nearly horizontal and form small scattered outcrops, but along the southern edge of the plateau the Cretaceous beds are nearly 300 m. thick and plunge steeply southwards below the Tertiaries or into the alluvium at the foot of the plateau.

Burma

In the Arakan Yoma of Burma, and in the southward continuation of the same strike in the Andaman Islands, is found a large thickness of beds which are at least in part Cretaceous. Owing to the paucity of fossils and lack of detail maps of Arakan Yoma country, the classification of these beds in uncertain. The *Mai-i series* is largely sandstone including an argillaceous limestone with *Schloenbachia inflatus*. The *Negrais series* includes sandstones and shales, somewhat metamorphosed, evidently a flysch deposit recalling that of Spiti in Northern Himalayas. The uppermost Cretaceous contains *Cardita beaumonti*, also characteristic of beds in Sind and Baluchistan.

Among the intrusive Cretaceous rocks of Burma are masses of serpentines traversed by veins of jadeite, which yield the jadeite of commerce for which Burma is famous (pp. 457-458).

Cretaceous rocks have lately been found in the Irrawaddy river defile near Yanbo in Upper Burma, containing species of *Orbitolina* allied to those occurring in the Cretaceous of Eastern Tibet. This suggests an extension of the Cretaceous sea of the Tibetan zone of the Himalayas into Burma.[1]

---

[1] *Rec. G.S.I.* vol. lxxi. pp. 350-375, 1937.

REFERENCES

W. T. Blanford, Ancient Geography of Gondwanaland, *Rec. G.S.I.* vol. xxix. pt. 2, 1896.

A. von Krafft, Exotic Blocks of Malla Johar, *Mem. G.S.I.* vol. xxxii. pt. 3, 1902.

H. B. Medlicott, Shillong Plateau, *Mem. G.S.I.* vol. vii. pt. 1, 1869.

H. H. Hayden, Geology of Spiti, *Mem.* G.S.I. vol. xxxvi. pt. 1, 1904.

E. Spengler, *Pal. Indica*, N.S. vol. viii. mem. 1, 1923.

H. L. Chhibber, *Geology of Burma* (Chapter on "Igneous Activity in Burma"), (Macmillan), 1934.

E. Vredenburg, Geology of Some Baluchistan Districts, *Rec. G.S.I.* vol. xxxviii. pt. 2, 1910.

H. Douville, *Cardita Beaumonti* Beds, *Pal. Ind.*, N.S. vol. x, 1929.

PLATE XIIA.

CRETACEOUS-EOCENE FLYSCH DEPOSITS OF KARGIL LADAKH

CHAPTER XV

The Cretaceous System (*Continued*)

PENINSULA

Upper Cretaceous of the Coromandel coast—Upper Cretaceous rocks of the south-east coast of the Peninsula form one of the most interesting formations of South India, and have been studied in great detail by many geologists and palaeontologists. They are a relic of the great marine transgression of the Cenomanian age, whose records are seen in many other parts of the world, besides the coasts of the Gondwana continent in India as well as Africa. Three small inliers of these rocks occur among the younger Tertiary and post-Tertiary formations which cover the east coast of the Peninsula. Their bottom beds rest either upon a basement of the ancient Archaean gneisses or upon the denuded surface of some division of the Upper Gondwana. As is usual with deposits formed during transitory inroads of the sea, as mentioned in a previous chapter, the dip of the strata is towards the sea; hence the outcrops of the youngest stage occur towards the sea, while the older beds are seen more towards the interior of the mainland.

Interest of the south-east Cretaceous—South of Madras these rocks are exposed in three disconnected patches, in which all the divisions of the Cretaceous from Cenomanian (Lower Chalk) to Danian (uppermost Cretaceous) are present. The most southerly outcrop, *viz*. that in the vicinity of Tiruchirapalli, has an area of from five to eight hundred square km., while the other two are much smaller. But the fauna preserved in these outcrops is of remarkable interest and of inestimable value alike on account of the multitude of genera and species of old-world invertebrata that are preserved, and for the perfect state of their preservation. Sir T. H. Holland speaks of these three small patches of rocks as forming a little museum of palaeozoology, containing more than 1000 species of extinct mollusca, including forms which throw much light on the problems connected with the distribution of land and sea during the Cretaceous. Their distribution and their relations to the Cretaceous fauna of the other Indian and African regions, from Tamilnadu to Madagascar and Natal, have much to tell about the geography of the Gondwana continent at this epoch, and of the barriers to inter-oceanic migrations of life which it interposed.

The Cretaceous rocks of South India are classified into three stages in the order of superposition :

>Ariyalur,
>Trichinopoly,
>Utatur.

Utatur stage—The lowest, Utatur, stage rests upon an ancient land-surface of the Archaean gneisses or on Upper Gondwanas. It is mostly an argillaceous group about 300 to 600 m. in thickness. At the base it contains as its principal member a coral limestone (an old coral reef) succeeded by fine silts, clays, and gritty sandstones. The Utatur outcrop is the westernmost, and is continuous through the whole Cretaceous area along its western border. At places its width is greatly reduced by the overlapping of the next stage, the Trichinopoly. The Utatur fossils are all, or mostly, littoral organisms, such as wood-boring molluscs, fragments of cycadaceous wood, and numerous ammonites. The preponderance of the latter at particular horizons enables the series to be minutely subdivided into sub-stages and zones. The genus *Schloenbachia* occurs largely at the base, and gives its name to the lowest subdivision of the Utaturs, followed by the *Acanthoceras* zone, etc.

Trichinopoly stage—The next group is distinguished as the *Trichinopoly stage*, and comes somewhat unconformably on the last. This group is also 300 m. in thickness, but in lateral extent is confined to the outcrop in the vicinity of Tiruchirapalli only. Both the composition of this group as well as the manner of its stratification show it to be a littoral deposit from top to bottom. The rocks are conspicuously false-bedded coarse grits and sands, clays and shelly limestones, with shingle and gravel beds. Granite or gneiss pebbles are abundantly dispersed throughout the deposits. The proximity of the coasts is further evidenced by the large pieces of cycad wood, sometimes entire trunks of trees, enclosed in the coarser sandstone and grits. The shell-limestone has compacted into a beautiful, hard, fine-grained, translucent stone which is much prized as an ornamental stone, and used in building work under the name of Tiruchirapalli marble. Fossils are many, though not so numerous as in the Utatur division. They indicate a slight change in the fauna.

Ariyalur stage. Niniyur stage—The Trichinopoly is conformably overlain by the *Ariyalur stage*, named from the town of Ariyalur in the Tiruchirapalli district. It consists of about one thousand meters of regularly bedded sands and argillaceous strata, with, towards the top, calcareous and concretionary beds full of fossils. The Ariyalur stage occupies by far the largest part of the Cretaceous area, the breadth of its outcrop exceeding 25 km. The Ariyalur fauna exceeds in richness that of the two preceding stages, the gastropods alone being represented by no less than one hundred and forty species. Besides these, reptilian and fish remains, cephalopods, lamellibranchs, echinoderms, worms, orbitoids, etc. are present in large numbers of species. The uppermost beds of this stage are sharply marked off from those below and form a distinct subdivision, known as the *Niniyur stage*, and distinguished from the remainder on palaeontological grounds, though there is no stratigraphic break visible. The *ammonites* have disappeared from this division, and with them also many lamellibranch genera, while the proportion of gastropod species shows a marked increase. Numerous beds of algal[1] and foraminiferal limestones are enclosed among argillaceous and gritty sediments. The following genera of fossil marine algae are common: *Dissocladella, Indopolia, Acicularia* and several *Lithothamnia*. Milioline foraminifers are associated with these. The fossils of the Niniyur beds reveal a Danian affinity; according to Vredenburg, these beds are equivalent to the *Cardita beaumonti* beds of Sind and Baluchistan. The decline of the ammonites and the increase in the families and orders of the gastropods are a very significant index of the change of times: the Mesozoic era of the earth's history has well-nigh ended, and the third great era, the Cainozoic, is about to commence.

Eocene (Palaeocene) of Pondicherry—Over the Ariyalur beds in the Pondicherry area occur beds with a foraminiferal fauna comparable with that of the Ranikot stage of the Eocene formation of Sind. Species of *Nummulites, Discocyclina* and *Cibicides* are found in a limestone bed and are identical with the forms observed in the basal Eocene of Pakistan.[2]

Fauna of the south-east Cretaceous—The following list shows the distribution of the more common genera in the three stages:

Utatur Stage :

    Brachiopods : *Kingena, Terebratula* (many species), *Rhynchonella* (many species).
    Corals : *Trochosmilia, Stylina, Caryophyllia, Isastrea, Thamnastrea.*
    Gastropods : *Fusus, Patella, Turritella.*
    Cephalopods : *Schloenbachia, Acanthoceras, Hamites, Mannites, Turrilites, Nautilus neocomiensis.*
    Lamellibranchs : *Exogyra, Gryphaea, Inoceramus, Tellina, Opis, Nuculana, Nucula, Arca, Aucella, Radula, Pecten, Spondylus, Lima, Pinna, Trigonoarca.*

---

[1] L. R. Rao and Julius Pia, *Pal. Indica* N.S., vol. xxi. mem. 4, 1936.
[2] L. R. Rao, *Curr. Sc.*, 1939.

Trichinopoly Stage :
  Ammonites : *Placenticeras, Pachydiscus, Heteroceras, Holcodiscus, Scaphites.*
  Lamellibranchs: *Pholadomya, Modiola, Ostrea, Corbula, Mactra, Cyprina, Cytherea, Trigonia, Trigonoarca, Pinna, Cardium, Pecten.*
  Reptiles : *Ichthyosaurus, Megalosaurus* (Dinosaur).

Ariyalur Stage :
  Ammonites: *Pachydiscus, Baculites, Sphenodiscus, Desmoceras, Puzosia, Anisoceras.*
  Lamellibranchs : *Cytherea, Cardium, Cardita, Lucina, Yoldia, Nucula, Axinea, Modiola, Radula, Gryphaea, Radiolites, Trigonoarca, Exogyra, Plicatula, Anomia.*
  Gastropods : *Voluta, Cypraea, Aporrhais, Alaria, Pseudoliva, Cancellaria, Cerithium, Turritella, Solarium, Patella, Nerita, Nerinea, Phasianella, Rostellaria.*
  Reptiles : *Ichthyosaurus,* ? *Titanosaurus, Megalosaurus* and other theropod and sauropod dinosaurs[1].
  Corals : *Stylina, Caryophyllia, Thamnastrea, Cyclolites.*
  Echinoids: *Epiaster, Cardiastar, Holaster, Catopygus, Holectypus, Salenia, Pseudodiadema, Cyrtoma.*
  Crinoids : *Marsupites, Pentacrinus.*
  Polyzoa : *Discopora, Membranopora, Lunulites, Cellepora, Entalophora.*

Niniyur Stage : *Nautilus danicus,* large specimens of *Nerinea* and *Nautilus* with *Orbitoloides, Cyclolites, Nummulites.* Many gastropods and foraminifers, and algae and other plant remains.

The above list gives but an imperfect idea of the richness of the fauna and of its specific relations. All the groups of the Invertebrata are represented by a large number of genera, each genus containing sometimes ten or even more species. The mollusca are the most largely represented group, and of these the cephalopods form the most dominant part of the fauna. There are one hundred and fifty species of cephalopods, including three species of *Belemnites,* twenty-two of *Nautilus,* ninety-three of the common species of *Ammonites,* and three species of *Scaphites,* two of *Hamites,* three of *Baculites,* eight of *Turrilites,* eleven of *Anisoceras,* and three of *Ptychoceras.* The gastropods and lamellibranchs number about two hundred and forty species each. The next group is corals, represented by about sixty species, echinoids by forty-two species, polyzoa twenty-five and brachiopods twenty.

---
[1]*Rec. G.S.I.* vol. lxi. pt. 4, 1929.

Of Vertebrata there occur seventeen species of fishes, and two or three of reptiles, one of *Megalosaurus* and one of *Ichthyosaurus* and ? *Titanosaurus*, relatives of the giant reptiles of the European and American Cretaceous. No fossil mammals belonging to the Cretaceous age have yet been discovered in any part of India.

Marine Cretaceous of the Narmada Valley: Bagh Beds

The Narmada valley Cretaceous—A number of small detached outcrops occur along the Narmada valley, extending along an east-west line from the town of Bagh in the Gwalior region to beyond Baroda, stretching as far west as Wadhwan in Saurashtra. They cover an extensive area of the Panch Mahals and Broach districts of N. Gujarat, generally underlying the Deccan Traps. The rocks are characterised by a heterogeneous composition including cherts, impure shelly limestones, quartzitic sandstones and shales. In most cases they occur around inliers of older rocks in the Deccan Trap, by the denudation of which these beds are laid bare. They are the much worn relics of another of the incursions of the sea (this time it is the sea to the north—the Tethys) during the Cenomanian transgression and, therefore, of the same age as the Utatur beds described above. The fossiliferous portion of the Bagh Cretaceous comprises only a very small thickness, 18–20 m., of limestone and marls, which may be classified into three sections:

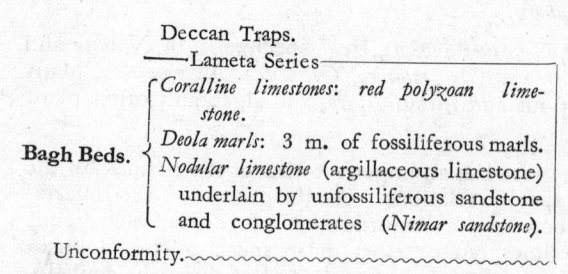

Deccan Traps.
———Lameta Series———

Bagh Beds. { *Coralline limestones*: red polyzoan limestone.
Deola marls: 3 m. of fossiliferous marls.
Nodular limestone (argillaceous limestone) underlain by unfossiliferous sandstone and conglomerates (*Nimar sandstone*). }   Senonian to Cenomanian

Unconformity.

Gneisses, Middle Gondwana rocks, etc.

The lower beds are nodular argillaceous limestones, of a wide extension horizontally, met with in the majority of the outcrops between Bagh and Baroda, followed by richly fossiliferous marls—the Deola and Chirakhan marls—and by a coralline limestone formed of the remains of polyzoa. The last two zones do not extend much westwards. The fossils are numerous, the chief genera being (Ammonites) *Placenticeras*, *Namadoceras*; (Lamellibranchs) *Ostrea*, *Inoceramus*, *Pecten*, *Pinna*, *Crasinella*, *Grotriana*, *Protocardium*, *Cardium*; (Echinoids) *Salenia*, *Cidaris*, *Echinobrissus*, *Hemiaster*,

Opisaster, Cyphosoma; (Polyzoa) *Escharina, Eschara*; (Coral) *Thamnastrea*; (Gastropods) *Triton, Turritella, Natica, Cerithium*.

The sandstones underlying the Bagh beds, the *Nimar beds*, have thickened to several thousand metres to the west; these sandstone strata of the western inliers, particularly near Baroda, have furnished to this region large quantities of an excellent building stone of very handsome appearance and great durability[1] (*Songir sandstone*[2]). The Nimar sandstone, unconformably overlying the Archaeans, contains fossil *Ptilophyllum* and *Sphenopteris* of Up. Gondwana (Up. Jurassic) affinities.

Conclusions from the Bagh fauna—The Bagh fauna covers but a small part of geological time—some of the chalk (Cenomanian to Senonian). The main interest of the fauna is the contrast which it offers to the fauna of the Trichinopoly Cretaceous, from which it differs as widely as it is possible for two formations of the same age to differ. The Bagh fauna, as a whole, bears much closer affinities to the Arabian and European Cretaceous than to the former. This is a very significant fact, and denotes isolation of the two seas in which they were deposited by an intervening land-barrier of great width, which prevented the inter-sea migrations of the animals inhabiting the two seas. The one was a distant colony of the far European sea, connected through the Tethys, the other was a branch of the main Southern Ocean. The two areas, though so adjacent to each other, were in fact two distinct marine zoological provinces, each having its own population.[3] The barrier was no other than the Gondwana continent, which interposed its entire width between the two seas, *viz*. that which occupied the Narmada valley and that which covered the south-east coast.

While the difference between these two Cretaceous provinces is of such a pronounced nature, it is interesting to note that there exists a very close agreement, both lithological as well as faunal, between the Trichinopoly Cretaceous and the Assam Cretaceous described in the last chapter. This agreement extends much further, and both these outcrops show close relations to the Cretaceous of Central and South Africa. These facts point to the inference that it was the same sea which covered parts of Africa,

---

[1] The appearance of the stone is greatly improved by the abundant diagonal bedding, made conspicuous by the inclusion of red and purple laminae in the white or cream-coloured general mass of the rock.

[2] The Songir sandstone of Gujarat is probably the same as the *Ahmednagar sandstone* of the Idar region.

[3] Later discovery of some fossil forms related to the Upper Cretaceous species from the Tiruchirapalli area has somewhat reduced this distinctiveness of the Bagh fauna from the Coromandel fauna.

the Coromandel coast and Assam, in which the conditions of life were similar and in which the free intercourse and migrations of species were unimpeded. These series of beds must therefore show very wide faunal discrepancies from the deposits that were laid down in an arm of the great northern sea, Tethys, which was continuous from West Europe to China, and was peopled by species belonging to a different marine zoological province.

Lameta Series : Infra-Trappean Beds

Age of the Lameta series. Metasomatic limestones—*Lameta series* is the name given to a fairly widely distributed series of estuarine or fluviatile deposits of the same or a slightly newer stratigraphic position than that of the Bagh beds of the Narmada. Outcrops of the series are found scattered in Gujarat, central India, Madhya Pradesh, and also in many parts of the Deccan, directly underlying the Deccan Traps. They generally appear as thin narrow discontinuous bands round the borders of the trap country, particularly the north-east and east borders. The name is derived from the Lameta ghat near Jabalpur, where they were first noticed. The Lameta group is not of any great vertical extent in comparison to its wide horizontality. The constituent rocks of the series are cherty or siliceous limestones, earthy sandstones, grits and clays, attaining in all from 6 to 30 m. in total thickness. The limestones form the most characteristic part of the series, and in some places they contain a few badly preserved fossils.[1] The sandstones and clay beds of the Lameta series have yielded a few land or fresh-water shells and the remains of numerous reptiles; among the former are species of *Bulinus, Melania, Corbicula, Paludina*, etc. which are readily recognised as fresh-water, or at the most estuarine, species. The vertebrate fossils include Dinosaurian reptiles, turtles (*Chelonia*) and some fish remains. The latter are valuable as having yielded conclusive evidence with regard to the stratigraphy of the Lameta series. The fishes were obtained from Dongargaon in Madhya Pradesh. They include some species of *Eoserranus, Lepidosteus*, and *Pycnodus*. The first of these belongs to the order *Teleostei* of bony fishes; the latter two belong to the less highly organised order of *Ganoidei*. Sir Arthur Smith Woodward has, from the evidence of these fish remains, determined the age of the Lameta series to be between Danian and Lower Eocene. Von Huene, on the evidence of fossil Dinosaurs, places the Jabalpur Lametas in the Turonian (base of the Upper Cretaceous).

The recent discovery of remains of Cretaceous dinosaurs from Jabalpur and Pisdura (Chanda district) has greatly increased our knowledge of the fossil Dinosauria of India. Twelve new genera

---

[1]Dr. C. A. Matley, *Rec. G.S.I.* vol. liii. pt. 2, 1921.

have been added to the known Indian fossil dinosaurs; these include the first records of the *Stegosauria* and the *Coelurosauria*. The dinosaurs reached their highest development in India during the Lameta epoch. The twelve genera have been identified from the vertebrae, skulls and limb-bones, armour-plates, teeth and coprolites. The following are the principal genera: *Titanosaurus*, three species; *Antarctosaurus*, two species; *Indosuchus*, two species; *Lametosaurus*, *Laplatasaurus*; *Jubbulporia*; *Megalosaurus* and some carnivorous dinosaurs. Prof. von Huene states that Madhya Pradesh fossil dinosaurs are closely allied to those occurring in the Cretaceous of Madagascar and also with those found in Patagonia and Brazil. This would suggest land-bridges in the existing Indian and Atlantic oceans, or the persistence of large remnants of the old Gondwana continent. (See p. 169.)

The Lameta series everywhere rests with a great unconformity over the older rocks, whether they are Archaean gneisses or some member of the Gondwana or the Bagh beds. As a rule it is conformably overlain by the earliest lava-flows of the Deccan Traps series of volcanic eruptions, which began at this time and the geology of which now claims our attention. At a few places, however, the lowest Traps exhibit discordant relations to the Lametas, denoting that a considerable interval of time elapsed before the volcanic cycle began. It is quite probable, however, that the discordant relations may be only apparent and may be due to the fact that in these particular cases the supposed Lameta limestone is only a metasomatic limestone,[1] which Fermor and others have found so commonly between the Traps and the Archaeans and which has in the past been so often mistaken for Lameta limestone. This we must now discuss.

Investigations by Fermor have revealed that many of the supposed Lameta limestones are metasomatic in origin, and have resulted from the calcification of the underlying Archaean gneisses and schists through the process of molecular transformation, effected by the agency of percolating waters. The metasomatic changes are seen in all stages of progress, from unaltered gneisses through partly calcified rock to siliceous limestone resembling the Lameta beds. The calcification and silicification have affected all kinds of underlying rocks, gneisses, granites, and hornblende and other schists.

Western India

The *Himmatnagar sandstone*, a massive and horizontally bedded group of red and brown sandstones with shales in the Idar region, has recently yielded a small but interesting flora including *Weichselia* and *Matonidium*, two extinct genera of ferns which are of con-

---
[1]See Chapter III, p. 79.

siderable stratigraphical value. The former genus is represented by *W. reticulata*, a very characteristic Wealden species. The *Matonidium* (*M. indicum*) is closely allied to the well-known European species *M. goepperti*. This genus reached its maximum development in the Lower Cretaceous, though it also occurs in the Jurassic.[1] The Himmatnagar (Ahmednagar) sandstone is newer than the Dharangadhra sandstone (p. 191) and of the same age as the Songir sandstone of Baroda and the Barmer (Balmir) sandstone of Western Rajasthan, all of which are extensively used as building stones.

REFERENCES

F. Kossmatt, Cretaceous Deposits of Pondicherry, *Rec. G.S.I.* vol. xxviii. pt. 2, 1895, and xxx. pt. 2, 1897.

F. Stoliczka and H. F. Blanford, Cretaceous Fauna of Southern India, *Pal. Ind. Sers.* I., III., V., VI. and VIII., 1861–1873.

L. L. Fermor, *Rec. G.S.I.* vol. xlvii. pt. 2, 1916.

W. T. Blanford, Geology of the Taptee and Narbada Valleys, *Mem. G.S.I.* vol. vi. pt. 3, 1869.

F. von Huene, Cretaceous Reptiles of the Central Provinces, *Pal. Ind.* N.S., vol. xxi. pt. 1, 1933.

C. A. Matley, Relationships of the Lameta beds, *Rec. G.S.I.* vol. liii. pt. 2, 1921.

B. C. Gupta and P. N. Mukherjee, Geology of Gujarat and Southern Rajputana, *Rec. G.S.I.* vol. lxxiii. pt. 2, 1938.

L. Rama Rao, Cretaceous Rocks of South India, *Ind. Acad. Sc.*, 44, 4, 1956; The Cretaceous-Tertiary Boundary, xxii. *Int. Geol. Cong.*, Delhi, 1964. Cretaceous-Tertiary Formations of South India, *Geol. Soc. Ind*, 1966.

---

[1] *Rec. G.S.I.* vol. lxxi. pt. 2, 1936.

CHAPTER XVI

Deccan Trap

The great volcanic formation of India—Towards the close of the Cretaceous, subsequent to the deposition of the Bagh and the Lameta beds, a large part of the Peninsula was affected by a stupendous outburst of volcanic energy, resulting in the eruption of a thick series of lava and associated pyroclastic materials. This series of eruptions proceeded from fissures and cracks in the surface of the earth from which highly liquid lavas welled out intermittently, till a thickness of some thousands of metres of horizontally bedded sheets of basalts had resulted, obliterating all the previously existing topography of the country and converting it into an immense volcanic plateau. That the eruptions took place from fissures such as those which arise when the surface of the earth is in a state of tension, and not from the more localised vents of volcanic craters, is evident from a number of circumstances, of which the entire absence of any traces, even the most vestigial, of volcanoes of the usual cone-and-crater type, and the almost perfect horizontality of the lava-sheets in the immense basaltic region, are the most significant.

This great volcanic formation is known in Indian geology under the name of the Deccan Traps. The term "trap" is a vague, general term, which denotes many igneous rocks of widely different nature, but here it is used not in this sense but in its Swedish meaning of "stairs" or "steps", in allusion to the usual steplike aspect of the weathered flat-topped hills of basalts which are so common a feature in the scenery of the Deccan.

Area—The Deccan Traps encompass to-day an area of about 500,000 square km., covering a large part of Kutch, Saurashtra, Gujarat, Deccan, central India, Madhya Pradesh, the Hyderabad region, etc., but their present distribution is no measure of their past extension, both areally and vertically, since denudation has been at work for ages, cutting through the basalts and detaching a number of outliers, separated from the main area by wide distances. These outliers, which are scattered over the whole ground from W. Sind to Rajahmundry on the east coast, therefore, must testify to the original extent of the formation, which at the time of its completion could not have been much less than one and a quarter million square kilometres.

Thickness—The maximum thickness attained by the Deccan Traps is a matter of conjecture, but it is possible that it might have been as much as 3,000 m. along the coast of Bombay. The thickness, however, rapidly becomes less farther east, and varies much at different places. Towards its southern limit it is between

FIG. 32.—View of Deccan Trap country (Oldham).

600 and 800 m; at Amarkantak, the eastern limit, the thickness is 150 m., while in Sind, *i.e.* the northern limit, it dwindles down to a band of only 30 or 60 m. In Kutch the Traps are about 800 m. in thickness. The individual lava-flows are about 5 m. on an average, but occasionally some flows are seen reaching 15 to 30 m. in thickness. In a boring near Bhusawal 370 m. deep, 29 distinct flows were encountered. The successive sheets of lava are often separated by thinner partings of ashes, scoriae and green earth, and in very many cases by true sedimentary beds, which are hence called inter-trappean beds. The ash and tuff beds are pretty uniformly distributed throughout, but they are scarcer towards the lower part.

The presence of volcanic ashes and tuffs suggests explosive action of some intensity. This might have been the case at certain local vents along the main fissures, where a few subsidiary cones may have been raised. The eruption of the main mass of the lava was, however, of a quiet, non-explosive kind, as is the case with fissure-eruptions.

Horizontality of the lavas—A very remarkable character of the lavas of the Deccan Trap, having an important bearing on the question of their mode of origin, is their persistent horizontality throughout their wide area. It is only in the neighbourhood of Bombay that a marked departure from horizontality appears and a gentle dip is perceptible, of about 5° (in the top beds as much as 15°), towards the sea. Other localities, where a slight but appreciable inclination and even gentle folding of the lava-sheets are noticeable, are the Western Satpuras, Kandesh and the Rajpipla hills, near Broach, but these dips are believed to be due to the effects of late disturbances of level due to tectonic causes rather than to an original inclination of the flows.[1]

Petrology—In petrological composition the Deccan basalts are singularly uniform. The most common rock is a normal augite-basalt, of mean specific gravity 2.9. This rock persists, quite undifferentiated in composition, from one extremity of the trap area to the other. The only variation is in the colour and texture of the rock; the most prevalent colour is a greyish-green tint, but a perfectly black colour or lighter shades are not uncommon. A few, especially those of trachytic or more acid composition, are even of a rich brown or buff colour; less common are red and purple tints. The texture varies from a homogeneous, crypto-crystalline, almost vitreous basalt, through all gradations of coarseness, to a coarsely crystalline dolerite. The rock is often vesicular and scoriaceous, the amygdaloidal cavities being filled up by numerous secondary minerals like calcite, quartz, and zeolites. Porphyritic close-grained varieties with phenocrysts of glassy felspar (a medium labradorite) have an almost semivitreous lustre, a dark lustrous colour, and conchoidal fracture. Owing to the high basicity, and consequent fluidity of the lavas, crystallisation was a comparatively rapid process, for which reason basalt-glass or *tachylite* is quite rare, except in some "chilled edges", where a vitreous glaze appears.

Over enormous extents of the trap area there is no evidence at all of any magmatic differentiation or variation indicated by the presence of acidic or intermediate varieties of lavas. Some notable exceptions, however, have been observed in Kutch, the Pawagarh hills, the Girnar hills, and the Satpura where rocks of more acid or basic composition (rhyolite, granophyre, monzonite, andesite, monchiquite, limburgite and gabbro) are found associated with the basalts. Their occurrence in close association with the ordinary basalts suggests that they were local differentiation products of the same magma, erupted from differing depths, the more mafic types being derived from the deeper Sima layers. The most common of these acid lavas are *rhyolites*, approaching *dacites* and *quartz-andesites*, *pitchstones* and *pumice* found at Pawagarh.[2] The

---

[1] *Rec. G.S.I.* vol. xlvii. pt. 2, 1916.
[2] Fermor, *Rec. G.S.I.* vol. xxxiv. pt. 3, 1906.

gabbroid complex of the Girnar hills is more noteworthy. Here are masses of gabbros and allied basic intrusives occupying a large tract of hilly country rising abruptly from the level trap-built plains of Saurashtra. The relations of the plutonic masses with one another and with the surrounding country-rocks, which are Deccan Trap flows of usual composition, suggest some post-trappean intrusion, or series of intrusions, proceeding from the same magma reservoir as that of the basalts. Subsequent differentiation of the intruded magma by prolonged segregative processes appears to have given rise to several interesting types ranging in basicity from gabbro, lamprophyre, limburgite, diorite, and syenite to granophyre, exposed in the vicinity of Junagadh town. Clusters of dykes and sills are found in Kutch, in the Satpura area, in Rewah and parts of Maharastra and Gujarat. The basic varieties are of dioritic or doleritic composition, while acidic dykes are composed of trachytes or rocks of allied composition and character. Other types from the Saurashtra peninsula are *monchiquite, nepheline syenite, rhyolite, monzonite, oceanite, anka ramite*. Acid differentiates of the Deccan Traps, *trachytes, granophyres*, and *rhyolites*, are found on Maharashtra and Gujarats coast. Similar acid rocks also occur in the Narmada valley and Porbandar. Some acidic varieties may have arisen from assimilation of Sialic rocks during ascent of the basaltic magma; others may be of hybrid origin. Xenoliths of granite have been observed in them.

Of these rocks, the ultrabasic types occur in dykes and small stocks along the west edge of the trap outcrop from Kutch to the Bombay region in all three phases, volcanic, hypabyssal and plutonic. The acidic types show a more extensive distribution, but individual occurrences are small and their total volume is insignificant in proportion to the vast bulk of the plateau basalts.

As we have observed in Chapter XIV, there is a much greater diversity of petrological composition among the eruptive and intrusive products of the extra-Peninsula, which are in all probability the representatives of the Deccan Trap of the Peninsula.

Microscopic character of the Deccan basalts—In microscopical characters, the basalts are augite-basalts, generally free from olivine (Tholeitic). The mineral olivine is locally abundant in some places. The bulk of the rock is composed of a fine-grained mixture or ground-mass of plagioclase and augite. Besides abundant plagioclase (labradorite or anorthite) prisms, which are often corroded at the edges, there occur sometimes large tabular crystals of clear glassy labradorite of medium composition as phenocrysts in the ground-mass. But porphyritic structure is not common. The augite, often enstatitic, the next important constituent, is present in small grains, very rarely with any crystalline outline. Magnetite is abundantly disseminated through the ground-mass either as idiomorphic crystals or grains, or as secondary dendritic aggregates. In the ordinary grey or green basalts there is very little glass, or isotropic residue, left, it being all devitrified; but in the black dense specimens there is a large quantity of glass present, of a green or brown colour. In some cases the peculiar amorphous isotropic product *palagonite*[1] is seen infilling cavities and interstices of the rock.

The relation of the plagioclase to augite crystals, when apparent, is of a modified *ophitic* type, the latter having a tendency partially to enclose the former. Primary accessory minerals like apatite are few, but secondary minerals, produced by the widespread meteoric and chemical changes that the basalt has undergone, are many, *viz*. calcite, quartz, chalcedony, glauconite, prehnite, zeolites, etc., filling up the steam-cavities as well as the interstices of the rock. A host of other secondary minerals have been described from the basalts of different localities—chlorophaeite, delessite, celadonite, serpentine, chlorites, iddingsite and lussatite. By the discoloration attending these changes the original black colour of the basalts is altered to a grey or greenish tint (glauconitisation). Glauconite is a very widely distributed product in the basalts of the Deccan Trap, both in the body of the rock as well as coating the amygdaloidal secretions. The basalt-tuffs are composed of the usual comminuted lava-particles, with fragments of pumice, crystals of hornblende, augite, felspar, etc. They are usually finely bedded, and have a shaly aspect.

Petrography of the Traps—The detailed petrography of the Deccan Trap is based on the work of L.L. Fermor on the cores of a boring at Bhusawal which penetrated 29 horizontally bedded flows of an aggregate thickness of 357 m., the thicknesses of individual flows varying from 1·5 m. to 30 m. His descriptions of the rocks encountered in this thick succession are regarded as typical of the greater portion of the flows of the Deccan trap, the predominant type being a basalt of specific gravity 2.91, consisting essentially of labradorite ($Ab_1 An_2$), enstatite-augite, glass and iron-ore, olivine occurring in most of the Bhusawal flows, but not universally. A host of secondary minerals are found as alteration-products of the glassy base, or of some primary minerals of the rock, *e.g.*, palagonite, chlorophaeite, celadonite, chabazite, iddingsite, delessite, or as late secretions filling the amygdaloidal cavities of the lava—zeolites, chalcedony, opal, delessite, calcite, quartz and lussatite. Fermor has shown that some ultra-basic modific-

---

[1]Palagonite is the name given to a peculiar green or brown amorphous alteration-product met with in basic volcanic rocks, resulting from change of their ferromagnesian constituents as well as from residual glass. Much of it is analogous to *chlorophaeite*. Its exact origin is not known with certainty. See *Rec. G.S.I.* vol. lviii. pt. 3, 1925.

ations of the basalt may have originated by gravity differentiation, *i.e.* by the sinking of olivine and basic felspar phenocrysts to the base of thick lava-flows which remained fluid enough after eruption for a longer period than other flows. This, however, is not generally the case, though they may have originated thus in special cases.

[CHEMICAL COMPOSITION: Eleven specimens of Deccan traps, collected from widely scattered localities, have been chemically analysed in detail by H. S. Washington. The most striking feature of these analyses is the uniformity of composition of the majority of the basalts, with variation in silica from 48.6 to 52 per cent.

This chemical constitution of the traps, expressed in terms of standard normative minerals calculated from the composition, gives the following result as the norm of the Deccan Traps:[1]

Quartz	4.14
Orthoclase	4.45
Albite	22.01
Anorthite	23.07
Diopside	17.41
Hypersthene	17.78
Olivine	—
Magnetite	4.64
Ilmenite	3.65
Apatite	1.01]

The basalts exhibit a tendency to spheroidal weathering by the exfoliation of roughly concentric shells, hence rounded weathered masses are everywhere to be seen in the exposed outcrops, whether in the field or in stream-courses or on the sea-coasts. Prismatic jointing, or columnar structure, is also observed in the step-like series of perpendicular escarpments which the sheets of basalt so often present on the hill-sides or slopes. At some places beautiful symmetrical prismatic columns are to be seen; this is especially observed in some dykes, in the Bombay area and Kutch. It is the tendency to this kind of jointing, giving rise to the landing-stair-like or "ghat"-like aspect of the basalt hills of the Deccan, that has given the name of the Deccan Trap to the formation.

Recent studies of the traps of the Bombay islands indicate that the westerly dipping tholeiitic lava flows belong to the Upper Traps of Eocene age and are partly sub-serial and partly sub-aqueous, the latter grading into spilite with well-developed pillow structures. Intercalated with the basalt flows are fossiliferous estuarine inter-Trappean sediments, mainly of tuffaceous origin, with

---

[1] H. S. Washington, *Bulletin, Geological Society of America*, vol. xxxiii., 1922.

small intrusives of highly mafic rocks, monchiquite and ankaranite and some felsic extrusives of rhyolite type.

An interesting occurrence of a large deposit of carbonatites (the first discovery of the type in India) has been reported from a Deccan Trap area in Varodara (Baroda) district, Gujarat. The carbonatite is part of a ring complex of alkaline, nepheline-aegerine rocks, intruded in a dome in the upper Deccan Trap flows, penetrating through the underlying Cretaceous limestones of *Bagh beds*. The carbonatite is closely associated with a large deposit of economically workable fluorite, together with some radioactive minerals, sovite, felspathoid and alkaline minerals.

Among the abundant secondary minerals, due to hydrothermal activity during cooling of the lava-sheets, that are found as kernels in the amygdaloidal cavities, the most common are the zeolites, stilbite, apophyllite, heulandite, scolecite, ptilolite, laumontite; also thomsonite and chabazite; calcite, crystalline quartz, or rock-crystal and its cryptocrystalline varieties, chalcedony, agates, carnelian, heliotrope, bloodstone, jasper, etc. Glauconite is abundant as a coating round the kernels. A quantity of bitumen and asphalt, filling large cavities in the lavas near Bombay, was found in 1919. This may have originated by distillation of organic matter contained in the associated sedimentary beds by the heat of the lavas.

Stratigraphy of the Deccan Trap—The following table shows the stratigraphic relations of the Deccan Traps among themselves, and also with the overlying and underlying rocks:

Nummulitics of Surat and Broach; Eocene of Kutch; laterite.
Unconformity.

Upper Traps 450 m.	Of Bombay region and Saurashtra. Lava flows with numerous ash-beds; sedimentary inter-Trappean beds of Bombay with large number of fossil vertebrata and molluscan shells.
Middle Traps 1,200 m.	Of Malwa and central India. Lavas and ash-beds forming the thickest part of the series. No fossiliferous inter-Trappean beds.
Lower Traps 150 m.	Of Madhya Pradesh, Narmada, Berar, etc. Lavas with few ash-beds. Fossiliferous inter-Trappeans numerous.

Slight unconformity.
Lameta or Infra-Trappean series; Bagh beds; Jabalpur beds and older rocks.

Inter-Trappean beds—At short intervals the lava-flows are separated by sedimentary beds of small vertical as well as horizontal extent, of lacustrine or fluviatile deposition, formed on the irregularities of the surface during the eruptive intervals. These sedimentary beds, known as Inter-Trappean beds, are fossiliferous, and are valuable as furnishing the history of the periods of eruptive quiescence that intervened between the successive outbursts, and of the animals and plants that again and again migrated to the quiet centres. Usually they are only 1 to 3 m. in thickness, and are not more than five to seven km. in lateral extent, but they are fairly regularly distributed throughout the *lower* and *upper traps*, being rarely absent for any distance in them. The rocks comprising these beds are a black, cherty rock resembling lydite, stratified volcanic detritus, impure limestones and clays. Many plant-remains and fresh-water molluscan shells are entombed in these, together with insects, crustacea, and the relics of fishes, frogs, tortoises, etc. The most common shell, which is also the most characteristic fossil of the inter-Trappean beds, wherever they have been discovered, is *Physa (Bulinus) prinsepii*—a species of fresh-water gastropod; other fossils are *Lymnaea, Unio, Paludina, Valvata, Melania, Natica, Vicarya, Cerithium, Turritella, Pupa*; the crustacean *Cypris*, some insects, and bones, scales, scutes and teeth of vertebrate animals, *e.g.* fish, frogs (*Rana* and *Oxyglossus*) and tortoises (*Hydraspis, Testudo,* etc.). The flora is very rich in palms, of which numerous stems have been found as well as leaves and fruits; several species of dicotyledonous trees are also present. In places a rich aquatic flora including the fresh-water alga *Chara*, the water-fern *Azolla* and other aquatic plants has been found beautifully preserved in a cherty rock which is probably the silicified mud of lakes.

Inter-Trappean beds are exposed in good sections at Bombay (Malabar hill and Worli), where about 30 m. of well-bedded shales are seen between two lava-flows, containing numerous carbonised plants, many frogs, a tortoise and *Cypris* shells.

A prolific area for fossiliferous inter-trappeans is Chhindwara in Madhya Pradesh, where beautifully silicified leaves, flowers, fruits, seeds and wood of many species of plants are preserved in abundance.

A type-section through a portion of the basalts will show the relations of the traps to these sedimentary intercalations as well as to the infra-Trappean Lametas.

1. Bedded basalts, thick. Individual flows often marked on upper and lower surfaces by steam-holes.
2. Cherty beds, lydites, with *Unio, Paludina, Cypris*, fossil wood, 1.5 m.
3. Bedded basalts, very thick.
4. Impure limestone, stratified tuffs, etc., with *Cypris, Physa (Bulinus)*, and broken shells, 2 m.
5. Bedded basalts, thick.
6. Siliceous limestones with sandstone (Lametas), with a few shell fragments, 6 m.

The mode of eruption of the Deccan Trap—The lowermost trappean beds rest upon an uneven floor of older rocks, showing that the eruptions were subaerial and not subaqueous. In the latter case, *i.e.* if the eruptions had taken place on the floor of the sea or lakes, the junction-plane between the two would have been quite even, from the depositing action of water. As already alluded to, the actual mode of the eruptions was discharge through linear fissures, from which a highly liquid magma welled out and spread itself out in wide horizontal sheets. This view is abundantly borne out by the monotonous horizontality of the traps everywhere, and the absence of any cone or crater of the usual type as the foci of the eruptions, whether within the trap region or on its periphery. The most gigantic outpourings of lavas in the past, in other parts of the world, the "Plateau Basalts", have all taken place through fissures, *viz*. the great basaltic plateau of Idaho in the U.S.A., the Abyssinian plateau and the sheet-basalts of Antrim, etc. A recent analogy, though on a very much smaller scale, is furnished by the Icelandic type of eruptions, *i.e.* eruptions from a chain of craters situated along fissure-lines. (Cf. the Laki eruption of 1783.)

Fissure-dykes in the traps—For any proof of the existence of the original fissures which served as the channels of these eruptions we should look to the peripheral tracts of the Deccan Traps, as it is not easy to detect dykes and intrusions, however large, in the main mass of the lavas unless the former differ in petrological characters from the latter, which is rarely the case actually. Looked at in this way, some evidence is forthcoming as to the original direction and distribution of the fissures. Dykes of large size, massive irregular intrusions, and ash-beds are observed at a number of places in the neighbourhood of the Trap area around its boundary.[1] The most notable of these is the Rajpipla hill tract near Broach. In Kutch likewise there are numerous large dykes and complex ramifications of intrusive masses visible, along the edge of the Trap country, among the Jurassic rocks. The trap area of Saurashtra is traversed by a large number of dykes intruded into the main mass of the lavas. They are of all sizes, from thin veins to masses hundreds of metres wide and some kilometres in length, and follow different directions. The dykes of Saurashtra are composed either of an acid, trachytic rock or of a coarse-grained dark doleritic or dioritic mass. Similar fissure-dykes occur in the Narmada valley and Satpura area among the Gondwana rocks; they are likewise seen in the Konkan, while ash-beds are of very frequent occur-

---

[1] These dykes, intrusions and ash-beds must naturally abound in the vicinity of an eruptive site, and thus help to indicate the location of the fissure and its probable direction in the interior.

rence near Pune (Poona); all these are evidences of the vicinity of an eruptive focus. It is clear that the foregoing instances of dykes, etc. are only the starting-points of the linear fissures which extended a great way into the interior.

It is possible that all these dykes may not have been feeders to the lava eruption. Some evidence has been provided lately to suggest that a number of dykes, especially those observed within the body of the trap-flows, were of later age than the lava-flows, belonging to a subsequent hypogene phase.[1]

Age of the Deccan Traps—There is no conclusive internal evidence in the Deccan Traps with regard to their age. The inter-trappean fossils do not throw any certain light on the age of the beds in which they are entombed. To establish an accurate correlation of the great volcanic series in terms of the standard stratigraphic sequence, we must look to external evidence furnished by the underlying and overlying marine and estuarine beds. The eruptions were certainly subsequent to the Bagh beds (Cenomanian) which they overlie at some places, and to the Lameta series which they overlie at others. Another indication of the age is provided by the interstratification of a few flows of the traps with the *Cardita beaumonti* beds of Sind, whose horizon is fixed as Danian or somewhat newer. At one or two places on the west coast the traps are seemingly unconformably overlain by small outliers of Nummulitic beds, as at Surat and Broach. Here the apparently unconformable junction, denoting an appreciable lapse of time between the last eruptions and the submergence of the area, is quite marked. At Rajahmundry, on the Godavari delta, a distant outlier of the traps occurs resting on the top of a small thickness of marine Cretaceous sandstone of Ariyalur age. In the midst of the trap series in the last-named locality are found sedimentary beds of estuarine and marine deposition containing fossils such as *Physa* (*Bulinus*) *prinsepii*, *Turritella*, *Nautilus*, *Certhium*, *Morgania Potamides*, *Corbula*, *Hemitoma*, *Tympanotomus*. These fossils, however, do not lead to any definite inference, as the affinities of the species and genera are not very pronounced. Examination by Prof. Birbal Sahni of the rich fossil flora from the base of the trappean series of the Nagpur-Chhindwara area, containing an abundance of fossil palms, the occurrence among them of *Nipadites*, a characteristic Eocene genus, and the presence of numerous fertile specimens of *Azolla* (a modern genus of floating water-ferns of which all the previous fossil records are post-Cretaceous), lead him to infer an early Tertiary age for the traps. According to Sahni, the inter-trappean flora finds its clearest affinities with the *London*

---

[1] J. B. Auden, *Trans. Hat. Inst. Sc. Ind.*, vol. iii. 3, 1949.

clay flora. This conclusion seems to find support from later finds by L. R. Rao and others of foraminifers of the families Rotaliidae, Lagenidae, and Miliolidae, of *charophytic* remains from marl beds, and of *Acicularia* and other *algae* from an inter-trappean limestone occurring in the small Trap outcrop near Rajahmundry.

If Sir A. Woodward's inference of the age of the fish fossils from the Lameta series (which is distinctly *infra*-Trappean in position) is accepted (p. 272), the base of the Trap would be positively Eocene. An Eocene age is also supported by the study of some fossil fish-scales from the *inter*-Trappean beds of Betul district, Madhya Pradesh. Dr. S. L. Hora recognises in these scales representatives of an osteoglossid genus *Musperia* and several species of the genus *Clupea*, with some percoid fishes the fossil members of which family carry it only as far back as the Eocene.[1]

The present position may be thus summarised: from external evidence it is quite apparent that the Deccan Traps cannot be older than the Danian stage of the uppermost Cretaceous, while from the internal evidence of fossil fishes, palms and foraminifers, etc. they cannot be much younger than the Eocene.

Except for the Tertiaries at Surat and Broach noted above, together with the alluvial deposits of river-valleys, by far the largest area of the Traps is not covered by any later formation. The peculiar subaerial alteration-product known as *laterite* surmounts the highest flow of the Traps everywhere as a cap, having been produced by a slow meteoric alteration of the basalts.

Economics—The basalts are largely employed as road-metal, in public works, and also to a certain extent as a building stone in private dwellings. From their prevailing dark colour and their generally sombre aspect, however, the rocks are not a favourite building material, except some light-coloured varieties, *e.g.* the buff trachytes of Malad, near Bombay. The large kernels of chalcedony often yield beautiful agates, carnelians, etc., worked into various ornamental articles by the lapidaries, for which there was once a large market at Cambay. These are obtained from a Tertiary conglomerate, in which pebbles of chalcedony, derived from the weathering of the Traps, were sealed up. The sands of some of the rivers and some parts of the sea-coast are magnetitic, and when sufficiently concentrated (as on some sea-beaches) are smelted for iron. Conditions of underground water storage and supply in the Deccan Trap areas are of interest. The vesicular parts of the bedded lavas make good aquifers and yield fair supplies of underground water. These together with the numerous joints and fissures are the only means of water storage in this otherwise impervious and massive formation, containing but few stratification-planes or porous layers. The soil produced by the de-

---

[1] *Rec. G.S.I.* vol. lxxii. pt. 2, 1937.

composition of the basalts is a rich agricultural soil, being a highly argillaceous dark loam, containing calcium and magnesium carbonates, potash, phosphates, etc. Much of the well-known "cotton-soil", known as the "black-soil", or *regur*, is due to the subaerial weathering of the basalts *in situ* and a subsequent admixture of the weathered products with iron and organic matter.

REFERENCES

W. T. Blanford, Traps and Inter-trappean Beds of Western and Central India, *Mem. G.S.I.* vol. vi., 1867–1869.

L. L. Fermor and C. S. Fox, Deccan Trap Flows of Chhindwara District, *Rec. G.S.I.* vol. xlvii. pt. 2, 1916, and *Rec. G.S.I.* vol. xxxiv. pt. 3, 1906.

L. L. Fermor, *Rec. G.S.I.* vol. lviii. pt. 2, 1925.

J.W. Evans, A Monchiquite from Girnar, Q.J.G.S., lvii. pp. 38–54, 1900.

H. S. Washington, Deccan Traps and other Plateau Basalts, *Bull. Geol. Society America*, 33, 1922.

K. K. Mathur *et. al.*, Magmatic Differentiation in the Girnar Hills, *Journ. of Geology*, vol. 34, 1926.

H. Crookshank, *Mem. G.S.I.* vol. lxvi. pt. 2, 1936.

B. Sahni, *Proc. 21st and 24th Ind. Sc. Congr.*, 1934 and 1937.

M. S. Krishnan, Petrography of Girnar Rocks, *Rec. G.S.I.* vol. lviii. pt. 3, 1926.

W. D. West, Deccan Traps, *Trans.*, N.I.S.I., 4, 1958.

R.N. Sukheswala & Poldenvaart, *Geol. Soc. America Bull.*, v., 1958.

G. R. Udas & R. N. Sukneshwala, Carbonatite in Deccan Traps, XXII *Int. Geol. Congr.*, 1964.

CHAPTER XVII

The Tertiary Systems

INTRODUCTORY

General—In Europe the upper limit of the Cretaceous is marked by an abrupt *hiatus* between it and the overlying Eocene group of deposits. A sudden and striking change of fauna takes place in the latter system of deposits, whole families and orders of animals die out, and new and more advanced types of creatures make their appearance. The class of reptiles, the pre-eminent vertebrates of the Cretaceous period, undergo a serious decline by the widespread extinction of many of the orders of the class, and mammals begin to take precedence. The earliest mammals are of a simple or generalised type of organisation, but they soon increase in complexity, and are differentiated into a large number of genera, families and orders. Among the invertebrata the cephalopod class suffers widespread extinction of its species with the advent of the new era; the ammonites and belemnites are swept away altogether. They are now merely items of geological history, like the trilobites of the palaeozoic era. The place of the cephalopods is taken by the gastropods, which enter on the period of their maximum development.

In India these changes in the history of life are as well marked as in the other parts of the world, although there is not any sharply marked stratigraphical break perceptible as in Europe.

Physical changes—The Tertiary era is the most important in the physical history of the whole Indian region, the Himalayas as well as the Peninsula. It was during these ages that the most important surface-features of the area were acquired, and the present configuration of the country was outlined. With the middle of the Eocene, an era of earth-movements set in which materially altered the old geography of the Indian region. Two great events of geodynamics stand out prominently in these readjustments: one the final breaking up of the old Gondwana continent by the submergence of large segments of it underneath the sea,[1] the other

---

[1] It is probable that the disruption of Gondwanaland was not single even but that it proceeded in stages. The first part to separate was Australia and the Malay Archipelago. The next severance took place between South Africa and South America; and the last act was the foundering of Lemuria (the land-bridge between India and Madagascar), which brought into existence the Arabian sea.

the uplift of the Tethyan geosynclinal tract of sea deposits to the north into the lofty chain of the Himalayas.

The prodigious outburst of igneous forces at the very end of the Cretaceous seems explicable when viewed in connection with these powerful crust-movements and deformations. The close association of periods of earth-movements with phenomena of vulcanicity in the records of the past lends support to the inference that the late-and post-Cretaceous igneous activity was in some way antecedent to these earth-movements.

The transfer of large masses of magmatic matter, as we have seen in the last chapters, from the inner to the outer zone of the earth's sphere could not but be accompanied by marked effects on the surface, chiefly of the nature of subsidence of crust-blocks and, secondly, wrinkles and folds of the superficial crust. It is now a growing belief among geologists that the major crust movements, viz. block-subsidences, uplifts and deep corrugations of the crust, are caused by movements and tranfer of large magma masses, convection currents, etc., taking place in the hot, semi-liquid Mantle layer underneath the Crust. The exact nature of this interaction between the exterior and the interior of the earth is not understood. But there can be no doubt regarding the collateral and consequential nature of the displacements occurring in the plastic Mantle and the surface crustal movements.

The elevation of the Himalayas—The pile of marine sediments, that was accumulating on the border of the Himalayas and in Tibet since the Permian period, began to be upheaved by a slow secular rise of the ocean-bottom. From Mid-Eocene to the end of the Tertiary this upheaval continued, in several intermittent phases, each separated by long periods of time, till on the site of the Mesozoic sea was reared the greatest and loftiest chain of mountains of the earth. The last signs of the Tethys, after its evacuation of the Tibetan area, remained in the form of a few straggling basins. One of these basins occupied a large tract in Ladakh north of the Zanskar range, and another occurred in the Hundes province of Kumaon; on their floors were laid down the characteristic deposits of the age, including among them the Nummulitic limestone—that indubitable and unfailing landmark of Tertiary geological history. These sedimentary basins are of high value, therefore, in fixing the date of commencement of the uplift of the Himalayas in the time-scale of geology.

Three phases of upheaval of the Himalayas—There appear to have been three important phases of the upheaval of this mountain system. The first of these was post-Nummulitic, culminating in the Oligocene; this ridged up the central axis of ancient sedimentary and crystalline rocks. It was apparently followed by a movement of greater intensity about the middle of the Miocene. The most important phase elevated the central part of the range together with the outlying zone of Siwaliks into the vast range of mountains which have since been reduced by denudation to form the present Himalayas. This last stage was mainly of post-Pliocene, later than the deposition of the greater part of the Siwaliks, later than the appearance of Man on earth, and did not cease till after the middle of the Pleistocene.[1] There is some proof that the elevatory movement has not entirely disappeared even within recent times.

After the final breaking up of Gondwanaland, the most prominent feature of the earth's Mesozoic geography, the Peninsula of India, acquired its present restricted form. Incidental to this change, a profound redistribution of land and sea must have taken place in the southern hemisphere. Few geographial changes of any magnitude have occurred since these events, and the triangular outline of South India acquired then has not been altered to any material extent.

Distribution of the Tertiary systems in India—Tertiary rocks, from the Eocene upwards to the Pliocene, cover very large areas of India, but in a most unequal proportion in the Peninsula and the extra-Peninsula. In the Peninsula a few insignificant outcrops of small lateral as well as vertical extent are exposed in the vicinity of the coast of Kerala, Malabar, Gujarat and Saurashtra. A larger area is covered on the east coast by marine coastal deposits of variable horizon, from Eocene to Miocene and Pliocene, the *Cuddalore sandstone*, outcrops of which are found from Orissa to the extremity of the Peninsula. A third and more connected sequence of Tertiary deposits is in Kutch, where a band of these rocks overlies the south border of the Deccan Trap.

Tertiary systems of the extra-Peninsula—The Tertiary rocks of the extra-Peninsula are much more important, and occupy an enormous superficial extent of the country. They are most prominently displayed in a belt running along the foot of the mountainous country on the western, northern, and eastern borders of the country. The Tertiary rocks are essentially connected with these mountain-ranges, and enter largely into their architecture. The geological map of India depicts an unbroken band of Tertiary development running from the southernmost limit of Sind and Baluchistan along the whole of the west frontier of India, through the trans-Indus ranges, to Hazara-Kashmir where it attains its greatest width; from there the Tertiary band continues eastward, though with a diminished breadth of outcrop, flanking the foot of

---

[1] In the Potwar geosyncline 1,500 m. of Up. Siwalik boulder-conglomerates (Pliocene-Pleistocene) have been tilted up to a vertical position for many miles. In the upper valley of the Sutlej in Ngari Khorsum, Pleistocene ossiferous alluvium rests unconformably on tilted Pliocene strata. The tilted Upper Karewa of Kashmir carry sub-Recent semi-tropical plant fossils.

the Punjab, Kumaon, Nepal and Assam Himalayas, up to their termination at the gorge of the Brahmaputra. Thence the outcrop continues southward with an acute bend of the strike. It is here that the Tertiary system attains its greatest and widest superficial extent, expanding over eastern Assam and Upper and Lower Burma to the extreme south of Burma.

Dual facies of Tertiary deposits—In all these areas the Tertiary system exhibits a double facies of deposits—a lower marine facies and an upper fresh-water or subaerial. The exact horizon where the change from marine conditions to fresh-water takes place cannot be located with certainty at all parts, but from Sind to Burma everywhere the Eocene is marine and the Pliocene fluviatile or even subaerial. The seas in which the early Tertiary strata were laid down were gradually driven back by an uprise of their bottoms, and retreated southwards from the two extremities of the extra-Peninsula, one towards the Bay of Bengal and the other towards Sind and the Rann of Kutch, giving place, in their slow regression, to gulf, estuarine and then to fluviatile conditions.

There were, however, two periods at which important changes took place throughout the greater part of the area of Tertiary deposition. The first was at the end of the Oligocene, when there was a temporary but widespread retreat of the sea, so that nowhere in the Tertiary outcrops that have been studied in detail has there been found an unbroken succession from the Oligocene to the Miocene. The break is greatest in the northern part of Pakistan, where the Oligocene is completely absent, and in northeastern Assam where some of the Oligocene and much of the Lower Miocene are missing. The second important break, marked by local folding movements in Assam and Burma, occurred late in the Miocene, before the deposition of the Pontian. During this break thousands of metres of Miocene deposits were removed from the uplifted areas.

The backbone of Tertiary India—its main water-shed—was the Vindhyan mountains and the Kaimur ridge, continued northeast by the Hazaribagh-Rajmahal hills and the Assam ranges. This divide separated the northerly drainage, flowing into the remnant of the Tethys (left after the first, mid-Eocene uplift of the Himalayas), from the southward-flowing drainage into the Indian Ocean. There were then two principal gulfs : the Sind gulf extending through Kutch, Western Rajasthan, Punjab, Simla and Nepal; and the Eastern gulf, subdivided into two by the ridge of the Arakan Yoma into the Assam gulf and the Burma gulf. The Gangetic plains then were a featureless expanse of rocky country sloping northwards from the central highlands towards the narrow eastward extension of the Sind gulf.

The whole Tertiary history of India is exhaustively recorded in the deposits filling up these two gulfs. As the seas dwindled and receded, they were replaced by the broad estuaries of the rivers succeeding them, *i.e.* the Indus in Sind, the Ganges-Brahmaputra system in the case of the Assam gulf (p. 51), and the Irrawaddy in Burma; their earlier marine deposits were succeeded, as the heads of the gulfs were pushed outwards, by the growing estuarine and deltaic sediments of the rivers superseding them.

TERTIARY SYSTEMS OF PENINSULAR INDIA
Gujarat

Tertiaries of Surat and Broach—Two small exposures of Eocene rocks, also underlying the laterite cap, are seen as inliers in the alluvial country between Surat and Broach.[1] The component rocks are thick beds of ferruginous clay, with gravel beds, sandstones and limestones, from 150 to 300 m. in thickness, resting with a distinct unconformity on the underlying Traps. These beds are well exposed at Bodhan, near Surat, on the Tapi (Tapti), and extend northwards and westwards for many miles. The gravels are wholly composed of rolled basalt-pebbles and some agates derived from the disintegration of the Traps. Limestone strata are found in the lower part of the exposure, and are full of foraminifers, belonging to several species of the genus *Nummulites* and also *Ostrea, Tibia, Natica*, etc., from the evidence of which the lowest part of the Gujarat Tertiaries is correlated with the Kirthar series of Sind. The highest beds, which contain the foraminifer *Pellatispira*, are, according to S. R. Narayana Rao[2] and F. E. Eames, equivalent to the uppermost Eocene. The name *Tapti series* has been suggested. Above these beds comes a great thickness, 1,200–1,500 m., of conglomeratic gravel beds and clayey and ferruginous sandstones well exposed at Ratanpur, near Broach. The gravel and shingle beds contain many waterworn pebbles of chalcedony derived from the underlying amygloidal Deccan Traps. The pebbles are extracted from the conglomerate, for working them for agates by lapidaries of Cambay. The age of the upper group is estimated as equivalent to the Gaj series of Sind.

This great thickness of "Agate conglomerate", overlying the Nummulitics of Surat and Broach, is well exposed in sections on the banks of the Tapti near Tarkeshwar; the prevailing fossil is *Lepidocyclina*, characteristic of the Nari and Gaj series of the Sind Oligocene-Lower Miocene.

Extensive areas of N. W. Gujarat are thus covered by the Lr. Tertiaries overlying the Deccan Traps. It is probable that the superficial alluvial mantle spreading from Surat to Palanpur is of estuarine as well as of marine origin, filling up a broad arm of the sea which connected the Gulf of Cambay with the Rann

---
[1] W. T. Blanford, *Mem. G.S.I.* vol. vi. pt. 3, 1869.
[2] *Journ. Mysore Univ.* vol. ii. pt. 2, 1941.

of Kutch—an inland sea which persisted up to early Pleistocene times (p. 375). Rocks of the Gaj Series probably extend underneath the alluvium as far as Ahmednagar and westward beneath the coastal alluvium of the shallow waters of the Cambay Gulf to Bhawnagar in Saurashtra underlying the newer Tertiary deposits of this coast. Between the Saurashtra peninsula and Ahmadabad there is a long depressed tract containing a large shallow brackish-water lake (*Nal*), which confirms the probability of this tract being an old marine inlet.

Saurashtra

Perim Island Tertiary—At the extreme east and west points of the Saurashtra peninsula, Tertiary strata ranging from Oligocene to Pliocene age are found overlying the traps. The western outcrop is known as the *Dwarka beds*, and consists of soft gypsiferous clays overlain by sandy limestone containing many foraminifers. The other occurrence is near Bhavnagar, a detached outlier of which crops out in the Gulf of Cambay as the island of Perim. The Perim island was a famous locality for the collection of Tertiary mammalian fossils, and has yielded in past years many perfect fossil specimens of several varieties of extinct quadrupeds. The rock is a hard ossiferous conglomerate, enclosing many skulls, limb-bones, jaws, teeth, etc. of mammals like goats (*Capra*), pigs (*Sus*), *Dinotherium*, *Rhinoceros*, *Mastodon*, etc., of Middle and even Upper Tertiary affinities (Miocene to Pliocene). Many of these relics were found among the beach-shingles produced by wave-action on the conglomerate coasts.

Nummulitic and later strata of Eocene-Miocene age (Nummulitic to Gaj horizon) probably exist on both sides of the Gulf of Cambay, buried under post-Tertiary alluvia; this fact is presumed from the existence of sporadic reservoirs of natural oil and gas underground in parts round Baroda and the east coast of Saurashtra.[1] The fact that the chief petroliferous horizons of the Punjab, Assam and Burma are restricted to rocks of this system (Eo-Miocene) lends colour to the supposition that the Gulf of Cambay was a subsidiary branch of the Sind gulf, and locally afforded conditions suitable for the deposition of oil-forming material. Underneath the Gulf of Cambay, it is probable that suitable structure or disposition of anticlinal or dome folds may exist favouring the storage of oil in commercial quantity.

[With the exceptions of the rather large Jurassic inlier around Dhrangadhra, a few small Cretaceous outcrops near Wadhwan, and the Tertiary development described above, by far the largest surface-extent of the Saurashtra peninsula is occupied by the basaltic traps. It is only in the peripheral parts of the province, in the immediate vicinity of the coast, that rocks of different com-

---

[1] P. K. Ghosh, *Rec. G.S.I.* vol.lxix. pt. 4, 1936.

position are met with, composed of marine coastal accumulations of later ages. Of these the deposits known as the Porbander sandstones (Miliolite) are the most important, and will be described later.]

Kutch

Tertiaries of Kutch—The Tertiary area of Kutch is on a larger scale than those last described. It is seen bordering the Trap and the Jurassic area of Kutch proper, in two long bands parallel with the coast. The older, inner, band abuts upon the Traps directly, while the outer, newer, band runs parallel with the latter, but approaches the Traps by overlapping successively the different members of the older Tertiaries. To the east it encroaches still further north, and comes to rest unconformably on the Jurassic beds by overlapping the Traps in turn.[1]

The bottom beds are argillaceous, with bituminous gypseous and pyritous shales, which by their constitution recall the Laki series of the much more perfectly studied Tertiary sequence of Sind. They are succeeded by about 220 m. of impure, sandy limestones with *Nummulites, Alveolina*, corals, echinoderms, etc., representing the massive Nummulitic limestone of the Kirthar horizon. Above these comes a thick succession of clays, marls, and calcareous shales, crowded with fossils of mollusca, corals and echinoderms, *e.g. Turritella, Venus, Corbula, Breynia*, etc. This part of the sequence corresponds to the Gaj (Miocene) horizon of Sind. It is succeeded by a large development of Upper Tertiary strata representing the Manchar series of Sind and the Siwalik of the Himalayas. The greater part of the latter formation, however, is concealed under recent alluvium, blown sand, etc.

The accompanying table gives a general idea of the Tertiary system of Kutch, correlated with the European Tertiary:

Recent alluvium: blown sand, etc.		Recent and Pleistocene.
Unconformity		
Ferruginous conglomerates, sandstones and clays (*Manchar of Sind*).	152 m.	Pliocene.
Richly fossiliferous shales, clays, and marls with sandstone beds (*Gaj series*).	366 m.	Lower Miocene.
Unconformity		
Impure Nummulitic limestone (*Kirthar series*).	213 m.	Upper and Middle Eocene.
Bituminous, gypseous and pyritous shales, etc. (*Laki series*).	61 m.	Middle Eocene.
Unconformity		
Basalts of the Deccan Trap.		

[1] Wynne's Map of Cutch, *Mem. G.S.I.* vol. ix. pt. 1, 1872; also Geological Map of India (1925), scale 1 in. = 32 miles.

Rajasthan

Rocks of the Tertiary (Eocene) system occur in connection with the Jurassic and Cretaceous inliers of Bikaner, Jodhpur and Jaisalmer regions in the desert tract of Rajasthan, west of the Aravallis. The characteristic Nummulitic limestone is readily recognised in them by means of its *Foraminifers* and other fossils. The nummulitic strata are underlain by a group of shaly beds, the shales enclosing some seams of bituminous coal and lignite. Gypsum in considerable amounts is interbedded, and the series suggests the Laki facies of the Sind Tertiary. Some beds of yellow and brown earthy shale belonging to this series are quarried for the use of the material as fuller's earth, while the lignite and gypsum are capable of further economic development. The Palana coal-field of the Bikaner region, situated on an outcrop of this series, produces at present less than 50,000 tonnes of brown coal per year. The Rajasthan Tertiaries were laid down in a northward extension of the Kutch Eocene sea, which was probably a branch of the main Sind Gulf.

The Coromandel Coast

Cuddalore series—A fairly widely developed series of Tertiary fossiliferous rocks is found along the east coast, underlying the post-Tertiary or Quaternary formations and overlying the various Mesozoic coastal deposits. These formations range from Eocene to Pliocene. A fossiliferous Lower Eocene limestone occurs near Pondicherry and near-by borings have found Middle Eocene sands. Near Tanjore, richly fossiliferous beds of Pliocene age, known as *Karikal beds*, occur in the coastal belt at a depth of 107 m. The principal formation is named the *Cuddalore series*, from the town of that name. Outcrops of the Cuddalore series commence as far north as Orissa and Midnapur, from whence they extend in a number of more or less disconnected inliers through Karikal, hidden under the alluvium of the Kaveri, and the whole length of the coast to the extremity of the Peninsula. A related formation, but of somewhat older age, is also met with on the west coast, extending through Kerala, fringing the coast as far north as Ratnagiri. Throughout this extent the deposits are of irregular distribution and of variable composition. A variously coloured and mottled, loose-textured sandstone is the principal component of the Cuddalore series. It is often ferruginous, argillaceous and gritty. It rests everywhere unconformably on the older deposits of various ages, in one instance overlying the Ariyalur stage of the Trichinopoly Cretaceous. At some places it is covered by a laterite cap, at others by later alluvium. Some sandstones attributed to the Cuddalores abound in fossils, mainly gastropods, *e.g. Terebra, Conus, Cancellaria, Oliva, Mitra, Fusus, Buccinum, Nassa, Murex, Triton*, etc. *Ostrea* and *Foraminifers* of several species are also present.

A great part of the Cuddalore sandstones is believed to be of Miocene-Pliocene age.

Important deposits of lignite have been found, interbedded with the Cuddalores of S. Arcot at Neyveli in seams 6–25 m. thick. The lignite occurs with soft water-logged sandstones and shales and is of good quality and usable as fuel. The *Baripada beds* of Mayurbhanj, marine fossiliferous limestones, over 45 m. thick, are of the same or slightly older age, hidden under the laterite cap and have been only known from well-borings. It is probable that similar marine Miocene-Pliocene sediments form extensive beds, completely hidden under later coastal alluvium or laterite, along the whole coast line.

Malabar Coast

A series of small outcrops of Upper Tertiary strata are found along the Quilon coast of Kerala beneath the superficial cover of laterite (*Quilon* and *Warkalli* beds). A few bright-coloured sands and clays, enclosing bands of lignite with lumps of fossil resin (amber), and pyritous clays occur over the limestones. The limestone strata are full of fossil molluscs, corals and foraminifers. The most abundant are gastropods, *e.g. Conus, Strombus, Voluta, Cerithium, Natica, Rimella, Murex, Terebra, Turritella,* etc. A species of foraminifer, *Orbitolites,* is also present in the limestone. The fauna of the Quilon beds indicates approximately an Upper Gaj horizon (Middle Miocene). On the whole, the Malabar coast Tertiaries denote an older stratigraphic horizon than the *Cuddalore* and *Karikal beds* of the east coast, described above, which are regarded as of Upper Miocene to Pliocene age.

A very similarly constituted outcrop of Tertiary rocks is seen further north at Ratnagiri, on the Malabar coast, underneath the laterite.

Sri Lanka

The large outcrop of horizontally bedded richly fossiliferous limestone seen along the coastal strip of N.W. Sri Lanka—the *Jaffna beds*—is probably a south-east extension of the same formation. The Jaffna limestone is several hundred metres thick and on palaeontological evidence is considered to be Middle to Upper Miocene, homotaxial with the Travancore beds (the species *Orbiculina malabarica* is common to both), but older than the Karikal beds.

TERTIARY SYSTEMS OF EXTRA-PENINSULAR INDIA

The Tertiary development of the extra-Peninsula is far more extensive, and in it all the stages of the European Cainozoic from Eocene to Pliocene are developed on a scale of great magnitude. It has again been more closely studied, and its stratigraphy as well as palaeontology form the subject of several voluminous memoirs published by the Geological Survey of India. The palaeontological evidence available enables us to make a correlation of the different exposures with one another in the immense region which they cover, and also to determine approximately the correspondence of the Indian divisions with the stages of the standard Tertiary scale.

Until very much more work has been done on the Tertiary palaeontology of India it is hardly possible to put forward a completely satisfactory classification, and no scheme has yet been devised to which all Indian palaeontologists agree. The classifications here adopted are from the writings of Vredenburg, Pilgrim, and more recent authors as best suited to the purposes of the student.

The following are the principal localities where the system is well developed: Sind, the Salt-Range and Potwar, the outer Himalayas, Assam, and Burma.

Sind

The great series of Tertiary deposits of Sind are typically exposed in the hill ranges, Kirthar, Laki, Bugti, Sulaiman, etc., which separate Sind from Baluchistan. The Tertiary sequence of Sind is, by reason of its exceptional development, taken as a type for India for systematic purposes. The following table gives an idea of the chief elements of the sequence:

Series	Description	Age
Manchar Series (3,050 m.)	Lower and upper beds, grey sandstones with conglomerates; middle part, brown and orange shales and clays, unfossiliferous. Lower Manchar conglomerates containing teeth of *Mastodon, Dinotherium, Rhinoceros*.	Lower Pleistocene or Upper Pliocene to Middle Miocene.
Gaj Series (455 m.)	Marine yellow limestones and shales, fossiliferous.	Lower Miocene (Burdigalian).
Nari Series (1,830 m.)	Upper Nari, thick sandstones, unfossiliferous and partly of fluviatile origin. Includes the *Bugti beds* of Baluchistan, freshwater, with mammalian fossils. Unconformity.	Lower Miocene (Aquitanian).
	Lower Nari, fossiliferous marine limestone.	Oligocene (Stampian).
Kirthar Series (900–2,800 m.)	Massive nummulitic limestones forming all the higher ranges in Sind, richly fossiliferous.	Upper and Middle Eocene
Laki Series (150–250 m.)	Argillaceous and calcareous shales with coal-measures. Alveolina limestones. Thickness varying.	Middle Eocene (Lutetian).

Unconformity	~~~~~~~~~~~~~~~~~~~~~	Lower Eocene
	Upper, fossiliferous brown limestone and shales.	(Thanetian).
Ranikot Series (610 m.)	Lower, variegated shales and sandstones, gypseous and carbonaceous, fluviatile.	
	Cardita beaumonti beds.	Danian.

Salt-Range and Potwar

The north-western part of the Punjab contains, in the Salt-Range and the plateau country to the north, a very important development of Tertiary rocks, and one which has received much attention. The uppermost scarp of the Salt-Range is a prominent cliff of limestone which has generally been termed the nummulitic limestone. This has developed along the whole length of the range from the eastern spurs near Jhelum almost to the Indus near Kalabagh. Although at the eastern end of the Salt-Range the limestone lies wholly within the Laki stage, towards the western end of the range a lower limestone of Ranikot age develops and reaches a considerable thickness. Above the Laki series there is a pronounced unconformity, the whole of the Oligocene being absent. The limestones and associated marls are overlain by Upper Tertiary rocks, the unconformity being clearly visible in sections at the head of the Nilawahan. In the eastern part of the range the lowest beds above the unconformity belong to the *Murree series*, but further west the overlying *Kamlial stage* rests upon the Eocene. Above the Kamlials, there is developed a complete sequence of the Siwalik system; this is seen not only in the Salt-Range itself, but also in the large plateau to the north known as the Potwar. This comprehensive development of the Siwalik system constitutes the type area for India. The abundance and wide distribution of its mammalian fauna have enabled a very careful and detailed zoning to be established by Dr. Pilgrim, and this affords a basis for the correlation of the Siwalik deposits of the various different areas in India.

The succession in the Salt-Range is as follows:

Upper Siwalik (1,830 m.)	*Boulder conglomerate zone*: conglomerates, sands and clays. *Pinjor zone* : pebbly sandstones. *Tatrot zone*: sandstones and conglomerates.	Lower Pleistocene to Pliocene.

Middle Siwalik (1,830 m.)		*Dhok Pathan zone*: light grey and white sandstones and pale-coloured shales, containing a rich Pontian (Pikermi) fauna. *Nagri zone*: grey sandstones and red and pale-coloured shales.	Upper Miocene (Pontian) to Middle Miocene.
Lower Siwalik (1,525 m.)		*Chinji stage*: bright red nodular shales and clays with grey soft sandstones and pseudo-conglomerates. *Kamlial stage*: hard dark-coloured sandstones, red shales and pseudo-conglomerates.	Middle Miocene (Helvetian).
Murree Series (up to 610 m.)		Light-coloured and purple sandstones, pseudo-conglomerates, red and purple shales.	Lower Miocene (Burdigalian and Aquitanian).
Unconformity			
Laki Series (120 m.)		*Bhadrar beds*: marls and limestones. *Sakesar limestone*: massive limestone forming the summit of the Salt-Range scarp. *Nammal limestone-shale*: bedded limestone, marls, and thin shales.	Middle Eocene (Lutetian).
Unconformity			
Ranikot Series (15–300 m.)		*Patala stage*: shale with thin limestones and impersistent sandstone; coal seam at the base. *Khairabad limestone*: brown nummulitic limestone of very variable thickness with calcareous shale. *Dhak Pass beds*: Shale with pisolitic ferruginous beds at the base.[1]	Lower Eocene (Landenian and Thanetian).

The succession in the Potwar differs somewhat; the gap between the Eocene and the Miocene is reduced both by the development of the lower beds of the Kirthar series and by a great increase in the thickness of the Murree rocks. The succession here merges into that of the Kashmir Himalayas and is given in the table on pp. 299–300.

Himalayas

Tertiary rocks enter preponderatingly into the composition of the outer, lower, ranges of the Himalayas, *i.e.* the ranges lying outside (south of) the central zone of crystalline and metamor-

---

[1] L. M. Davies and E. S. Pinfold, Eocene Beds of the Punjab Salt Range, *Pal. Ind.* N.S. xxiv. mem. 1, 1937.

THE TERTIARY SYSTEMS

phosed sedimentary rocks. In fact, the whole of the outer stratigraphic zone, which is known as the sub-Himalayan zone,[1] is almost exclusively constituted of Lower and Upper Tertiary rocks. With the exceptions noted below, Tertiary rocks are absent from the ranges to the north of the sub-Himalayas. In the Kashmir and Simla Himalayas, where these rocks have been studied, they are disposed in two broad belts, an outer belt and an inner, formed respectively of the Upper Tertiary and the Lower Tertiary. These strata continue eastwards with much the same disposition, but greatly reduced in width of outcrop, along the Kumaon, Sikkim and still more eastern Himalayas, forming the outermost foothills of the mountains, separating them from the plains of Uttar Pradesh, Bengal, and northern Assam.

At this place must be mentioned the Pliocene and Eocene occurrences belonging to the Tibetan zone in localities north of the central crystalline axis of the Himalayas. Two or three such have been observed, *e.g.* North Kashmir (Ladakh) and the Hundes province of Kumaon. Of these the Ladakh exposure is the best known. In the upper Indus valley in Ladakh, to the north of the Zanskar range, there is a narrow elongated outlier composed of marine sedimentary strata, with nummulites and other fossils associated with peridotite intrusions and contemporaneously erupted lava-flows, ash-beds and agglomerates. The sedimentary part of this outlier resembles in some measure the Subathus of the outer Himalayas. This outcrop is described below somewhat more fully. No marine strata of younger age than these have been discovered in any part of the Northern Himalayas.

The succession is given in the following table:

W. Punjab and Kashmir Himalayas and northern Part of the Potwar.	Kumaon and Simla Himalayas.	
Upper Siwalik: Boulder-conglomerates, clays, sands and grit, 1,830 m.	**Upper Siwalik** Soft earths: clays and boulder-conglomerates, 1,800–3,050 m.	Pleistocene to Lower Pliocene.
Middle Siwalik: Massive grey sandstone with pale or drab shales, 1,830 m.	**Middle Siwalik:** Massive sand-rock, clays and shales, fossiliferous at the base, 1,200 (?) m.	Upper to Middle Miocene.
Lower Siwalik: *Chinji:* bright red nodular shales with fewer grey sandstones, 915 m. *Kamlial:* hard brown sandstones and purple shales, 305 m.	**Lower Siwalik:** (*Nahan*): Grey micaceous sandstones and red shales, generally unfossiliferous, 900–1,200 m.	Middle Miocene.

(Siwalik System, 4,880 m.)

---

[1] Chapter I.—The Geological Classification of the Himalayas, pp. 9, 10.

Upper Murree: Sandstones, soft, pale and coarse-grained, with purple splintery and nodular shales, 915 m.	Murree Series, 2,440 m.	**Kasauli:** Lacustrine, coarse, soft, grey or green coloured sandstones.	Lower Miocene.
Lower Murree: Indurated dark sandstones, deep red and purple-coloured splintery shales, 1,525 m; at the base the *Fatehjang zone* of ossiferous sandstones and conglomerates (Upper Nari).		**Dagshai:** Brackish-water or lagoon, bright red and purple nodular clays overlain by fine sandstones.	Lower Miocene.

Unconformity

Chharat: *Nummulite* shale, variegated shales, gypseous marls and thin-bedded limestone, 150–275 m. (Kirthar).	Eocene, 610 m.	**Subathu:** Grey and red gypseous shales with subordinate lenticular nummulitic limestone with pisolitic limonite (laterite?) at base.	Eocene.
Hill limestone: Massive well-bedded nummulitic limestone, some shale and thin coal 60–500 m. (Ranikot to Laki).			

Kashmir

The Tertiaries of Kashmir call for notice because of a few local peculiarities which they exhibit. The Tertiary band at the Jhelum stretches eastwards through the Kashmir area, preserving all its geological characters and relations unchanged, to the Ravi and thence to the Sutlej, where it merges into the much better explored country of the Kangra Himalayas. There is one tectonic distinction in the Kashmir Tertiary belt—the gradual disappearance of the bounding faults separating the Murrees from the successive Siwalik divisions to the west of the river Chinab. However, the more important northerly fault-junction, separating the Siwaliks and the older Tertiaries of the foot-hills Zone from the older rocks of the Mid-Himalaya Zone, persists as the most characteristic feature of Himalayan tectonics from the Indus to the Brahmaputra.

The Tertiary outcrop is the widest where it crosses the Jhelum but is much constricted in its eastward extension beyond the Sutlej, though the broad features of its stratigraphy and lithology are readily recognisable as far as Dehra Dun. Beyond Simla, the Murree horizon is missing, the foot-hills Tertiaries being confined to the Siwalik horizons in Kumaon, Nepal and Bhutan. Equivalent of the Murree series reappears in *Assam (Surma series)*.

Tertiaries of the Inner Himalayas—the Indus Valley Tertiaries—The retreating waters of the Tethys have left an enormous volume of Lr. Tertiary, mainly Eocene, deposits to the N. of the axial Himalaya Range, constituting the *Kampa system* of Tibet (Ranikot-Laki). A long belt of these strata is seen in S. Ladakh along with the Cretaceous Volcanics, the Jurassic, Triassic and the Permo-Carboniferous, described in earlier chapters. Another wide outcrop of the Eocene is met with in the Hundes province of N. Kumaon, in continuation of the Kailas and Mansarovar tertiaries. These are the newest deposits of the Tethyan geosyncline. The Ladakh Lr. Tertiary extends from Dras, Kargil and Leh to beyond Hanle, north of Spiti. From such fossil evidence as is available its age is determined to be Up. Cretaceous to Oligocene.

The Tertiaries of Ladakh rest unconformably over gneissic and metamorphic rocks. The base is of coarse felspathic grits and conglomerates, followed by brown calcareous and green and purple shales. The shales are overlain by a thick band of blue shelly limestone, containing ill-preserved *Nummulites*. This nummuliferous limestone is succeeded by a coarse limestone-conglomerate. On either extremity of this sedimentary basin there is a large development of igneous rocks of acid as well as extremely basic composition. They include both contemporaneously erupted dark basalts, with ash and tuff-beds, as well as dykes and sills of intrusive granite and quartz- and augite-porphyries together with peribotites and gabbros. In the north-west prolongation of the Kargil band of Eocene volcanics, in Dras, there is a close association of tuffs, volcanic ash-beds, lavas and augite-porphyries, with limestones, containing *Alveolina*, *Dictyoconoides*, *Nummulites* and gastropods.

The sedimentary part of this group has preserved a few fossils, besides the *Nummulites* noticed above, but owing to the great deal of folding and fracturing which they have undergone the fossils are mostly deformed and crushed beyond recognition. The following genera are identified, with more or less certainty: *Unio* and *Melania* in the lower part (which bear witness to estuarine conditions), and *Hamites*, *Hippurites*, *Conus*, etc. which show that besides the Lower Tertiaries there are Cretaceous beds present.

Tertiaries of the Sub-Himalayas—The Tertiary zone of the Outer Himalayas is disposed in three or four parallel belts conforming to the general strike, the oldest, Eocene, abutting on the S. W. flank of the Pir Panjal, while the newer ones build the successive outer low ranges of the foot-hills zone. This zone is traceable from Hazara to as far as Assam, though greatly modified, along its course in the relative proportions of its constituent units—Nummulitics, disappearing East of the Kosi. At the N.W. extremity, where the Jhelum river emerges from the mountains, there is a deep inflexion in the strike of the mountains, the entire

Himalayan system undergoing a profound bending inwards of over 150 km. The acute-angled re-entrant bay thus produced is filled with Up. Tertiary sediments. The significance of this feature is dealt with on p. 398, where it is explained as probably due to some crustal obstruction which has deflected the main axes of the fold-systems and converged them in a knot (Syntaxis).

The table on pp. 299–300 shows the relations of the Tertiaries of the Jammu hills to the corresponding rocks of the Kumaon and Simla Himalayas.

Assam

In Assam the Tertiary deposits reach a very great thickness, probably exceeding that of any other part of India; where fully developed the sediments are more than 15,000 metres thick. Despite this, there are several gaps in the succession, the most important one being the absence of the top part of the Oligocene. Owing to the extreme paucity of fossils in the greater part of Assam, it is impossible to give very accurate correlations with other areas or with the standard time-scale, but the following table summarises the results of investigations[1] by P. Evans and colleagues, and indicates an approximate correlation with the Tertiary of Northwest India. Evans's survey of Assam has laid the foundation of the stratigraphic classification of the Tertiaries in that difficult and inhospitable geological terrain.

Alluvium	Alluvium of the Brahmaputra and Surma valleys, high-level alluvium, river-terraces, gravels, etc.	Recent and Pleistocene.
Dihing Series 1,525 m.	(*Upper Siwalik*). Thick pebble-beds with clays and sands.	Pliocene.
Unconformity.	~~~~~~~~~~~~~~~~~~~~	
Dupi Tila Series, 2,440 m.	(*Middle Siwalik*). Coarse ferruginous sands, mottled sands and clays, fossil wood.	Upper Miocene (Pontian).
Unconformity.	~~~~~~~~~~~~~~~~~~~~	
Tipam Series, 2,440 m.	(*Lower Siwalik*). Thick, coarse, ferruginous sandstones, mottled sandy clays, shales, fossil wood and lignite.	Middle Miocene.
Surma Series 4,000 m.	(*Murree*; *Upper Nari* and *Gaj*). Sandy shales and sandstones, conglomerates.	Lower Miocene.
Unconformity.	~~~~~~~~~~~~~~~~~~~~	
Barali Series, 4,575 m.	(*Lower Nari*). Sandstones, shales, and carbonaceous shales.	Oligocene to Upper Eocene.

---

[1] P. Evans, Tertiary Succession in Assam, *Trans. Geol. Min. Inst. Ind.* vol. xxvii., 1935.

Jaintia Series, 900 m.	(*Ranikot* to *Kirthar*). Alternating sandstones and shales with coaly beds, including the Sylhet limestone—the *Nummulitic limestone* of Assam: equivalent to part of the *Disang Series*.	Middle Eocene.

Unconformity. ~~~~~~~~~~~~~~~~~~~~~~~

Cretaceous and older rocks.

The above classification refers mainly to the western part of Assam. In the eastern portion of the province, the succession below the Oligocene-Miocene unconformity is:

Barail Series. 4,575 m.	(*Lower Nari*). Sandstones, shales, clays, with thick coal seams in Upper Assam.	Oligocene to Upper Eocene.
Disang Series (very thick).	(*Ranikot* to *Kirthar*). Thick series of grey splintery shales with fine sandstones, partly equivalent to the Jaintia series. Base not seen.	Middle to Lower Eocene.

Burma

The Tertiary system of Burma is composed of rocks which differ considerably in lithological characters from the standard sections of North-West India, but as fossils are abundant, an approximate correlation is not difficult, although much remains to be done in the investigation of the details. As might be expected, the Burma succession shows more resemblance to the succession in the neighbouring province of Assam. The Eocene beds reach a great thickness and, although foraminifers are found in some beds there are no thick developments of nummulitic limestone such as those seen in Sind, Baluchistan, and the Punjab. The middle part of the succession, composed of Oligocene and Lower Miocene strata, is distinguished as the *Pegu system* and is approximately correlated with the Nari and Gaj series. It has recently been established that a break occurs in the middle of the system, approximately at the boundary between the Oligocene and Miocene, so that the Pegu system really consists of two separate units. The uppermost beds (known as the *Irrawaddy system*) form a great thickness of fluviatile strata corresponding both in lithological aspects as well as in organic characters to the upper parts of the Manchars of Sind and of the Siwaliks of the Punjab and sub-Himalayas. In central Burma they lie with marked unconformity on the Pegus.

The Tertiary history of Burma is largely the history of the filling up of a north and south geosynclinal basin, 970 km. long and 240 km. wide—the basin of the old gulf of Pegu lying between the Arakan Yoma and the Shan Plateau—which was filled up by the deltaic deposits of the Irrawaddy gradually pushing southward into the gulf and ultimately replacing it by the present valley of the Irrawaddy. Hence a marine facies of deposits preponderates towards the south and characterises all the stages till as late as the Pliocene, while in the north the same stages show a terrestrial facies of deposits, it being a common feature of many of the stages that when traced laterally from north to south they show a variation from fluviatile to estuarine and brackish-water, passing thence into marine further south, in which direction the gulf-conditions persisted till the beginning of the Pliocene.

The following table is based on the work of Cotter and Vredenburg, combined with that of the Burmah Oil Company geologists as described by G. W. Lepper.

Plateau gravels and red earth.

	Irrawaddy System 1,525 m. Unconformity.	Fresh-water sandstones with abundant fossil wood, mammalian fossils.	Pleistocene. Pliocene to Upper Miocene (Pontian).
PEGU SYSTEM.	Upper Pegu, 3,050 m. Unconformity.	Sandstones, clays, and shales, with many fossils.	Middle and Lower Miocene.
	Lower Pegu, 1,500–3,050 m.	Mainly sandstones above, shales in the middle, and shallow-water sandstones with coal-seams at the base; fossiliferous.	Oligocene.
EOCENE SYSTEM.	Yaw Stage, 610 m.	Shaly clays, marine, with *Nummulites*.	
	Pondaung Stage, 1,980 m.	Marine sandstones and clays passing up into fluviatile sandstones and deeply coloured clays containing the earliest mammalian fauna: Anthracotheroids, Rhinoceratoids and Titanotheres.	Upper Eocene.
	Tabyin Clay, 1,525 m.	Green shales with thin coal-seams.	Middle Eocene.
	Tilin Sandstone, 1,200 m.	Marine sands and sandstones with *Nummulites*.	
	Laungshe Shales, 3,050 m.	Shales containing *Orbitoides* and *Gastropoda*.	Lower Eocene.
	Paunggyi conglomerate, 900 m.	Conglomerates containing *Orthophragmina*. Basal unconformity.	

CORRELATION OF TERTIARY FORMATIONS

		KASHMIR AND POTWAR	SIMLA HIMALAYAS	SIND AND KUTCH	PENINSULA (SOUTH)	ASSAM	BURMA
Recent		U. Karewas	Sutlej alluvium	Indus alluvium	Newer river deposits; deltas	Brahmaputra alluvium	Irrawaddy alluvium laterite
Pleistocene		III & IV Glacial — 3rd Himalayan Upheaval — II Glacial — I Glacial		Rann of Kutch	Older human artifacts; older alluvium		
Pliocene		L. Karewas and U. Siwalik	U. Siwalik	Manchar	Karikal (?) Cuddalore and Warkalli	Dihing	Irrawaddy
Miocene	U.	M. Siwalik	M. Siwalik	Gaj	Quilon	Dupi Tila Tipam	U. Pegu
	M.	L. Siwalik — 2nd Himalayan Upheaval —	L. Siwalik				
	L.	Murree	Kasauli Dagshai	U. Nari	Baripada	Surma	
Oligocene		Absent — 1st Himalayan Upheaval (Intrusive Granite) —	Absent	L. Nari Tapti	Widespread Unconformity	Barail	L. Pegu
Eocene	U.	Kirthar (Chharat)	Subathu	Kirthar	Eocene of Pondicherry	Jaintia and Disang	Eocene
	M.	Laki		Laki			
	L.	Ranikot		Ranikot	Deccan Trap (top part)		

The Tertiary basin of Burma is separated from the Palaeozoic and Mesozoic highlands of the Shan Plateau to the east by a great north and south boundary fault. On the west, the Pegu and Eocene rocks outcrop in the form of a large monoclinal fold running north and south through the foothills of the Arakan Yoma. This range is still largely a *terra incognita* to geologists, and thus it is impossible to say what the nature of the western limit of the Burma Tertiaries is.

From the above résumé of the stages of Tertiary history of North India, it must have been gathered that the Tertiary records of India are far fuller than the Primary and Secondary ones. It was entirely within the Tertiary that the geomorphic evolution of India, as a separate entity, was initiated and completed, for, as we have seen in the preceding pages, in the Mesozoic age even the skeletal outlines of this area could not be discerned. All the earth-features north of the Vindhyas came to be stamped upon it during the latter half of the Tertiary. Its physical isolation from the Asiatic continent was brought about by the emergence of the great mountain barriers of the west, north and east. Concomitantly with these was produced the extraordinary trough or sunken-valley region of India—a depression 3,000 km. long and 300 km. broad in its narrower parts, separating Northern from Peninsular India—two distinct crust-segments. The geological history of this vast sunken tract, now filled up by the river-deposits of the Indo-Ganges systems, does not commence till the very end of the Tertiary. Thus out of the three great geomorphic divisions of the Indian region two owe their evolution to processes operating during or subsequent to the Tertiary era of the earth's history.

REFERENCES

H. B. Medlicott, Geological Structure and Relations of the Himalayas, *Mem. G.S.I.* vol. iii. pt. 2, 1864.

G. E. Pilgrim, Tertiary Fresh Water Deposits of India, *Rec. G.S.I.* vol. xl. pt. 3, 1910.

E. W. Vredenburg, A Review of the Tertiary Sequence of Sind, *Pal. Indica*, New Series. vol. iii. pt. 1, 1909.

G. de P. Cotter, Geology of the Attock District, *Mem. G.S.I.* vol. lv. pt. 2, 1933.

D. N. Wadia, Tertiaries of Jammu Hills and N. W. Punjab, *Mem. G.S.I.* vol. li. pt. 2, 1928; Tertiary Geosyncline of N. W. Punjab, *Q.J.G., M.S.* vol. iv., 1932.

P. Evans, Tertiary Succession in Assam, *Trans. Min. Geol. Inst. Ind.* vol. xxvii. pt. 3, 1932.

Duncan and Sladen, Tertiary and Up. Cretaceous Fauna of Western India, *Pal. Ind.*, Ser. vii. and xiv., 1871--1885.

H. M. Sale and P. Evans, Geology of the Assam-Arakan Oil Region (India and Burma), *Geol. Mag.* vol. lxxvii., 1940.

P. Evans and C.A. Sansom, The Oilfields of Burma, *Geol. Mag.* vol. lxxviii., 1941.

CHAPTER XVIII

The Eocene System

THE Eocene system includes three divisions: the lowest, known as the *Ranikot series*, directly overlies the *Cardita beaumonti* beds. Its typical development is in Sind, but the horizon has also been recognised in many other parts of North-west India, Assam, Tibet and in Burma. The middle division, the *Laki series*, is composed chiefly of richly fossiliferous nummulitic limestones, green shales, variegated shales and marls, while the upper, designated the *Kirthar series*, includes the bulk of the nummulitic limestone of Sind and of some of the extra-Peninsular hill-ranges. The names of the series are derived from hill-ranges in Sind. After summarising each of the three series of the Eocene we shall describe the developments in the more important areas in which the rocks have been studied.

Ranikot Series

This series is typically developed at Ranikot, on the Laki range, and occupies a considerable tract in Sind. The distribution of this series is somewhat more limited than that of the other members of the Tertiary, but fossiliferous Ranikot beds have been recognised in Sind, Kohat, the Salt-Range, Hazara, Pir Panjal, Tibet, Assam and Burma, and it is probable that unfossiliferous representatives occur elsewhere.

The series, which in Sind lies with apparent conformity on the *Cardita beaumonti* beds, includes in most of the Pakistan exposures a lower division of sandstones, clays, and shales, and an upper division of limestones and shales. The Ranikot series includes the coal-measures of the N.W. Punjab.

Fossils of the Ranikot series—The leading fossils of the Ranikot series are: (Echinoids) *Conoclypeus, Cidaris, Salenia, Phymosoma, Dictyopleurus, Paralampas, Hemiaster, Schizaster*; (Corals) *Trochosmilia, Stylina, Cyclolites, Montlivaltia, Feddenia, Isastraea, Astraea, Thamnastraea, Litharaea*; (Gastropods) *Tibia, Nerita, Terebellum, Velates, Crommium*; (Foraminifers) *Lockhartia, Alveolina, Nummulites* (*N. nuttalli*). The species *N. nuttalli* is characteristic of the Ranikot horizon.

Laki Series

Although of no great vertical extent, this series is of wide geographical prevalence in India. It includes a considerable thickness of nummulitic limestones and in places these are associated with oil-bearing beds. The series is well developed in Sind, Baluchistan, Kohat, the Salt-Range, the north-western part of the Punjab, Jammu, Bikaner, Assam and Burma. The rocks show numerous local variations; there is an essentially calcareous facies which is seen in the Salt-Range, a gypseous shaly facies which is found in Baluchistan, whilst in Assam and Burma there is a very thick development of dark shales at this horizon. The salt and gypsum of Kohat and perhaps of the north-western part of the Salt-Range belong to the Laki series, though much controversy still exists regarding the Eocene age of the Saline series of the eastern part of the Range. Laki (and Ranikot) horizons occur, moreover, in the group of massive limestones known as the *Hill limestone* in the Kala Chitta and Potwar area; it is there overlain by the lower part of the Chharat series (p. 313), which is Kirthar in age.

The important fossil organisms contained in the Laki strata may be recorded as *Nummulites atacicus, Assilina granulosa, Alveolina oblonga,* some species of *Nautilus, Echinolampas, Metalia, Blagraveia, Corbula, Gisortia,* etc., with numerous leaf impressions, fruits, seeds, etc. of plants belonging to the *angiospermous* division of the flowering plants.

Kirthar Series (Chharat Series)

Like the Ranikot and Laki, this series derives its name from a range in western Sind.

The Kirthar nummulitic limestone forms a conspicuous group of rocks in many parts of extra-Peninsular India, particularly Sind, Baluchistan, Kohat and Hazara, and to a more limited extent in the outer parts of the Himalaya, the central Assam range, and Burma. The prominent nummulitic limestone scarps of the Salt-Range are older, being of Laki age. In its type-area, Sind, it attains a great thickness of massive homogeneous limestone, capping all the high ranges of the Sind-Baluchistan frontier. There is no doubt that the nummulitic limestone of India is an eastern continuation of the same formation of Europe, a direct connection being traceable between these two regions through the nummulitic limestone formations of Baluchistan, Iran, Asia Minor, North Africa, Turkey and Greece to the west of Europe up to the Pyrenees. It thus forms a conspicuous landmark in the stratigraphical record of the whole world.

Fossils of the Kirthar series—The fossils include species of *Nummulites* etc., of which *N. laevigatus, N. perforatus, N. gizehensis,*

Assilina spira, *A. exponens*, *Alveolina elliptica*, *Discocyclina* sp., *Dictyoconoides* sp. are the most common. Other foraminifers are *Lockhartia*, *Orbitolites*, etc. Gastropods are present in large numbers; the genera *Conus*, *Cypraea*, *Cerithium*, *Strombus* and *Turritella* are very frequent. Portions of the coronae and spines of echinoids of large size, such as *Cidaris*, *Phymosoma*, *Echinolampas*, *Micraster*, *Hemiaster*, *Schizaster*, *Conoclypeus*, are common. The lamellibranchs are represented by the genera *Cytherea*, *Astarte*, *Cardita*, *Lucina*, and *Pholadomya*.

Important Areas

The following are the principal localities where Eocene rocks (Ranikot, Laki, and Kirthar series) are found: Sind, Baluchistan, the Salt-Range, Kohat, the Potwar, Hazara, Kashmir, the outer Himalayas, Assam and Burma.

Sind and Baluchistan

Ranikot series. Laki series—The Sind exposures of the Eocene provide the chief type for India. The Ranikot beds consist of soft sandstones, clays and carbonaceous and lignitic shales, containing pyrites in the lower part, succeeded by highly fossiliferous limestones and calcareous shales. The lower beds, both in their mineral composition as well as in the few dicotyledonous plants and fragmentary fossil bones that they contain, bear the impress of undoubted fluviatile origin. The overlying limestone with intercalated shales is about 200–250 m. in thickness, and abounds in fossil *echinoidea*, by means of which the series has been classified into zones. The Laki of Sind has been subdivided by W.L.F. Nuttall into a basal laterite, the *Meting limestone*, the *Meting shales* and the *Laki limestone*. The two limestones are lithologically similar but the lower bed is thinner and more fossiliferous, and contains a slightly different fauna. In Baluchistan the *Dunghan limestone*, which varies in thickness up to hundreds of metres, includes both Ranikot and Laki beds; and the overlying *Ghazij shales*, about 450 m. thick, are Laki.

Kirthar series—The Laki series is overlain by the Kirthar beds, a slight unconformity occurring in some sections, where the lowest bed is often a dark conglomeratic limestone. For the most part the lower Kirthar rocks are thick, massive, white, rather sparsely fossiliferous limestones with occasional shales. Higher in the sequence, shales form a much greater proportion of the beds, but in the highest beds white limestones are again abundant.

The Salt-Range

Ranikot—The lowest Eocene bed in the Salt-Range is a ferruginous pisolite, which passes laterally into a haematite clay and haematitic sandstone. This is in many places overlain by a thin gypseous and carbonaceous shale. These beds have received the name *Dhak Pass* beds. They are in turn followed by the *Khairabad limestone*, a brown and grey nummulitic limestone which shows very great variations in thickness. In the eastern part of the range the Ranikot series exhibits a predominantly shaly facies—the *Patala beds*, which overlie the attenuated Khairabad limestone and include thin coal seams. All these beds belong to the Ranikot and not to the Laki as was previously supposed. The coal is worked in a number of small mines in the face of the Salt-Range, and the pyritous shales have provided a source of alum.

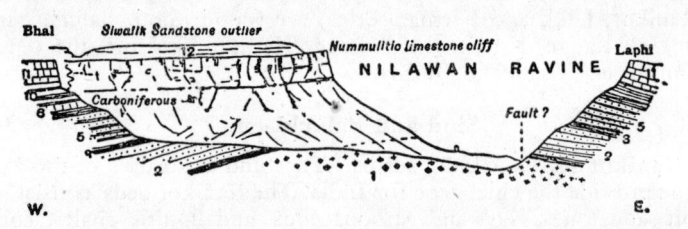

Fig. 33.—Sketch and section to show the nummulitic (Laki) limestone scarp in the Salt-Range. (Wynne, *Mem. G.S.I.* xiv.)

Laki—The lowest Laki beds of the Salt-Range tend to be somewhat shaly and are known as the *Nammal limestone-shales*; above them there is a more uniformly calcareous development, the *Sakesar limestone*, a light-coloured, somewhat cherty limestone which covers a large area of the Salt-Range. It has a well-defined series of joints and consequently a tendency to weather in cliffs having the aspects of "mural escarpments", presenting from a distance the general appearance of ruined walls or fortifications. Some of the finest cliffs of the range are produced in this manner by the action of the weathering agents. The mass of the rock is nearly pure calcium carbonate, made up almost wholly of foraminiferal shells, mostly of *Nummulites*, which on weathered surfaces of the rock stand out as little ornamented discs, flat or edgewise. In microscopic sections of this rock the internal structure of the *Nummulites*, as well as of other fossils, is clearly revealed, where crystallisation has not destroyed the organic structures. There are a large number of other fossils present as well, but they are difficult to extract from the unweathered rock. Large chert or flint nodules are irregularly dispersed in the limestone. The uppermost beds (*Bhadrar beds*) are more argillaceous. They are found on the plateau at the top of the range but seldom enter into the southward-facing scarp.

No Kirthar beds are known in the main part of the Salt-Range but they may occur in the north-eastern spurs towards Jhelum.

In the north-western portion of the Salt-Range a few kilometres south-east of Kalabagh, E. R. Gee has described a remarkable passage of the Sakesar limestone into massive gypsum. In the same neighbourhood the Bhadrar beds are also associated with gypsum, and show considerable resemblances to the lower parts of the Chharat series of the northern portion of the Potwar. This change from limestone into gypsum is evidently a secondary alteration phenomenon. Slightly further north-west of the point where this change takes place the gypsum is intimately associated with red marl and salt, and a little further north-west at Mari-Indus and Kalabagh the salt is worked on a large scale. Recent work by Gee has shown the complicated nature of the structure in the central part of the Salt-Range; and he believes the red marl, gypsum and salt to be of Cambrian age, and not Laki, as might be expected from the occurrence of salt and gypsum of Laki age to the north-west in Kohat.

[**Saline series of central Salt-Range of disputed age**—A reference has already been made to the Saline series of the central part of the Salt-Range (Khewra, Warcha, etc.). The gypsum, salt, and red marl underlie the Cambrian sequence, but the junction is demonstrably an irregular one and there has been much discussion about the relations between the Cambrian Purple sandstone and the underlying beds (p. 135). It is thought by some geologists that the disturbance is of small extent and is merely an expression of the difference in competence of the beds above and below. This explanation necessitates the assumption that there are two separate Saline series in the same neighbourhood, one of Cambrian and one of Eocene age. This seems an improbability, especially in view of the very close similarity of the supposed two sets of beds, and other geologists have interpreted the disturbed junction between the Saline series and the overlying Cambrian rocks as a thrust-fault of major importance which has brought the Cambrian strata on to the Eocene Saline series. So far no definite evidence has been forthcoming to show that the Saline series of the central portion of the Salt-Range is really of Eocene age, but the Salt-Range is known to include a series of important over-thrust faults and it does not seem improbable that there is a thrust-plane running along the base of the range. Further evidence of movement between the Saline series and the overlying beds is provided by the sections near Musa Khel; here the red marl and gypsum are overlain not by Cambrian beds but by the Talchir boulder-bed, and at many places along the junction the boulders have been greatly sheared.

B. Sahni has in voluminous papers described the nature and mode of occurrence of numerous plant micro-fossil remains (angiosperms, see page 135) in various rocks of the Saline series, obtained from deep borings (to eliminate chances of any adventitious introduction of microscopic air-borne plant-débris). The subject is still a major controversy among geologists in India and conflicting evidences continue to appear, supporting one side or the other, from natural rock-cuttings, thrust-planes, micro-fossils, and from deep borings put down for oil-wells[1].]

Kohat

The Rock-salt deposits of Kohat—A short distance to the north-west of the Salt-Range, at Bahadur Khel and elsewhere in the hills of the Kohat district, there are outcrops of Kirthar rocks, which are remarkable for being underlain by a great thickness of Laki beds composed of Alveolina limestone, which has been mostly altered to massive gypsum[2], and beds of rock-salt. At Bahadur Khel about 300 m. of these beds are laid bare in a perfect anticlinal section; the beds of rock-salt, which are seen at the centre of the anticline, are overlain by gypsum, the upper part of which is interbedded with green clay and shale. These beds are in turn succeeded by red clays and by Kirthar limestones containing *Nummulites, Alveolina* and other fossils. In lateral extent the outcrop of rock-salt is traceable for several kilometres. The salt is chemically pure crystalline sodium chloride with some admixture of calcium sulphate, but with no associated salts of potassium or magnesium as in the Salt Range deposits of the same mineral, from which also the Kohat salt differs in its prevailing dark-grey colours and in being slightly bituminous. It has been suggested that the two salt-deposits, in spite of these slight differences, have had a common origin, and are of the same age, and that the apparent infra-Cambrian position of the Salt-Range salt-deposits 150 km. to the S.E., as seen near Khewra, is due to overthrust faulting.

Samana Range—In the extreme north-west of Kohat the Ranikot beds are strongly developed in the Samana range and have been studied in considerable detail by Col. L. M. Davies. The basal *Hangu beds* are sandstones and shales with an abundant molluscan fauna; these constitute the lowest Eocene horizon found in India; the higher beds are mainly limestones (*Lockhart limestones*). A more normal facies of the Laki beds (as compared with the Bahadur Khel development) is exposed near Kohat itself. The lowest beds are green clays and shales; these are overlain by the *Shekhan limestone*. Above this limestone there is the Kirthar series with a gypseous red clay at the base, followed by the *Kohat shales*, limestones and shales, the *Nummulite shale* and the *Kohat limestone*. The Laki and Kirthar limestones and accompanying shales have been traced eastwards and north-eastwards, through the Margala and Kala Chitta hills and the Hazara mountains, to Muzaffarabad on the Jhelum and thence into Kashmir.

---

[1] *Proc. Nat. Acad. Sc. Ind.* vols. 14 and 16, 1944-46.
[2] D. N. Wadia and L. M. Davies, *Trans. Min. Geol. Inst. Ind.* vol. xxiv., 1929.

THE EOCENE SYSTEM

Sulaiman range—The Eocene succession in the hills west of Dera Ghazi Khan (West Punjab) may be referred to here as it has recently been described by F. E. Eames.[1] The Ranikot beds are of variable character with much shale; rather over 300 m. occur near the road to Fort Munro. The Laki series, about 1,200 m. thick, is mainly shale, with one thick limestone in the upper part. The Kirthar series, about 500 m., is largely chocolate-coloured clay with subordinate marl and limestone. In the uppermost 180 m. of Eocene beds *Pellatispira* is common and these beds are correlated with the Tapti series.

Potwar

East of Kohat, in the Kala Chitta hills in the northern part of the Potwar, the Eocene beds present a somewhat different facies. There is a strong development of limestones which include both Ranikot and Laki beds, and a thin coaly horizon presumably corresponding to the coal of the Salt-Range. These limestones, which are not strikingly fossiliferous, have been termed *Hill Limestone*. They are overlain by gypseous limestones which are followed by variegated shales (*Chharat series*) 100–150 m. thick, containing fossil vertebrate bones and a few fresh-water fossils. This horizon, known as the *Planorbis beds*, appears to be the base of the Kirthar series, and is associated with seepages of oil. The Kirthar series is represented also by the Kohat shales and the *Nummulite* shales, about 60–100 m. which come above the variegated shales (age approximately Middle Kirthar), the higher beds having been removed during the Oligocene denudation. The range of beds from the Planorbis beds upwards is known as the *Chharat series*.

In the southern part of the Potwar the Eocene has a development similar to that in the Salt-Range. Oil has been found at Khaur, Dhulian, Joya Mair, and Balkassar in the Sakesar limestone of Laki age. Drilling near Chakwal has shown the presence of the Ranikot (Khairabad and Dhak Pass beds) beneath the Laki.

Hazara

Eocene rocks, principally composed of massive, dark nummulitic limestone, play a prominent part in the geology of Hazara and, indeed, of the whole country around the N.W. frontier. At the base, the coal-bearing Ranikot series is identified, though it does not possess any economic resources, the quantity as well as quality of the coal being very inferior. The nummulitic limestone is a grey or dark-coloured massive rock of great thickness interbedded with nummulitic shale beds, and is thus somewhat different

---

[1] F. E. Eames, *Q. J.G.S.* vol. cvii. pt. 2, 1952.

from the equivalent beds in the Salt-Range. The limestone passes upwards into the Chharat series of shales and limestones which are unconformably overlain by the Murree series of fluviatile deposits.

The Eocene of Hazara extends eastwards beyond the Jhelum into Kashmir, following the great bend of the mountains and, as mentioned in the next section, it joins up with the nummulitic border fringing the south-western foot of the Pir Panjal.

Kashmir

The strong band of nummulitic limestones and associated rocks, belonging to the Ranikot and Laki series, which stretches from Kohat through the Hazara mountains to Kashmir, persists as a narrower band across the Jhelum, where it turns abruptly round the great syntaxial re-entrant of the mountains and runs along the foot of the Pir Panjal for more than 400 km. The width varies greatly, the band widening and narrowing between the two Panjal thrusts (page 392) which bound it on either side. The Eocene is also associated with the large inliers of Permo-Carboniferous limestones (page 216) in the younger Tertiaries of the Jammu hills, and includes deposits of coal, and aluminium and iron ores. The largest of these inliers are near Riasi and Poonch. Among these the Laki horizon is recognised by the presence of species of *Assilina*, *Alveolina* and *Nummulites* in the nummulitic limestone. Fossiliferous *Nummulites*-bearing rocks occur in Ladakh, closely associated with Cretaceous Flysch and igneous rocks (ophiolite), in a long band along the Indus from Drass to Rupshu and thence to Hundes and as far as Mt. Kailas. They form the north margin of the Tibetan zone of marine sediments of the Himalayas, representing the youngest formation of the Tethyan geosyncline. Detailed study of this area is still lacking, but the relationship of the Ladakh Eocene with the Eocene of Hazara, Kashmir and the Salt Range is beyond doubt.

The Eocene of Kashmir exhibits a double facies—one analogous with the Nummulitics of Hazara and the N.W. Punjab, the other recalling the Subathu facies of the type area in the Simla hills. The former type is well developed in the south-west flank of the Pir Panjal wherein, along its whole length from the Jhelum to the Ravi, it constitutes a remarkably consistent and characteristic belt of altered, obscurely nummulitic limestone of the "Hill Limestone" facies, overlain by a thick series of variegated shales with coal seams at the base (Chharat Series). Its width varies from a few metres to about 7 km. Lydekker ascribed these rocks to an indefinite age between the Carboniferous and Trias, and named them the "Kuling" and "Supra-Kuling" series; later work in the

Pir Panjal range, however, has established the Eocene age of these rocks beyond doubt.[1]

The Subathu series of Jammu—The Sind facies of the Nummulitics mixed with the Subathu type of the Eocene is met with in a number of inliers exposed in the Murree zone lying to the south of the Pir Panjal. The most important of these inliers occurs as a narrow rim bordering the outcrop of the Permo-Carboniferous (Krol) limestone, Sirban limestone (p. 216), exposed as the core of an anticlinal near Riasi, north of Jammu. Another is seen in Poonch exhibiting like relations.

The section given below illustrates the sequence of formations in the Eocene:

Lower Murree (some thousands of metres)
{ Purple and grey sandstones and shales of great thickness, underlain by ossiferous pseudoconglomerates.

Unconformity.

Subathu (100–200 m.)
{ Nummulitic limestone, thin-bedded, nodular, bituminous.
Olive shales, papery.
Nummulitic limestone, up to 100 m. thick.
Grey and olive shales.
Pyritous shales.
Ironstone shales, carbonaceous.
Coal seams (15 cm. to 3 m.) in pyritous shales.
Pisolitic bauxite and aluminous clays, 2 m.
Dykes of ultra-basic intrusive.
Chert breccia, 2 m. to 3 m.

Unconformity.

Permo-Carboniferous.
{ White cherty and silicified dolomitic limestone, unfossiliferous, thickness over 300 m., interbedded with Agglomeratic slate, near Sumlar, Kotli. (Sirban)

These inliers are exposed as the cores of faulted anticlinals in the Murree series, the north limb of which shows an apparently conformable passage of the Eocene into Murrees, while the south limb is generally missing as the result of strike-faulting.

Eocene bauxite and coal—The basal beds of the Eocene are highly interesting as containing evidence of an extensive laterite formation, which appears variably at different places, either as a pisolitic limonite, as highly aluminous clays, or as a pure bauxite. The laterite or bauxite covers an old land surface of the pre-Tertiary limestone, and marks a great erosional unconformity. In the valley of the Poonch, near Kotli, the base of the Eocene rests on the truncated edges of nearly vertically inclined strata of the

---
[1] D. N. Wadia, *Mem. G.S.I.* vol. li. pt. 2, 1928.

"Great Limestone", but this discordant junction is not equally apparent everywhere.

The pisolitic limonite and ironstone of Riasi and Poonch have been largely drawn upon in the past to support a flourishing industry of iron-smelting, while the associated bauxite deposits of these localities form large potential reserves of a high-grade ore of aluminium. At Riasi the overlying coal-measures, containing seams of anthracitic coal up to 6 metres in thickness, have been found to be workable and capable of supporting remunerative mining, but further westwards the coal is excessively friable, and distributed in very thin and inconstant seams which are severely crushed and in part graphitised.

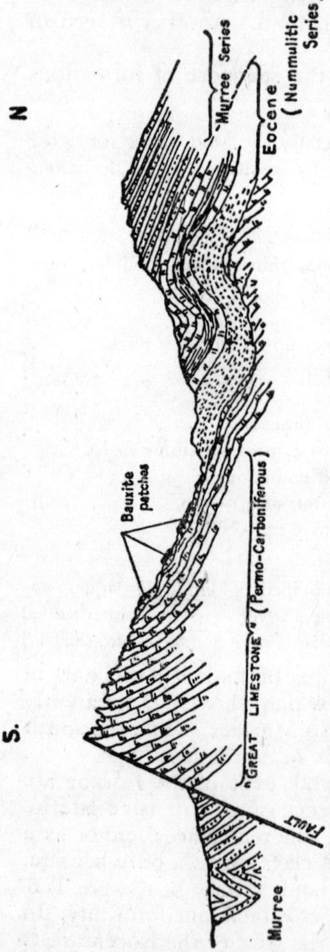

The nummulitic limestone is thin-bedded and black-coloured; it has a tendency to assume greater proportions as it is traced westwards of the Jhelum, in which direction the constitution of the whole series changes materially. The coal-seams become thinner and then disappear; the pisolitic iron-ore and bauxite are barely seen, while the nummulitic limestone steadily increases in bulk, becoming a massive monotonous formation of white or pale colour, whose aggregate thickness is over 500 metres. The species of foraminifers so far identified in these rocks are *Nummulites beaumonti, N. atacicus,* and *Assilina granulosa.*

Eocene of the Pir Panjal— The Eocene of the Pir Panjal probably belongs to a lower horizon, though its base is not exposed anywhere. The limestones are about 60–120 m. thick, generally thin-bedded and lenticular, containing obscure tests of *Nummulites* and *gastropoda*; they are greatly compressed, inverted

Fig. 34.—Section showing the relation of Permo-Carboniferous and Eocene, Jammu Hills.

THE EOCENE SYSTEM

and overthrust along both their inner and outer margin (see p. 399, *Murree* and *Panjal* thrusts), but they show a general resemblance to the "Hill Limestone" of the Punjab and Hazara (Laki-Ranikot age). A typical section shows:

Murree Series

———————————————————————Thrust-plane.

Laki	Variegated red and green shales with quartzose sandstones. - - - - -	250 m.
	Thin-bedded lenticular, black bituminous limestones with *Nummulites, Operculina, Assilina* and *Ostrea*. - - - -	30 m.
	Coaly and pyritous shales with ironstone shales and jasperitised beds. - - - -	15 m.
Ranikot	Massive, pale, grey-coloured, cherty, generally thin-bedded limestones, with badly preserved *Nummulites* and *gastropoda*. - Shale partings very few.	100–120 m.

Unconformity and thrust ~~~~~~~~~~~~~~~~~~~~~~~~~~~~~~ *Panjal thrust*

Panjal Trap and Permian or Trias limestones.

It is probable that this limestone group extends in a continuous outcrop along a general north-west direction from the northerly termination of the Panjal chain near Uri, along the Jhelum valley to Muzaffarabad, and thence to Hazara, merging into the wider Eocene zone of that region.

Punjab Himalayas

The extent and boundary of the Eocene sea of North India, a gulf of the Tethys, can be roughly judged by the extent and distribution of the outcrops of Nummulitic limestone preserved to-day—a more or less continuous belt extending from Sind and the Sulaiman hills, where it attains its maximum development, through Hazara, Muzaffarabad and the Pir Panjal chain to beyond Dalhousie and Subathu, and thence with decreasing width and some intermissions to Naini Tal.

The Eocene of the Outer Himalayas of Simla is distinguished as the *Subathu series*, which is collateral with part of the Kirthar series, possibly with the underlying Laki series. The Subathu series is typically developed near Simla. Subathu is a military station, from which the group takes its name. The rocks are red and grey, gypsiferous and calcareous shales, with some interbedded sandstones and subordinate limestones in which *Nummulites*, and other fossils, are found. This development differs from the more usual Laki and Kirthar beds in the small proportion of limestone.

There is also a difference in colour and texture, the Subathu lime-stone being grey to black in colour, very compact and thinly bedded. The lower beds are very variable and inconstant; there is a workable coal-seam in one locality, but this is missing from the type area, the lower beds being instead ferruginous sandstone and grits containing pisolitic haematite and limonite.

Assam

The Eocene rocks occupy a large area in Assam, and offer several points of interest. The lowest beds exhibit two sharply contrasted facies, one in the east of the province and the other in the west. In the Naga hills (in the eastern area) the lowest Eocene beds are the *Disang shales*—a great thickness of very well-bedded dark-grey shales with thin well-cemented sandstones. Towards the interior of the hills separating Assam from Burma the shales become hardened and slaty, and are associated with quartz veins and serpentine. It is just possible that in this area some of the beds referred to the Disang series are of pre-Tertiary age.

In the western part of Assam, there is developed a calcareous facies of the Eocene; this occupies a large area in the Shillong plateau (Garo hills, Khasi and Jaintia hills). The lowest beds here have recently been termed by C. S. Fox the *Tura stage*; they include sandstones and shales and thin seams of coal. This stage is now believed to include the *Cherra sandstone*, a band of hard coarse sandstone, and the various outcrops of thin coal occurring in and near the Garo hills; these were previously thought to be of Cretaceous age. These beds rest with no marked discordance on the Cretaceous but overlap on to the gneiss and other metamorphic rocks. At the base is commonly found kaolin and occasionally laterite. The Tura beds are followed by nummulitic limestones (*Sylhet limestone stage*) which show considerable lateral variation; shales and even sandstones are locally developed. The fauna is fairly rich and shows affinities with the Kirthar fauna of Pakistan. Above these limestones are the *Kopili alternations* including shale, thin coal, thin limestone, and thin sandstone; these beds are also fossiliferous. The range of beds including the Tura, Sylhet limestone, and Kopili alternations stages is known as the *Jaintia series*, corresponding in age to the upper part of the Disang series.

Recent work by the Burmah Oil Company geologists suggests that some revision of this classification is desirable. It appears that the Tura beds and Sylhet limestone are facies divisions characteristic of different conditions of deposition, and not successive stages of the Eocene. The following divisions are proposed, approximately correlated with the well-known sequence of Sind.

Prang limestone	Kirthar
Narpuh beds	Laki
Unconformity or palaeontological break	
Lakadong beds ⎫	
Therria beds ⎭	Ranikot

The Tura or Cherra sandstone of the Therriaghat-Cherrapunji neighbourhood is equivalent to the Therria beds, of Ranikot age, but farther west, near Tura, the Tura stage is younger, of Laki or even Kirthar age. The Sylhet limestone includes Lakadong to Prang. The coal of the Cherrapunji district is limited to the Lakadong series, and is thus of Ranikot age. In a north-westerly direction the various divisions of the Eocene become more arenaceous and thinner.

The Therria beds contain few fossils, but contain some algae which are elsewhere associated with Ranikot beds. The Lakadong beds contain foraminifers which indicate a Ranikot age. The Prang limestone includes many fossils known from the Kirthars of Pakistan, such as *Assilina, Discocyclina, Nummulites,* etc. In the Kopili alternations reticulate *Nummulites* and *Pellatispira* appear, and the fauna suggests a correlation with the uppermost Kirthars and the Tapti beds (pp. 291–92).

Both the Jaintia series and Disang series are overlain by the very thick *Barail series* which is of considerable economic importance as it contains thick seams of coal. This series includes thick hard sandstones which give rise to the Barail range which is the "backbone" of Assam; in addition there is a fairly large proportion of argillaceous beds which increases slightly in a north-eastern direction. The Barail series has been subdivided into stages on lithological grounds, but as fossils are extremely rare it has not been possible to correlate these stages precisely with those in other areas. The Barail beds show an important lateral variation; when traced from south-west to north-east, the carbonaceous material very steadily increases in amount and the carbonaceous shales pass into coaly shales, shaly coals, and thence into thin coals and so in Upper Assam into thick coals of good quality. In this area, the upper portion of the Barail series forms the *Coal-measure sub-series* but with a few thick, workable seams restricted to a small portion of the sequence. A few fossils discovered near the coal horizon suggest an uppermost Eocene or Oligocene age. Oil-shows are found in association with the Barail series in the Surma valley and in Upper Assam. There is no separation of "oil-measures" and "coal-measures"; oilsands often occur in between the coal seams and petroliferous coal seams.

Bengal

Underneath the alluvium of Bengal and the Gangetic delta

borings for petroleum deposits have revealed a thick series of Eocene strata, over 1,000 m. thick, resting over a SE shelving platform of Rajmahal trap and some Cretaceous beds and underlying a thick succession of estuarine and marine Oligocene to Pliocene formations, aggregating 1,700 m. This Tertiary series becomes thicker in a southerly direction, reaching over 4,600 m. near Port Canning. (B. Biswas, *Min. Res. Ser*, ECAFE, 1963).

Andaman Islands

These islands lie on the southern continuation of the orographic axis of the Arakan Yoma and are mainly composed of Tertiaries, resting upon a substratum of Cretaceous with its ultrabasic intrusives. A fairly full Cainozoic sequence is seen, commencing with Nummulite-bearing group of beds (*Port Blair sers*), through the Oligocene and Miocene to the Pleistocene.

Burma

The Eocene rocks in Burma are developed on a large scale, reaching a thickness of well over 6,000 m. They show a facies of deposits very different from that of their equivalents in Pakistan (Sind, Baluchistan, North-West Frontier Province, the Salt-Range) and Kashmir, but have considerable resemblances to the Eocene of Assam. Cotter has divided them into six stages (see Table on page 304). The Tertiary sequence commences with a basal conglomerate, *Paunggyi conglomerate*, of Ranikot age, resting over a somewhat obscure group of rocks which forms a large part of the Arakan Yoma from Cape Negrais northwards. These are known as the *Axial, Mai-i* and *Negrais* groups which probably include beds from Triassic to Cretaceous age. The greater part of the Lower Eocene is made up of the thick *Laungshe shales* which probably correspond to the Disang shales of Assam and are mainly of Laki age. A few thin seams of coal are met with in the overlying *Tabyin clays*. The Upper Eocene beds are of great interest. The *Pondaung sandstones*, about 1,500 m. thick, mark a temporary retreat of the nummulitic sea which was thrown back by thick deltaic accumulations, in which are preserved the earliest fossil mammals of the Indian region. These belong to the *Amynodonts, Metamynodonts* and *Titanotheres*, the ancestral forms of rhinoceroses, and the highly generalised extinct group of ungulates (the Anthracotheres)—*Anthracotherium, Anthracohyus,* and *Anthracokeryx*. Marine conditions were soon resumed, however, before the Eocene period came to a close, and in the *Yaw stage* there is a considerable development of marine beds containing foraminifers.

REFERENCES

W. T. Blanford, Geology of Western Sind, *Mem. G.S.I.* vol. xvii. pt. 1, 1879.

P.M. Duncan and W. P. Sladen, Tertiary Fauna of Sind, *Pal. Indica*, ser. vii. and xiv. vol. 1. pts. 2, 3 and 4, 1882–1884.

E. W. Vredenburg, Tertiary System in Sind, *Mem. G.S.I.* vol. xxxiv. pt. 3, 1906.

E. S. Pinfold, Structure and Stratigraphy of N.W. Punjab, *Rec. G.S.I.* vol. xlix. pt. 3, 1918.

E. H. Pascoe, Petroleum Deposits of Punjab and N.W. Frontier Provinces, *Mem. G.S.I.* vol. xl. pt. 3, 1921.

W. L. F. Nuttall, Stratigraphy of the Laki Series, *Q.J.G.S.* vol. lxxxi. pt. 3, 1925.

L. M. Davies, Ranikot Beds at Thal, *Q.J.G.S.* vol. lxxxiii. pt. 2, 1927; Fossil Fauna of the Samana Range, *Pal. Indica*, N. S. vol. xv., 1930; Tertiary Echinoidea of the Kohat-Potwar Basin, *Q.J.G.S.* vol. xcix. pt. 1, 1943.

L. M. Davies and E. S. Pinfold, Eocene Beds of the Punjab Salt-Range, *Pal. Indica*. N.S. vol. xxiv. mem. 1, 1937.

P. Evans, Tertiary Succession in Assam, *Trans. Min. Geol. Inst. Ind.* vol. xxvii., 1932.

E. R. Gee, Geology of the Salt-Range, *Proc. Nat. Acad. Sci. Ind.*, 14, 16, 1947.

F. E. Eames, Study of the Eocene in Western Pakistan and Western India: B. The Lamellibranchia, *Phil. Trans. Roy. Soc.* B. vol. 235, 1951; C. The Scaphopoda and Gastropoda, *ibid.* vol. 236, 1952; A. Geology of Standard Sections, *Q.J.G.S.* vol. cvii. pt. 2, 1952; D. Faunas of Standard Sections, *ibid.*

D. N. Wadia, Permo-carboniferous Limestone Inliers in the Sub-Himalayan Tertiary zone of Jammu, Kashmir Himalaya, *Rec. G.S.J.*, vol. 72. pt. 2, 1937.

CHAPTER XIX

The Oligocene and Lower Miocene System

OLIGOCENE

Restricted Occurrence—The Oligocene system is very poorly represented in India and it seems that during a part of this period a considerable amount of what is now the Tertiary outcrop was undergoing denudation, which resulted in the removal of such Oligocene deposits as had been formed, as well as some of the Eocene. The fullest developments of Oligocene rocks are in Sind and Burma. Rocks which are probably of Oligocene age occur also in Assam. In the few areas in which it is developed, the Oligocene appears to lie conformably upon the Eocene, although it is not impossible that there is a palaeontological break at this horizon. The Oligocene system is usually separated from the overlying Miocene beds by an unconformity or at least a palaeontological break.

The Oligocene system appears to be absent from Kohat, the Punjab, Kashmir and the North-West Himalaya.

Baluchistan

Flysch—In Baluchistan there is a great thickness of shallow-water sandstones and green arenaceous shales, with only subordinate limestone; this closely resembles the Flysch of Switzerland and covers a wide tract north of the Mekran coast. This formation is designated the *Kojak shales* succeeded by thick masses of sandstone and shales, *Hinglaj stage*. Fossils are rare but a few gastropods have been found, indicating the Oligocene age of at least some part of these beds.

Sind

Nari Series—The Oligocene beds in Sind are more interesting than the almost unfossiliferous Baluchistan Oligocene. They are part of the Nari series and overlie the Kirthar limestone with apparent conformity.

The name is derived from the Nari river, along the banks of which a section of the series is seen. The lower part of the Nari is composed of limestones, marls, shales and sandy limestones. Blanford pointed out the sharp distinction betwen these beds and the overlying arenaceous beds, and described an angular unconformity between them south of Sehwan. Later work by W. B. Metre and other geologists of the Burmah Oil Company has confirmed the distinction between the Upper and Lower Nari, and has shown that the Nari series is not a single unit, but is divisible into two parts separated everywhere by a break, the lower part being Oligocene and the upper part Miocene. This important break in the Tertiary sequence corresponds to the break found in Assam and Burma.

The Lower Nari attains a thickness of 550 m. near the Gaj and Nari rivers, but it thins in both directions away from the Kirthar range and is absent in the lower ground west of the Indus near the Hyderabad region. In the lowest part of the series whole beds are made of the tests of *Nummulites* and *Lepidocyclina*, especially *L. dilatata*, specimens of which (previously called *Orbitoides papyracea*) upto 5 or 7.5 cm. in diameter are not uncommon. *Nummulites intermedius-fichteli* is an index species for the Lower Nari, and other common forms are *N. nanggoelani* and *N. vascus*. Other fossils are *Montlivaltia, Schizaster, Eupatagus, Clypeaster, Lucina, Clementia, Corbula, Ostrea, Natica, Voluta*, etc.

Assam

Barail Series—It is probable that the greater part of the Barail series of Assam is of Oligocene age, but the extreme rarity of fossils makes it impossible to establish the age with certainty. The Barail series is overlain with marked unconformity by lower Miocene rocks.

Burma

Pegu Series—A very fossiliferous development of the Oligocene occurs in Burma and reaches a thickness of as much as 3,050 m. These beds form the lower half of the Pegu system, the upper half of the system being of Miocene age and separated from the underlying beds by an unconformity. The Lower Pegus have been subdivided into three stages. The lowest, the *Shwezetaw sandstone*, locally contains a few thin seams of impure coal; when traced from south to north the group shows a passage from marine beds into a continental facies. The Shwezetaw sandstones are overlain by the *Padaung clays* with a characteristic Middle Oligocene fauna. Above the Padaung clays, there is the *Singu stage* of Vredenburg including both sandstones and shales, but the Burmah Oil Company geologists have put forward a different classification for all the beds above the Padaung clays and they have termed the highest Oligocene rocks the *Okhmintaung sandstone*—a formation which shows great variation in thickness. The Okhmintaung sandstone is separated by a marked palaeontological break from the overlying Upper Pegu rocks.

Petroleum—While petroleum, like coal, can occur in rocks of any geological age from Cambrian to Pliocene, the most productive petroliferous strata in Asia are of Jurassic to Miocene age. The Pegu system of Burma has yielded large quantities of petroleum and its associated products. The oil of Yenangyaung has been known from ancient times. It was formerly obtained from wells dug by hand to considerable depths and was used as a preservative of wood-work, as a medicine, for lubricating, and as an illuminant. The most important oil-fields are Yenangyaung, Singu (Chauk), Lanywa, Yenangyat, and Minbu, all of which are in a small area in central Burma. The oil is found at the summits of anticlines and is obtained by drilling to very considerable depths. In Assam, the oil is of Barail-Surma age, while in Punjab it is of Upper Eocene-Oligocene horizon. In the Gujarat area it is also Nummulitic to post-Nummulitic.

Petroleum—Petroleum is a liquid hydrocarbon of complex chemical composition, of varying colour and specific gravity (0.8–0.98). Crude petroleum consists of a mixture of hydrocarbons—solid, liquid and gaseous. These include compounds belonging to the paraffin series (C_nH_{2n+2}) and also some unsaturated hydrocarbons and a small proportion belonging to the benzene group. Petroleum accumulations are usually asssociated with some gas (methane, ethane, etc.) called *natural gas*.

Origin of Petroleum—The origin of petroleum has been much debated; at one time it was thought that it had an igneous origin and the action of steam on metallic carbides was cited as an example of a possibly analogous process. It is now generally held that oil has an organic origin. This has been established not only by careful consideration of the circumstances in which oil is found throughout the world but also by the presence of optically active constituents in petroleum. The oil occurrences in India support the view of the organic origin of petroleum from animal or vegetable matter contained in shallow marine sediments, such as sands, silts and clays, deposited during periods when land and aquatic life was abundant in various forms, especially the minor microscopic forms of plants and animals. The history of lower and middle Tertiary sedimentation in certain deposition-centres in India shows that conditions for petroleum formation were favourable. Dense forests and a rich plankton flourished

---

[1] The impure bituminous substance sold in the bazaars as a drug of many virtues (*Salajit*) is a solid hydrocarbon found in the more exposed parts of the higher Himalayas as a superficial deposit. This substance, however, has nothing in common with petroleum, being of entirely different, and recent, organic origin.

in profusion in the gulfs, estuaries and deltas, and the lands surrounding them, during this period. Deposits of this organic muddy sediment in the land-locked sea or estuary or marsh must have precluded oxidation and decomposition of the organic matter, and promoted bacterial and biochemical action leading to the formation of various hydrocarbons.

Researches show that 60% of organic matter in modern marine sediment is derived from vegetation. The material which contributes most of this is the shallow-water plankton (floating or free-swimming algae, weeds and other organisms in fresh or sea water). Thousands of metres of diatomaceous beds have been met with in some of the oil-fields of California, and equally thick masses of foraminiferal limestones form productive oil-beds in Iran and Iraq.

The degree of porosity of reservoir rocks plays an important part in the underground storage of petroleum. The porosity of rocks may vary from 1 to 5%, in compact strata, and increase to 30 to 40% in some sands and sandstones. A porosity of 20% in a rock would mean a storage capacity of 2,462 cubic m. of oil per acre-metre. If, as is sometimes the case, the reservoir rocks are up to hundreds of feet thick, the oil stored would aggregate up to 1,000,000 barrels per acre.

Mode of Occurrence—Petroleum occurs in the pores and minute interstices of sands and in crevices in limestones, and is always closely associated with sediments which are of shallow water, usually marine, origin. The oil is derived from decomposition of the organic matter contained in the sediments, but the method by which the transformation into petroleum takes place is not yet completely known. It is evident that there must be special conditions in which there is incomplete oxidation of the carbon and hydrogen, and it has been suggested that the action of bacteria is a factor in these processes, especially in the elimination of the nitrogen of the animal tissues. It is possible that the change takes place in different stages.

At first the petroleum is disseminated throughout the geological formation in which it originated, but the pressure of overlying beds forces it to migrate into the most porous rocks and consequently it is generally found in sand beds and sandstones intercalated amongst clays and shales, although in some areas it occurs in the fissures and crevices of limestones. It is rarely found without gas, and saline water is likewise often present, associated with the oil. Oil in commercial quantities is not usually found where the component strata are horizontal, but in inclined and folded strata the oil and gas are found collected in a sort of natural chamber or reservoir, in the highest possible situations, *e.g.* the crests of anticlines. In such positions the gas collects at the summits of the anticlines, with the oil immediately below it. This follows of course from the lower density of the oil as compared with the water saturating the petroliferous beds. "In all cases there must apparently be an impervious bed above to prevent an escape of the oil and gas, and in this there is a certain similarity to the conditions requisite for artesian wells, but with the difference that the artesian wells receive their supplies from above and must be closed below, while the oil and gas wells receive the supplies from below and must be closed above. Both require a porous bed as a reservoir, which in the one case, ideally, but not always actually, forms a basin concave above, in the other concave below."[1] Where the rocks are not saturated with water, oil may occur in different circumstances, for example in the bottoms of synclines, but this type of accumulation is unknown in India. The porous sand beds, sandstones, conglomerates or fissured limestones which contain the oil must be capped by impervious beds in order that oil be not dissipated by percolation in the surrounding rocks.

Gas—The oil usually contains a large proportion of hydrocarbons which under normal pressure would be gaseous, but the pressures at great depths below the surface are sufficient to liquefy these hydrocarbons. In addition, other hydrocarbons (such as marsh gas) which are not liquefied by pressure are readily soluble in petroleum under pressure; in consequence, when the puncturing of an oilsand by drilling into it brings about a great local reduction in pressure there follows a brisk evolution of gas. This gas, escaping towards the well through the minute crevices in the sand or limestone, carries the oil with it. In this way the oil reaches the well and, if the pressure is sufficient, it will come up to the surface—sometimes with great force. Occasionally a well on reaching an oilsand may get out of control, and the oil flows high above the ground, but in India, Pakistan and Burma care is taken to avoid waste both of the oil and of the gas which plays so important a part in bringing about the production of the oil.

Oil-springs—In the search for oil in India a great deal is made of the existence of surface oil and gas springs. The presence of petroleum springs in an area, while it indicates the existence of subterranean oil, is not necessarily a proof of its existence in quantity. It may as likely prove the reverse. A single oil-spring discharging only half a litre of petroleum in a day may have during its whole existence dissipated at least 12 million tonnes of petroleum. A multiplicity of oilshows and springs, therefore, may be indicative more of the quantity of the oil and gas that has escaped than of what remains underground after the oozing or leakage of ages since its accumulation in the early or middle Tertiary.

Migration—The oil and gas are usually not indigenous to the rocks containing them but have been concentrated from a fairly large area by the combined effects of gravitation, capillary action and percolation, and underground water. In some cases the oil occurs a considerable distance away from, or above, the original source.

Petroleum Areas in India—Pascoe has drawn attention to the analogy between the petroleum areas of India, Burma, Assam, and N.W. Punjab, which appear to have been gulfs or arms of the nummulitic sea which were filled up by sedimentation. The Cambay gulf and the Gujarat oil-fields lie in a southern extension of the Punjab gulf.

G. W. Lepper has pointed out that the most prolific fields in Burma are situated on the eastern margin of a broad syncline corresponding approximately to the Chindwin-Irrawaddy valley. He suggests that the bulk of the oil-forming sediments were de-

---

[1] Chamberlin and Salisbury, *Geology*, vol. ii., 1909.

posited in a shallow marine environment, and that most of the oil of the Burma oil-fields has migrated into them from the sediments of this long and broad syncline.

LOWER MIOCENE

Distribution—Unlike the Oligocene, the Miocene system is very fully developed in India, being found in all the Tertiary areas of the extra-Peninsula. It is convenient to deal separately with the Lower Miocene, since in several areas this presents a notably different development from the upper portions of the system. In Sind the Lower Miocene rocks include the Upper Naris and also the *Gaj series*; in the Potwar and Kashmir they consist of the *Murree series*. In the Simla Himalayas the term *Sirmur* was applied to a group ranging from Eocene to Miocene but this term is inconvenient as it does not represent a natural unit, the Oligocene being absent. Consequently the term Sirmur system is now seldom used. The upper part of this group of rocks includes two series—*Dagshai* and *Kasauli*, which belong to the Lower Miocene. In Assam the Barail series is unconformably overlain by the *Surma series* which is of Lower Miocene age. In Burma the *Upper Pegus* are important, since they contain a petroliferous horizon and are very fossiliferous.

Sind

Gaj Series—The Upper Naris are overlain with apparent conformity by the Gaj series; this consists of richly fossiliferous dark-brown coral limestone, with shales, distinguished from the underlying Naris by the absence of *Nummulites*. The higher beds are red and olive shales which are sometimes gypseous; these in turn pass up into a series of clays and sandstones whose characters suggest deposition in an estuary or the broad mouth of a river. This shows a regression of the sea-border and its replacement by the wide basin of an estuary. Fossils are very numerous in the marine strata, representing every kind of life inhabiting the sea. The commonly occurring forms are *Ostrea* (spp. *O. multicostata* and *O. latimarginata*), *Tellina*, *Brissus*, *Breynia*, *Echinodiscus*, *Clypeaster*, *Echinolampas*, *Temnechinus Eupatagus*, *Lepidocyclina* and *Orbitoides*. The species *Ostrea latimarginata* is highly characteristic of the Gaj horizon, it being met with also in the parallel group of deposits within the upper part of the Pegu system of Burma. It is evident from the estuarine passage-beds that the Upper Gaj was the time for the expiry of the marine period in Sind and the beginning of a continental period. On the land which emerged from the sea, a system of continental deposits began to be formed, which culminated in an alluvial formation of great thickness and extent enclosing relics of the terrestrial life of the time. *Rhi-*

noceros is the only land-mammal whose remains have been hitherto obtained from the Upper Gaj beds.

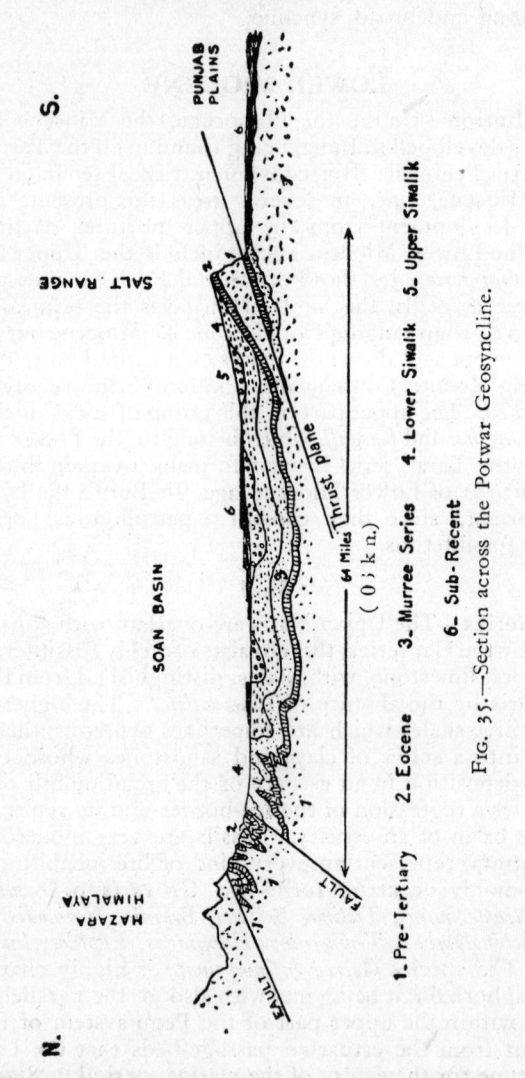

FIG. 35.—Section across the Potwar Geosyncline.

Bugti beds—In the Bugti hills of the Bugti country, in East Baluchistan, the fluviatile conditions had established themselves at an earlier date, the marine deposits in that country ceasing before the end of the Nari epoch. The overlying strata, *i.e.* Upper

Nari and the lower part of the Gaj, are fluviatile sandstones containing a remarkable fauna of vertebrates, of Upper Oligocene or Lower Miocene affinities. The leading fossils are the mammals *Anthracotherium, Cadurcotherium, Diceratherium, Baluchitherium* (a rhinoceratid, one of the largest land-mammals), *Brachyodus, Teleoceras* and *Telmatodon*, together with a few fresh-water molluscs, among which are a number of species of *Unio*. These beds are known as the *Bugti beds*.

Salt-Range, Potwar, Jammu Hills

The Potwar trough—One of the most perfect developments and exposures of the whole Tertiary sequence in India is observed in the geosynclinal trough of the Potwar, a plateau lying between the Salt-Range and the foot-hills of the Hazara district. In this area, with the exception of the Oligocene break, continuous sedimentation took place from the Ranikot stage onwards to late Pleistocene, resulting in deposits upto 7,600 m. thick, in which fossils belonging to most of the Tertiary time-divisions are recognised. On a floor constituted mainly of Mesozoic rocks there occur about 300 m. of the Nummulitics, overlain by 1,800 m. of the ferruginous, brackish-water sediments of Aquitanian and Burdigalian age, the *Murree series*, succeeded by over 5000 m. of the fluviatile and subaerial Siwalik strata. At the top, the Upper Siwaliks pass transitionally into the Older and Newer Pleistocene alluvia, loess, gravels, etc. The rock sequence in the Potwar basin-fold epitomises the Tertiary geology of Northern India. This syncline is 110 km. broad and 240 km. in length along its strike, tapering out east of the Jhelum into the Siwalik foothills zone. The Potwar basin is the smaller ramification of the Indo-Gangetic trough, the other southward and larger branch being the Rajasthan trough.

The southern edge of the Potwar basin is the great scarp of the Salt-Range mountains, a disrupted monocline; while its northern rim is the isoclinally folded Kala Chitta range at the south border of Hazara (Fig. 35).

The Potwar trough forms almost the north-western extremity of the much wider and larger Indo-Gangetic synclinorium, also filled up by Tertiary and post-Tertiary deposits, of which the Potwar may be regarded as a small-scale replica.[1]

Murree Series—In the eastern end of the Salt-Range, in the Potwar, and in the hills fringing the Jammu and Kashmir Himalaya, the various members of the Eocene are overlain by alternating sandstones and shales, the Murree series, very variable in thickness, but exceeding 2,450 m. where fully developed. At the

---

[1] *Memoirs, G.S.I.* vol. li. pt. 2, 1928; *Quart. Jour. Geol. Min. Met. Soc. Ind.* vol. iv. No. 3, 1931.

base of the series there is often a well-marked conglomeratic bed with bone fragments and derived nummulites. For some time the age of these beds was uncertain. The nummulites are of Eocene age (mainly Kirthar) and the other organic remains of Miocene (Gaj) age. It was therefore thought that the horizon represented a passage from the Eocene through the Oligocene to the Miocene. When it was recognised that the nummulites were all derived by erosion of the underlying Eocene rocks, the difficulty disappeared and the basal Murrees took their correct place in the succession.

Fatehjang zone—A few *palm* and *dicotyledon* leaf impressions and silicified wood remains, with very rare mammalian bones, fish and frogs, are all the fossils hitherto observed in the main body of the group. At the base, however, some 30 m. of ossiferous sandstone and conglomerate occur—the Fatehjang zone—containing *Anthracotherium*, *Teleoceras* and *Brachyodus*, which indicate close affinities with the Bugti beds fauna.

The Murree outcrop is over 40 km. wide where it crosses the Jhelum, but it thins eastwards rapidly, and where it intersects the valley of the Ravi it is only 5-7 km. across. At this point it merges into the Dagshai series of the Simla region. On lithological grounds the series is divisible into Lower and Upper stages of variable thickness:

Upper Murree	Soft, brown and buff, coarse sandstones, with inner cores of grey colour.
	Red and purple shales and nodular clays.
	Numerous *di-* and *mono-cotyledon* leaf impressions.
Lower Murree	Indurated, deep-coloured, at times inky purple and red sandstone, generally flaggy.
	Splintery, purple shales and deep red clays, with abundance of vein calcite.
	Numerous bands of pseudo-conglomerates.
	Unfossiliferous, except at base, where a few beds are ossiferous. Derived *Nummulites*.

Structurally and in their field relations the Upper Murrees present aspects of Siwalik type—open, broad folds weathered into strike-ridges and valleys with a succession of escarpments and dip-slopes, while the Lower Murrees show a far greater amount of compression, fracture and dislocation, being plicated in a series of tight isoclines and overfolds with repeated local faulting. They weather in the fashion of older rocks which are cleaved and jointed, and in which the alignment of the spurs and ridges has no close relation to the prevalent strike or "grain" of the country.

The range of age presented by the Murree series is difficult to determine, but it is clear that they are in the main Lower Miocene; it is not impossible that the lowest beds range down into the Oligocene. There is no sharp upward limit to the series, the passage into the overlying Kamlial stage being quite gradual.

The Murree series has a very restricted development in the Salt-Range, being absent from the greater part of the area. The gap between the Eocene and the overlying beds is thus greater than further north. In the western part of the Salt-Range the Eocene is followed by Kamlial beds, but in the trans-Indus area further west successively higher Siwalik horizons rest upon the Eocene.

It appears probable that unlike the Siwaliks, which are derived wholly from the denudation of the Himalayan granites and other rocks, the Murrees have originated from sediments whose source was the iron-bearing Purana formations of the Peninsular highlands to the south.

In Punjab, the inner limit of the Murree is a great thrust-plane—the *Panjal thrust*, where it abuts upon the Panjal range, a structural feature which persists further E'ds in the Simla Himalayas (*Nahan thrust*), and in the Kumaon Himalayas (*Krol thrust*). Further east, the equivalent of the Murrees are the Dagshais and Kasaulis.

Petroleum—In the Potwar, the Murree series occasionally contains petroliferous beds and has yielded at three localities, since 1916, considerable volume of liquid hydrocarbons. It is believed that the oil has migrated into the Murree series from the underlying Eocene.

Outer Himalayas

Dagshai and Kasauli Series—The outcrop of the Murree series in the Jammu hills, forming a belt 25-40 km. wide from the Indus to the Chenab, narrows towards the east and merges into the typical Dagshai-Kasauli band (*Sirmur series*) of the Simla area, a connection between the two being discernible in some plant-bearing beds in the valley of the Ravi. The Dagshai beds overlie the Subathus without any marked discordance, but there is nevertheless a large break, the whole of the Oligocene being absent. The lower part of the Dagshai series is made up of bright red nodular clay; the upper is a thickly stratified, fine-grained, hard sandstone which passes up, with a perfect transition, into the overlying Kasauli group of sandstones, which rocks are the chief components of the Kasauli series. No fossils are observed in the Dagshai group except *fucoid* marks and worm-tracks, fossils which are of no use for determining either the age of the deposit or its mode of origin. The Kasauli group also has yielded no fossils except a few isolated plant remains and a *Unionid*. The only traces of life visible in this thick monotonous pile of grey or dull-green coloured coarse, soft sandstones are some impressions of the leaves of the palm *Sabal major*. These are of importance because they enable the Kasauli horizon to be recognised further north-west in the Jammu hills.

Assam

Surma Series—The coal measures of Assam, belonging to the Barail series described in the last chapter, are unconformably overlain by the *Surma series*, equivalent to the Upper Nari and Gaj beds of Sind. The Surma series has a wide extent in the Naga hills, North Cachar hills, and Surma valley of Assam, and extends southwards through Chittagong to the Arakan coast of Burma. It is composed of sandstones and sandy shales, mudstones and thin conglomerates, generally free from carbonaceous content. In the Garo hills a small range of beds in the Surma series has yielded a large number of marine fossils, and another fossiliferous bed has been described from a slightly lower horizon in the Surma valley. Both faunas belong to the Lower Miocene; otherwise the series is remarkably unfossiliferous. Indications of petroleum are common in the Surma series in several localities.

Burma

Upper Pegu Series—The Upper Pegu rocks of Lower Miocene age form an important part of the Burma Tertiary sequence. As mentioned on pages 322-4 there is a break at the top of the Oligocene, and there is also a strong unconformity between the Pegus and the overlying Irrawaddy series. Consequently the thickness of the Upper Pegus is very variable. Petroleum is found in the Miocene beds but these are hardly as important a source of this mineral as the Oligocene. The abundant fossils of the Upper Pegus enable the age of the greater part of the group to be definitely identified as Lower Miocene; it is however probable that the uppermost Pegu beds are of Middle Miocene age. The Upper Pegus, like the Lower Pegus, show evidence of passing northwards into rocks deposited in more shallow water conditions. The extent of the unconformity between the Pegus and Irrawaddies varies considerably in different localities, and it has been suggested that in some parts of Burma there is very little break between the two sets of beds.

Igneous action during the Oligocene and Lower Miocene

The Middle Tertiary was the period for another series of igneous outbursts in many parts of extra-Peninsular India. The igneous action was this time mainly of the intrusive or plutonic phase. Unfortunately it is difficult to fix the precise age of these intrusions. The early Eocene rocks were pierced by large intrusive masses of granite, syenite, diorite, gabbro, etc. In the Himalayas, in Baluchistan and in Burma the records of this hypogene action are numerous and of a varied nature. Intrusions of granite took place along the central core of the Himalayas. In Baluchistan the plutonic action took the form of *bathyliths* of granite, augite-syenite, diorite, porphyrites, etc., while in Upper Burma and in the Arakan Yoma it exhibited itself in peridotitic intrusions piercing through the Eocene and possibly Oligocene strata. In the Myitkyina district of Upper Burma, a basaltic tuff appears to be interbedded with the Tertiary rocks, which are mainly of Eocene or Oligocene age.

Final Retreat of the Sea from N. India—In all the above Tertiary provinces of India that we have reviewed so far, from Sind to Burma, the transition from an earlier marine type of deposits to estuarine and fluviatile deposits of later ages must have been perceived. The passage from the one type of formation to the other was not simultaneous in all parts of the country, and marine conditions may have persisted in one part long after a fluviatile phase had established itself in another; but towards the middle of the Miocene period the change appears to have been complete and universal, and there was a final retreat of the sea from the whole of north India. This change from the massive marine nummulitic limestone of the Eocene age, containing abundant foraminifers, corals and echinoids, to the fluviatile deposits of the next succeeding age crowded with fossil wood and the bones of elephants and horses, deer and hippopotami, is one of the most striking physical revolutions in India. We must now turn to the great system of uppermost Tertiary river-deposits which everywhere overlies the Middle Miocene, enclosing in its rock-beds untold relics of the higher vertebrate and mammalian life of the time, comprising all the types of the most specialised mammals except Man.

REFERENCES

W. T. Blanford, Geology of Western Sind, *Mem. G.S.I.* vol. xvii. pt. 1, 1879.

G. E. Pilgrim, Tertiary Fresh-water Deposits of India, *Rec. G.S.I.* vol. xl. pt. 3, 1910.

E. H. Pascoe, Oil-fields of India, *Mem. G.S.I.* vol. xl. 1912–1920; *Rec. G.S.I.* vol. xxxviii., pt. 4, 1910.

P. M. Duncan and W. P. Sladen, Tertiary Fauna of Western India, *Pal. Indica*, ser. vii. and xiv., vol. 1, pt. 3.

G. E. Pilgrim, Vertebrate Fauna of the Bugti Beds, *Pal. Indica*, N.S., vol. iv. *Mem.* 2, 1912.

A. B. Wynne, Tertiaries of the Punjab, *Rec. G.S.I.* vol. x. pt. 3, 1877.
E. S. Pinfold, Stratigraphy of N.W. Punjab, *Rec. G.S.I.* vol. xlix., 1918.
G. de P. Cotter, Geology of the Attock District, *Mem. G.S.I.* vol. lv. pt. 2, 1933.
P. Evans, Tertiary Succession in Assam, *Trans. Min. Geol. Inst. Ind.* vol. xxvii., 1932.
G. W. Lepper, Geology of the Chindwin-Irrawaddy Valley of Burma and of Assam-Arakan, *Proc. World Petroleum Cong.* vol. i. pp. 15-25, 1933.
H. M. Sale and P. Evans, Geology of the Assam-Arakan Oil Region (India and Burma), *Geol. Mag.* vol. lxxvii., 1940.
P. Evans and C.A. Sansom, The Oilfields of Burma, *Geol. Mag.* vol. lxxviii., 1941.
E. V. Corps, Digboi Oilfield, Assam, *Bull. Amer. Assoc. Pet. Geol.* vol. 33, 1949.
H. R. Tainsh, Tertiary Geology and Principal Oilfields of Burma, *Bull. Amer. Assoc. Pet. Geol.* vol. 34, 1950.

CHAPTER XX

The Siwalik System

MIDDLE MIOCENE TO LOWER PLEISTOCENE

General—The newer Tertiaries occur on an enormous scale in the extra-Peninsula, forming the low, outermost hills of the Himalaya along its whole length from the Indus to the Brahmaputra. They are known as the Siwalik system, because of their constituting the Siwalik hills near Hardwar, where they were first known to science, and from which were obtained the first palaeontological treasures that have made the system so famous in all parts of the world. The same system of rocks, with much the same lithological and palaeontological characters, is developed in Baluchistan, Sind, Assam, and Burma, forming a large proportion of the foot-hill ranges of these provinces. Local names have been given to the system in the extra-Himalayan areas, *e.g.* the Mekran system in Baluchistan, Manchar system in Sind, the Tipam, Dupi Tila and Dihing series in Assam, and the Irrawaddy system in Burma, but there is no doubt about the parallelism of all these groups.

The Siwalik deposits—The composition of the Siwalik deposits shows that they are nothing else than the alluvial detritus derived from the subaerial waste of the mountains, swept down by their numerous rivers and streams and deposited at their foot. This process was very much like what the existing river-systems of the Himalayas are doing at the present day on their emerging to the plains of the Punjab and Bengal. An important difference is that the former alluvial deposits now making up the Siwalik system have been involved in the latest Himalayan systems of upheavals, by which they have been folded and elevated into their outermost foot-hills, although the oldest alluvium of many parts of northern India serves to bridge the gap between the newest Siwaliks and the present alluvium.

Geotectonic Relations of the Siwaliks—Some of these folds in the later phase of Himalayan orogeny were inverted and their middle limbs, reaching the limit of strength, have passed into highly inclined reversed faults, or *thrust-planes*, thus thrusting the older pre-Siwalik rocks of the Inner ranges over the younger rocks of the Outer ranges. These reversed over-thrust faults are a highly characteristic and significant feature of the Outer Himalayas; many of the reversed faults of the Siwalik zone can be traced over long distances. Wherever the Siwaliks are found in contact with older formations, the plane of junction is always a reversed fault with an apparent throw of many thousand metres, the younger Siwaliks dipping under the older Tertiaries or still older rocks of the Middle or even the Inner ranges of the Himalayas.

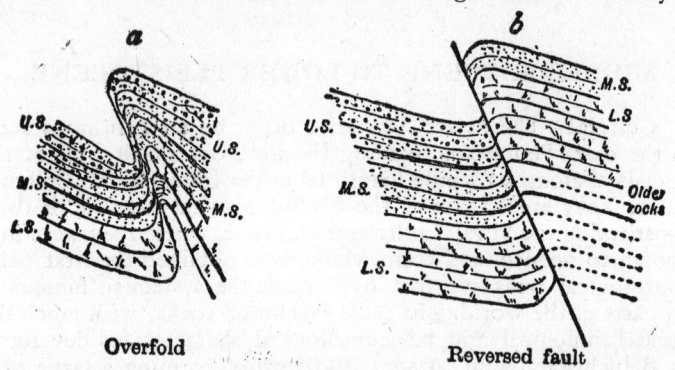

FIG. 36.—Diagrams to illustrate the formation of reversed faults in the Siwalik zone of the Outer Himalayas.

The Main Boundary Fault—This plane of contact is known as the Main Boundary Fault and is a noteworthy tectonic feature in the geology of the Siwalik foot-hill zone all along its length from the Indus to the Brahmaputra. The Main Boundary Fault again is not a single fault but is one of a series of more or less parallel faults among the Tertiary zone of strata, building the Outer Himalayan ranges, all of which exhibit similar tectonic as well as stratigraphic peculiarities of covering the Siwaliks under the older Tertiaries and the latter under the Puranas (slaty and schistose formations) of the Middle Himalayas.

The researches of Middlemiss and Medlicott tended to suggest that the Main Boundary is not an ordinary fault or dislocation which limits the boundary of the Siwaliks, but it marks the *original limit of deposition* of these strata against the cliff or foot of the then existing mountains, beyond which they could not extend. It was supposed that subsequently this limit had been further emphasised by some amount of faulting. The remaining boundary faults were considered to be the successive limits of deposition against the advancing foot of the Himalayas. This view of the nature of the Main Boundary faults will be made clearer by imagining that if the Indo-Gangetic alluvium, at present lying against the Siwalik foot-hills, were to be involved in a future Himalayan upheaval, they would exhibit much the same relations to the Siwaliks as the latter do to the older rocks of the Himalayas. These reversed faults thus were not "contemporaneous but successional". This hypothesis was based on the supposition that nowhere do the Siwaliks overstep the Main Boundary fault, or extend as outliers beyond it. Such outliers have in fact been found, and the faults, which are now proved to be of the nature of overthrusts, are definitely of later date than the deposition of the Siwaliks, and even subsequent to their plication.

In the Eastern Himalayas and particularly in the foot-hills south-east of the Brahmaputra valley, these faults that were previously regarded as Boundary faults, have been shown by detailed mapping, and in some cases by drilling, to be thrust-planes of moderate inclination along which older beds have been moved many miles across much younger beds. Furthermore in some cases, the rocks supposed to be bounded by the faults have been found to have extensive outcrops beyond the supposed "boundary faults". A striking example of this is provided by the Disang fault of Assam which was thought to be the boundary fault limiting the Barail coal-measures; large outcrops of Barails with coal-seams are now known beyond the fault. The mapping of a number of these over-thrust faults have now demonstrated that they may have throws of thousands of metres and consequently in overthrust areas the progress of denudation will have removed great thicknesses of strata from the up-throw side of the fault; it is thus evident that the above facts may account for the apparent restriction of the Siwaliks to a limited zone.

Geotectonic relations of the foot-hill zone of the Himalayas
—These reversed over-thrust faults giving rise to sheet-like recumbent folds (nappes) are a characteristic and highly significant feature of the Outer Himalaya. Many of the reversed faults we have discussed in the above paragraphs can be traced for hundreds of km. Wherever the Tertiary rocks of the outer ranges are found in contact with the older Himalayan formations, the plane of junction is always a low-angle reversed fault or thrust, along which the latter have moved several kilometres southwards as a sheet-fold—*nappe*—the Tertiary beds dipping under them. These have been designated, according to their locality, the *Murree*, and *Nahan* thrusts, followed inwards by the *Panjal* and *Krol* thrusts and these again in the interior by the *Zanskar*, *Giri* and the *Garhwal* thrusts, as the central ranges are approached[1].

---

[1] For geotectonics of the Tertiary belt and the Outer Himalayas reference may be made to Heim & Gansser, Assam & Nepal Himalayas: Swiss Nat. Sc. Soc., vol. lxxiii, 1, 1939; J.B. Auden, Garhwal Himalaya: *Rec. G.S.I.*, vol. lxvii, 4, 1934 and vol. lxxi, 4, 1937; W. D. West, Simla Himalaya; *Mem. G.S.I.*, liii, 1928; D. N. Wadia, Kashmir Himalaya; *Mem. G.S.I.*, vol. li, 2 1928 and *Rec.*, vol. lxv, 2, 1931.

The width of the foot-hill zone is thus determined by the outermost of these thrusts. In some parts the foot-hill zone is greatly constricted, sometimes even obliterated, by the trespass of its overthrust towards the south. This is very clearly seen in the Bhutan foot-hills, to the south of Darjeeling, where the thrust-sheet

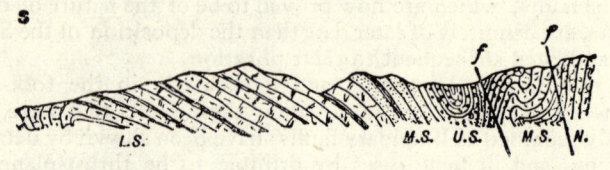

FIG. 37.—Section to illustrate the relations of the outer Himalaya to the older rocks of the mid-Himalaya (Kumaon Himalaya).

L.S. Lower Siwalik sandstones.   U.S. Upper Siwalik conglomerate.
M.S. Middle Siwalik sand-rock.   N. Older rocks. (C. S. Middlemiss).

originating in the Middle Himalayas, has moved southwards and over-riding the entire Tertiary Zone, including the Siwaliks, has come to lie against the edge of the alluvial plains of North Bengal. (Fig. 38b).

The palaeontological interest of the Siwalik system—The most notable character of the Siwalik system of deposits, and that which has invested it with the highest biological interest, is the rich collection of petrified remains of animals of the vertebrate sub-kingdom which it encloses, animals not far distant in age from our own times, and consequently, according to the now universally accepted doctrine of descent, the immediate ancestors of most of our modern species of land mammalia. These ancient animals lived in the jungles and swamps which clothed the outer slopes of the mountains. The more durable of their remains, the hard parts of their skeletons, teeth, jaws, skulls, etc., were preserved from decay by being swept down in the streams descending from the mountains, and entombed in rapidly accumulating sediments. The fauna thus preserved discloses the great wealth of the Himalayan zoological provinces of those days, compared to which the present world looks quite impoverished. Many of the genera disclose a wealth of species, now represented by scarcely a third of that number, the rest having become extinct. No other mammalian race has suffered such wholesale obliterations as the Proboscideans. Of the nearly thirty species of elephants and elephant-like creatures that peopled the Siwalik province of India, and were indigenous to it, only one is found living today. The first discovered remains were obtained from the Siwalik hills near Hardwar in 1839, and the great interest which they aroused is evident from the following popular description by Dr. Mantell: "Wherever gullies or fissures expose the section

THE SIWALIK SYSTEM

of the beds, abundance of fossil bones appear, lignite and trunks of dicotyledonous trees occur, a few land and fresh-water shells of existing species are the only vestiges of mollusca that have been observed. Remains of several species of river-fish have been obtained. The remains of elephants and of mastodontoid animals comprise perfect specimens of skulls and jaws of gigantic size. The tusks of one example are 2.895 m. in length and 68 cm. in circumference at the base.[1] This collection is invested with the highest interest not only on account of the number and variety of the specimens, but also from the extraordinary assemblage of the animals which it presents. In the sub-Himalayas we have, entombed in the same rocky sepulchre, bones of the most ancient extinct species of mammalia with species and genera which still inhabit India: *Eleurogale*, *Hyaenodon*, *Dinotheria*, mastodons, elephants, giraffes, hippopotami, rhinoceroses, horses, camels, antelopes, monkeys, struthious birds and crocodilian and chelonian reptiles. Among these mammalian relics of the past are the skulls and bones of an animal named *Sivatherium* that requires a passing notice. This creature forms, as it were, a link between the ruminants and the large pachyderms. It was larger than a rhinoceros, had four horns, and was furnished with a proboscis, thus combining the horns of a ruminant with the characters of a pachyderm. Among the reptilian

Fig. 38.—Section across the sub-Himalayan zone east of the Ganges river. (After Middlemiss.) L.S. Lower Siwalik. M.S. Middle Siwalik. U.S. Upper Siwalik. X. Nummulitic limestone. T. Tal series. N. Older rocks.

---

[1] This has been much exceeded in some later finds, e.g. a specimen discovered by the writer in the Upper Siwalik beds near Jammu, in which the left upper incisor of *Stegodon ganesa* was found intact with the maxillary apparatus and the upper molars. The tusk measured from tip to socket 3.226 m., the circumference at the proximal end being a little over 62 cm.

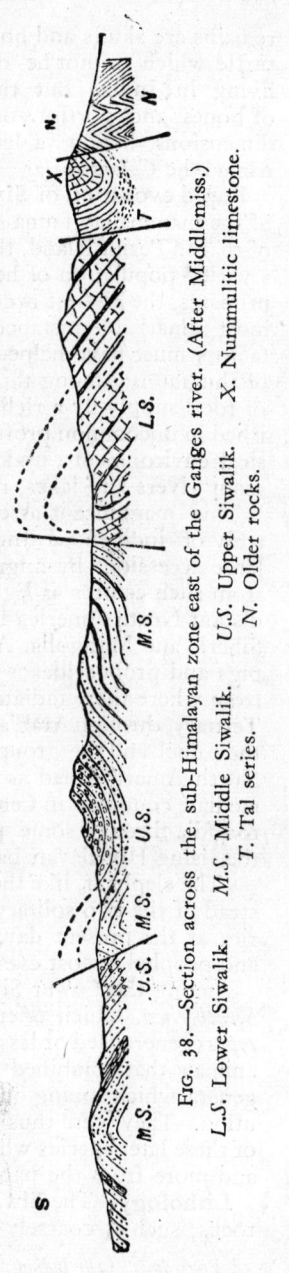

remains are skulls and bones of a gigantic crocodile and of a land turtle which cannot be distinguished from those of species now living in India. But the most extraordinary discovery is that of bones and portions of the carapace of a tortoise of gigantic dimensions, having a length nearly 20 feet. It has aptly been named the *Colossochelys Atlas.*"

Rapid evolution of Siwalik fauna—After the first few glimpses of the mammalian fauna of the Tertiary era in the Bugti beds and of that in Perim Island, this sudden bursting on the stage of such a varied population of herbivores, carnivores and rodents and of primates, the highest order of the mammals, must be regarded as a most remarkable instance of the rapid evolution of species. Many factors must have helped in the development and differentiation of this fauna; among those favourable conditions the abundance of food-supply by a rich angiospermous vegetation, which flourished in uncommon profusion, and the presence of suitable physical environments, under a genial climate, in a land watered by many rivers and lakes, must have been the most prominent.

This magnificent assemblage of mammals, however, was not truly of indigenous Indian origin; it is certain that it received large accessions by migration of herds of the larger quadrupeds from such centres as Egypt, Arabia, Central Asia and even from distant North America by way of the land-bridge across Alaska, Siberia and Mongolia. According to Pilgrim[1] our hippopotamus, pigs and proboscideans had their early origin in Central Africa, from where they radiated out and entered India during the late Tertiary, through Arabia and Iran; while the rhinoceros, the horse, the camel and the group of Primates, probably all originating in North America, had as their evolutionary centres various intermediate countries in Central and Western Asia, and were migrants to India through some passes on the north-west or north-east of the rising Himalayan barrier.

The elephant, like the horse, has been a world traveller and instead of the two solitary species inhabiting India and South Africa at the present day, it had in late Tertiary times spread to and peopled almost every country of the world except Australia.

Among the Lower Siwalik mammals there are forms, like the *Sivatherium,* which offer illustrations of what are called *synthetic types* (generalised or less differential types), *i.e.* the early primitive animals that combined in them the characters of several distinct genera which sprang out of them in the process of further evolution. They were thus the common ancestral forms of a number of these later species which in the progress of time diverged more and more from the parent type.

Lithology—The Siwalik system is a great thickness of detrital rocks, such as coarsely-bedded sandstones, sand-rock, clays and

---

[1] *Proceedings,* 12*th Indian Science Congress,* Geology Section, Benares, 1925.

conglomerates, measuring between 4,500 and 5,200 m. in thickness. The bulk of the formation, as already stated, is very closely similar to the materials constituting the modern alluvia of rivers, except that the former is somewhat compacted, has undergone folding and faulting movements, and is now resting at higher levels, with high angles of dip. Although local breaks exist here and there, the whole thickness is one connected and complete sequence of deposits, from the beginning of the Middle Miocene to the close of the Siwalik epoch (Lower Pleistocene). The lower part, as a rule, consists of fine-grained micaceous sandstones, more or less consolidated, with interbedded shales of red and purple colours: silicified mono-and dicotyledonous wood and often whole tree-trunks are most abundant throughout the Siwalik sandstones, and leaf-impressions in the shales. The upper part is more argillaceous, formed of soft, thick-bedded clays, capped at places, especially those at the debouchures of the chief rivers, by an extremely coarse boulder-conglomerate, consisting of large rounded boulders of siliceous rocks.

The lithology of the Siwaliks suggests their origin; they are chiefly the water-worn débris of the granitic core of the central Himalaya, deposited in the long and broad valley of the "Siwalik" river (p. 51). The upper coarse conglomerates are the alluvial fans or talus-cones at the emergence of the mountain streams; the great thickness of clays and sands represents the silts and finer sediments of the rivers laid down in flood-plains; while it is probable that the lower, *e.g.* Kamlial, beds were formed in the lagoons or estuaries of the isolated sea-basins that were left by the retreating sea as it was driven back by the post-Murree upheavals. These lagoons and estuaries gradually freshened and gave rise to fluviatile and then to subaerial conditions of deposition.

The composition as well as the characters of the Siwalik strata everywhere bears evidence of their very rapid deposition by the rejuvenated Himalayan rivers, which entered on a renewed phase of activity consequent on the uplift of the mountains. There is very little lamination to be seen in the finer deposits; the stratification of the coarser sediments is also very rude; the great thickness of clays and sands represents the silts and finer sediments of the rivers laid down in flood-plains; while current-bedding is universally present. There is again little or no sorting of grains in the sandstones, which are composed of unassorted sandy detritus derived from the Himalayan gneiss, in which many of its constituent minerals can be recognised, *e.g.* quartz, felspar, micas, hornblende, tourmaline, magnetite, epidote, garnet, rutile, zircon, ilmenite, etc.

[Under the direction of P. Evans a great deal of detailed examination of heavy mineral constituents of the Upper Tertiary sediments of India has been carried out.] The results of several thousand analyses have afforded useful data regarding the distribution of hornblende, epidote, kyanite, stauroite, etc., which are likely to be of value for correlation purposes where other means such as fossils or stratigraphic proofs are not available.]

The idea of the older geologists that the whole Siwalik system of rocks was deposits of the nature of alluvial fans, talus slopes, etc., at the debouchures of the Himalayan rivers very much along the sites of their present-day channels, does not appear to be tenable on the ground of the remarkable homogeneity that the deposits possess. Not only do they show on the whole uniformity of lithological composition at such distant centres as Hardwar, Simla hills, Kangra, Jammu and the Potwar, but also there is a striking structural unity of disposition along a definite and continuous line of strike. This negatives any theory of the deposition of these rocks in a multitude of isolated basins.

The periodic uplift of the Himalayas, accompanied by the encroachment of the mountain-foot gradually towards the rapidly filling depression to the south, resulted in the main drainage channels being pushed southwards. As the uplift proceeded, each periodic uprise of the mountains rejuvenated the vigorous young streams from the north while the drainage from the south became enfeebled and disorganised, so that in the building up of the Siwalik pile the sediments from the Gondwana mainland had but little share. How far southwards the Siwaliks extended is not certain, but it is highly probable that a considerable breadth of the Siwaliks lies buried under the alluvium of the Ganges.

Classification—On palaeontological grounds the system is divisible into three sections, the passage of the one into the other division being, however, quite gradual and transitional:

Upper Siwalik, 1,800–2,750 m.	Boulder-conglomerate zone : *Elephas namadicus, Equus, Camelus, Buffelus palaeindicus.* Pinjor zone: *E. planifrons, Hemibos, Stegodon.* Tatrot zone : *Hippohyus, Leptobos.*	Coarse boulder-conglomerates, thick earthy clays, sands, and pebbly grit, passing up into older alluvium. Richly fossiliferous in the Siwalik hills.	Lower Pleistocene to Lower Pliocene.

Middle Siwalik, 1,800– 2,500 m.	Dhok Pathan zone: *Stegodon, Mastodon,* large *Giraffoids, Sus, Merycopotamus.*	Grey and white sandstones and sandrock with shales and clays of pale and drab colours. Pebbly at top. The richest Siwalik fauna occurs in the Salt-Range.	Pontian to Middle Miocene.
	Nagri zone: *Mastodon, Hipparion, Prostegodon.*	Massive, thick, grey sandstones with fewer shales and clays, mostly red coloured.	
Lower Siwalik (*Nahan*), 1,200– 1,500 m.	Chinji stage: *Listriodon, Amphicyon, Giraffokeryx, Tetrabelodon.*	Bright red nodular shales and clays with fewer grey sandstones and pseudo-conglomerates. Unfossiliferous in the Siwalik hills (Nahans).	Middle Miocene. Tortonian.
	Kamlial stage: *Aceratherium, Telmatodon, Tetrabelodon, Anthropoids, Hyoboops.*	Dark, hard sandstones and red and purple shales and pseudo-conglomerates. Fossiliferous in the Punjab.	Helvetian.
Upper Murree.	Conformable passage downwards into Upper Murree sandstones and shales.		Burdigalian.

The top beds of the Upper Siwaliks—the boulder-conglomerate stage—probably mark the beginning of the Ice Age in N.W. India; this conglomerate carries some mingled glacial débris, though the majority of the boulders show no sign of ice action. An interesting occurrence of a true glacial boulder-bed is observed at Bain, in the Marwat hills, Shekh Budin range of Waziristan, among Upper Siwaliks. The Bain boulder-bed (about 20 m. thick) overlies strata containing *Elephas planifrons, Equus siwalensis* and *Bos*, and it underlies beds containing *Elephas planifrons, E. hysudricus, Equus* and *Bos*. This may be considered the earliest Pleistocene glacial deposit in India.[1]

---

[1] T. Morris, *Q.J.G.S.* vol. xciv., London, 1938.

The Potwar terrain immediately north of the Salt-Range and the Kangra-Hardwar tract may be regarded as type-areas of the Siwaliks both as regards stratigraphy and faunas.

Siwalik Zone of Kashmir Sub-Himalayas—For a brief description of the character and disposition of the main divisions of the Siwalik zone in N.W. India we might select the Siwalik belt of Jammu as a type area.

Rocks of the Siwalik system are disposed in parallel folded zones constituting the outermost foot-hills, which have a width of some 40 km. The Siwalik system of the Jammu hills does not differ in any essential respect from that developed in the rest of the Himalayas from Afghanistan to Assam. Structurally, stratigraphically, as well as palaeontologically, it exhibits similar characters, broadly speaking, to those found in the better-known areas of Kangra to the east and the Potwar to the west of the Jammu hills, which have yielded relics of the highest value, bearing on the problem of the phylogeny of Mammals.

On the whole, while the Upper and Middle Siwaliks of the Jammu hills show a more or less close lithological analogy with those of the adjacent Salt-Range and Potwar areas, the lower division exhibits marked local variations, which relate it more nearly to the Murrees than to the typical Kamlial or Chinji facies. This persistence of Murree conditions of deposition during Lower Siwalik time becomes more marked nearer the Jhelum valley, in the Poonch area, where between the Upper Murrees and the basal beds of the Lower Siwaliks there is no difference whatever of rock-facies, save the local occurrences of fragmentary bones of fresh-water reptiles and mammals in the latter group.[1]

Lower Siwalik—Petrologically the Lower Siwaliks are composed, from the bottom upwards, of indurated brown sandstones liberally intercalated with thick strata of red and purple semi-nodular clays, having a general resemblance with the Upper Murrees on the one hand towards the west and the typical Nahans of the Simla hills towards the east. The lower, harder and more purple coloured beds, about 600–900 m. in thickness, possess a fauna of Kamlial age, though of a very meagre description. The upper, scarcely less indurated, but more shaly division is of like vertical extent, and is characterised by a newer fauna of Chinji type, in the few localities from which fossils have been collected. Fossil plants and woody tissue are met with abundantly in the lower part, together with bones of a varied reptilian population of *Chelonia*, *Crocodilus*, *Gavialis*, fishes and snakes, mixed with gastropod shells and their opercula. The upper division has yielded numerous *Mastodon*, *Dinotherium*, *Microbunodon*,

---

[1] Geology of Poonch State, Kashmir, *Mem. G.S.I.*, vol. li. pt. 2, 1928.

Dorcatherium, Giraffokeryx, Aceratherium, several species of Anthropoid apes,[1] Antelopes, Giraffes, and several genera of the *Suidae* and *Anthracotheriidae*.

Middle Siwalik—Overlying this group there comes the Middle Siwalik group of thick massive beds of coarse micaceous sand-rock, at times too incoherent to be termed sandstone. Clays and shales are sparingly developed in these, and they have not the bright vivid coloration of the shales of the lower division. The prevalent colour of the sand-rock is pepper-and-salt grey. Its cementation is very unequal, much of the cement being concentrated in large, hard, fantastically shaped concretions which at times enclose fossil teeth, skulls or bones, leaving the main part of the rock a crumbling mass of sand. There is a well-marked Dhok Pathan stage, underlain by the Nagri zone in the Udhampur *Dun*. Pebbles are found, and increase in numbers and size as the upper limit of the Middle Siwalik series is reached, till they form enormous beds and lenticles of coarse bouldery conglomerates. The Dhok Pathan stage is recognised by *Hipparion, Bramatherium*; several suidae, *e.g. Potamochoerus, Listriodon* and *Tetraconodon*; *Tragocerus, Hippopotamus, Stegodon* and *Rhinoceros*.

Upper Siwalik—The Upper Siwaliks consist lithologically either of very coarse conglomerates, the boulder-conglomerates, or massive beds of sand, grit and brown and red earthy clays. The former occur at the points of emergence of the large rivers—the Ravi, Tavi, Chenab and Jhelum—and of their chief tributaries, whereas the latter occupy the intervening ground. The clays in the upper part of the series are indistinguishable from the alluvial clays of the Punjab plains into which they pass by an apparently conformable passage upwards.

Fossils are numerous in the Upper Siwaliks at some localities. This area appears to have been a favourite haunt of a highly diversified elephant population, as is evident from the profusion and wide distribution of their skeletal remains. Incisors of *Elephas, Stegodon, Mastodon*, their molars, skull plates, mandibles, maxillae, limb-bones, etc., are commonly found in the sands and conglomerates. Other fossils are referable to *Bubalus, Bos, Hippopotamus, Rhinoceros, Sus, Equus, Cervus*, Apes, *Gavialis* and numerous *Chelonian* bones.

The precise boundaries of the various Siwalik divisions described above cannot be delimited in the absence of positive or sufficient fossil evidence, nor is more minute subdivision into stages and zones possible. The inner boundary of the Siwaliks is, as stated above, a faulted one only as far as the Chenab, beyond which, westwards, the fault gradually diminishes and is replaced by an

---

[1] From a locality near Ramnagar village, 32 km. north of Jammu, species of *Sivapithecus* and *Dryopithecus* have been found.

anticlinal flexure. It is well-marked and typical at Udhampur, but has lost its significance at Kotli, where Siwalik outliers are found inside the boundary, in synclinal troughs of the Murrees. The parallel boundary faults within the Siwalik zone of the eastern Himalayas (Figs. 37 and 38) are not observed in the foothills west of Udhampur; the system of strike-faults that is met with in this area is of the nature of ordinary dislocations, which have no significance as limits of deposition.

Physiography of Siwalik country—The weathering of the Siwalik rocks has been proceeding at an extraordinarily rapid rate since their deposition, and strikingly abrupt forms of topography have been evolved in this comparatively brief period. Gigantic escarpments and dip-slopes, separated by broad longitudinal strike-valleys and intersected by deep meandering ravines of the transverse streams—surface features which are the most common elements of Siwalik topography—give us a quantitative measure of the subaerial waste that has taken place since the Pleistocene. The strike is remarkably constant in a N.W.-S.E. direction, with only brief local swerves, while it is almost always in strict conformity with the axes of even the subordinate ridges and elevations. The only variations in strike-direction from this course are the ones already referred to.

Although the Siwalik strata are often highly inclined, especially towards their inner limits, they are never contorted or overfolded, as is the case with the Murrees.

Siwalik Fauna

The Siwalik deposits enclose a remarkably varied and abundant vertebrate fauna in which the class *Mammalia* preponderates. The first collections were obtained from the neighbourhood of the Siwalik hills in the early thirties of the last century, and subsequent additions were made by discoveries in the other Himalayan foothills. They have been recently considerably enriched by discoveries in the Potwar and Kangra areas by Dr. Pilgrim. He has brought to light, in a series of brilliant palaeontological researches, a number of rich mammaliferous horizons among these deposits, which are of high zoological and palaeontological interest. These have established the perfect uniformity and homogeneity of the fauna over the whole Siwalik province, and have enabled a revised correlation of the system. In very suggestive papers Pilgrim has discussed the problems of the phylogeny, interrelations and migrations of the various groups of prehistoric mammals into and out of India during the Siwalik epoch, when India's population of the higher mammals was far greater than it is to-day. An important element in the mammalian fauna of the Siwaliks consists of the remains of creatures belonging to the most highly developed order, the primates, these including some fifteen genera of anthropoid apes, extending in stratigraphic range from Middle Miocene to early Pleistocene. The fossil primates so far discovered are, however, unfortunately very fragmentary, and in the present stage of our knowledge no definite conclusions as to the probable lines of descent of these forms and their position with respect to the line of human ancestry in India can be safely drawn, yet the proof of the presence of a vigorous and highly differentiated family of the anthropoid apes (*Simiidae*), in an epoch directly anterior to that of man, suggests that Upper Siwalik Man may have existed in India and that his fossil remains may some day be found. The following is a list of the more important genera and species of *Mammalia* classified according to Dr. Pilgrim.

Upper Siwalik:
  Primates: *Simia, Semnopithecus, Papio.*
  Carnivores: *Hyaenarctos sivalensis, Mellivora, Mustela, Lutra, Canis, Vulpes, Hyaena, Crocuta, Panthera, Ursus, Hystrix, Viverra, Machaerodus, Felis cristata.*
  Elephants[1]**:** *Mastodon sivalensis, Stegodon ganesa, S. clifti, S. insignis, Elephas planifrons, E. hysudricus, E. namadicus.*
  Ungulates: *Rhinoceros palaeindicus, Equus sivalensis, Sus falconeri, Hippopotamus, Camelus antiquus, Giraffa, Indratherium, Sivatherium giganteum, Cervus, Moschus, Buffelus palaeindicus, Bucapra, Anoa, Bison, Bos, Hemibos, Leptobos.*

Middle Siwalik:
  Primates: *Palaeopithecus, Semnopithecus, Dryopithecus, Ramapithecus, Sugrivapithecus, Cercopithecus, Macacus.*
  Carnivores: *Hyaenarctos, Indarctos, Palhyaena, Mellivorodon, Lutra, Amphicyon, Machaerodus, Felis.*
  Rodents: *Hystrix.*

---

[1] The observed succession of fossil elephants in the Upper Siwaliks (Pliocene to Pleistocene) of India is:

Mastodon cautleyi	Lr. Pliocene	
Mastodon sivalensis	Mid. Pliocene	Tatrot stage
Elephas (Stegodon) clifti	Mid. Pliocene	Tatrot stage
Elephas (Stegodon) bombifrons	Mid. Pliocene	Pinjor stage
Elephas (Stegodon) insignis	Up. Pliocene	Pinjor stage
Elephas (Stegodon) ganesa	Up. Pliocene	Pinjor stage
Elephas hysudricus	Up. Pliocene	Pinjor stage
Elephas planifrons	Lr. Pleistocene	Blder-Congl. stage
Elephas namadicus (syn. *antiquus*)	Mid. Pleistocene	Blder-Congl. stage
Elephas primigenius (the mammoth)	Mid. to Up. Pleist.	Plateau gravels
Elephas maximus	Sub-Recent & Recent River alluvium	

Elephants: *Dinotherium, Tetrabelodon, Prostegodon cautleyi* and *latidens, Stegodon clifti, Mastodon hasnoti.*

Ungulates: *Teleoceras, Aceratherium, Hipparion* (very common), *Merycopotamus, Tetraconodon, Hippohyus, Potamochoerus, Listriodon, Sus punjabiensis, Hippopotamus irravaticus, Dorcatherium, Tragulus, Hydaspitherium, Vishnutherium, Cervus simplicidens, Gazella, Tragocerus, Anoa.*

Lower Siwalik:

Primates[1]**:** *Sivapithecus indicus, Dryopithecus, Indraloris, Bramapithecus, Palaeosimia.*

Carnivores: *Dissopsalis, Amphicyon, Palhyaena, Vishnufelis.*

Proboscidians: *Dinotherium, Trilophodon.*

Ungulates: *Aceratherium, Hyotherium, Anthracotherium, Dorcabune, Dorcatherium, Hemimeryx, Brachyodus, Hyoboops, Giraffokeryx, Conohyus, Sanitherium, Listriodon, Telmatodon.*

Besides these the lower vertebrate fossils are :

Birds: *Phalacrocorax, Pelecanus, Struthio, Mergus.*

Reptiles: (Crocodiles) *Crocodilus, Gavialis, Rhamphosuchus*; (Lizard) *Varanus*; (Turtles) *Colossochelys atlas, Bellia, Trionyx, Chitra*; Snakes, Pythons.

Fish: *Ophiocephalus, Chrysichthys, Rita, Arius,* etc.

Special interest attaches to the occurrence of about eleven genera of fossil primates in the Siwalik group. These fossils furnish important material for the study of the evolution of the highest order of Mammals, the phylogeny of the living anthropoid apes, and the probable lines of human ancestry.

A most interesting and representative collection of the Siwalik fossils of India is arranged in a special gallery, the Siwalik gallery, in the Indian Museum, Calcutta.

Age of the Siwalik system—From the evidence of the stage of evolution of the various types composing this fauna, and from their affinity to certain well-established mammaliferous horizons of Europe, which have furnished indubitable evidence of their age because of their interstratification with marine fossiliferous beds, the age of the Siwalik system is considered to extend from the Middle Miocene to the Lower and even Middle Pleistocene. The Middle Siwaliks are believed to be homotaxial with the well-known Pikermi series of Greece, of Pontian, *i.e.* uppermost Miocene, age.

A parallel series of deposits is developed in other parts of the extra-Peninsula, as already alluded to. These have received local names but they are in most cases also fluviatile or subaerially deposited sandstones, sand-rock, clays and conglomerates, con-

---

[1] Catalogue of the Fossil Anthropoids of India, *Rec. G.S.I.* vol. lxxii. pt. 4, 1938.

taining abundant fossil wood and (in some regions) mammalian remains agreeing closely with those of the Siwaliks.

Mekran system—The Mekran system in Baluchistan, representative of the Siwalik system, differs from the equivalents in other areas in having marine fossils. It comprises a thick series of shales and sandstones with shelly limestone intercalations containing a copious marine molluscan fauna. The fauna bears some resemblance to the Cuddalore and Karikal beds of the Madras coast and a stronger resemblance to the marine Mio-Pliocene of Java. They fall into two divisions—the lower *Talar stage* is of Mid. Siwalik and the overlying *Gwadar* of Up. Siwalik horizon.

Manchar system—In Sind the Manchar system has been divided into a lower group which is fossiliferous and is equivalent to the fossiliferous beds of the Potwar from the base of the Siwaliks to the Dhok Pathan zone, and an upper group which probably corresponds to the uppermost portion of the Upper Siwaliks. The whole group is 3,050 m. thick and is only occasionally fossiliferous. As the Upper Manchars are followed southwards, they become estuarine at first and then marine, resembling the Mekrans.

Tipam, Dupi Tila and Dihing series—In Assam the Siwalik system is approximately equivalent to the Tipam, Dupi Tila, and Dihing series. In the southern part of Assam and near Chittagong, and southwards almost to Akyab, the lower part of the *Tipam series* consists typically of coarse ferruginous sandstones and sandy shales, and has some marine fossils. In the upper part of the series mottled clays are prominent. It is unconformably overlain by the *Dupi Tila series* which includes sandstones with fossil wood, mottled clays and mottled sands and occasional conglomerates; this series is correlated with the Upper Miocene (Pontian) Irrawaddy sandstone of Burma. In the north-eastern part of Assam the Tipam series is entirely non-marine, consisting of ferruginous sandstones, mottled clays, and mottled sands. The lower beds, in which sandstones preponderate, are associated with oil-shows in Lakhimpur district, and at Digboi contain oilsands which have been worked since 1890. The highest beds contain abundant pebbles of lignite. The highest Tertiaries of Assam, the *Dihing beds*, consist mainly of pebble beds resting unconformably on the Tipam and Dupi Tila series, and presumably corresponding to the Upper Siwaliks and to the highest part of the Irrawaddies.

Structure of the Assam Tertiary outcrops—In most of the Surma Valley the Tertiary beds are folded into sharp anticlines separated by broad synclines. In the North Cachar Hills the anticlines open out against a belt of thrust-faults which can be followed for about 500 km. to the extreme north-east of Assam. This belt widens in the Naga Hills (which form the south-eastern side of the broad alluvial valley of the Brahmaputra in Upper Assam), where at least eight separate thrust-sheets can be distinguished. The younger Tertiaries are found in the thrust-sheets nearest to the alluvium and the older Tertiaries farther into the hills. One of the most prominent of the thrusts has long been known as the *Disang thrust*; it lies from 3 to 25 km. from the edge of the hills and for much of its length forms the north-western boundary of the outcrop of the Eocene Disang shales (p. 319). The outermost thrust has been termed the *Naga thrust*. A similar type of structure is found on the other side of the Assam valley in the Eastern Himalayas, but the over-thrusting here is of more pronounced type, Tertiary beds being found beneath Gondwana rocks (Fig. 38a).

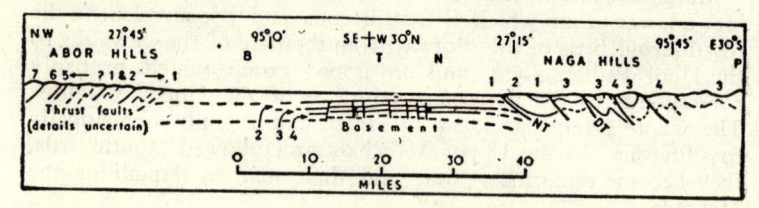

FIG. 38a—Diagrammatic section across the Brahmaputra Valley, Upper Assam. Scale 25 miles to an inch.

1 = Dihing series (with overlying alluvium)
2 = Tipam series
3 = Barail series
4 = Disang series and Jaintia series
5 = Gondwana beds
6 = Baxa series
7 = Daling series
NT = Naga thrust
DT = Disang thrust
B = Brahmaputra River near Dibrugarh
T = Tengakhat
N = Nahorkatiya
P = Patkai Range

Stratigraphy and Structure of Upper Assam—Between the two sets of overthrusts lies the Upper Assam valley deeply covered under a blanket of alluvium. Geophysical surveys have shown that the concealed Tertiary rocks are cut by numerous criss-cross faults. Beneath the alluvium comes the Dihing series which rests unconformably on the Tipam series. Both divisions of the Tipam have been recognised in bore-holes put down in this area for petroleum exploration (p. 437). The Tipam in turn rests on the Barail series (p. 320) which passes down into Eocene beds (Jaintia Series).

Test-drilling for petroleum has not yet penetrated below the Tertiaries, but seismic, gravity, and magnetic evidence agree in showing that the metamorphic basement, which appears at the surface in the Mikir Hills, falls north-eastwards to a depth exceeding 6,000 m.

Irrawaddy system—In Central Burma the lower portion of the Siwalik system appears to be missing, and there is a pronounced break between the Upper Pegus of Lower Miocene age and the overlying Irrawaddy system of Upper Miocene to Pliocene age. The Irrawaddy system is made up largely of coarse, current-bedded sands and occasional beds of clay and conglomerate, with locally at the base a conspicuous "red-bed" of lateritic origin. The total thickness may reach 3,000 m. Two fossiliferous horizons occur in this series, separated by about 1,200 m. of sands. The lower, containing *Hipparion* and *Aceratherium*, denotes the Dhok Pathan horizon of the Salt-Range, while the upper, characterised by species of *Mastodon*, *Stegodon*, *Hippopotamus* and *Bos*, is akin to the Tatrot zone of the Upper Siwaliks. The sediments are remarkable for the large quantities of fossil wood associated with them and they were originally known as the "fossil-wood group". Hundreds and thousands of entire trunks of silicified trees and huge logs lying in the sandstones suggest the denudation of thickly forested eastern slopes of the Arakan Yoma. Further north in Burma it is probable that the Irrawaddy system extends to somewhat lower horizons than in Central Burma, and the boundary between the Pegu and Irrawaddy rocks is often difficult to fix.

REFERENCES

H. Falconer and P. T. Cautley, *Fauna Antiqua Sivalensis*, 1846, London.

G. E. Pilgrim, Correlation of the Siwalik Mammals, *Rec. G.S.I.* vol. xliii. pt. 4, 1913 ; Tertiary Fresh-water Deposits of India, *Rec. G.S.I.* vol. xl. pt. 3, 1910.

R. D. Lydekker, Siwalik Fossils, *Pal. Indica*, series x. vols. i. ii. iii. iv., 1874–1887.

G. E. Pilgrim, *Pal. Indica*, N.S. : Fossil Giraffidae, vol. iv. mem. 1, 1911; Fossil Suidae, vol. viii. mem. 4, 1926; Fossil Carnivora, vol. xviii., 1932 ; Fossil Bovidae, *Pal. Ind.* vol. xxvi., 1939.

C. S. Middlemiss, Geology of the Sub-Himalayas, *Mem. G.S.I.* vol. xxiv. pt. 2, 1890.

A. B. Wynne, Tertiary Zone of N. W. Punjab, *Rec. G.S.I.* vol. x. pt. 3, 1877.

D. N. Wadia, Siwaliks of Potwar and Jammu Hills, *Mem. G.S.I.* vol. li. pt. 2, 1928.

E. H. Colbert, Siwalik Mammals, *Trans. Amer. Phil. Soc.*, N.S., vol. xxvi., 1935.

352 GEOLOGY OF INDIA

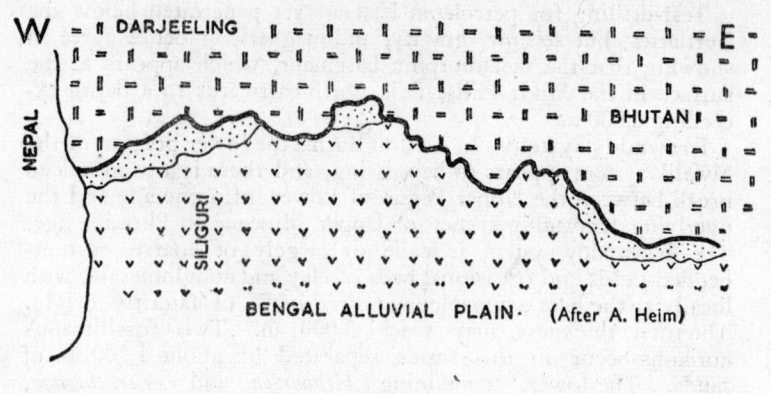

FIG. 38b.—Section showing trespass of the main boundary thrust southwards over Siwaliks.

FIG. 38c.—Siwalik Elephants.

CHAPTER XXI

The Pleistocene System

THE Pleistocene period of geology is in many ways the most fascinating, though the briefest, of earth history. It was during this period that the geography and topography of most parts of the world acquired their final outlines, and their floras and faunas their present distribution. The Pleistocene system in India has a fuller and more varied development than all the preceding systems save the Archaean; it covers 650,000 sq. m. of North India under river deposits; there are long stretches of contemporaneous ice-deposits in the Middle and Inner Himalayas, and desert, lateritic, littoral, lacustrine and subaerial accumulations in other parts of the country. An occurrence of sub-Recent ossiferous gravels over 1,000 m. deep in the Upper Sutlej basin gives us a clue to the varied geological, biological and meteorological conditions prevailing in India during the period. Extensive linear faulting along the west coast and tectonic disturbance of gravel beds, Karewas and Upper Siwalik strata are of further interest, showing that this last and sub-Recent epoch of earth history was not free from orogenic movements of a significant nature.

THE GLACIAL AGE IN INDIA

The Pleistocene Glacial Age of Europe and America—The close of the Tertiary era and the commencement of the Quaternary are marked in Europe, North America, and the northern world generally, by a great refrigeration of climate, culminating in what is known as the Ice Age or Glacial Age. The glacial conditions prevailed so far south as 39° latitude north, and countries which now experience a temperate climate then experienced the arctic cold of the polar regions, and were covered under ice-sheets radiating from the higher grounds. The evidence for this great change in the climatic conditions of the globe is of the most convincing nature, and is preserved in the *physical* records of the age, *e.g.* in the characteristic glaciated topography; the "glacial drift" or moraine-deposits left by the glaciers; and the effects upon the drainage system of the countries, as well as in the *organic* records, *e.g.* the influence of such a great lowering of the temperature on the plants and animals then living, on the migration or extinction of species and on their present distribution.

A modified Glacial Age in India—Whether India, that is, parts lying to the south of the Himalayas, passed through a Glacial Age is an interesting though an unsettled problem. In India, it must be understood, we cannot look for the actual existence of ice sheets during the Pleistocene glacial epoch, because a refrigeration which can produce glacial conditions in Northern Europe and America would not, the present zonal distribution of the climate being assumed, be enough to depress the temperature of India beyond that of the present temperate zones. Hence we should not look for the evidence in moraine-débris and rock-striations (except in the Himalayas), but in the indirect *organic* evidence of the influence of such a lowering of the temperature and the consequent increase of humidity on the plants and animals then living in India. Humidity or dampness of climate has been found to possess as much influence on the distribution of species in India as temperature. From this point of view sufficient evidence exists of the glacial cold of the northern regions being felt in the plains of India, though to a much less extent, in times succeeding the Siwalik epoch after the Himalayan range had attained its full elevation. The great Ice Age of the northern world was experienced in the southerly latitudes of India as a succession of cold pluvial epochs. The fluvio-glacial deposits of the Potwar, described on p. 385, and the boulder-bed referred to on p. 343 as within the Upper Siwaliks of the Shekh Budin hills in the Trans-Indus Salt-Range, are the only instances of actual glacial deposits recorded in India in latitudes so far south as 33° N.

The nature of the evidence for an Ice Age in the Peninsula—This evidence, derived from some peculiarities in the fauna and flora of the hills and mountains of India and Sri Lanka, is summarised by W. T. Blanford—one of the greatest workers in the field of Indian geology and natural history.

"On several isolated hill ranges, such as the Nilgiri, Animale, Shivarai and other isolated plateaus in Southern India, and on the mountains of Ceylon there is found a temperate fauna and flora which does not exist in the low plains of Southern India, but which is closely allied to the temperate fauna and flora of the Himalayas, the Assam Range, the mountains of the Malay Peninsula and Java. Even on isolated peaks such as Parasnath, 1,372 m. high, in Behar, and on Mount Abu in the Aravalli Range, Rajputana, several Himalayan plants exist. It would take up too much space to enter into details. The occurrence of a Himalayan plant like *Rhododendron arbireum* and of a Himalayan mammal like *Martes flavigula* on both the Nilgiris and Ceylon mountains will serve as an example of a considerable number of less easily recognised species. In some cases there is a closer resemblance between the temperate forms found on the Peninsular hills and those on the Assam Range than between the former and Himalayan species but there are also connections between the Himalayan and the Peninsular regions which do not extend to the eastern hills. The most remarkable of these is the occurrence on the Nilgiri and Animale ranges, and on some hills further south, of a species of wild goat, *Capra Hylocrius*, belonging to a sub-genus (*Hemitragus*) of which the only known species, *Capra Jeemlaica*, inhabits the temperate regions of the Himalayas from Kashmir to Bhutan. This case is remarkable because the only other wild goat found completely outside the palaearctic region is another isolated form in the mountains of Abyssinia.

"The range in elevation of the temperate flora and fauna of the Oriental regions in general appears to depend more on humidity than temperature, many of the forms which in the Indian hills are peculiar to the higher ranges being found represented by the allied species at lower elevations in the damp Malay Peninsula and Archipelago, and some of the hill forms being even found in the damp forests of the Malabar coast. The animals inhabiting the Peninsula and Ceylonese hills belong for the most part to species distinct from those found in the Himalayan and Assam ranges, etc., in some cases even genera are peculiar to the hills of Ceylon and Southern India, and one family of snakes is unrepresented elsewhere. There are, however, numerous plants and a few animals inhabiting the hills of Southern India and Ceylon which are identical with Himalayan and Assamese hill forms, but which are unknown throughout the plains of India.

"That a great portion of the temperate fauna and flora of the Southern Indian hills has inhabited the country from a much more distant epoch than the glacial period may be considered as almost certain, there being so many peculiar forms. It is possible that the species common to Ceylon, the Nilgiris and the Animale may have migrated at a time when the country was damper without the temperature being lower, but it is difficult to understand how the plains of India can have enjoyed a damper climate without either depression, which must have caused a large portion of the country to be covered by sea, a diminished temperature, which would check evaporation, or a change in the prevailing winds. The depression may have taken place, but the migration of the animals and plants from the Himalayas to Ceylon would have been prevented, not aided, by the southern area being isolated by the sea, so that it might be safely inferred that the period of migration and the period of depression were not contemporaneous. A change in the prevailing winds is improbable so long as the present distribution of land and water exists, and the only remaining theory to account for the existence of the same species of animals and plants on the Himalayas and the hills of southern India is depression of temperature."

Ice Age in the Himalayas—When, however, we come to the Himalayas, we stand on surer ground, for the records of the glacial age there are unmistakable in their legibility. At many parts of the Himalayas there are indications of an extensive glaciation in the immediate past, and that the present glaciers, though some of them are among the largest in the world, are merely the shrunken remnants of those which flourished in the Pleistocene age. Enormous heaps of terminal moraines, now grass-covered, and in some cases tree-covered, ice-transported blocks, and the smoothed and striated hummocky surfaces and other indications of the action of ice on the land surface are seen at all parts of the Himalayas that have been explored from Sikkim to Kashmir, at elevations thousands of metres below the present level of descent of the glaciers. On the Haramukh mountain in Kashmir a mass of moraine is described at an elevation of 1,675 m. Grooved and polished rock surfaces have been found at Pangi in the Upper Chenab valley, and at numerous localities in the Sind and Lidar valleys on cliffs at the 2,300 m. level. In the Pir Panjal, above 2000 m., the mountains have a characteristic glaciated aspect, while the valleys are filled with moraines and fluvio-glacial drift. On the southern slopes of the Dhauladhar range an old moraine (or what is believed to be such) is found at such an extraordinarily low altitude as 1,433 m., while in some parts of Kangra, glaciers were at one time believed, though not on good evidence, to have come below the 1,000 m. level. In southern Tibet similar evidences are numerous at the lowest situations of that elevated plateau. Equally convincing proofs of ice-action exist in the interruptions to drainage courses that were caused by glaciers in various parts of the mountains. Numerous small lakes and rock-basins in Kashmir, Ladakh and Kumaon directly or indirectly owe their origin to the action of glaciers now no longer existing. A more detailed survey and exploration of the Himalayas than has been possible hitherto will bring to light further proofs.

The ranges of the Middle Himalayas, which support no glaciers to-day, have, in some cases, their summits and upper slopes covered with moraines. The ice-transported blocks of the Potwar plains in Attock and Rawalpindi (referred to on page 385) also furnish corroborative evidence to the same effect. (Note also the testimony of some hanging valleys (p. 28), and of the well-known desiccation of the Tibetan lakes (p. 30).)

The extinction of the Siwalik mammals—one further evidence—Further evidence, from which an inference can be drawn of an Ice Age in the Pleistocene epoch in India, is supplied by the very striking circumstances to which the attention of the world was first drawn by the great naturalist Alfred Russel Wallace. The

TABLE SHOWING SUGGESTED CORRELATION OF GLACIAL STAGES WITH THE UPPER SIWALIKS OF N.W. INDIA (After de Terra)

Period.		Stage.	Fauna.	Glacial Cycle in Kashmir.
Pleistocene	Upper	Re-deposited silt.	Living species.	4th ice advance. Terminal moraine at 2,400 to 3,050 m.
		Erosion.		3rd Interglacial : erosion.
	Middle	*Potwar*: yellow, loess-like silt, and gravel.	Narmada fauna in "upper group".	3rd ice advance : 3–4 recessional moraines. Terminal moraine at 2,000 m.
		Erosion.		Long 2nd Interglacial: Upper Karewa beds: erosion.
		Boulder-conglomerate Stage.	Narmada fauna. *Equus, Bubalus, Hippopotamus, Elephas namadicus.*	2nd ice advance : boulder clay and gravel in Karewa beds.
	Lower	*Pinjor* Stage.	*Equus, El. namadicus, Bos, Sus Rhinoceros, Cervus, Felis, Sivatherium.*	1st Interglacial ; Lower Karewa beds, birch, oaks, pine-forest.
		Tatrot Stage.	*Stegodon bombifrons, Hippolyus, Hexaprotodon, Pentalophodon* Falc.	1st ice advance. Terminal moraine at 1,675 m.
Pliocene.		*Dhok Pathan* Stage.	*Hipparion, Tragocerus, Stegolophodon, Bramatherium.*	

sudden and widespread reduction, by extinction, of the Siwalik mammals is a most startling event for the geologist as well as the biologist. The great Carnivores, the varied races of elephants belonging to no less than twenty-five to thirty species, the *Sivatherium* and numerous other tribes of large and highly specialised Ungulates which found such suitable habitats in the Siwalik jungles of the Pliocene epoch, are to be seen no more in an immediately succeeding age. This sudden disappearance of the highly organised mammals from the fauna of the world is attributed by the great naturalist to the effect of the intense cold of a Glacial Age. It is a well-known fact that the more highly specialised an organism is, the less fitted it is to withstand any sudden change in its physical environments; while the less differentiated and comparatively simple organisms are more hardy, and survive such changes either by slowly adapting themselves to the altered surroundings or by migration to less severe environments. The extinction of the large number of Siwalik genera and species, and the general impoverishment of the mammalian fauna of the Indian region, therefore, furnish us with an additional argument in favour of an "Ice Age" (though, of course, greatly modified and tempered in severity) in India, following the Siwaliks.

Interesting glaciological investigations have been made in the Kashmir Himalaya and in the Karakoram by Dainelli, Grinlinton and de Terra. Dainelli records four distinct phases of glaciation in the N. W. Himalaya recognised by their moraines. Some indications of the oscillation of glacial and interglacial periods have been recognised in the heavy Pleistocene drift filling the Sind and Lidar valleys of Kashmir. De Terra has attempted a correlation of the moraines of successive glaciations with the Upper Siwalik stages of the Punjab. He believes that the terminal and ground moraines of the Kashmir glaciers merge into the boulder-conglomerate of the foot-hills and with the system of river terraces of the main valleys of Kashmir.

The system of lacustrine and river deposits known as *Karewas* in Kashmir contain many terminal moraines embedded in them. The moraines at some places contain finely laminated "varved" glacial clays.

PLEISTOCENE ICE AGE DEPOSITS OF KASHMIR

Pleistocene or post-Pliocene deposits of the nature of fluviatile, lacustrine or glacial have spread over many parts of Kashmir and occupy a wide superficial extent. Of these the most interesting as well as conspicuous examples are the fresh-water (fluviatile and lacustrine) deposits, found as low flat mounds bordering the slopes of the mountains above the modern alluvium of the Jhelum. In these, re-sorted terminal moraines of the glaciers from the higher ground have furnished a large constituent.

THE PLEISTOCENE SYSTEM

Karewa series—These are known as *Karewas* in the Kashmiri language. The Karewa formation occupies nearly half the area of the valley; it has a width of from 13 to 26 km. along its south-west side and extends for a length of some 80 km. from Shopyan to Baramula. The present view regards the Karewas as the surviving remnants of deposits of a lake or series of lakes which once filled the whole valley-basin from end to end. The draining of the lake or lakes, by the opening and subsequent deepening of the outlet at Baramula, has laid them bare to denudation which has dissected the once continuous alluvium into isolated mounds or platforms. The highest limit at which the Karewas have been observed on the N. E. slopes of the Pir Panjal is 3,800 m., more than 2,000 m. above the level of the Jhelum bed. At the height of the Ice Age this Karewa lake must have been no less than 7,800 sq. km. in area.

Lithologically the Karewa series consists of blue, grey and buff silts, sand, partly compacted conglomerates and embedded moraines. The series is divisible into two stages, separated by an unconformity representing an erosive interval during which some 600 m. of the Lower Karewas were denuded from the tops of two flat anticlines. Moraines of all periods are found interstratified with the finer lake sediments of the Karewas at different levels. The aggregate thickness of the Karewas exceeds 2,000 m.; it is difficult to estimate the exact thickness, due to the folding and unequal erosion. In stratigraphic range, the base of the Karewas is probably Pliocene, touching as low a horizon as the *Dhok Pathan* stage of the Middle Siwaliks; the top is upper Pleistocene, being conformably overlain by the sub-Recent Jhelum alluvium. What

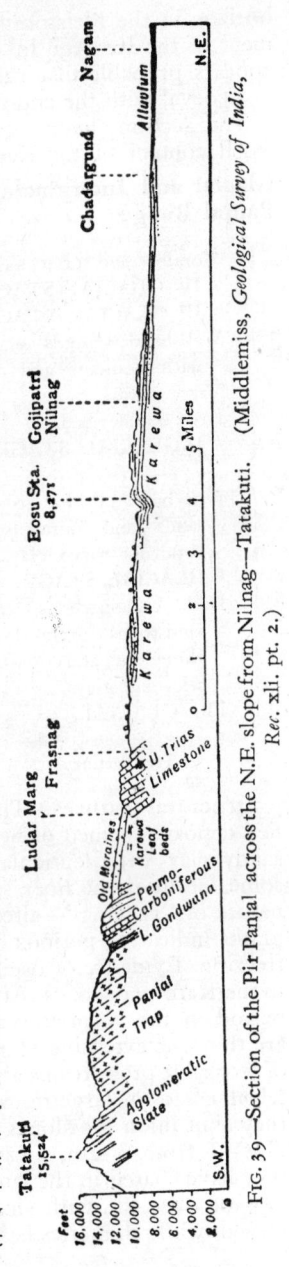

FIG. 39.—Section of the Pir Panjal across the N.E. slope from Nilnag—Tatakuti. (Middlemiss, *Geological Survey of India*, Rec. xli. pt. 2.)

horizon in the Pleistocene of Kashmir represents the commencement of the Ice Age in the Himalaya is not certain. Indeed it appears probable that the onset of the Ice Age in Kashmir was not coeval with the end of the Pliocene but was later in date.

The section below gives an idea of the stratigraphy and fossil content of the Karewa series.

Glacial and Interglacial Deposits in the Karewa Series, Pir Panjal Range

Upper	Moraines and terraces of IV GLACIAL STAGE. III GLACIAL STAGE. Well-bedded sands and clays with boulders and erratics. Varve clays. Basal boulder-bed. II GLACIAL STAGE.	Plant fossils locally abundant. Many gastropod and other land molluscs.
Lower	Fine buff and blue-grey shales, sands and gravels, cross-bedded, varve clays. I GLACIAL STAGE. Dark carbonaceous shales and sandstones with thick conglomerates and lignite seams. Silts and clays. PRE-GLACIAL.	Fossil leaves, fruit and spores, of rose, cinnamon, oak, maple, walnut, trapa; diatoms; landshells. *Elephas hysudricus, Rhinoceros* sp., *Cervus* sp. Schizothoracine fish remains, bird and ungulate bones, teeth, etc. Sub-tropical lowland plants.

Pre-Tertiary.

Structural features—The Karewas are mostly horizontally stratified deposits formed of beds of fine-grained sand, loam, and blue sandy clay with lenticular bands of gravelly conglomerate. At some localities the finer sands and clays show lamination of the nature of "varving"—alternating laminae of different colours and grains indicating periods of summer melting of ice and of winter freezing. Evidence of oscillation of the glacial climate is recorded in the Karewa deposits. At the end of the ice age there was a forest period in the Kashmir valley. Interstratified with the top beds are thin but extensive seams of lignite or brown coal which are of workable proportions at two or three localities in the Hundawar tehsil, enclosing large reserves of a medium-grade fuel. Only when they abut upon the slopes of the Pir Panjal do the Karewas show dips of from 5°-20°, away from the mountains, indicating that they have shared in the later upheaval of the Panjal range; locally dips of over 40°, with sharp monoclinal folding, have been observed, while the series has been traced continuously up to almost the summit of the Pir Panjal. This fact establishes the inference that the Pir Panjal has undergone considerable elevation since the material of the Karewas was laid down on its slopes.

Fossil leaves and wood of 120 recent species, *e.g.* birch, beech, willow, oak, walnut, trapa, rose, holly, various pines, together with land and fresh-water shells and some fish and other vertebrate remains, including *Elephas*, *Cervus* and a species of *Rhinoceros*, are found at places.

Glacial Moraines—Pleistocene and later glacial deposits are of wide distribution in Kashmir. Two or more distinct sets of moraines are observed—one at high level, which is of more recent accumulation by existing glaciers, and the others at considerably lower situations, whose age is Lower or Middle Pleistocene. The glaciation of the tributary valleys of Kashmir, the Sind, Lidar and Lolab, presents features of great interest. According to some observers[1] this part of the Himalayas underwent four distinct glaciations, separated by interglacial warm periods. Indications

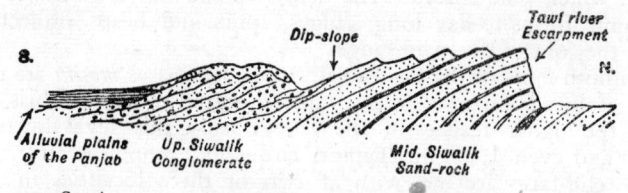

FIG. 40.—Section across the outermost hills of the sub-Himalayas at Jammu. Note the succession of dip-slopes and escarpments.

of these successive glaciations, according to them, are present in the glacier moraines and drifts which fill these ice-eroded, characteristically shaped tributary valleys and in the system of river-terraces in the upper reaches of the main valley of the Jhelum, into which the moraine deposits gradually merge. Moraines belonging to three or four successive glacial advances, interbedded with the Karewa deposits at various levels, have been recognised by de Terra. The terminal moraines of the latest glacial period are seen capping the top beds of the Upper Karewas. While moraines of the existing small glaciers of Kashmir rarely occur below 4,300 metres the older moraines are generally buried under either contemporaneous or later Karewa deposits. Four such moraines can be distinguished on the wide smooth glacis of the Pir Panjal range between Baramula and Banihal Pass:

---

[1] For glaciological studies of the Kashmir mountains and valleys, reference may be made to the published works of Dainelli and de Terra. See also Grinlinton, *Mem. G.S.I.* vol. xlix. pt. 2, 1928, and *Rec.* vol. xxxi. pt. 3, 1904.

IV GLACIATION moraines—elevation 3,600 m. to 2,700 m. bare, or at places covered under lichen, grass, or juniper shrubs, etc. (? 20,000-10,000 years ago).

III GLACIATION moraines—elevation up to 2,100 m. covered under thick pine forests.

II GLACIATION moraines—partly buried under the upper strata of the Lr. Karewas; few exposures seen. 2,000 m.

I GLACIATION moraines—buried under the Lr. Karewas or river-terraces of the Jhelum and Chenab valleys. 1,600 m. and lower.

At Baramula and Ganderbal some of the earliest moraines in Kashmir are seen. Moraines of the third glaciation are seen over a wide area on the top of the Lr. Karewas, themselves covered under thick pine forests. The débris of the last ice advance in Kashmir forms to-day long ridges, spurs and heaps projecting from the foot of the main range.

On both faces of the Pir Panjal, moraine masses *in situ* are met with at levels above 2,000 m. while re-sorted moraine débris has filled the higher reaches of the valleys below this level down to 1,500 and even 1,200 m. Typical cirque-like amphitheatres with steep cliff-faces are met with at two or three localities in the Poonch Pir Panjal.

High-level river-terraces—Among later deposits than these are the high-level river-terraces, 300 m. or more above the streambed, sub-Recent river alluvia, levees, and flood plains; the enormous "fan-taluses" in the Nubra and Changchenmo valleys of Ladakh; cave-deposits such as those of Harwan; travertine, etc.

The great thickness of gravel and pebble-beds, resting unconformably over the subjacent Upper Siwalik boulder-conglomerate, which fringes the outermost foothills in Jammu and near Dalhousie is likewise of the same age.

FOSSIL MAN IN GLACIAL AGE

The problem of the existence of Glacial Man in India in the Up. Siwalik Boulder Conglomerate stage, equivalent to II Glacial, is yet unsolved. While numerous flake implements (Acheulian) are found on the outcrops of the Boulder Conglomerate of Potwar and on the surface of the Upper Karewa deposits of Kashmir, and Chellean implements in the Narmada valley, there is no evidence to date them precisely. No skeletal remains of early Man have been found *in situ* in the Mid. Pleistocene either in the Karewas or in caves, or the older terraces of the Deccan rivers.

In other parts of the world, the existence of Glacial Man in the II and III epochs is well established. The following table shows the succession of fossil man and man-like apes so far known:

Homo neanderthalensis (Prussia)—III Ice age (100,000 yrs.)
Homo heidelbergensis (Germany)—II Ice age (500,000 yrs.)
Sinanthropus (China)   —I Ice age (600,000 yrs.)
Pithecanthropus (Java)   —I Ice age (600,000 yrs.)
Australopithecus (South Africa)—? Pre-glacial (1,000,000 yrs.)

Australopithecus { Eoanthropus / Zinjanthropus } — ,, ,,

Dryopithecus (India)   —Pliocene (10,000,000 yrs.)
Sivapithecus (India)   —Miocene (15,000,000 yrs.)
Ramapithecus (India)   —   ,,   ,,   ,,

REFERENCES

W.T. Blanford, *Geology of India*, vol. i., Introduction, 1879.
W. Theobald, Extension of Glaciers within Kangra District, *Rec. G.S.I.* vol. vii., 1874.
R. D. Oldham, Glaciation of the Sind Valley, Kashmir, *Rec. G.S.I.* vol. xxxi. pat. 3, 1904.
J. L. Grinlinton, Glaciation of the Lidar Valley, Kashmir, *Mem. G.S.I.* vol. xlix. pt. 2, 1928.
G. Dainelli, *Italian Expedition to the Himalaya* (1913-14), vols. i-xiii. Bologna, 1923-1935.
H. de Terra, Studies on the Ice Age in India, *Carn. Inst.*, Washington, No. 493, 1939.
D. N. Wadia, Pleistocene Ice Age Deposits of Kashmir, *Proc. Nat. Inst. Sc. Ind.* 7, 1941; Pliocene-Pleistocene Boundary in N.W. India, xviii. *Intern. Geol. Cong.*, London, 1948.

CHAPTER XXII

The Pleistocene System (*continued*)

THE INDO-GANGETIC ALLUVIUM

The plains of India—The present chapter will be devoted to the geology of the great plains of North India, the third physical division of India which separates the Peninsula from the extra-Peninsular regions. It is a noteworthy fact that these plains have not figured at all in the geological history of India till now, the beginning of its very last chapter. What the physical history of this region was during the long cycle of ages, we have no means of knowing. That is because the whole expanse of these plains, from one end to the other, is formed, with unvarying monotony, of Pleistocene and sub-Recent alluvial deposits of the rivers of the Indo-Gangetic system, which have completely shrouded the old land-surface to a depth of thousands of metres. The solid geology of the country, consisting of rock-systems at the south foot of the Himalayas as well as those along the north edge of the Peninsula is thus totally buried. The deposition of this alluvium commenced after the final phase of the Siwaliks and has continued all through the Pleistocene up to the present. The plains of India thus afford a signal instance of the imperfection of the geological record as preserved in the world, and of one of the many causes of that imperfection.

Nature of the Indo-Gangetic depression—In the Pleistocene period, the most dominant features of the geography of India had come into existence, and the country had then acquired almost its present form, and its leading features of topography, except that the lands in front of the newly-upheaved mountains formed a depression, which was rapidly being filled up by the waste of the highlands. The origin of this depression, or trough, lying at the foot of the mountains, is doubtless intimately connected with the origin of the latter, though the exact nature of the connection is not known and is a matter of discussion. The great geologist, Eduard Suess, has suggested, as we have already seen, that it is a "fore-deep" in front of the high crust-waves of the Himalayas as they were checked in their southward advance by the inflexible solid land-mass of the Peninsula. On this view the depression is of a synclinal nature—a *synclinorium*. From physical and geodetic considerations, Sir S. Burrard considers that the Indo-Gangetic plains occupy a "rift-valley", a portion of the Earth's surface sunk in a huge crack in the subcrust, between parallel faults on its two sides. This view has got few geological facts in it support and is not adopted by geologists, who conceive that the Indo-Gangetic depression is a true fore-deep, a downwarp of the Himalayan foreland, of variable depth, converted into flat plains by the simple process of alluviation. On this view, a long-continued vigorous sedimentation, loading a slowly sinking belt of the Peninsular shield from Rajasthan to Assam (under its cover of Aravallis, Vindhyans, and probably also Gondwanas and Cretaceous, with their varied topography and tectonics) the deposition keeping pace with subsidence, has given rise to this great tectonic trough of India. The exact depth of the alluvium has not been ascertained, but recent gravity, magnetic and seismic exploration shows that it is variable, from less than 1,000 to over 2,000 metres. Underlying the alluvium are unconsolidated Siwalik and older Tertiary sediments of the Himalayan piedmont and below these are more consolidated older formations, such as the Gondwanas and the Cretaceous, the presence of which is indicated by good reflections of the seismic wave and also by borings. The depth is not even—it is greater in the northern than in the southern sector. The northern rim of the basin, where it joins the foot-hill zone of the mountains, is one of considerable faulting and structural strain. It is also probable that the alluvium conceals two or three transverse ridges and three or four pre-Tertiary basins due to the crumpling and dislocation of the basement floor. Overall there prevailed here most favourable conditions for the quick accumulation of sediments in the zone of lodgment at the foot of the mountains. (Cf. Figs. 35 and 41.)

Extent and thickness—The area of these alluvial plains is 777,000 square km. covering the largest portion of Sind, Northern Rajasthan, almost the whole of the Punjab, Uttar Pradesh, Bihar, Bengal and half of Assam. In width they vary from a maximum of 500 km. in the western part to less than 150 km. in the eastern. Some indication of the basement configuration of the Gangetic trough is obtained from recent exploratory borings put down for water or oil. The floor is not structurally uniform but is segmented by ridges and hollows due to faulting. Magnetic surveys reveal local highs and lows, all of which dip steeply to the north. In 130 borings, the depth from surface to bed-rock was found to range between 400 m. and 100 m.

Oldham postulated the depth of the alluvium from geological considerations to be about 4,600 m. near its northern limit, from which the floor slopes upwards to its southern edge where it merges with the Vindhyan uplands of the Deccan. Recent calculations

366 GEOLOGY OF INDIA

from geodetic surveys, however, give a much greater thickness for these lighter deposits resting on the pre-Tertiary bed-rock.[1] How far southwards the Murree and Siwalik deposits of the foot-

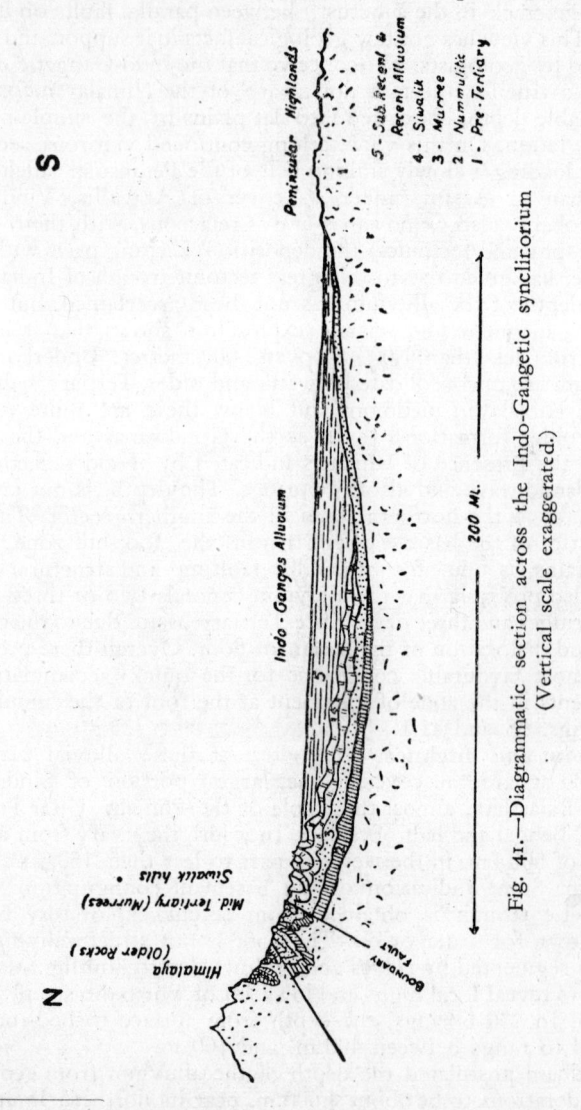

Fig. 41.—Diagrammatic section across the Indo-Gangetic synclinorium. (Vertical scale exaggerated.)

---

[1] E. A. Glennie, Gravity Anomalies and the Structure of the Earth's Crust, *Survey of India*, Dehra Dun, 1932.

hills zone extend underneath the alluvium we have no means of determining, except perhaps by geophysical surveys. The depth of the alluvium is at a maximum between Delhi and the Rajmahal hills, and it is shallow in Rajasthan and between Rajmahal and Assam. The Rajmahal-Assam gap is of recent origin, the two being connected underground at a small depth. The floor of the Gangetic trough is thus not an even plane, but is corrugated by inequalities and buried ridges. Two such ridges have been indicated by geodetic surveys: an upwarp of the Archaean rocks in structural prolongation with the Aravalli axis, between Delhi and Hardwar; and a ridge, submerged under the Punjab alluvium, striking north-west from Delhi to the Salt-Range.

The downwarp which produced the Gangetic geosyncline (in this case the subsidence was not deep enough to carry its surface beneath sea-level, except temporarily at the beginning) must have started as a concomitant of the Himalayan elevation to the north somewhere in the mid-Eocene. The deposition of the débris of the newly rising mountains and sinking of the trough must have proceeded *pari passu* all through the Tertiary up to late-Pleistocene and sub-Recent times.

Folding and faulting at the north margin—There is evidence of a considerable amount of flexure and dislocation at the northern margin of the trough, passing into the zone of the various boundary faults at the foot of the Himalaya; it is possible that a certain amount of folding and faulting extends southwards to the bottom of the downwarp. At any rate, it is clear that the northern rim of the great trough is under considerable tectonic strain due to the progressive downwarping, with the greatest subsidence where it merges into the foot of the mountains. The seismic instability of this part of India is well known, it being the belt encompassing the epicentres of the majority of the known earthquakes of Northern India. Field investigations on the late Bihar earthquake of January 1934 point to some crustal dislocation below the Gangetic valley between Motihari and Purnea as the cause of the disaster.[1]

The southern limit of the trough shows no structural peculiarities or features of any importance.

Changes in rivers—The highest elevation attained by the plains is about 275 m. above sea-level; this is the case with the tract of country between Saharanpur, Ambala, and Ludhiana, in the Punjab. The above tract is thus the present watershed which divides the drainage of the east, *i.e.* of the Ganges system, from that of the west, *i.e.* the Indus and rivers of the Punjab. There exists much evidence to prove that this was not the old water-parting. The courses of many of the rivers of the plains have

---

[1] *Mem. G.S.I.*, vol. lxxiii., 1938.

undergone great alterations. Many of these rivers are yearly bringing enormous loads of silt from the mountains, and depositing it on their beds, to raise them to the level of the surrounding flat country, through which the streams flow in ever-shifting channels. A comparatively trifling circumstance is able to divert a river into a newly scoured bed. The river Jumna, the sacred Saraswati of the Hindu Shastras, in Vedic times flowed to the sea through the Eastern Punjab and Rajasthan, by a channel that is now occupied by an insignificant stream which loses itself in the sands of the Bikaner desert. In course of time, the Saraswati took a more and more easterly course and ultimately merged into the Ganges at Prayag. It then received the name of Jumna.[1]

Most of the great Punjab rivers have frequently shifted their channels. In the time of Akbar, the Chenab and Jhelum joined the Indus at Uch, instead of at Mithankot, almost 100 km. downstream, as at present. Multan was then situated on the Ravi; now it is 60 km. from the confluence of that river with the Chenab. 250 years ago the Beas deserted its old bed, which can still be recognised between Montgomery and Multan, and joined up with the Sutlej near Ferozepur several hundred miles upstream.[2]

The records of the third century B.C. show that the Indus flowed more than 130 km. to the east of its present course, through the now practically dry bed of a deserted channel, to the Rann of Kutch,[3] which was then a gulf of the Arabian Sea. The westering of the Indus is thus a very pronounced phenomenon, for which different causes have been suggested. An old river bed, the Hakra, Sotra (Ghaggar), or Wahind, more than 1,000 km. in length, the channel of a lost river, is traceable from Ambala near the foot of the Himalayas through Bhatinda, Bikaner and Bahawalpur to Sind.[4] It is probably the old bed of the Saraswati (the Jumna when it was an affluent of the Indus) at a time when it and the Sutlej flowed independently of the Indus to the sea, *i.e.* the Rann of Kutch. The present dry river bed to the east of Sind, known as the Eastern Nara, is either the old bed of the Indus or, more probably, the channel of the Sind portion of the Sutlej after the river had deserted it.

---

[1] *Quart. Journ. Geol. Society*, xix. p. 348, 1863. The above example illustrates what, in a general manner, was the behaviour of the majority of the rivers of this tract, including the Indus itself, which is supposed to have been originally confluent with the Ganges. See also Pascoe, *ibid.* vol. lxxv. pp. 138–155 (1919); and Pilgrim, *Journ. Asiat. Soc. Bengal*, vol. xv. (1919), pp. 81–99.

[2] General Cunningham, *The Ancient Geography of India*, London, 1871.

[3] The Rann of Cutch, *Journ. Roy. Geog. Soc.*, vol. xl., 1870.

[4] Maj. C. F. Oldham, On the Lost River of the Indian Desert, *Calcutta Review*, 1874.

[The famous cities of Mohenjo Daro, situated on the Indus in Sind, and Harappa, on one of its affluents in the Punjab, were probably abandoned at a much earlier date due to the vagaries of the shifting rivers and also to their recurring flood deposits, which eventually buried them.]

Great changes have likewise taken place in Bengal and in the Gangetic delta since 1750; and hundreds of square kilometres of the delta have become habitable since then. In 1785 Rennel, the great geographer of Bengal, observed the Brahmaputra flowing through Mymensingh; now it flows 64 km. westwards. Moreover, the Tista flowed southward through Dinajpur and joined the Ganges; now it has a south-easterly course and discharges into the Brahmaputra.[1]

Old maps of Bengal show that hardly 250 years ago the river Brahmaputra, which now flows to the west of Dacca, and of the elevated piece of ground to its north, known as the Madhupur jungle, then flowed a great many kilometres to the east of these localities. This change appears to have been accomplished suddenly, in the course of a few years.

Lithology—The rocks are everywhere of fluviatile and subaerial formation—massive beds of clay, either sandy or calcareous, corresponding to the silts, mud, and sand of the modern rivers.

Gravel and sand become scarcer as the distance from the hills increases. At some depth from the surface there occur a few beds of compact sands and even gravelly conglomerates. A characteristic of the clayey part of the alluvial plains, particularly in the older parts of the deposits, is the abundant dissemination of impure calcareous matter in the form of irregular concretions—*Kankar*. The formation of Kankar concretions is due to the segregation of the calcareous material of the alluvial deposits into lumps or nodules, somewhat like the formation of flint in limestone. The alluvium of some districts contains as much as 30 per cent calcareous matter. Some concretionary limonite occurs likewise in the clays of Bengal and Bihar.

Classification—With regard to the geological classification of the alluvial deposits, no very distinctly marked stages of deposition occur, the whole being one continuous and conformable series of deposits whose accumulation is still in progress. But the following divisions are adopted for the sake of convenience, determined by the presence in them of fossils of extinct or living species of mammals:

3. Deltaic deposits of the Indus, the Ganges, etc. Recent.
2. Newer alluvium : *Khadar* of the Punjab.
   Fossils, chiefly living species, including relics of Man.

---

[1]*Physical Geography of Bengal*, from the Maps and Writings of Maj. J. Rennel, 1764–1776. Calcutta, Bengal Secretariat, 1926.

1. Older alluvium :   *Bhangar* of the Ganges valley.
                             Fossils of *Elephas antiquus*, *Equus namadicus*, *Mans gigantea*, extinct species of *Rhinoceros*, *Hippopotamus*. etc.

Unconformity. ~~~~~~

    Rocks of unknown age: possibly extensions of the *Archaean*, *Purana* and *Gondwanas* of the Peninsula and of *Nummulitic*, *Murree* and *Siwalik* of the sub-Himalayas.

The Bhangar—The *Bhangar*, or older alluvium of Bengal and of Uttar Pradesh, corresponds in age with the Middle Pleistocene, while the *Khadar* gradually passes into the Recent. The former generally occupies the higher ground, forming small plateaus which are too elevated to be flooded by the rivers during their rise. As compared to the Bhangar, the Khadar, though newer in age, occupies a lower level than the former. This, of course, happens in conformity with the principle that as a river becomes older in time, its deposits become progressively younger; and if the bed of the river is continually sinking lower, the later deposits occupy a lower position along its basin than the earlier ones. Such is the case with all old river deposits (*e.g.* river-terraces and flood-plains). Remnants of the Bhangar land are being eroded by every change in the direction of the river channels, and are being planed down by their meandering tendencies.

The Khadar. The Ganges delta—The Khadar deposits are, as a rule, confined to the vicinity of the present channels. The clays have less Kankar, and the organic remains entombed in them all belong to still living species of elephants, horses, oxen, deer, buffaloes, crocodiles, fishes, etc. The Khadar imperceptibly merges into the deltaic and other accumulations of the prehistoric times. The delta of the Ganges and the Brahmaputra is merely the seaward prolongation of the Khadar deposits of the respective river-valleys. It covers an area of 130,000 square km. composed of repeated alternations of clays, sands and marls with recurring layers of peat, lignite and some forest-beds.

    Southern Bengal has been reclaimed from the sea at a late date in the history of India by the rapid southward advance of the Ganges and Brahmaputra delta through the deposition of enormous loads of silt. J. Fergusson has stated that only 5,000 years ago the sea washed the Rajmahal hills and that the country round Sylhet was a lagoon of that sea, as was also a part of the province of Bengal at a later date. The cities of lower Bengal became established as the ground became desiccated enough to be habitable, only about 1,000 years ago. The diversion of the Brahmaputra to the east of Madhupur some centuries ago and its later deflection again to the west in the middle of the nineteenth century are well-recorded events. This diverted portion which broke away from its course to join the Ganges was named the *Jamuna*. The eastern sea-face of the delta is changing at a rapid rate by the formation of new ground and new islands, while the western portion of the deltaic coast-line has remained practically unchanged since Rennel's surveys of the 1770's.

The Indus delta—Similarly the Indus delta is a continuation of the Khadar of the Indus river. This delta is a well-defined triangle with its apex at Tatta; it is of much smaller area than the Ganges delta, since it is probable that the present delta is not of a very old age, but is of comparatively late formation. From old maps of Sind it is found that the delta has grown in size considerably during late historic times, and that the river has swung from the Gulf of Cambay in the south-east to Cape Monze in the north-west, frequently changing the character of the coast-line. It is inferred from various evidences that the Indus, within historic times, had a very much more easterly course, and discharged its waters at first into the Gulf of Cambay and then into the Rann of Kutch. Both in Sind and Kutch there exist popular traditions, as well as physical evidence, to support the inference. (See pp. 368–9).

Observation of the Khadar deposits of the Lower Indus basin of Sind shows that this strip of country is being *aggraded* by the deposition of silt by the river, till at places the Indus bed is nearly 20–22 m. higher than the level of the surrounding country. The river thus is in danger of leaving its bed in flood-time. The sub-Recent history of the river proves that such desertion of the channel has not been uncommon and that the Indus has wandered over the plains of eastern Sind and N.W. Kutch over a wide amplitude of territory, raising the level of the invaded country by the annual deposit of silt.

A few other vernacular terms are employed to denote various superficial features of geological importance in this area:

Bhaber denotes a gravel talus with a somewhat steep slope fringing the outer margins of the Siwalik hills. It resembles the alluvial fans or dry deltas. The rivers in crossing them lose themselves by the abundant percolation in the loose absorbent gravels. The student will here see the analogy of this Bhaber gravel with the Upper Siwalik conglomerates. The latter are, in fact, old *Bhaber* slopes sealed up into conglomerates by the infiltration of a cementing matrix.

Terai is the densely forested and marshy zone below the Bhaber. In these tracts the water of the Bhaber slopes reappears and maintains them in a permanently marshy or swampy condition.

The term *Bhur* denotes an elevated piece of land situated along the banks of the Ganges and formed of accumulated wind-blown sands, during the dry hot months of the year.

In the drier parts of the alluvial plains, a peculiar saline efflorescent product—*Reh*[1] or *Kalar*—is found covering the surface and destroying in a great measure its agricultural fertility. The Reh salts are a mixture of the carbonate, sulphate and chloride of sodium together with calcium and magnesium salts, derived originally from the chemical disintegration of the detritus of the mountains, dissolved by percolating waters and then carried to the surface by capillary action in the warm dry weather. (See p. 473.)

The *Dhands* of Sind are small, shallow, alkaline or saline lakes formed in hollows of the sand-dunes. The salts, carbonate, chloride and sulphate of sodium, are brought here by water percolating through the blown sands and accumulated in the basins, which form important concentrations of natron at some places.[1]

In the alluvial tract lying between south-east Sind and Kutch, there are likewise found fair-sized beds and lenses of pure rock-salt buried in the sand deposits. The total quantity of salt so buried is of the order of several million tonnes.

Economics—Though not possessed of any mineral resources, these alluvial plains are the highest economic asset of India because of their agricultural wealth. The clays are an unlimited store for rude earthenware and brick-making material, which is the only building material throughout the plains, while the Kankar is of most extensive use for lime- and cement-making and also for road construction. These plains are an immense reservoir of fresh sweet water, stored in the more porous, coarser strata, beneath the level of saturation, which is easily accessible by means of ordinary borings in the form of wells. The few deep borings that have been made have given proof of the prevalence of artesian conditions in some parts of the plains, and in a few cases artesian borings have been made with successful results. A considerable amount of success has attended tube-well boring experiments in the plains at many places; wells of large calibre and of a depth of 60-120 m. are supplying water for agricultural use in lands unprotected by irrigation.

Rajasthan desert—Of the same age as, or slightly newer than, the alluvial formation just described are the aeolian accumulations of the great desert tract of India, known as the Thar. The Thar, or Rajasthan desert, is one wide expanse of wind-blown sand and bare rock stretching from the west of the Aravallis to the basin of the Indus, and from the southern confines of the Punjab plains, the basin of the Sutlej, to as far south as lat. 25°, occupying an area 650 km. long by 160 km. broad, concealing beneath it much of the solid geology of the region. The desert is not one flat level waste of sands, since there are numerous rocky projections of low elevation in various parts of it, and its surface is further diversified

---

[1] *Rec. G.S.I.* vol. xiii. pt. 2, 1880 ; vol. lxxvii., *Prof. Paper* 1, 1942.
[2] *Mem. G.S.I.* vol. xlvii. pt. 2, 1923.

by the action of the prevailing winds, which have heaped up the sands in a well-marked series of ridges, dunes and hillocks. The rocky prominences which stand up above the sands belong to the older rocks of the country, presenting in their bare, bold and rounded outcrops, and in their curiously worn and sand-blasted topography, striking illustrations of the phenomena of desert-erosion. The aspect presented by the sand-hills resembles that of a series of magnified wind-ripples. Their strike is generally transverse to the prevailing winds, though in a few cases, *e.g.* those occurring in the southern part of the desert, the strike is parallel to the wind-direction. In both cases the formation of the sand-ridges is due to wind-action, the longitudinal type being characteristic of parts where the force of the wind is great, the transverse type being characteristic of the more distant parts of the desert where that force has abated. The windward slope is long, gentle and undulatory, while the opposite slope is more abrupt and steep. In the southern part of the desert these ridges are of much larger size, often assuming the magnitude of hills 100 to 200 m. high. All the dunes are slowly progressing inland.

Composition of the desert sand—The most predominant component of the sand is quartz in well-rounded grains, but felspar and hornblende grains also occur, with a fair proportion of calcareous grains. The latter are only casts of marine foraminiferal shells, and help to suggest the site of origin of the sands with which they are intimately mixed.

As is characteristic of all aeolian sands, the sand-grains are well and uniformly rounded, by the ceaseless attrition and sorting they have received during their inland drift. In other respects the Rajasthan sand is indistinguishable from the sand of the seashore.

Origin of the Rajasthan desert—The origin of the Indian desert is attributed, in the first instance, to a long-continued and extreme degree of aridity of the region, combined with the sand-drifting action of the south-west monsoon winds, which sweep through Rajasthan for several months of the year without precipitating any part of their contained moisture. These winds transport inland clouds of dust and sand-particles, derived in a great measure from the Rann of Kutch and from the sea-coast, and in part also from the basin of the Lower Indus. There is but little rainfall in Rajasthan—the mean annual fall being not much above 12 cm. —and consequently no water-action to carry off the detritus to the sea, and hence it has gone on accumulating year after year. A certain proportion of the desert sand is derived from the weathered débris of the rocky prominences of this tract, which are subject to the great diurnal as well as seasonal alternations of temperature characteristic of all arid regions. The daily variation of heat and cold in some parts of Rajasthan often amounts to

37.5° C in the course of a few hours. The seasonal alternation is greater. This leads to a mechanical disintegration and *desquamation* of the rocks, producing an abundance of loose débris, and there is no chemical or organic (*i.e.* humus) action to convert it into a soil-cap.

The desert is not altogether, as the name implies, a desolate treeless waste, but does support a thin scrubby vegetation here and there, which serves to relieve the usually dreary and monotonous aspects of its limitless expanses; while, in the neighbourhood of the big Rajasthan cities, the soil is of such fertility that it supports a fairly large amount of cultivation. Wells of good water abound in some places, admitting of some measure of well-irrigation.

Besides the above-described features of the great Indian desert, the Thar offers instructive illustrations of the action of aeolian agencies. As one passes from Gujarat or even central India to the country west and south of the Aravallis one cannot fail to notice the striking change in the topography that suddenly becomes apparent, in the bare and bold hill-masses and the peculiar sand-blasted, treeless landscapes one sees for kilometres around under a clear, cloudless sky. Equally apparent are the abundance of mechanical débris produced by the powerful insolation, the disintegration of the bare rock-surface by desquamation, the saline and alkaline efflorescences of many parts, and the general absence of soil and humus. A more subtle and less easily understood phenomenon of the Rajasthan desert is the growing salinity of its lake-basins by wind-borne salt dust from the sea-coasts.[1]

As stated earlier (on p. 4), the Rajasthan desert is of recent origin within historic times. There is evidence that the region north of Kutch and south of the Punjab was a fertile and forested tract of country, supporting well-populated cities, even so late as the time of invasion by Alexander the Great (323 B.C.). Old rivers flowing through Rajasthan (pp. 367-8) have fought a losing battle against the inroads of the desert and are now traceable only through their dry, forsaken channels.

The Rann of Kutch—This vast desiccated plain terminates to the south-west in the broad depression of the Rann of Kutch, another tract of the Indo-Gangetic depression which owes its present condition to the geological processes of the Pleistocene age. This tract is a saline marshy plain scarcely above sea-level, dry for one part of the year and covered by water for the other part. It was once an inlet of the Arabian Sea, but it has now been silted up by the enormous volume of detritus poured into it by the small rivers discharging into it from the east and north-east. From November to March, that is, during the period of the north-

---

[1] *Rec. G.S.I.* vol x. pt. 1, 1877, and *Mem. G.S.I.* vol. xxxv. pt. 1, 1902.

east or retreating monsoons, the Rann is a barren tract of dry salt-encrusted mud, presenting aspects of almost inconceivable desolation. "Its flat unbroken surface of dark silt, baked by the sun and blistered by saline incrustations, is varied only by the mirage and great tracts of dazzlingly white salt or extensive but shallow flashes of concentrated brine; its intense silent desolation is oppressive, and save by chance a slowly passing caravan of camels or some herd of wild asses, there is nothing beyond a few bleached skeletons of cattle, salt dried fish, or remains of insects brought down by floods, to maintain a distant and dismal connection between it and life, which it is utterly unfit to support."[1] During the other half of the year it is flooded by the waters of the rivers that are held back owing to the rise of the sea by the southwest monsoon gales. A very little depression of this tract would be enough to convert Saurashtra and Kutch into islands. On the other hand, if depression does not take place, the greater part of the surface of the Rann will be gradually raised by the silts brought by the rivers with each flood, and in course of time converted into an arable tract, above the reach of the sea, a continuation of the alluvial soil of Gujarat.

REFERENCES

J. Fergusson, Delta of the Ganges, *Quarterly Journal of the Geological Society* xix., 1863.

T. G. Carless, Delta of the Indus, *Journal of the Royal Geographical Society*, viii., 1838.

A. B. Wynne, Geology of Cutch, *Mem. G.S.I.* vol. ix. pt. 1, 1872.

H. B. Medlicott, The Plains of the United Provinces, *Rec. G.S.I.* vol. vi. pt. 1, 1873.

T. H. D. La Touche, *Mem. G.S.I.* vol. xxxv. pt. 1, 1902. (See the Plates at the end of the Memoir illustrating features arising from desert-erosion).

R. D. Oldham, Structure of the Gangetic Plains, *Mem. G.S.I.* vol. xlii. pt. 2, 1917

D. N. Wadia, Evolution of the Deserts of Asia, Their Post-Glacial Origin, *Monogr. Natl. Inst. Sc. Ind.*, 1960.

---

[1] Wynne, *Mem. G.S.I.* vol. ix. 1872.

Chapter XXIII

The Pleistocene System (*continued*)

LATERITE

Laterite, a regolith peculiar to India—In this chapter we shall consider laterite, a most widespread Pleistocene formation of the Peninsula and Burma, a product of subaerial alteration highly peculiar to India. Laterite is a form of regolith peculiar to India and a few other tropical countries. Its universal distribution within the area of the Peninsula, and the economic considerations that have of late gathered round it, no less than its obscure mode of origin, combine to make laterite an important subject of study in the geology of India.

Composition—Laterite is a kind of vesicular clayey rock composed essentially of a mixture of the hydrated oxides of aluminium and iron with often a small percentage of other oxides, chief among which are manganese and titanium oxides. The two first-named oxides are present in variable ratios, often mutually excluding each other; hence we have numerous varieties of laterite which have bauxite at one end and an indefinite mixture of ferric hydroxides at the other. The iron oxide generally preponderates and gives to the rock its prevailing red colours; at places the iron has concentrated in oolitic concretions, at other places it is completely removed, leaving the rock bleached, white or mottled. At some places again the iron is replaced by manganese oxides; in the lateritic cap over the Dharwar rocks this is particularly the case. Although the rock originally described as laterite by Buchanan from Malabar does contain clay and considerable amounts of combined silica, in the wide terrains of what is obviously the same rock in other parts of India there is no clay (kaolin) and the silica present is colloidal and mechanically associated. According to present usage it is the latter, clay-free rock which has come to be regarded as typical laterite. According to the preponderance of any of the oxides, iron, aluminium, or manganese, at the different centres, the rock constitutes a workable ore of that metal. Usually between the lateritic cap and the underlying basalt or other rocks over which it rests, there is a lithomarge-like rock, or bole, a sort of transitional product, showing a gradual passage of the underlying rock (basalt or gneiss) into laterite.

Laterite has the peculiar property of being soft when newly quarried, but becoming hard and compact on exposure to the air. On account of this property it is usually cut in the form of bricks for building purposes. Also loose fragments and pebbles of the rock tend to re-cement themselves into solid masses as compact as the original rock.

Distribution of laterite—Laterite occurs principally as a cap on the summit of the basaltic hills and plateaus of the highlands of the Deccan, central India and Madhya Pradesh. In its best and most typical development it occurs on the hills of Bombay region of the Deccan. In all these situations it is found capping the highest flows of the Deccan Traps. The height at which laterite is found varies from about 600 m. to 1,500 m. and considerably higher, if the ferruginous clays and lithomarges of the Nilgiri mountains are to be considered as one of the many modifications of this rock. In thickness the lateritic caps vary from 15 to nearly 60 m.; some of these are of small lateral extent, but others are very extensive and individual beds are often seen covering an immense surface of the country continuously. Laterite is by no means confined to the Deccan Trap area, but is found to extend in isolated outcrops from as far north as the Rajmahal hills in Bihar[1] to the southern extremity of the Peninsula. It extends to Sri Lanka where it forms a thick cap covering the gneiss and khondalite. In these localities the laterite rests over formations of various ages and of varying lithological composition, *e.g.* Archaean gneiss, Dharwar schist, Gondwana clays, etc. Laterite is of fairly wide occurrence in parts of Burma also.

High-level laterite and low-level laterite—The laterite of the above-noted areas is all of high level, *i.e.* it never occurs on situations below about 600 m. above sea-level. The rock characteristic of these occurrences is of massive homogeneous grain and of uniform composition. This laterite is distinguished as *high-level laterite*, to differentiate it from the *low-level laterite* that occurs on the coastal lowlands on both sides of the Peninsula, east and west. On the Malabar side its occurrences are few and isolated, but on the eastern coast the laterite occurs almost everywhere rising from beneath the alluvial tracts which fringe the coast. Laterite of the low-level kind occurs also in Burma, in Pegu and Martaban. Low-level laterite differs from the high-level rock in being much less massive and in being of detrital origin, from its being formed of the products of mechanical disintegration of the high-level laterite. As a rock-type, laterite cannot be said to constitute a distinct petrological species; it shows a great deal of variation from place to place, as regards both its structure and its composi-

---

[1] These hills are for the most part composed of Jurassic traps, in addition to a substratum of Gondwana rocks; the summit of the traps is covered with laterite.

tion, and no broad classification of the varieties is possible; but the above distinction of the two types of high and low level is well established, and is based on the geological difference of age as well as the origin of the two types.

Theories of the origin of laterite—The origin of laterite is intimately connected with the physical, climatic and denudational processes at work in India. The subject is full of difficulties, and although many hypotheses have been advanced by different geologists, the origin of the (high-level) laterite is as yet a much-debated question. One source of difficulty lies in the chemical and segregative changes which are constantly going on in this rock, and which obliterate the previously acquired structures and produce a fresh arrangement of the constituents of the rock. It is probable that laterites of all the different places have not had one common origin, and that widely divergent views are possible for the origins of the different varieties.

From its vesicular structure and its frequent association with basalts, it was at first thought to be a volcanic rock. Its subaerial nature was, however, soon recognised beyond doubt, and later on it was thought to be an ordinary sedimentary formation deposited either in running water or in lakes and depressions on the surface of the traps. Still later views regard the rock as the result of the subaerial decomposition *in situ* of basalt and other aluminous rocks under a warm, humid and monsoonic climate. Under such conditions of climate the decomposition of the silicates, especially the aluminous silicates of crystalline rocks, goes a step further, and instead of kaolin being the final product of decomposition, it is further broken up into silica and the hydrated oxide of aluminium (bauxite). The vital action of low forms of vegetable life was at one time suggested as supplying the energy necessary for the breaking-up of the silicates to this last stage. The silica is removed in solution, and the salts of alkalis and alkaline earths, derived from the decomposition of the ferromagnesian and aluminous silicates, are dissolved away by percolating water. The remaining alumina and iron oxides become more and more concentrated and become mechanically mixed with the other products liberated in the process of decomposition. The vesicular or porous structure, so characteristic of laterite, is due to molecular segregation taking place among the products left behind. For a summary of views on the laterite and bauxite of India see C. S. Fox's memoir.[1]

Mr. J. M. Maclaren[2] declared that laterite deposits are due to the metasomatic replacement (in some cases the mechanical replacement) of the soil or subsoil by the agency of mineralised solutions,

---

[1] Bauxite and Aluminous Laterite Occurrences of India, *Mem. G.S.I.* vol. xlix. pt. 1, chap. 1. 1923.

[2] *Geological Magazine*, Dec. V. vol. iii., 1906.

brought up by the underground percolating waters ascending by capillary action to the superficial zone.

From the highly variable nature of this peculiar rock, it is possible that every one of the above causes may have operated in the production of the laterites of different parts according to particular local conditions, and that no one hypothesis will be able to account for all the laterite deposits of the Indian Peninsula.

Laterite rock-bodies are subject to secondary changes, a fact which introduces further complexity. "Under conditions of free drainage and high rainfall (2,500 mm. per year, or more) the laterite may accumulate without much further change, the soluble products of hydrolysis being rapidly lost by leaching. On the other hand, under impeded drainage conditions and alternations of wet and dry seasons, the fluctuating ground water, carrying dissolved silica and bases, may effect a complete change in the laterite, whose gibbsite component, according to Harrison, is converted into secondary kaolins, stained red by hydrous iron oxide residues." In this manner some authorities have explained the formation of the vast masses of red earth capping igneous rock-terrains of humid tropics, such as the gneissic areas of Madras. This implies a resilicification of the bauxitic or gibbsitic base of laterite into secondary clays.[1]

The age of laterite—The age of the existing high-level laterite cap is not determinable with certainty; in part it may be Pliocene, or even older, in part its age is post-Tertiary (Pleistocene) or somewhat later, and it is probable that some of it may still be forming at the present day. The age of the low-level, coastal laterite must obviously be very recent. The earliest relics of prehistoric man in the shape of stone implements of the *Palaeolithic* type are found embedded in large numbers in the low-lying laterite.

There is evidence, however, that important masses of laterite were formed in the Eocene, and even in earlier ages. A thin but persistent substratum of pisolitic haematite, red earth, or of bauxite occurs at the base of the Nummulitic series in Pakistan. Its subaerial mode of origin under the above conditions being granted, there is no reason why it should be restricted to any particular age only. According to several authorities laterite is seen at several other horizons in the stratigraphical record of India, especially those marking breaks or unconformities when the old land-surfaces were exposed for long durations to the action of the subaerial agents of change. A ferruginous lateritic gravel bed among the rock-records of past ages is, therefore, held to be of the same significance as an unconformity conglomerate.

---

[1] F. Hardy, Some Aspects of Tropical Soils, *Trans. Third Int. Cong. Soil Sc.* vol. ii., 1935.

Economics—As stated above, laterite is at times, according to conditions favouring the concentration of any particular metallic oxide, a valuable ore of iron or an ore of aluminium and manganese. The use of laterite as an ore of iron is of very old standing, but its recognition as a source of alumina is due to Sir T. H. Holland, and of manganese to Sir L. L. Fermor. In several parts of southern India and Burma laterite is quarried for use as a building stone from the facility with which it can be cut into bricks. In fact the term laterite originally has come from the Latin word *later*, a brick.

Laterite does not yield good soil, being deficient in salts as well as in humus.

REFERENCES

L. L. Fermor, Laterite, *Geol. Magazine*, Dec. V. vol. viii., 1911.
C. S. Fox, *Mem. G.S.I.* vol. xlix. pt. 1, 1923; *Mining Magazine*, vol. xxvi. pp. 82–96, and *Records G.S.I.* vol lxix. pt. 4, 1936; *Aluminous Laterite and Bauxite*, London, 1932.
P. Lake, Geology of S. Malabar, *Mem. G.S.I.* vol. xxiv. pt. 3, 1890.
Sir T. H. Holland, *Geol. Magazine*, Dec. IV. vol. x., 1903.
J. Harrison, The Katamorphism of Igneous Rocks under Tropical Conditions, *Imp. Bur. Soil. Sc.* London, 1934.

References to laterite in *G.S.I.* publications are too numerous to quote. The earlier *Mems.*, vols. i. ii. ix. and x., may be consulted for descriptive purposes.

CHAPTER XXIV

Pleistocene and Recent

Examples of Pleistocene and Recent Deposits—Among the Pleistocene and Recent deposits of India are the following, each of which in its respective locality is a formation of some importance: the high-level river-terraces of the Upper Sutlej and other Himalayan rivers, and of the Narmada, Tapti and Godavari among the Peninsular rivers; the lacustrine deposits (*Upper Karewa*) of the Upper Jhelum valley in Kashmir and the similar accumulations (*Tanr*) in the Nepal valley[1]; the foraminiferal sandstone (*Porbander stone*) of the Saurashtra coast and the *Teris* of the Tinnevelli and Kerala coasts; the aeolian deposits of the Godavari, Krishna and Cauvery banks (resembling the *Bhur* of the Ganges valley), and the *loess* deposits of the Salt-Range, Potwar, and of Baluchistan; the fluvio-glacial deposits of the Potwar plateau; the stalagmitic cave-deposits of the Kurnool district; the black cotton-soil or *Regur* of Gujarat and the Deccan; the great gravel-slopes (*daman*) of the Baluchistan hills. These are examples, among many others, of the Pleistocene and later deposits of India each of which requires a brief notice in the present chapter.

Alluvium of the Upper Sutlej—Ossiferous clays, sands and gravels, the remains of the Pleistocene alluvium of the Upper Sutlej,[2] are found in the Hundes province of the Central Himalayas covering several hundreds of square km. and resting at a great height above the present level of the river-bed. These deposits were laid down in the broad basin of the Upper Sutlej while it was at a considerably higher level, enclosing numerous relics of the living beings that peopled this part of the Himalayas. The old alluvium of the river is now being deeply trenched by the very Sutlej which has already cut out of it a picturesque and deep, narrow gorge some 1,000 m. in depth. The chief interest of the Hundes deposits attaches to the mammalian fossils preserved in the horizontally bedded gravels. These deposits have so far not been investigated systematically, and only *Rhinoceros*, *Pantholops*, *Equus*, *Bos* and *Capra* have so far been known from isolated specimens.

[1] *Rec. G.S.I.* vol. viii. pt. 4, 1875.
[2] *Rec. G.S.I.* vol. xiv. pt. 2, 1881.

Tapti and Narmada—In the broad basins of many of the Peninsular rivers large patches of ancient alluvium occur, characterised by the presence of fossils belonging to extinct species of animals. Of these the old alluvial remains of the Narmada and Tapti are remarkable as lying in deep rock-basins, at considerable elevations, over 150 m., above their present bed. Among other vertebrate and mammalian fossils,[1] these ancient river sediments have preserved the earliest undoubted traces of man's existence. Scattered in their alluvia are the stone knives, hatchets, arrows and other implements which prehistoric man manufactured out of any hard stone that he came across, whether it was Cuddapah quartzite, or a Vindhyan sandstone, or the amygdaloidal agates.

There is some proof that the Narmada in those days was confluent with the Tapti, and that its separation into a distinct channel was effected at a comparatively late date by earth-movements. That the course of the Narmada has undergone a serious disturbance during late geological time is corroborated by another piece of evidence, namely the precipitous falls of this river at Jabalpur.

The Karewas of Kashmir—The valley of Kashmir is an alluvium-filled basin, a large part of which is of recent formation by the river Jhelum. More than half of its area, however, is occupied by outliers of a distinctly older alluvium, which forms flat mounds or platforms, sloping away from the high mountains that border the valley on all sides. These deposits, known in the Kashmiri language as *Karewas*,[2] are composed of fine silty clays with sand and bouldery gravel, the coarse detritus being, as a rule, restricted to the peripheral parts of the valley, while the finer variety prevails towards the central parts. The bedding of the Karewas is for the greater part almost horizontal, but where they abut upon the Pir Panjal, or the mountains of the south-west border of the valley, they show evidence of a good deal of upheaval, dipping sometimes as much as 40° at some places, the direction of the dip being towards the valley.[3]

Middlemiss's work in the Pir Panjal and elsewhere has greatly modified the views regarding the age and thickness of these deposits. He has shown that their thickness amounts to 1,400 m.

---

[1] *Crocodilus, Trionyx, Pangshura, Ursus, Bubalus, Bos, Equus, Sus, Cervus, Elephas, Hippopotamus* and *Rhinoceros*. Besides these, shells of land molluscs such as *Melania, Planorbis, Paludina, Lymnaea, Bulinus, Unio* are found in the alluvium of the Narmada.

[2] F. Drew, *Jammu and Kashmir Territories*, p. 210, 1875.

[3] Later investigations have revealed some Karewa deposits even on the summit of the Pir Panjal (3,660 m.), thus proving that the latter mountains have been elevated nearly 1,500 m. since the Karewas were deposited. *Rec. G.S.I.* vol. xliv. pt. 1, 1914.

at least, and that the lower part of the Karewa deposits is considerably older than any of the glacial moraines on the Pir Panjal and may be of Middle Siwalik age. In the Upper Karewas several successive terminal glacial moraines, composed of boulders, pebbles and sands, separated by fine clays (some of them of the type of *varved* clays), denoting the deposits of the warm interglacial periods of melting ice, have been observed. In some sections of the Karewas, according to some observers, deposits of three or four distinct glacial periods can be made out (p. 360).

The Karewas, in their upper part at least, are supposed to be the relics of old extensive lake-basins, which intermittently came into existence during the warm interglacial periods of melting ice and which periodically filled the whole valley of Kashmir from end to end to a depth of more than 300 m. This old alluvium has been subsequently elevated, dissected, and in a great measure removed by subaerial denudation as well as by the modern Jhelum, leaving the Karewa outliers of to-day. For further information regarding the Karewas, see p. 359.

Old alluvial deposits, to which a similar origin is ascribed, are found in the Nepal valley, and are known there under the local name of *Tanr*. They contain a few peat and phosphatic beds enclosing mammalian relics.

Coastal alluvial deposits—In a previous chapter it was mentioned that all along the eastern coast of India, from the Ganges delta to the extremity of the Peninsula, there is a broad strip of Tertiary and post-Tertiary alluvium containing marine shells and other fossils. The Tertiary part of these deposits has been described already under the title of the Cuddalore series, in Chapter XVII; the remaining younger part occupies small tracts both on the east and west coasts. That on the east coast, however, assumes a considerable width and forms many tracts of fertile country from the Mahanadi to the Cape. On the Malabar coast this alluvial belt is very meagre and is confined to the immediate vicinity of the coast except at its north end, where it widens out into the alluvial flats of Gujarat. On the Saurashtra coast at some places a kind of coastal deposit occurs known as the *Porbander stone* (sometimes also as *Miliolite*), which is noteworthy. It is composed of calcareous wind-blown sand, the sand grains being largely made up of the casts of foraminifers, the whole compacted into a white or cream-coloured, rudely bedded freestone. The rock known as Junagarh limestone is a typical aeolian limestone, situated 50 km. inland from the sea coast and 60 m. thick. It is mainly composed of fragments of calcareous shells (most of them of living species) cemented by lime. About 6 to 12 per cent of foreign particles of the Girnar igneous rocks enter into the composition. It is believed that the Saurashtra peninsula stood 45 m. lower than at present and was probably in Pleistocene time an island or group of islands. From their softness and the ease with which they receive dressing and ornamental treatment, these limestones are a favourite material for architectural purposes in many parts of Gujarat, Maharashtra and the Deccan.

Sand-dunes—Sand-dunes are a common feature along the Indian coasts, particularly on the Malabar coast, where they have helped to form a large number of lagoons and backwaters, which form such a prominent feature of the western coast of India. In Orissa there are several parallel ridges of sand-dunes on the plains fronting the coast which are held to indicate the successive positions of the coast-line. Sand-loving grasses and other vegetation help to check the further progress of the dunes inland.

Sand-dunes are also met with in the *interior* of the Peninsula, in the broad valleys of the Krishna, Godavari, etc., and occupy a wide stretch of the coastal terrain of Orissa. They are also common in the lower Indus valley, in Rajasthan, Kutch and for some distance inland on the Mekran coast. The sand is blown there by the strong winds which blow through these valleys during the hot-weather months. A large volume of sand is thus transported and accumulated along the river courses, which are unable to sweep it away (cf. *Bhur* land of the Ganges valley).

The peculiar form of sand-hills known as *Teri* on the Tinnevelli coast is also of the same origin.

Loess—In the localities east of the Indus, in the N. W. Punjab and on the Salt-Range, there are subaerial Pleistocene accumulations of the nature of loess, a loose unstratified earthy or sandy deposit but little different in composition from the alluvium of the plains. Loess, however, differs from the latter in its situation at all levels above the general surface of the plains and in its being usually traversed by fine holes or tubes left by the roots of the grasses growing upon it. The lower parts of Baluchistan are largely covered with wind-blown, more or less calcareous and sandy earth, unstratified and loosely consolidated. On the flat plateau top of the Salt-Range loess is a very widespread superficial deposit, and on many plateaus, which form the summit of this range, the accumulation of loess from the dust and sand blown from the Punjab plains is yet in progress. The inequalities of the surface, produced by its irregular distribution, are the cause of the numerous shallow lakes[2] on the summit of the Salt-Range. Loess is also a prevalent superficial formation in the country to the north (Potwar), where its dissection into an intricate system of branching ravines has produced *bad land* tracts.

The conditions that have favoured the growth of loess in these parts are their general aridity and long seasons of drought. These

---

[1]Fedden, *Mem. G.S.I.* vol. xxi., 1885.
[2]*Rec. G.S.I.* vol. xl. pt. 1, 1910.

give rise to dust-storms of great violence in the hot-weather months preceding the monsoons, which transport vast clouds of dust and silt from the sun-baked plains and dried-up river-basins, and heap them on any elevated ground or accidental situation. The isolated dust-mounds one notices in some parts of the Punjab are attributable to this cause.

Potwar fluvio-glacial deposits—The Potwar[1] is an elevated plain lying between the northern slopes of the Salt-Range and the Rawalpindi foot-hills. A few metres below the ordinary surface alluvium of some parts of these plains is found a curious inter-mixture of large blocks of rocks up to 15 m. in girth, with small pebbles and boulders, the whole embedded in a fine-grained clayey matrix. The material of the blocks suggests their derivation from the high central ranges of the Himalayas, while their size suggests the action of floating ice, the only agency which could transport to such distances such immense rock-masses. Scattered moraine and erratic blocks, assigned to the action of floating ice during the last glacial period, are found between Attock and Campbellpur on the surface of the Potwar plateau. The Indus river is noted for floods of extraordinary severity (owing to accidental dams in the upper narrow gorge-like parts of its channel or those of any of its tributaries).[2] Many such floods have been known in historic times, and some have been recorded in the chronicles. The water so held up by the dam spreads out into a wide lake-like expanse in the broader part of the valley above the gorge. In the Pleistocene times, when, as has been shown in a previous chapter, the Himalayas were experiencing arctic conditions of climate, the surface of the lake would be frozen. The sudden draining of the lake, consequent on the removal of the obstacle by the constantly increasing pressure of the waters resulting from the melting of the ice in springtime, would result in the tearing off of blocks and masses of rock frozen in and surrounded by the ice. The rushing debacle would float down the ice-blocks with the enclosed blocks of rock, to be dropped where the ice melted and the water had not velocity enough to carry or push them further. This would, of course, happen at the site where the river emerged from its mountain-track and entered the plains. The above is regarded as the probable explanation of the origin of the Potwar deposits. It thus furnishes us with further cogent evidence of the existence of glacial conditions, at any rate in the Himalayas.

---

[1] *Rec. G.S.I.* vol. x. pt. 3, 1877, *Rec. G.S.I.* vol. xiii. pt. 4, 1880, and vol. lxi. pt. 4, 1929.

[2] For an interesting account of some of the disastrous floods of the Indus and their cause, obtained from eye-witnesses and from personal observations, see Drew, *Jammu and Kashmir Territories*, London, 1875.

Cave deposits—*Caves*.[1] But few caves of palaeontological interest exist in India, and of these only one has received the attention of geologists. The caves in other countries have yielded valuable ossiferous stalagmitic deposits, throwing much light on the animal population, particularly the cave-inhabiting larger mammals, of late geological times, their habits, mode of life, etc. During Pleistocene times caves were used as dwellings by prehistoric man, and important relics of his handiwork, art and culture are sometimes preserved on the walls and floors of the caves. The only instances of the Pleistocene caves are a few caverns in the Kurnool[2] district, in the neighbourhood of Banaganapalli, in a limestone belonging to the Kurnool series. In the 7-9 m. thick stalagmite at the floor, there occurs a large assemblage of bones belonging to a mixture of Recent and sub-Recent species of genera such as *Cervus, Viverra, Hystrix, Sus, Rhinoceros* (extinct), and *Cynocephalus, Equus, Hyaena, Manis* and (living species) of bat, mice, squirrel and porcupine.

A small cave in a limestone belonging to the Triassic age, occurring in the neighbourhood of Srinagar, near Harwan, was recently found to contain mammalian bones on its floor. They included remains of sub-Recent species such as *Cervus aristotelis* (sambur), *Sus scropha* (European pig), and an unknown antelope. A number of small caves are found in the great Trias limestone cliffs of Kashmir, but they have not been investigated. Scores of caves, large and small, found in limestone and calcareous rocks in various parts of the country require systematic examination of their floor deposits.

Regur—Among the residual soils of India there is one variety which is of special agronomic and geological interest. This is the black soil, or *Regur*[3] (Chernozem), of many parts of Gujarat, Madhya Pradesh and other "cotton districts" of the Deccan. Regur is a highly argillaceous, somewhat calcareous, very fine-grained black soil. It is extremely sticky when wetted and has a capacity for retaining a large proportion of its moisture for a long time. Among its accessory constituents are a high percentage of iron oxide and calcium and magnesium carbonates, the $CaCO_3$ disseminated as *kankar*, and a very varying admixture of organic matter (humus) ranging from one to as much as 10 per cent. It is probably to its iron and humus content that the prevailing dark, often black, colour is due. The black cotton soil is credited with an extraordinary degree of fertility by the people; it is in some cases known to have supported agriculture for centuries

---

[1] *Rec. G.S.I.* vol. xix. pt. 2, 1886 ; *Pal. Indica*, sers. x. vol. iv. pt. 2, 1886.

[2] *Mem. G.S.I.* vol. iv. pp. 183 and 357, and vol. vi. p. 235 ; *Rec. G.S.I.* vol. iv. p. 80, 1871.

[3] From Telugu word *Regada*.

without manuring or being left fallow, and with no apparent sign of exhaustion or impoverishment.

The origin of Regur—The origin of this soil is yet not quite certain. It is generally ascribed to long-continued surface action on rocks like the Deccan Trap and Peninsular gneisses of a basic composition. The decomposition of the basalts *in situ*, and of aluminous rocks generally, would result in an argillaceous or clayey residue, which, by a long cycle of secondary changes and impregnation by iron and decomposed organic matter (humus) resulting from ages of jungle growth over it, would assume the character of Regur.

The thickness of the regur soil-cap is highly variable, from 30 cm. to 15 m., while the composition of the soil shows considerable variation with different depth horizons, especially in its clay content and lime segregation. The clay-fraction of black cotton soil is very rich in silica, 60 per cent, and iron 15 per cent, with only 25 per cent of alumina.

"Daman" slopes—Alluvial fans or taluses fringing the mountains of Baluchistan, and known as *Daman*, are another example of Pleistocene deposits. These are a very prominent feature of the hilly parts of Waziristan and Baluchistan where the great aridity and drought favour the accumulation of fresh angular débris in enormous heaps at the foot of the hills. Wells that are commonly excavated in these gravel slopes (and which are known as *Karez*) illustrate a peculiar kind of artesian action. The Karez is merely a long underground, almost horizontal, tunnel-like bore driven into the sloping talus till it reaches the level of permanent saturation of water, which is held in the loose porous gravel. The water is found at a sufficient pressure to make it flow at the mouth of the well. The underground tunnel may be several miles in length and connected with the surface by bore-holes.[1]

The Human epoch—In the foregoing account of the later geological deposits of India there is everywhere a gradual passage from the Pleistocene to the Recent, and from that to Prehistoric. These periods overlap each other much as do the periods of human history. As in other countries, the Pleistocene in India also is marked by the presence of Man and is known as the Human epoch.

Man's existence is revealed by a number of his relics preserved among the gravels of such rivers as the Narmada and Godavari, and the Soan, or in other superficial alluvia, both in South and North India. On the surface of the Potwar plateau there are found scores of sites containing flint artifacts of ? Chellean industry in hundreds of flakes and cores. Stratigraphically these implements are dated by being preserved in a few cases in the topmost beds

---

[1]Vredenburg, *Mem. G.S.I.* vol. xxxii. pt. 1, 1901 ; Oldham, *Rec. G.S.I.* vol. xxv. pt. 1, 1892.

of the Upper Siwalik boulder-conglomerate and in the older alluvium of the Narmada, Godavari and the Soan. A chipped stone hatchet of quartzite, found near Godawara village on the Narmada, is believed to be of pre-Chellean age (Lr. Pleistocene), the earliest prehistoric relic of Man in India. From the evidence of the few artifacts found in the older Jumna-Ganges alluvium as well as that of fossil bones, it seems that this alluvium is intermediate in antiquity between the Narmada-Godavari beds and the Mid Pleistocene. This suggests human settlements in the latter valleys. of an earlier Palaeolithic race. These archaic human relics consist of various stone implements that prehistoric man used in his daily life, ranging from rude stone-chippings, cores and flakes to skilfully fashioned and even polished instruments like knives, celts, scrapers, arrow-heads, spears, needles, etc., manufactured out of stone or metal or bone. These instruments ("artifacts") become more and more numerous, more widely scattered, and evince an increasing degree of skill in their making and in their manipulation as we ascend to newer and younger formations. This evidence of man's handiwork furnishes us with the best basis for the classification of this period into three epochs.

F. E. Zeuner considers that, though no classification applicable to the whole of India is possible, a rough guide is:

Neolithic—passing upwards into the Asoka period (274 B.C.) and downwards into the Indus Valley culture (about 2,500 B.C. and beyond).

Microlithic—from late Pleistocene to pre-historical.

Palaeolithic—from about 500,000 years ago to the end of the Pleistocene.

Palaeoliths were first found in India by R. B. Foote; his collection is in the Madras Museum. The Palaeolithic in relation to geological deposits has lately been examined in Kashmir, Punjab, Gujarat (Sabarmati Valley), Bombay (Kandivli), Nellore etc. The earliest finds seem to be large flakes made from pebbles in the boulder-conglomerates of the Potwar, correlated by de Terra with part of the Karewas of Kashmir. The later Punjab Palaeolithic is named after the Soan Valley. Farther south the Palaeolithic is increasingly characterised by Acheulian implements. The Sabarmati Palaeolithic is based on river pebbles, and according to Zeuner of penultimate glaciation age. The Palaeolithic of Madras and the Deccan is still more markedly of Acheulian type and associated either with river gravels or with derived laterites.

The Microlithic industries, consisting of very small tools of chalcedony, jasper and quartz, are comparable with the European Mesolithic but chronologically are probably later. They include arrow-heads, barbs and long parallel-sided blades. Polished axes very similar to those of the European Neolithic appear in India with the later Microlithic, by which time metal was known but not generally used. Besides bronze, the primitive Indian made implements of native copper which he found in south India.[1]

The existence of Man in an age earlier than the older alluvia of the Narmada and Godavari is a matter of conjecture only. No signs of the existence of human beings are observed in the Upper Siwalik, except perhaps in the topmost strata (pp. 362–3). Whether he was a witness of nature's last great phenomenon, the erection of the Himalayan chain to its present height, or whether he was a contemporary of the *Sivatherium* or the *Stegodon*, is a profoundly interesting speculation but one for which no clue has been so far discovered. The question has hardly received any attention in India in the past due mainly to the paucity or absence of cave-deposits. It is, however, possible that valuable geological and anthropological data may be obtained by search in the Upper Siwalik, in the older alluvia and river terraces, the travertine deposits of springs, loess caps and mounds, etc.

Here, however, we reach the limits of geological inquiry. Further inquiry lies in the domains of anthropology and archaeology.[2]

Few changes of geography have occurred in India since the Pleistocene. After the great revolutions at the end of the Pliocene, the present seems to be an era of geological repose. A few minor warpings or oscillations in the Peninsula;[3] the extinction of a few species; the migration and redistribution of others; some changes in the courses of rivers, the degradation of their channels a few feet lower, and the extension of their deltas; the silting up of the Rann of Kutch; a few great earthquakes; the eruptions of Barren Island and other minor geological and geographical changes are all that a geologist notes since the advent of man in India.

REFERENCES

References to the various subjects treated in this chapter have been given against each.

---

[1] *Prehistoric and Protohistoric Relics of Southern India*, by R. B. Foote, Madras, 1915 ; *Old Chipped Stones of India*, by A. C. Logan, Calcutta, 1906 ; *Prehistoric India*, P. Mitra, Calcutta University, 1927.

[2] W. J. Sollas, *Ancient Hunters* (Macmillan), 1924 ; F.E. Zeuner, *Dating the Past* (Methuen), 1950.

[3] E. Vredenburg, Pleistocene Movement in the Indian Pensinula, *Rec. G.S.I.* vol. xxxiii. pt. 1, 1906.

CHAPTER XXV

Physiography

In the light of what we have seen of the geological history of India, a brief re-examinaton of the main physiographic features of the region will be of interest. Every geological age has its own physiography, and, therefore, the present surface features of India are the outcome, in a great measure, of the latest chapters of its geological history.

Principles of physiography illustrated by India—Physiography is that branch of geology which deals with the development of the existing contours of the land part of the globe. In the main, dry land owes its existence *en masse* to earth-movements, while the present details of topography, its scenery and its landscapes, are due to the action of the various weathering agents. In the case of elevated or mountainous regions of recent upheaval, the main features are, of course, due to underground forces, hypogene agencies; but in old continental areas, which have not been subject to crustal deformation for long ages, the epigene or meteoric forces have been the chief agents of earth-sculpture. Land areas of great antiquity, therefore, possess earth-features of a subdued relief; ultimately it is the fate of the centres of ancient continents to be overspread by deserts. In this latter class of earth-features there is no correspondence observable between the external configuration of the regions and their internal geological structure. Here the high ground does not correspond to anticlinal, or the hollows and depressions of the surface to synclinal, folds. The accumulation of the eroded products, derived from the degradation of the elevated tracts by the subaerial, meteoric agencies, in a low, broad zone of lodgment gives rise to a third order of land-forms —the plains of alluvial accumulation.

The three physiographic divisions of India afford most pertinent illustrations of the main principles of physiography stated above. The prominent features of the extra-Peninsula, the great mountain border of India, are those due to upheaval of the crust in late-Tertiary times, modified to some extent by the denuding agents which have since been operating on them; those of the Peninsula are mainly the results of subaerial denudation of a long cycle of geological ages, modified in some cases by volcanic and in others by sedimentary accumulations; while the great plains of India, dividing these two regions, owe their formation to sedimentary deposition alone, their persistent flatness being entirely due to the aggrading work of the rivers of the Indus-Ganges system during comparatively recent times.

All through geological history, the Earth's crust has been subject to deformation, displacements, folding, faulting, inversion and overthrusts, through the play of many hypogene forces—viz. contraction following cooling, up-warps and down-warps in the *sima*, accumulation of radio-active heat, unequal distribution of heat in the dense, basic Mantle layer beneath the crust. Whatever be the causes of crustal movements giving rise to the major features of its relief, the great ocean-basins, continents, mountain-chains, movements of depression must be in excess of elevation. Uplifts take place when two master-segments of the Earth's sphere, such as the high block of Central Asia and the horst of Gondwanaland in their subsidence squeeze between them the intervening Tethyan geosyncline and ridge it up into a mountain-range by the enormous tangential thrusts involved in their sinking.

According to current belief, subcrustal currents maintained by thermal processes, especially liberation of radio-active heat, play an important role in the development of major earth-features and in sustaining their isostatic balance.

The main elements of the physiography of a country are five:
(1) Mountains. (2) Plateaus and plains. (3) Valleys.
(4) Basins and lakes. (5) Coast-lines.

1. MOUNTAINS

Mountains may be (1) original mountains, or (2) subsequent or relict. The student already knows that these two types characterise the two major divisions of India. Original mountains include (*a*) accumulation-mountains and (*b*) deformation-mountains. Volcanoes, dunes or sand-hills and moraines are examples of the former, while mountains produced by the deformation or wrinkling of the earth's crust are examples of the latter. In the latter the relief of the land is closely connected with its geological structure, *i.e.* the strike, or trend, of these mountains is quite conformable with their axis of uplift. They are divisible into two classes: (i) folded mountains, and (ii) dislocation-mountains. Of these, the first are by far the most important, comprising all the great mountain-chains of the earth. The Himalayas, as also all the other mountain-systems of the extra-Peninsular area, are of this type.

The Structure of the Himalayas—The structure of the outer or sub-Himalayan ranges is generally of great simplicity; they are made up of a series of broad anticlines and synclines of the normal type, a modification of the Jura type of mountain structure. These outer ranges, dissected into a series of escarpments and dipslopes, are separated by narrow, longitudinal tectonic valleys or depressions, called *Duns*. The reversed strike-faults mentioned on pp. 337-8 are a characteristic feature in the tectonics of these sub-Himalayan ranges. The most prominent of these is the Main Boundary Fault, which extends along the length of the mountains from the Punjab to Assam. We have seen on page 338 the true nature of these faults and the significance attached to them.

Many of the ranges of the outer Himalayas and several of the middle Himalayas as well are of the *orthoclinal* type of structure, *i.e.* they have a steep scarp on the side facing the plains and a gentle inclination facing Tibet. It is a characteristic of the folds of this part of the Himalayas that the anticlines are often faulted steeply in their outer or southern limbs, the fault-scarp lying in juxtaposition with much younger rock-zones.

This zone is succeeded by a belt of more compressed isoclinal folds, which are strictly autochthonous in their position. It is followed, in the Pir Panjal range and in the Simla-Kumaon area, by a system of overfolds of the recumbent type, severed by reversed faults that have passed into thrust-planes, along which large slices of the mountains have moved bodily southwards—the *Nappe zone* of the Himalayas[1]. Two more or less parallel and persistent planes of thrust have been traced at the foot of the Pir Panjal range along its whole length from the Jhelum to the Beas in Kulu. The outer of these (the *Murree thrust*) has thrust the autochthonous Carboniferous-Eocene belt of rocks over the Mid-Tertiary Murree series, while the inner thrust (the *Panjal thrust*) has driven the older Purana schists and slates of the central mountains over the autochthonous Carboniferous-Eocene rocks along an almost horizontal plane of thrust (Kashmir nappe).

In the Krol belt of Simla-Kumaon-Nepal, a similar tectonic sequence is observed, revealing at least two nappes of older rocks overriding the autochthonous fold-belt of the Tertiary rocks of the outer Himalayas. These are the Krol nappe and the Garhwal nappe, separated by two distinct thrust-planes. In the neighbourhood of Solon and Subathu, Nummulitic and Dagshai strata crop out as *windows* from beneath rocks of the Krol nappe.[1]

The structure of the inner Himalayas has not yet been the subject of such intensive study and investigation as that which has so far unravelled the inner architecture of the Alps. A great deal

---

[1]Pilgrim and West, Structure of the Simla Rocks, *Mem. G.S.I.* vol. liii., 1928; Wadia, *Rec. G.S.I.* vol. lxv. pt. 2, 1931 and vol. lxviii. pt. 2, 1934; Auden, *Rec. G.S.I.* vol. lxvii. pt. 4, 1934.

of investigation in the central ranges, especially the zone of most complex folding and intrusion, remains to be done before it is possible to say anything regarding the structure of these mountains except in very general terms. East of Kumaon systematic geological work has now begun. The evidence so far obtained, however, tends to show that large areas of the Western Himalayas possess a comparatively simple type of mountain tectonics, and the piles of nappes, their complex re-folding, digitation and inversions such as those to which modern theory ascribes the formation of the Swiss Alps have not been observed on the same scale of intensity or order of magnitude. The thrusts in the Himalayas that have driven sheets of older rocks over the newer recall rather the thrust-planes of the Scottish Highlands. The great sedimentary basins of Hazara and Kashmir, lying between the crystalline axis and the zone of the great thrusts (the nappe zone), show a system of normal open anticlines and synclines without shearing, or reduplication, indicating that the nappes have undergone no subsequent body deformation.

As we approach the central crystalline axis of the Himalayas, there is manifested a puzzling monotonous uniformity of rock-facies—a uniformity that is only apparent—induced by the regional and thermal metamorphism to which the rocks have been subjected. The Central zone, however, is not wholly crystalline. Parts of it are capped by unmetamorphosed Palaeo-Mesozoic members of the Tibetan zone of geosynclinical facies, lying to its North. The Tibetan zone, confined almost wholly to the north in the Eastern Himalaya, crosses over the Central axis west of Spiti into Kashmir and Hazara; it then turns westwards, through the Salt Range, to join on to the Iranian Arc of mountains of Alpine-Himalayan orogeny.

From a tectonic point of view, according to present data, we may divide the Western Himalayas into the following structure-zones:

The Foreland. North fringe of Gondwanaland, covered under Tertiary sediments:
1. *Siwalik belt*—Jura type of folds of Upper Tertiary river-deposits.
2. *Sirmur belt*—more compressed isoclinal folds of lagoon sediments.

Autochthonous Fold Zone.
3. *Carboniferous-Eocene belt*—recumbent folds of Eocene with cores of Carbon-Trias rocks, Panjal volcanics, Blaini, or Krol series.

Nappe Zone.
4. *Purana Slate belt*—unfossiliferous slates containing Palaeozoic and Mesozoic outcrops which have expanded out in the Hazara and Kashmir sedimentary basins.

5. *Crystalline belt*—of the central axial chain of metamorphic rocks with granite intrusions, a geanticline within the main geosyncline.
   6. *Tibetan belt*—marine sediments of Cambrian to Eocene age in the Himalaya geosyncline, crossing the axial range.

The Nappe Zone of Kashmir

In these mountains the *nappe zone* of inner Himalayan rocks has travelled far along a horizontal thrust (Panjal thrust) so as to lie fitfully sometimes against a wide belt of the autochthon, at other times almost against the foreland. The Kashmir nappe is composed mostly of pre-Cambrian sediments (Salkhala series) with a superjacent series (Dogra Slate), forming the floor of the Himalayan geosynclinal that has been ridged up and thrust forward in a nearly horizontal sheet-fold. On this ancient basement lie synclinal basins containing a more or less full sequence of fossiliferous Palaeozoic and Triassic marine deposits in various parts of Kashmir. The latter are detached outliers of the Tibetan marine zone, which in the eastern Himalayas is confined to the north of the central Himalayan axis.

In the nappe zone to the north are more thrusts, not easily recognisable in the crystalline complex which builds the Great Himalayan range of the centre. These thrusts, however, are not of wide regional or tectonic significance. As a tectonic unit, the Great Himalayan range is made up of the roots of the Kashmir nappe, the principal geanticline within the main Himalayan geosyncline, consisting of the Archaean and pre-Cambrian sedimentary rocks together with large bodies of intrusive granites and basic masses. Several periods of granitic intrusions have been observed, the latest being post-Cretaceous, or still later, connected with the earlier phases of the Himalayan uplift. A subordinate element of the Great Himalayan range is formed by the southward extensions of the representatives of the Tibetan belt of marine formations belonging to the Palaeozoic and Mesozoic.

The Nappes of the Simla Himalaya

Detailed mapping and study of the metamorphic gradations in ancient rock-complexes have led G. E. Pilgrim and W. D. West to conclude that the rocks of the Simla-Chakrata area, lying to the north of the Tertiary belt (Outer Himalayas), are not in the normal position as previous observers had believed, but have undergone complex inversions and thrusting. Four overthrusts are noted which have trespassed over the 100 km. broad Upper Tertiary area of Kangra and constricted it to barely 26 km. at Solon. The thrusts represent flat recumbent folds of great amplitude, showing bodily displacement from the north towards the autochthonous belt of the southwest. The pre-Cambrian (Jutogh and Chail series) is piled up on the Carboniferous and Permian systems (Blaini and Krol series), the entire sequence being totally unfossiliferous. Evidence of the superposition of the highly metamorphosed pre-Cambrian (Jutogh and Chail series), building some of the conspicuous mountain tops of the area (klippen), over the less altered lower Palaeozoics and Blaini beds (Upper Carboniferous), is obtained by a study of relative metamorphism and the structural relations of thrusts and discordances. The older rocks, now isolated, were once part of a continuous sheet over this area, but are now separated from the roots in the north by the deep valley of the Sutlej. To the south of the thrust zone, in the foothills, the older Tertiaries (Nummulitics) are separated from newer Tertiaries of the foothills by the series of parallel reversed faults which have been designated as boundary faults: (1) separating the Upper Tertiary from the lower Tertiary, and (2) separating the Lower Tertiary from pre-Tertiary rocks. This last "boundary" fault is really an overthrust, corresponding with the Murree thrust of the Kashmir mountains. Medlicott, Oldham and Middlemiss saw these faults and thrusts not only as dislocations but also as limits of deposition, no Upper Tertiary occurring north of the outer fault and no Upper or Lower Tertiary occurring north of the inner fault. Though this conception may still hold true to some extent, there are exceptions here as in the other parts of the Himalayas, *viz*. the occurrence of Nummulitic and later Tertiaries to the north of the inner line of faulting.

The nappe zone of the Simla region makes a more striking feature than in Kashmir. It commences some kilometres north of Solon and follows a meandering E.S.E. course, separating the Krol (Permo-Carboniferous) belt by the two great thrusts, Jutogh and Giri, which correspond with the Panjal thrusts of the western Himalaya. The outer limit of the Krol belt is the Krol thrust, corresponding to the Murree thrust of Kashmir. As shown by West and Auden, the Krol thrust itself is steeply folded by later disturbances which have plicated the Krol belt. This Krol belt, which tectonically corresponds with the Panjal range of the Kashmir Himalayas, runs along the Outer Himalayas for about 300 km. south-east of Solon in a tightly compressed sequence of Permo-Carboniferous strata. Near Solon, Tertiary rocks crop out as windows from under the Krols.

East of Nahan the Krol thrust transgresses southwards and overlaps the main boundary fault. Broadly speaking, the Krol zone of Simla corresponds with the authochthonous fold-belt of Kashmir, but as with the latter area the autochthon is often greatly narrowed and at places obliterated by the approach of the nappe front of the gently inclined overthrust slices from the north. Here and there, as at Solon, the Krol zone itself is deformed and thrust forward over the Nummulitics.

Massive porphyritic granite is intruded on a large scale into the pre-Cambrians. This granite is part of the central crystalline axis of the Himalaya, as in Kashmir and Hazara.

The Superposed Nappes of the Garhwal Himalaya

The tectonics of this part of the Himalayas are discussed in a recent paper by J.B. Auden.[1] Two nappes, the Krol and the Garhwal nappe, are superposed one on the other and thrust forward to the obliteration of the auto-

---

[1] *Rec. G.S.I.* vol. lxxiii. pt. 4, 1937.

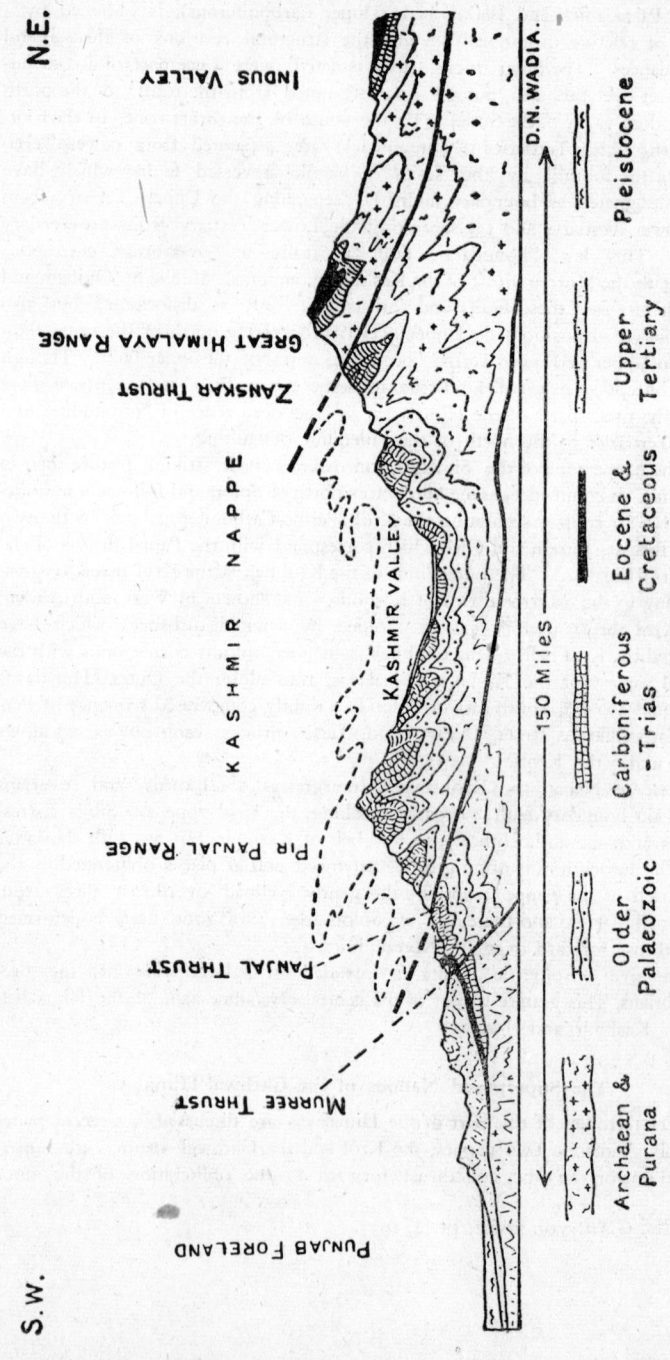

FIG. 42.—Diagrammatic section across the Kashmir Himalaya, showing the broad tectonic features.

chthon at places. Middlemiss's and Griesbach's previous studies of this section of the Himalaya had given, in accord with the tectonic ideas prevalent then, a simple interpretation to the profile across the Garhwal Himalaya, involving no horizontal displacements.

Proceeding north-east from the Sub-Himalayan Upper Tertiary zone (Siwalik and Dagshai), there are encountered, according to Auden, the following well-defined units:

(1) The autochthonous fold-belt, comprising a substratum of Simla Slates folded in with the Eocene, Dagshai and Siwalik series.

(2) The Krol nappe, comprising a thick succession of rocks in the Krol series (probable Permo-Carboniferous) overthrust upon the Nummulitics and Dagshai of (1).

(3) The Garhwal nappe superposed on the Krol nappe, the relations being such that the Nummulitic, Jurassic and Krol rocks belonging to the underlying Krol nappe completely surround older Palaeozic metamorphosed and schistose series of rocks of the superincumbent nappe and dip below them in a centripetal manner.

(4) The Great Himalayan range of crystalline phyllites and schists, together with the numerous para-gneisses and also intrusive granite bodies.

(5) The Tibetan zone of fossiliferous sediments rang-

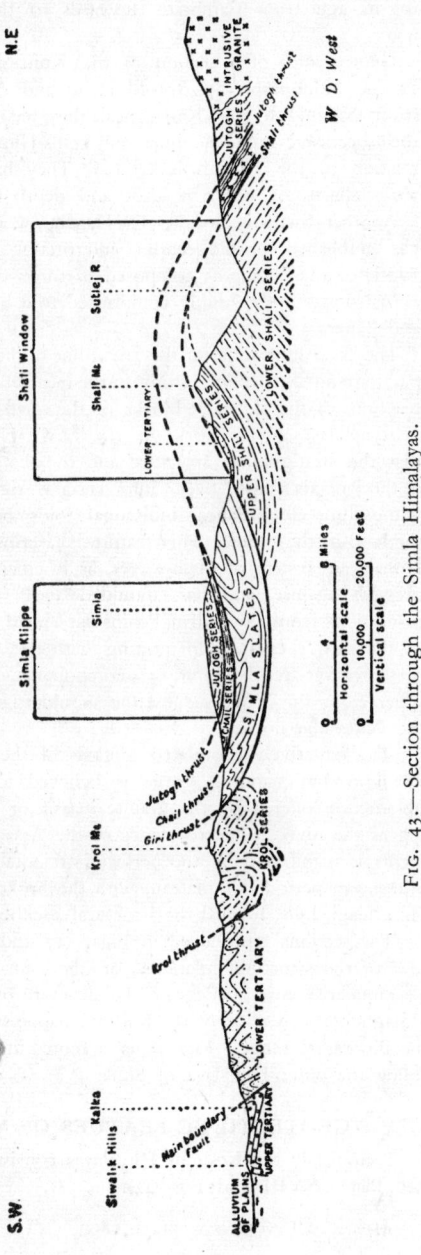

Fig. 43.—Section through the Simla Himalayas.

ing in age from Cambrian upwards to the Cretaceous (see Fig. 44, p. 400).

Geotectonics of a section of the Kumaon Himalayas is well disclosed in the monograph by Arnold Heim and A. Gansser. In their traverses from Nainital to the Kailas peak they recognised a number of nappes in the successive middle, inner and trans-Himalaya zones and their genetic relation to the Tibetan hinterland. They have exposed several compressed thrust-sheets of different sizes and depth from north to south.

Another Swiss geologist, T. Hagen, working in the Nepal Himalayas, has published profile-sections interpreting the tectonics of the 160-km. width of the Range as composed of three or four superposed nappes or thrust-sheets in the highly compressed fold belt between the Siwalik foothills and Tibet.

The deep inflexions in the trend-line of the Himalayas noted in Chapter I, p. 7, are an interesting study in the mechanism of mountain-building and the reactions of the old stable blocks of the earth against the weaker zones, the geosynclines. Field work in the N.W. Himalayan syntaxis has proved that the stratigraphy, structure and rock-components on the Kashmir flank of the syntaxis pass over into Hazara right round the re-entrant angle without any discordance, individual folds being traceable from one side of the loop to the other. This feature is ascribed to the circumstance that the Himalayan system of earth-waves, as it emerged from the Tethys, has been pressed against and has moulded itself on the shape of a tongue-like projection from the Indian Peninsular shield, one of the most rigid segments of the earth's crust. On meeting with this obstruction the northerly earth-pressures were resolved into two components, one acting from the N.E. and the other from the N.W., against the shoulders of this triangular promontory of the Peninsular horst.

The tentatively postulated syntaxis of the Assam Himalayas beyond the Tsangpo (Brahmaputra) gorge is believed to have originated through the obstruction offered by the granite massif of the Assam Plateau functioning as the pivot. The resistance of the Assam Plateau to folding movements is manifested in the perfect horizontality of its strata. In the pre-Himalayan period this plateau, with the broken chain of the Rajmahal and Hazaribagh hills, formed the structural backbone of Northern India.

The sections reproduced in Figs. 37 and 38 from Middlemiss[1] give an idea of the structural relations of the sub-Himalayan belts. Figs. 42 and 43 summarise current ideas on the structure of the Himalayas. Fig. 42 gives a diagrammatic section of the Kashmir nappe superposed on the S.W. flank of the Pir Panjal range. Fig. 43 is a representation of the Simla nappe over-riding the outer Himalaya of Simla.

GEOTECTONIC FEATUES OF N.W. HIMALAYAS

Tectonically the Kashmir Himalayas consist of three structural elements (see Plate XVIII and Fig. 42).

---

[1] *Mem. G.S.I*, vol. xxiv. pt. 2, 1890.

(1) The tongue of the *Foreland*, its peneplaned surface being buried under a thick cover of Murree sediments.

(2) A belt of *autochthonous*, mainly recumbent, folds consisting of rocks ranging in age from Carboniferous to Eocene, thrust against and over the foreland covered under the Murree series—the Murree thrust. Southward overfolding and thrusting with a dominant north-east dip is the prevalent structural tendency of this region.

(3) The *Nappe zone* of inner Himalayan rocks which has travelled far along an almost horizontal thrust (the Panjal thrust) so as to lie fitfully sometimes against a wide belt of the autochthon, at other times almost against the foreland. The Kashmir nappe is composed mostly of pre-Cambrian sediments (Salkhala series), with a superjacent series (Dogra slate), forming the floor of the Himalayan geosyncline that has been ridged up and thrust forward in a nearly horizontal sheet-fold. On this ancient basement lie synclinal basins containing a more or less full sequence of Palaeozoic and Triassic marine deposits in various parts of Kashmir. The latter are detached outliers of the Tibetan marine zone, which in the eastern Himalayas is confined to the north of the central Himalayan axis.

The most important tectonic feature of this region is the occurrence of two great concurrent thrusts on the southern front of the Himalayas, delimiting the autochthonous belt, which have been traced round the syntaxial angle from Hazara to Dalhousie, a distance of 400 km. Of these two thrusts, the inner (Panjal thrust) is the more significant, involving large-scale horizontal displacements. The outer, the Murree thrust, shows greater vertical displacement and is steeper in inclination, but has an equal persistence over the whole region. In its geological constitution, the autochthonous zone between the two thrusts consists of a series of inverted folds of the Eocene (Nummulitic) rocks enclosing cores of the Permo-Carboniferous Panjal Volcanics, and Triassic, all closely plicated but with their roots *in situ*.

As a tectonic unit, the Great Himalayan range is made up of the crystalline complex, the roots of the Kashmir nappe, the principal geanticline within the main Himalayan geosyncline. Several large bodies of intrusive granite and basic rocks occur in this zone. The latest period of granite intrusion is post-Cretaceous, or still later, connected with the earlier phases of Himalayan uplift. A subordinate element of the Great Himalayan range of Kashmir is the southward extension of the Tibetan belt of marine formations belonging to the Palaeozoic and Mesozoic.

The Simla Himalayas: The rocks of the Simla-Chakrata area have trespassed across the broad Upper Tertiary area of Kangra and constricted it to a narrow strip near Solon. The thrusts (the *Nahan* and *Krol* thrusts excluded from the *Murree* and outer *Panjal* thrusts of the Kashmir Himalayas) represent flat, recumbent folds of great amplitude. The pre-Cambrian (Jutogh and Chail series) is piled up on the Carboniferous and Permian systems (Blaini and Krol series), the *Giri* thrust, the entire sequence here, however, being totally unfossiliferous. Evidence of the superposition of the highly metamorphosed pre-Cambrian Jutogh and Chail series, building some of the conspicuous mountain-tops of the area (klippen), over the less altered lower Palaeozoics and Upper Carboniferous (Blaini series), is obtained by a study of relative metamorphism and the structural relations of thrusts and unconformities. The older rocks, now isolated, were once part of a continuous sheet over this area but are separated from their roots in the north by the deep valley of the Sutlej. To the south of the thrust zone, the older Tertiaries (Nummulitics) are separated from the newer Tertiaries of the foot-hills by the series of parallel reversed faults which have been termed boundary faults.

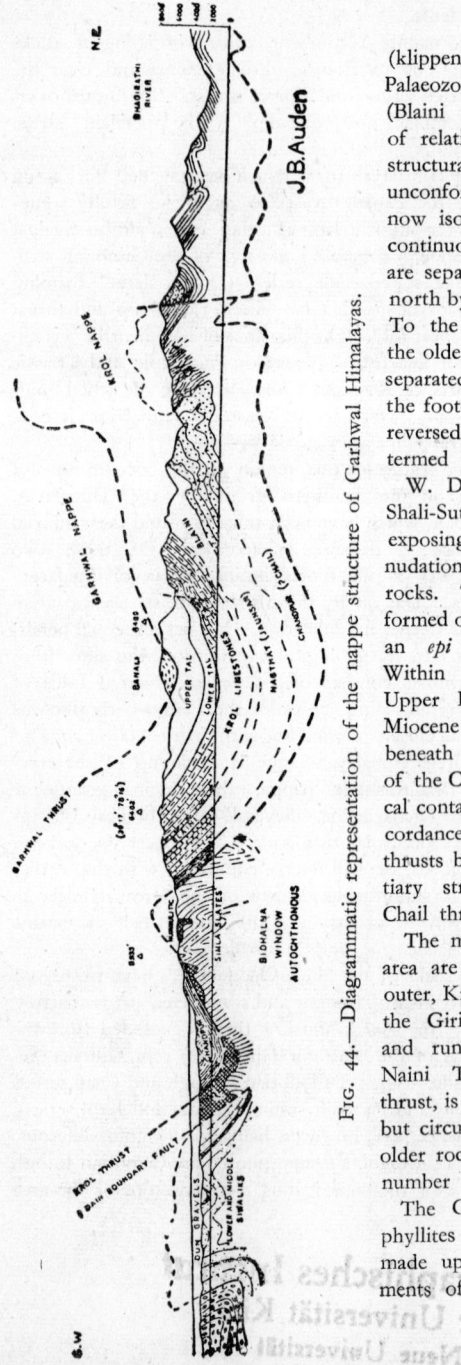

W. D. West has mapped in the Shali-Sutlej area a "window" exposing younger rocks by the denudation of the overlying older rocks. The sides of the window are formed of the Chail series showing an *epi* grade of metamorphism. Within the window there occur Upper Palaeozoic, Nummulitic and Miocene rocks, dipping centrifugally beneath the Chail cover. The base of the Chails is a plane of mechanical contact and one of marked discordance, some recumbent folds and thrusts being developed in the Tertiary strata immediately beneath the Chail thrust.

The major thrusts of the Garhwal area are shown in Plate XVII. The outer, Krol thrust, is continuous with the Giri thrust of the Simla hills and runs south-eastwards beyond Naini Tal. The inner, Garhwal thrust, is not one continuous plane, but circumscribes cake-like masses of older rocks lying over the Krols in a number of detached "outliers".

The Great Himalayan range of phyllites and crystalline schists is made up of the metamorphosed elements of the Garhwal nappe (which

FIG. 44.—Diagrammatic representation of the nappe structure of Garhwal Himalayas.

had its roots in this part), together with intrusive granite, para-gneisses and schists. This range denotes roughly the apex of the geanticline within the main geosyncline of the Himalayas.

The Tibetan zone of marine fossiliferous sediments, containing representatives of all ages from Cambrian to Cretaceous is confined to the north of the last zone in this part of the Himalayas, unlike Kashmir, where portions of it extend southwards of the crystalline axis. The high peaks of the central snowy range of the Himalayas, largely composed of granitic rocks, for the most part define the southern limit of the Tibetan zone, east of the Sutlej.

The Orographic Trends of North India and their relation to Central Asian Mountain-systems

Recent geological work in Ferghana (East Turkestan) and the Pamir region by the Russian geologists, and in the N.W. Himalaya by the Indian Geological Survey, tends to establish unity of structural plan and features, disclosing a common cause and origin for all the great mountain-systems of Central Asia, both of the Hercynian and Alpine age. It is probable, as Argand believes, that powerful crust-movements of the Tertiary and post-Tertiary Alpine orogeny superseded, and in a great measure altered, the old trend lines of Asia (Altaid orogeny) ; the existing alignment of all the ranges therefore, which meet in the Pamir knot is largely the work of the late-Tertiary diastrophism. The orientation of the Tien Shan-Alai-Kuen Lun system of radiating chains in the north, that of the Hindu Kush Karakoram arc of the middle, and of the deeply reflexed Himalayan arc in the south, all fuse in the Pamir vertex or knot, a crust segment possessing unique significance as having an equatorial strike orientation in the midst of numerous divergent trend-lines radiating away from it. To the south of the Pamirs is the Punjab wedge, the pivot on which are moulded the Himalayan syntaxis and Hindu Kush-Karakoram syntaxis. This N-S line of the crust connecting Pamir with Punjab is thus of critical importance in the orography of Asia, and will take a key position in future work on orogenesis and mechanics of crustal motion in mountain-building.

The knee-bends and festooning of the Himalayan arc, caused by the reaction of the plastic earth-folds of the newly rising mountains against the rigid Deccan horst, are of great interest. We have seen the two most prominent of these in the Punjab and Assam wedges causing acute looping of the mountain-arcs. An equally abrupt syntaxis of the mountain-folds which belong to the south-eastern flank of the Sulaiman bifurcation of the main Himalayan axis, where it joins on to the Iranian arc to the west, is seen near Quetta in Baluchistan. This comparatively minor but most spectacular re-entrant shows a bundling up of a multitude of normal anticlines and synclines of Tertiary, Cretaceous and Jurassic strata into a closely packed sheaf and forced out of straightness by two abrupt curves. These curves are the result of some crustal peg arresting the free movement of the folds towards the south, under pressures acting from the north and north-west.

From these considerations the view is expressed that the Great Himalaya Range from the gorge of the Brahmaputra to Nanga Parbat on the Indus denotes the Himalayan *protaxis*, the axis of original upwarp of the bottom of the Tethys geosyncline. At both ends it has undergone sharp southward deflections owing to the obstruction offered by the north edge of the Gondwana block of the Deccan and its projecting angles and capes.

With regard to the question of the *direction* of these great crustal movements and displacements in the North Indian region, two contrary views prevail and tectonic work in the field has provided no convincing data. From the markedly convex trends of the Malayan, Indo-Burman, Himalayan and Iranian ranges *away* from the centre of Asia, one view postulates a southward creep or drift of the crust from middle Asia towards Gondwanaland. The irregularities, deviations and convexities are, on this view, ascribed to the resistances offered by the irregular front of the rigid block of Peninsular India, Arabia and perhaps, to a less extent, the north-east corner of Africa. The second view regards the main direction of pressure as having come from the south. It does not regard Gondwanaland as a passive resistant block only, but as an active agent in pressing the plastic geosynclinal strata northwards. It does not credit a single, united, southward movement as competent to produce the convexities of island and mountain arcs (*e.g.* the Pacific arc facing east and the Himalayan facing south-west), but regards underthrusting of the ocean floor and a positive northward drift of the Indian foreland as more probable sources of crustal compression. The latter view has the plausible support of isostasy. For, on a lateral or radial spreading out of Central Asia, there ought to be a defect of matter in the Tibet-Mongol region, which on the contrary shows a positive gravity anomaly (a superficial excess of matter) and contains the greatest amount of land mass protruding above mean sea-level of any region in the world. A northward drift of the Indian continent, however, still remains to be proved. Measurement of the astronomical latitude at Dehra Dun and elsewhere in India has been repeated at different times during the last hundred years; results have shown a few irregular changes, but there is no suggestion of any continuous drift of the land-mass either to the north or any other direction.

On the whole, the evidence from structures lends support to the view which postulates tangential pressures from the north. Suess, J. W. Gregory, Bailey Willis, and also Mushketov and other Russians who have worked in the Pamir-Turkestan field have all suggested this as the main direction of earth-pressure in the post-Tertiary earth-movements of Central Asia and India.[1]

The Age of the Tectonics of North India

Evidence of the extreme youth of Himalayan orogeny has multiplied of recent years. The tilting and elevation of the Pleistocene lake deposits

---

[1] D. N. Wadia, Trend-line of the Himalayas, *Himal. Journ.* vol. viii., Oxford, 1936.

of the Kashmir valley (*Karewa series*), containing sub-Recent plant and vertebrate remains, to a height of 1,500–1,800 m. the dissection of river-terraces, containing post-Tertiary mammalia, to a depth of over 900 m., and the overthrusting of the older Himalayan rocks upon Pleistocene gravel and alluvium of the plains have been noted by various observers. The uplift of the Himalayas, initiated in post-Nummulitic time, continued late into the Pleistocene; strata containing *Bos*, *Elephas* and artifacts of prehistoric Man have been steeply folded in the foot-hills of Rawalpindi and Kangra. The main downwarp of the plains of Punjab, Bihar and Bengal was concomitant with this and even continued to later times.

In the Salt-Range and in the Assam Ranges, two periods of uplift are recognised: (1) post-Eocene and (2) Pliocene which continued into the Pleistocene. This latter diastrophism was probably a sympathetic movement accompanying the final Himalayan phase. The Assam plateau also received an epeirogenic uplift during the late Tertiary.[1]

Mountain ranges which are the result of one upheaval are known as *monogenetic*; those of several successive upheavals as *polygenetic*. The Tertiary deposits in the parallel belts flanking the Himalayas on the south side bear testimony to successive uplifts of these mountains.

The mountain arcs of Sind-Baluchistan and Burma, to the west and the east of the Himalayas, are of a more simple geological structure, and, in the succession of normal anticlines and synclines of which they are built up, recall the type of mountain-structure known as the Appalachian. In the former area, especially, the mountains reveal a very simple immature type of topography. Here the hill-ranges are anticlines with intervening synclines as valleys. The sides of the mountains, again, are a succession of dip-slopes.

In regions of more advanced topography, with greater rainfall and a consequently greater activity of subaerial denudation, *e.g.* the outer and middle Himalayas, this state of things is quite reversed, and the valleys and depressions are carved out of anticlinal tops while the more rigid, compressed synclinal systems of strata stand out as elevated ground.

While the broad features of these regions are solely due to movements of uplift, the characteristic scenery of the mountains, the serried lines of range behind range, separated by deep defiles and valleys, the bewildering number of watersheds, peaks and passes and the other rugged features, which give to the mountains their characteristic relief and outline, are the work of the eroding

---

[1] Structure of the Himalaya and of the North Indian Foreland, *Ind. Sc. Congr. Proc.* vol. xxv. pt. 2, Calcutta, 1938.

agents, playing on rocks of different structures and varying hardnesses.[1]

Among the mountains of the extra-Peninsula, the Salt-Range must be held as an illustration of a dislocation-mountain. Its orthoclinal outline, *i. e.* its steep southern scarp and the long gentle northern slope, suggests that the Salt-Range is the result of a monoclinal uplift combined with a lateral thrust from the north, which has depressed the southern part of the monocline under the Punjab plain, while the upper part has travelled some distance over it along a gentle plane of thrust[2]. The Assam range, on the other hand, has had a different origin and history, having as its backbone a granite *massif*. This plateau part of Assam, a horst uplift in the Miocene, has not experienced any folding movements, but there has been a tilting which has given the plateau a slope down to the north-east. On the south side the plateau is bounded by a zone of steep southerly dips associated with faulting. Eastwards in Upper Assam the gently dipping Tertiary rocks of the plateau have been overridden by the Haflong-Disang and Naga thrusts which have carried the Tertiary beds of the Patkai range many miles north-westwards. In the plateau portion of the Assam range there are also north-south faults, some of which showed a large movement during the earthquake of 1897. The Assam region's susceptibility to earthquakes is due to this late multiple faulting. These two ranges, at either extremity of the plains of India, share some common physical features and are unique in their physiography among the mountain-systems of India.

The Structure of the Peninsula

The geological structure of the Deccan Peninsula forms a single crust-block, the rigidity of which is not weakened by any overloading of large or deep belts of deposits, *e.g.* a recent geosynclinal mountain chain. This old crust-block, no doubt, was seamed with ancient mountains, now peneplaned and eroded almost to their roots and forming, in fact, the girders of its framework. These corrugations have given the Peninsula its main trend-lines or structural strikes—the Aravallis, and the Eastern Ghats strike-lines, roughly NNE.-SSW., and the Dharwarian strike extending under the Deccan Trap and joining with the Aravallis. Across

---

[1] The extremely rugged and serrated aspect of the lofty central ranges of the Himalayas, which are constantly subject to the action of snow and ice, contrasts strongly with the comparatively smooth and even outlines of the lesser Himalayas. The scenery of the outer Siwalik ranges is of a different description, the most conspicuous feature in it being a succession of escarpments and dip-slopes with broad longitudinal valleys in between.

[2] E. R. Gee.

PHYSIOGRAPHY 405

these structural lines is the less definite Satpura strike, nearly E.-W., extending from Assam to the west coast. Besides these

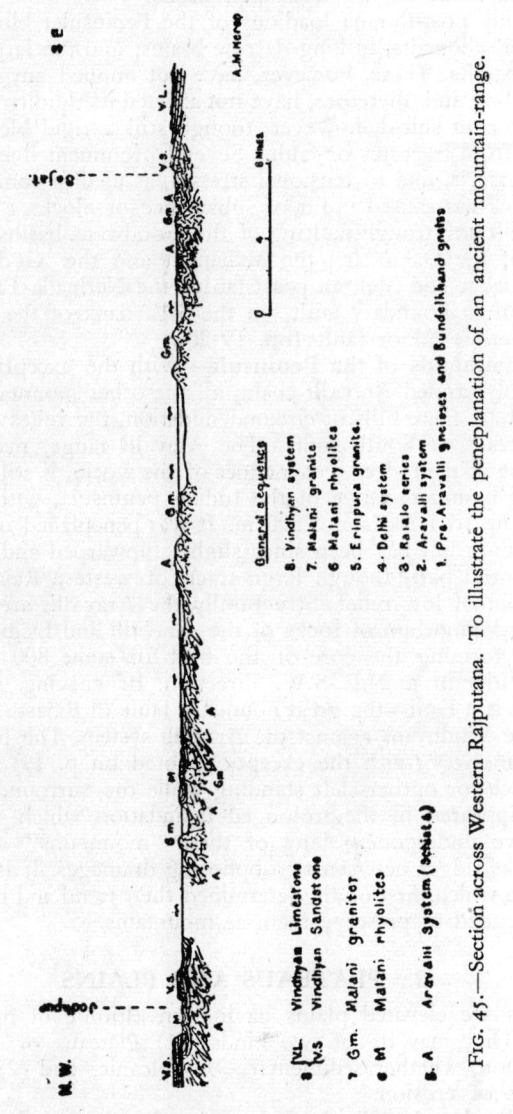

Fig. 45.—Section across Western Rajputana. To illustrate the peneplanation of an ancient mountain-range.

General sequence
8. Vindhyan system
7. Malani granite
6. Malani rhyolites
5. Erinpura granite
4. Delhi system
3. Raialo series
2. Aravalli system
1. Pre-Aravalli gneisses and Bundelkhand gneiss

V.L. Vindhyan Limestone
V.S. Vindhyan Sandstone
Gm. Malani granites
M Malani rhyolites
A Aravalli System (schists)

three prominent fold-axes of the Deccan, there are several minor axes of crumpling or folding, *e.g.* the Mahanadi axis; the NW.-SE.

trend-lines of the Khondalite rocks of South Nilgiri, Madurai and Tinnevelli, which persist over a large part of Sri Lanka and the crescentic strike of the Cuddapah basin.

The only post-Purana loadings of the Peninsular block are the Godwana sediments, in long narrow basins, and the large mass of plateau basalts. These, however, have not implied any crumpling or wrinkling and, therefore, have not affected its rigidity as a horst.

The Deccan shield, however, though still a rigid block, is not immune from fractures or faults. Several prominent lines of faulting traverse it, due to tensional stresses, as against compressional force. This has caused uplift or subsidence of blocks, *e.g.* Nilgiri-Palni uplift and trough-faulting of the Gondwana basins along the valleys of the Damodar, the Mahanadi and the Godavari; the great Malabar and Mekran coast faults; the Narmada-Tapti faults; the Rajasthan boundary fault, on the S.E. flank of the Aravallis; and numerous minor faults (pp. 19-20).

The mountains of the Peninsula—With the exception of the now deeply eroded Aravalli chain, all the other mountains of the Peninsula are mere hills of circumdenudation, the relics of the old high plateaus of South India. The Aravalli range, marking the site of one of the oldest geosynclines of the world, is still the most dominant mountain-range of the Indian peninsula, with summits reaching up to 1,200 and 1,500 m. It was peneplaned in pre-Cretaceous times but has been since slightly upwarped and dissected in the central part, though large tracts of western Rajasthan are a peneplain of low relief. Structurally the Aravallis are a closely plicated synclinorium of rocks of the Aravalli and Delhi systems, the latter forming the core of the fold for some 800 km. from Delhi to Idar in a N.E.-S.W. direction. Its curving south-east boundary is a fault—the great boundary fault of Rajasthan, which brings the Vindhyans against the Aravalli system. The hills south of the Vindhyas (with the exception noted on p. 19) are mere prominences or outliers left standing while the surrounding parts have disappeared in the prolonged denudation which these regions have undergone. Many of these "mountains" are to be regarded as ridges between two opposing drainages. It is this circumstance which, first of all, determined their trend and has subsequently tended to preserve them as mountains.

2. PLATEAUS AND PLAINS

Plateaus are elevated plains having an altitude of more than 300 m. They may be of two kinds: (1) Plateaus or plains of **accumulation**, whether sedimentary or volcanic, and (2) plateaus and plains of erosion.

Volcanic plateau—The best example of a plateau of accumulation in India is the volcanic plateau of the Deccan, built up of horizontal lava-sheets, now dissected into uplands, hills, valleys and plains. Its external configuration corresponds exactly with the internal structure, in the flat table-topped hills and the well-cut stair-like hill-sides. The Western Ghat country abounds in such plateaus.

Erosion plateau—Plateaus of erosion result from the denudation of a tectonic mountain-chain to its base-level and its subsequent upheaval. In them there is no correspondence at all between the external relief and geological structures. Some parts of Rajasthan and central India are an example of a plateau or plain of erosion (peneplain); parts of the Potwar plateau are another. The Archaean terrain of Chhota Nagpur received successive epeirogenic uplifts during the Tertiary of over 750 m. Today the Ranchi plateau contains peneplaned tracts studded with a few isolated worn hill-tops (*inselbergs*), detached by circumdenudation. The Parasnath hill and the numerous solitary eminences of southern Chhota Nagpur are good examples of such inselbergs rising over a peneplaned plateau country. The Assam plateau must be regarded as a plateau of erosion, a detached, outlying fragment of the Peninsula, connected with it through the intermittent Rajmahal hills. It has received some epeirogenic uplift since the early Tertiary.

Plains of accumulation—The great plains of the Indus and Ganges are plains of sedimentary accumulation. The horizontally stratified alluvium has the simplest geological structure possible, which is in perfect agreement with their flat level surface.

3. VALLEYS

A valley is any hollow between two elevated tracts through which a stream or river flows. Valleys are grouped into two classes according to their origin:

(1) Tectonic or Original Valleys.
(2) Erosion-Valleys.

Valley of Kashmir a tectonic valley—Tectonic valleys are exceptional features in the physiography of a country. They owe their origin to differential movements within the crust, such as trough-faulting, or the formation of synclines, which may be considered as the complementary depression between two mountain-chains. The valley of Kashmir, the Nepal valley and the many *Duns* of the sub-Himalayas are instances of tectonic valleys, these being synclinal troughs enclosed between two contiguous anticlinal flexures. This aspect is, however, somewhat modified by the deep alluvium which has filled up the bottom as well as that which rests on the slopes of the bordering mountains. Some valleys also result from the irregular accumulation of volcanic or morainic material or of dunes of sand. Valleys which run along fault-planes or fissures in the crust are also tectonic valleys, being determined by earth-movements; examples of such "fissure-valleys", according to some geologists, are afforded by the Upper Narmada and Tapti (pp. 25-26), which flow not in shallow base-levelled valleys of their own eroding, but in deep, linear fault-troughs filled with older alluvia. These faults were of post-Deccan Trap formation, caused by the tectonic movements in North India. Such valleys are of very rare occurrence, however, though it is possible that some of the deep "rifts" of Baluchistan have originated in this manner; all such valleys have been greatly modified by later erosion.

Erosion-Valleys

Valleys of the Himalayas. The transverse gorges—With the exceptions noted above, the great majority of the valleys of the Peninsula as well as of the extra-Peninsula are true erosion-valleys. The most prominent character of the major Himalayan valleys is their transverse course, *i.e.* they run across the strike of the mountains, in deep gorges or *cañons* that the rivers have cut for themselves by the slow, laborious process of vertical corrasion of their beds. The only exceptions are the head portions of the Indus, Brahmaputra, the Sutlej, and a few of their principal tributaries, which, for parts of their courses, form longitudinal streams and flow parallel to the mountain-strike. The cause of this peculiarity of the Himalayan system of valleys has already been explained in Chapter I, as arising from the situation of the watershed to the north of the main axis of uplift of the Himalayas. Hence the zone of the highest, snow-capped ranges is deeply trenched by all the rivers as they descend from their watershed to the plains of India. The "curve of erosion" of these valleys, which are yet in an immature stage of river-development, is, of course, most irregular and abounds in many inequalities. The most conspicuous of these is an abrupt fall, of nearly 1,500 m. which most rivers have as they cross from the axial Himalayan zone into the middle Himalayan zone, proving that the former zone is one of late special uplift. The same valleys, as they enter the end-portion of their mountain track—the Siwalik zone—cut through the very deposits which they themselves laid down at an earlier period of their history. Thus here also the apparently paradoxical circumstance is witnessed, that "the rivers are older than the hills which they traverse", which, on equally trustworthy evidence, is true for the greater part of the Himalayas. (See Chapter I.)

The configuration of the Himalayan valleys—The transverse gorges of the Himalayas, which are such characteristic features of the mountains, illustrate several interesting phases of river-action. In the first place their physical configuration in the eastern and western parts of the mountains is quite different. In the Kashmir Himalaya the upper courses of these streams show a series of abrupt alternations of deep precipitous U- or I-shaped gorges with broad, open V-shaped valleys, the latter always being found above the gorge-like portions. In the Eastern Himalayas of Sikkim and Nepal, on the other hand, the valley courses are uniformly broad, with gently sloping sides, and they do not exhibit these abrupt changes. This difference is due to the fact that the eastern part is a region of heavy rainfall, and hence the valley-sides are subject as much to erosion as the bed of the channel; here lateral corrasion is scarcely less marked than the vertical or downward corrasion of the bed. In the Western Himalayas, on the other hand, the rainfall is much smaller. River-erosion is the chief agent of denudation; hence deep defiles are cut out of the hard crystalline rocks, and broad V-like valleys from the softer clay rocks. The latter yield more readily to river-action because of the absence of any protective covering of vegetaion.

The Himalayan valleys are all in an early or immature stage of their development; they have been rejuvenated again and again with every upheaval of the inner higher ranges; hence the varying lithological characters and structures of the surface over which they flow have given rise to a number of waterfalls, cascades, and rapids in their courses. These will gradually disappear by the process of head-erosion, and in the later stages of valley-growth will be replaced by ravines and gorges. The narrow defiles of the Himalayan valleys are liable to be choked up by various accidental circumstances, such as landslips, glaciers, etc., and produce inundations of a terrific nature when the dam is removed. Several of these floods are recorded within recent times[1]. Many of the Himalayan valleys have been important high-roads of commerce with Tibet, China and Russian Turkestan, etc. since very ancient times.

The deep gorges and *cañons*, so characteristic of the Salt-Range and Baluchistan, are due chiefly to climatic causes. The river at the bottom is actively corrading and lowering its bed, while there is no denuding agency to lower the banks in these arid, rainless countries at an equal rate. The limestone rocks of the above districts have also been a factor in evolving these features.

Valleys of the Peninsula—The valleys of the Peninsula offer a striking contrast to those of the extra-Peninsula, for the former have reached the adult stage of development. Some of the principal valleys of the Peninsula are broad and shallow, their gradients low, and by reason of the levelling process being in operation for a long series of ages, they are near the attainment of their baser level. Their *curve of erosion* is, in the majority of cases, a regular curve from the source to the mouth.

---

[1] Chap. I. pp. 41-42.

One exception to the above general case is afforded by the falls of the Narmada near Jabalpur, and another by the falls of the Cauvery near Sivasamudram. Their existence in river channels of such great antiquity is surprising, and must be attributed to recent tectonic disturbances (see page 382). The physiographic features of the Narmada valley are of interest. To the north, the fault-trough is bounded by the great table-topped sandstone escarpments, 150 to 250 m., of the Vindhyan range, and to the south are the gentle slopes of the Satpura or the Mahadev hills, falling away in a series of abrupt scarps to the Tapti river at their south foot. The eastward extension of the Vindhyan scarps, the Kaimur hills, continues the range along the north flank of the Son Valley. The sources of the Son, Narmada and the Mahanadi lie around the trap plateau of Amarkantak.

The above remarks only apply to the valleys of the eastern drainage. The small but numerous streams that discharge into the Arabian Sea are all in a youthful state of development, being all actively eroding, torrential streams. Many of them abound in rapids and falls, of which the most famous are the Gersoppa Falls on the River Sharavati in the North Kanara district, but there are a number of other less-known instances. This greater activity of the westerly flowing streams, as compared to the opposite system of drainage is, of course, due to the former streams having to accomplish the same amount of descent to the coast as the latter, but within a far shorter distance from the watershed. Under such circumstances, a river performs much head-erosion, with the result that the watershed goes on continually receding. This process will continue till the watershed has receded to about the middle of the Peninsula and brought the grades of the channels on either side to an approximate equality.

4. BASINS AND LAKES

Basins and lakes. Functions of lakes—We have already considered the great troughs of India—the Indo-Gangetic and Potwar troughs, the West Rajasthan alluvial basin, all of which are parts of the great synclinorium of North India,[1] and the Irrawaddy trough of Burma; these are a part of the structural framework of India. We have here to consider the minor topographic depressions. Basins are larger or smaller depressions on the surface, the majority of which are filled with water, which, according to local conditions, may be fresh, brackish or salt. Lakes are of importance as regulators of the water-supply of rivers, ensuring for them a more or less even volume of water at all times and seasons, and

---

[1] R.D. Oldham, Structure of the Gangetic Plain, *Mem. G.S.I.* vol. xlii. pt. 2. 1917.

preventing sudden inundations and droughts. Their effect on the hydrography of a country like India would be very beneficial, but as stated before, in the chapter on Physical Features, there are very few lakes in India of any considerable magnitude. Hence basins as a feature in the physiography of India play but little part. The origins of lakes are diverse. The following are a few Indian examples:

Types of lakes—(1) **Tectonic lakes** are due to differential earth-movements, some of which are of the nature of symmetrical troughs, while others are due to fracture or subsidence of the underlying strata. The old Pleistocene lakes of Kashmir, whose existence is inferred from the Karewa deposits of the present day, were of this type.

(2) **Volcanic basins.** These are crater-lakes or explosion-crater lakes. The famous Lonar lake of salt water in the Buldana district, Berar, occupies a hollow which is supposed to have originated in a violent volcanic explosion—(*explosion-crater*).

(3) **Dissolution basins.** These are due to a depression of the surface by underground solution of salt-deposits, or of soluble rocks like gypsum and limestone. Some of the small lakes on the top of the Salt-Range may be due to this circumstance, aided by the irregular heaping of the loess deposits on its surface. Some of the Kumaon lakes also are of this nature.

(4) **Alluvial basins.** These are formed by the uneven deposition of sediments in deltas of rivers (Jhils); some lakes are formed of the deserted loops of rivers (bayou lakes), etc. The present lakes of the Kashmir valley are alluvial basins of this nature, while the Pangkong, Tsomoriri and the Salt-Lake of Ladakh in Kashmir territory are explained by Drew as having a somewhat different origin[1]. They have been formed by the alluvial fans from the side valleys (the tributaries) crossing the main valley and forming a dam which the waters of the main valley were unable to sweep away. A number of the lakes of Tibet have also originated in this manner, while some are supposed to have originated by differential earth-movements—tectonic basins.

(5) **Aeolian basins** are hollows lying among wind-blown sand-heaps and dunes. These are small and of temporary duration. Some of the Salt-Range lakes are aeolian basins; the numerous *Dhands*, small saline or alkaline lakes of Sind and W. Rajasthan, are other examples.

(6) **Rock-fall basins** are lakes produced by landslips or landslides, causing the precipitation of large masses of rock across the stream-courses. They are in some cases permanent. The small lakes of Bundelkhand are examples. The Gohna lake of Garhwal, formed by a huge landslip across a tributary of the Ganges in 1893, is a recent instance.

---

[1] *Jammu and Kashmir Territories*, London, 1875.

(7) **Glacial lakes**. They are often prevalent in districts which bear the marks of glaciation. In some cases the hollows are of glacial erosion (true rock-basins);[1] in other cases they are due to heaps of morainic débris constituting !a barrier across glacial streams. Numerous tarns and lakes on the north-east slopes of the Pir Panjal are examples. Some of the Kumaon lakes are ascribed to the latter origin. Old glacial basins, now altered to grassy meadows, and bounded by terminal moraines, are met with in front of some of the Himalayan glaciers (which are now retreating). Some of the *margs* of Kashmir are illustrations of such moraine-bound basins.

(8) **Lagoons**. The Chilka lake of Orissa and the Pulicat lake of Nellore are lagoon-like sheets of brackish water which owe their origin to the deposition of *bars* or *spits* of sand, drifted up along the coast by the action of oblique sea-currents, across the mouths of small bays or inlets. The lagoons of the Kerala coast (*kayals*) and of Sri Lanka are of this nature.

5. COAST-LINES

The coast-lines of a country are the joint product of epigene and hypogene agents. A highly indented coast-line is generally due to subsidence, while a recently elevated coast is fronted by level plains or platforms, cliffs and raised beaches.

In old lands, which have not undergone recent alteration of level, many of the features are the result of the combined marine and subaerial erosion.

Coast-lines—The coast-line of India is comparatively uniform and regular, and is broken by few indentations of any magnitude. For the greater part of its length a sandy and gently-shelving coast-strip is washed by a shallow sea. The proportion of the sea-board to the mean length of the sides of the Peninsula is very small. The western sea-board has, however, a number of shallow lagoons and backwaters, which constitute an important topographic feature of these coasts. This coast has no deltas, being exposed to the action of the persistent south-west monsoon gales which blow from May to October, and is, therefore, subject to a more active erosion by the sea-waves than the east coast. The rapidity of the coastal erosion is, however, in some measures retarded by the gently shelving nature of the sandy shores, and also by the lagoons and backwaters, both of which factors help to break the fury of the waves. The coasts are fronted by a low submarine plain or platform, where the sea is scarcely 100 fathoms deep. This "continental shelf" is much broader on the western

---

[1] The small lakes and tarns on the Pir Panjal are supposed to be of this description.

coast than on the eastern. On both the coasts there are "raised beaches" or more or less level strips of coastal detritus, situated at a level higher than the level of highest tides. This is a proof of a slight recent elevation of the coasts. For structural and tectonic features of the Bay of Bengal and Arabian Sea coast-lines, see p. 32.

The Arakan coast, with its numerous estuaries and inlets extending inland from a broad submarine shelf, is an excellent instance of an area that has undergone sub-Recent depression. Similarly it appears that the whole of the Malay region, which was once a continuous stretch of land from Assam to Indonesia, has been converted into a chain of islands and peninsulas by prolonged submergence of the land or a rise in sea-level.

For recent changes of level on the west coast and on the floor of the north Arabian Sea, see p. 43.[1]

REFERENCES

W. H. Hobbs, *Earth-Features and their Meaning* (Macmillan), 1912.

Sir S. G. Burrard and H. H. Hayden, *The Geography and Geology of the Himalaya Mountains*, 1907; Second Edition, revised by Sir S. G. Burrard and A.M. Heron, 1932.

H. H. Hayden, Relationship of the Himalaya to the Indo-Gangetic Plain and the Peninsula, *Rec. G.S.I.* vol. xliii., pt. 2, 1913.

R. D. Oldham, Support of the Mountains of Central Asia, *Rec. G.S.I.* vol. xlix. pt. 2, 1918.

J. W. Gregory, *The Structure of Asia* (Methuen), London, 1929.

D. N. Wadia, Syntaxis of the N. W. Himalaya, *Rec. G.S.I.* vol. lxv., 2, 1931.

Himalayan Geosyncline, NISI vol. 32, 5, 1966.

A. Gansser, Geology of the Himalayas, 1964, (John Wiley).

A. Heim and A. Gansser, Central Himalaya, Swiss Nat. Hist. Soc., vol. lxxiii., Mem. 1, Zürich, 1939.

A. M. Heron, The Physiography of Rajputana, *Proc. 25th Ind. Sc. Cong.* pt. 2, 1938.

M. S. Krishnan, Structural and Tectonic History of India, *Mem. G.S.I.* vol. lxxxi., 1953.

H. de Terra, Himalayan and Alpine Orogenies, *XVI Intn. Geol. Cong.*, vol. ii., 1937.

J. B. Auden, Traverses in the Himalayas, *Rec. G.S.I.* vol. lxix., 2, 1935.

---

[1] See also S. P. Chatterjee, *Annals of Geomorphology*, 1962.

CHAPTER XXVI

Economic Geology

In the preceding chapters we have dealt with the stratigraphical and structural geology of India. It is necessary for the student of Indian geology to acquaint himself with the various mineral products of the rock-systems of India and the economic resources they possess. In the following few pages we shall deal with the occurrence, the geological relations and some facts regarding the production of the most important of these products. For fuller details as well as for statistics, reference may be made to the excellent Reports of Mineral Production, published by the Bureau of Mines and Geological Survey of India[1].

For our purpose the various useful products, which the rocks and minerals of India yield, can be classified under the following heads :

(1) Water.
(2) Clays, Sands.
(3) Lime, Cements etc.
(4) Building-Stones.
(5) Coal, Lignite, Peat, Petroleum.
(6) Metals and Ores.
(7) Precious and Semi-precious Stones.
(8) Other Economic Minerals and Mineral Products.
(9) Soils.

An appraisal of the total mineral resources of India so far known to geologists brings home the fact that the mineral wealth of India is not inconsiderable for a country of her size and population, and that it encompasses a sufficient range of useful products that are necessary to make a modern civilised country more or less industrially self-contained. Except in the case of minerals such as iron-ore, aluminium-ore, titanium-ore, mica and a few other minerals, the resources in economic minerals and metals are, however, limited. Chances of discovery of new mineral deposits of any extent and richness by ordinary geological methods are

---

[1] J. Coggin Brown and A. K. Dey, *India's Mineral Wealth* (Oxford University Press), 1955; Dictionary of Economic Products of India, Counc. Sc.Ind. Res., New Delhi, 1949-51. Series of Bulletins and Prof. Papers in *Rec. G.S.I.* vols. lxxvi. and lxxvii., 1940-45.

not many, though the new geophysical methods of locating underground mineral occurrences by electrical, magnetic, gravimetric and seismic methods seem to offer possibilities of bringing to light hitherto undiscovered, but in some cases suspected, deposits of petroleum, coal-measures, natural gas, underground water, metallic lodes, etc.

Geological and Geographical distribution of India's minerals
—Barring coal and petroleum, and the somewhat disputed position of salt and gypsum due to their undecided age, the bulk of the valuable minerals and metals won in India are products of rocks of pre-Palaeozoic age and are confined to metamorphic rock-systems of either the Archaean or pre-Cambrian period. The principal ore and metal deposits, the precious and semi-precious stones, mica and a large number of industrially valuable minerals are derived from the Dharwar system. 98 per cent of the coal is of Lower Gondwana age, the remainder being Tertiary. The main petroleum horizons in India are Tertiary in Assam and Gujarat.

Nature has made a very unequal territorial distribution of minerals in the Indian region. The vast alluvial plains tract of Northern India is devoid of mines of economic minerals. The Archaean terrain of Bihar and Orissa possesses the largest concentration of ore-deposits such as iron, manganese, copper, thorium, uranium, aluminium, chromium; valuable industrial minerals like mica, sillimanite, phosphates; and over three-fourths of India's reserves of coal, including coking coal. The iron-ore reserves lying in one or two districts of Bihar and in the adjoining territories of Orissa are calculated at over 8 thousand million tonnes, surpassing in richness and extent those of any other known region. There are large reserves of manganese-ores; over 50 per cent of the world's best mica, block, splittings and sheet, is supplied by the mica mines of Kodarma and Gaya in Bihar. The second minerally rich province is Madhya Pradesh, carrying good reserves of iron and manganese, coal, limestone and bauxite. Madras has workable deposits of iron, manganese, magnesite, mica, limestone and lignite. Karnataka has yielded all the gold of India, besides producing appreciable quantities of iron, porcelain clays and chrome-ores. Andhra has good reserves of second-grade coal. Gujarat produces oil, manganese, salt, fluorite and bauxite worth 5 crores. Kerala possesses enormous concentrations of heavy-mineral sands of high strategic importance, calculated to contain, together with lesser deposits along the Malabar coast, some 200 million tonnes of ilmenite, besides monazite, zircon, rutile, and garnet in workable quantities. The provinces of U.P. and Eastern Punjab have been far less productive and have scarcely as yet figured in India's mineral statistics. Rajasthan, for a long time absent from India's annual mineral returns, is gradually becoming a productive centre, holding promise for the future in non-ferrous metals (copper, lead and zinc), phosphates, mica, steatite, beryllium and precious stones (aquamarine and emerald). Assam supplies about 2·5 million tonnes of much needed petroleum, besides carrying important reserves of Tertiary coal. West Bengal's mineral resources are confined to coal (annual production capacity about 16 million tonnes) and iron-ore. Of the vast extent of the Himalayan region, the only proved mineralised region of importance is the territory of Kashmir south of the Great Himalayan Axis, with its coal (some of it anthracitic), aluminium-ore, sapphires and some minor industrial minerals. The next mineralised terrain is Nepal, from where occurrences of cobalt, nickel and copper ore are reported, but which has scarcely yet been thoroughly explored. But for the partly-known copper and magnesite deposits of Sikkim and Kumaon and some fairly widespread iron-ore bodies in these areas, the Himalayan region is a veritable *terra incognita* as regards economic minerals.

Sind, N.W. Frontier and Western Punjab (Pakistan), and Bangladesh, have less development of the minerally productive geological rock-systems and have not in the past supported any considerable mining industry. Sind, N. W. Frontier and the rocky parts of Western Punjab have prospects of petroleum resources on a workable scale, which are being developed. This feature is shared also by Baluchistan, where two gas-fields of large capacity have been brought into production; it has in addition important chromium deposits of a high grade together with moderate reserves of low-grade lignite, coal and sulphur. Two or three oil-fields in Western Punjab produce a variable quantity of petroleum (some 273 to 409 million litres). Other minerals of Pakistan are the rock-salt (175,000 tonnes annually) from Khewra, the yet untapped reserves of millions of tonnes of brine in Baluchistan, and nitre and potash-salts, gypsum, and reserves of some 300 million tonnes of Tertiary coal of inferior but usable quality. Energetic prospecting for oil in Sind and in Bangladesh, at present in progress, may reveal new oil-fields of better productive capacity than the existing Punjab oil-fields which are small and variable in their output. Pakistan's metal resources do not extend beyond Baluchistan's high-grade chromite, Chitral antimony and arsenic, and the hitherto indifferently prospected iron-ore occurring in parts of Hazara and N. W. Frontier Province. The country is richly supplied with vast reserves of pure limestone and other raw materials for cement manufacture, together with extensive deposits of high-quality gypsum and ornamental marble. The following table gives the quantity and value of the annual production of leading economic minerals mined in India during recent years :

ECONOMIC GEOLOGY 417

MINERAL	QUANTITY		VALUE (Rupees)
Apatite	9,500	tonnes	7,00,000
Asbestos	10,000	,,	24,00,000
Barytes	60,000	,,	11,00,000
Bauxite	1,010,000	,,	1,20,00,000
(Aluminium	120,000	,,	55,00,00,000)
Building Stones	—		14,00,00,000
Chromite	220,000	,,	1,28,00,000
Clays (Industrial)	600,000	,,	51,40,000
Coal	75,000,000	,,	250,00,00,000
Copper	9,700	,,	8,40,00,000
Diamonds	11,900	carats	50,00,000
Dolomite	1,280,000	tonnes	1,93,00,000
Fluorite	1,870	,,	11,85,000
Glass-sand and Silica	390,000	,,	38,00,000
Gold		Kg.	
Gypsum	1,370,000	tonnes	1,33,00,000
Ilmenite	47,700	,,	19,00,000
Iron-ore	21,280,000	,,	23,37,00,000
Kyanite and Sillimanite	87,800	,,	1,74,33,000
Lead-concentrate	3,300	,,	25,36,000
Limestone	22,300,000	,,	18,83,50,000
Magnesite	293,130	,,	63,12,000
Manganese-ore	1,283,500	,,	7,29,16,000
Mica	17,620	,,	1,88,10,000
Ochre	31,600	,,	6,00,000
Petroleum (crude)	6,722,000	,,	62,73,53,000
Salt	5,550,000	,,	8,80,00,000
Silver	3,280	Kg.	14,91,000
Steatite	170,000	tonnes	49,44,000
Zinc-concentrate	21,620	,,	85,18,000

1. WATER

Wells. Springs. Artesian wells—Besides its use for domestic and agricultural purposes, water has many important uses in manufacturing and engineering operations, and the geologist is often called upon to face problems regarding its sources and supply. Porous water-bearing strata exist everywhere among the old sedimentary formations as well as among recent alluvial deposits, but a knowledge of the geological structure is necessary in order to tap these sources with the maximum of efficiency. A large part of the rain that falls in India is speedily returned to the sea, only a very small percentage being allowed to soak underneath the ground. This arises from the peculiar monsoonic conditions of the climate which crowd into a few months all the rainfall of the year, which rapidly courses down in flooded streams and rivers.

The small percentage which is retained soaks down and, flowing in the direction of the dip of the more pervious strata, saturates them up to a certain level (*level of saturation*) and, after a variable amount of circulation underground, issues out again on a suitable outlet being found, whether in the form of springs, wells or seepages. In India the great alluvial plains of the Indus, and Ganges are a great reservoir of such stored-up water, and yield large quantities of sweet water by boring to suitable depths below the surface. Wells, the most common source of water in India, are merely holes in the surface below the line of saturation, reaching the more porous of the rock-beds, in which water accumulates by simple drainage or by percolation. Springs are common in the rocky districts where pervious and impervious strata are interbedded, and inclined or folded, or where a set of rocks is traversed by joints, fissures or faults. If a porous water-bearing stratum with a wide outcrop is enclosed between impervious strata above and below it, and bent into a trough, conditions arise for artesian wells when a boring is made reaching the water-bearing stratum. Such ideal conditions, however, are rarely realised actually, but there are some other ways by which less perfect artesian action is possible.[1] The formation of an underground water-tight reservoir, either by the embedding of tongues of gravel and sand under impervious alluvial clays, the abutting of inclined porous strata against impervious unfissured rocks by means of faults, or the intersecting of large fissures in crystalline rocks, gives rise to conditions by which water is held underground under a sufficient hydrostatic pressure to enable it to flow out when an artificial boring is made reaching the water. Artesian wells are not of common occurrence in India, nor are conditions requisite for the formation of artesian areas of any magnitude often met with. The best known examples are those of Jaisalmer and Quetta along with the Karez already referred to (p. 387) in the great gravel slopes (or Daman) of Baluchistan. Artesian wells are possible in the alluvial districts of North India and in Gujarat,[2] by the embedding of pockets of loose gravel or coarse sands in the ordinary alluvium. Artesian wells in the arid parts of this country, suffering from irregular or scanty rainfall, would be of great utility for the purposes of irrigation, but a knowledge of the geological structure of the district is essential before any costly experiment can be undertaken in borings.

Tube-wells 60 to 125 metres deep are a simpler means by which

---

[1] Vredenburg, *Mem. G.S.I.* vol. xxxii. pt. 1, 1901.

[2] Instances of successful artesian borings in Gujarat are numerous. Artesian wells also exist in the alluvial tracts in Rawalpindi, Pondicherry, South Arcot Neyveli, Sylhet, and several other rock-bound depressions.

supplies of underground water of good quality can be tapped in the alluvial districts for domestic, industrial and for agricultural use. Thousands of tube-wells have been bored since 1947 in the Gangetic plains, many of which have a discharge capable or irrigating 162 hectares of agricultural land. Tube-wells of 15 to 20 cm. diameter yield as much as 60,000 gallons per hour. Wells of this calibre are, however, few and their discharge depends more on the water-bearing capacity of the aquifer tapped than on the diameter of the tube. Tube-well water, being derived from depth, is bacteriologically purer and freer from organic impurities than ordinary well or surface waters, though there may be a greater proportion of chemically dissolved salts in it.

In the drier parts of the country it is of the utmost importance that the subsoil water-level should be conserved by such devices as inverted wells, tanks or reservoirs, constructing of small dams across glens and ravines, so as to impound the run off from the rainfall for the benefit of wells situated downstream. Such dams, made of earth, masonry or loose rock-fill, in the foot-hills of mountainous country are of greater service in underground water conservation than large projects for damming rivers. In such districts it is often found that construction of new, or renovation of old, tanks or reservoirs improves the yield of surrounding wells. These devices replenish the underground water storage by diverting into the soil a part of the surface rainfall which would otherwise run away uselessly into the rivers. Surface running water is harnessed by engineers for irrigation and power development. India has made great advances since 1860 in irrigation and hydraulics, and made striking progress in the last few years in investigating a number of engineering schemes for river training, water conservation, flood protection, irrigation reservoirs and electric power generation. India today has its 28.3 million hectares of dry land (out of nearly 162 million hectares normally under cultivation) annually watered by canal irrigation, an area greater than the combined total of ten other leading countries of the world. Irrigation canals in India have a carrying capacity of 400,000 cubic feet per second and the total average consumption of water for agriculture is roughly 163,000 cubic feet per second. This amount is, however, barely 6 per cent of the available "water wealth" in the rivers of India, which is running to waste into the sea, often after doing extensive, recurring damage in floods and soil erosion. Indian engineers are nevertheless bringing us nearer to that yet far-off era dreamt of by the great Sinhalese king of old, who commanded his engineers to see that not a single measure of monsoon rain was lost by flowing away uselessly into the sea. Among the prominent hydro-electric, irrigation and storage reservoir projects in operation are the Bhakra-Nangal, Rihand, Kosi, Konar, Tilaiya, Maithon, Panchet Hill, Hirakud, Gandhi Sagar, Rana Pratap Sagar, Nagarjuna Sagar, Tungabhadra and Sharavathi. Projects under development are the Beas-Sutlej link, Tehri Dam, Ranganga, Jawahar Sagar, Srisailam, Upper Krishna and Malaprabha.

Mineral springs—Thermal[1] **and mineral springs** occur in many parts of India, especially in mountainous districts like the Punjab, Bihar, Assam, Salt-Range, in the foot-hills of the Himalayas, in Kashmir, etc. Among them are sulphurous (which are the most common), saline, chalybeate, magnesian and other springs according to the principal mineral content of the waters. There are over 300 such thermal and mineral springs known in India.[2] Thermal sulphurous springs are very numerous on the outcrops of Eocene Nummulitic rocks in Rawalpindi district and in Sind. Some springs in the latter area have a temperature over 48.9 C°. Chalybeate and sulphurous springs are common in the Himalayas. Springs or radio-active water are known, *e.g.* at Rajgir and Monghyr (Bihar) and at Ramkund (Gujarat), the latter with unusually high radio-activity. Fairly large radium-emanations are observed in the springs at Manikaran (Kulu), Vajrabai and Unai near Bombay.[3] From Ratnagiri to Gujarat there are a number of thermal springs probably suggesting the course of the Malabar Coast fault-line. Many medicinal virtues are ascribed to such springs in Europe. In India no such powers are recognised in them, and where, in a few cases, they are recognised, no economic benefit is derived from them.

1. CLAYS

China clay—Clay, that kind of earth which, when moistened, possesses a high degree of tenacity and plasticity, is of great industrial use in the making of various kinds of earthenware, tiles, pipes, bricks, etc., and when of sufficient purity and fine grain it is of use in the manufacture of glazed pottery and high-grade porcelain, for all of which an immense demand exists in the modern world.[4] Pure china clay, or kaolin, occurs in deposits of workable size among the Archaean of some parts of Bihar and Singhbhum, Karnataka, Kerala, Delhi, and in Jabalpur. China clay, which has resulted from the decomposition of the felspar of the gneisses, occurs in useful aggregates in some districts of the Madras Deccan. Recent investigations have revealed large workable deposits of

---

[1] There are several thermal springs in the Karakpur hills. One of these, the *Sitakund*, near Monghyr, is well known. At Gangotri, the source of the Ganges, there is another well-known spring of hot water. At the boiling springs of Manikarn (Kulu) people cook their food in the issuing jets of water. *Tatta pani* in Poonch is a thermal sulphurous spring with a large volume of discharge; temp. about 190°F. Rajgir (Patna), Thana (Bombay), Jwalamukhi (Kangra), Jamnotri (Tehri Garhwal) are other well-known examples.

[2] Thermal Springs of India, *Memoirs G.S.I.* vol. xix. pt. 2, 1882; P. K. Ghosh, Mineral Springs of India, *Indian Science Congress*, 1948.

[3] *Indian Medical Gazette*, vols. xlvi. and xlviii., 1911 and 1913.

[4] *Clays, their Occurrence, Properties and Uses*, H. Reis, 1906.

kaolin of varying degrees of purity, fit for ceramic uses, in W. Bengal, Bihar, Orissa, Saurashtra and Rajasthan.

China clay which is somewhat impure and coloured buff or brown is known as *terra-cotta*, which finds employment in the making of unglazed large-size pottery, statuettes, etc., and to some extent for architectural purposes. Terra-cotta clay deposits are of more common occurrence in India than pure kaolin. The deposits of the Rajmahal hills at Colgong (Pattarghatta) are of much interest, both as regards the quantity available and the purity of the material, for the manufacture of very superior grades of porcelain.[1] Similar deposits, though on a more restricted scale, are found in Bhagalpur, Gaya, and in many parts of Maharashtra Tamilnadu, Karnataka, Kerala and Orissa. *Ball clays*, fine-grained, highly plastic Clays of good binding power, are fairly widely distributed in the above-mentioned parts of India. Annual production of china clay, mostly for ceramic uses, has been around 300,000 tonnes.

Fire-clay—Fire-clay is clay from which most of the iron and salts of potassium and sodium are removed, and which, therefore, can stand the heat of furnaces without fusing. Fire-clay from which fire-bricks of high refractory quality can be manufactured occurs in beds on the western side of the Rajmahal hills, Jabalpur, near Dandot in the Salt-Range, and in the vicinity of Kolar, Karnataka. It also occurs as underclays in beds up to several metres in thickness in the Gondwana coal-measures and associated with other coal-bearing series, and is now raised for various manufactures in considerable amounts near Barakar.[2] Besides these localities, fire-clay of texture and refractoriness suitable for the manufacture of furnace-bricks is obtained from a number of localities in Madhya Pradesh, Bengal, etc., where its deposits are of fairly wide distributon. Annual producton approximates 435,000 tonnes.

Fuller's earth—Fuller's earth is a kind of white, grey or yellow coloured clay. It has a high absorbent power for many substances, for which reason it is used for washing and cleaning purposes. It is found, among many other places, in the Lower Vindhyan rocks of Jabalpur district (Katni). It is also obtained from some districts of Karnataka, from the Khairpur area in Sind and from the Eocene rocks of Jaisalmer and Bikaner in Rajasthan, where it is quarried and sold under the name of *Multani mattee*. A nearly 2 m. bed of fuller's earth occurs in the Salkhala series of Budil, Rajaori (Jammu Province), containing a large stock of the mineral. The variety known as bentonite, a plastic clay with large absorbent

---

[1] M. Stuart, *Rec. G.S.I.* vol. xxxvii. pt. 2, 1909.
[2] W. H. Bates, Indian Earths and Clays, *Trans. Min. and Geol. Inst. Ind.*, vol. xxxviii., 1933.

power, of use in petroleum and foundry industries, occurs in association with a Siwalik conglomerate near Bhimber, Jammu Province. The bed is 60 cm. thick and extends for many kilometres. Bentonite deposits are found in the Jodhpur area also.

Ordinary alluvial clay, mixed with sand and containing a certain proportion of iron, is used for brick-making and crude earthen pottery. Fine-grained clay, mixed with fine sand, is used in tile-making. Mangalore, together with some surrounding places, is the home of a flourishing tile industry, where tiles suitable for paving, roofing and ceilings are manufactured.

The total production of clays in India for industrial purposes is worth about Rs. 195 lakhs per annum on an average; this may be contrasted with the value of clays raised in the United States of America for various manufactures, which amounts to about Rs. 200 crores per year.

SANDS

Glass-sand—Pure quartz-sand, free from all iron impurities and possessing a uniform grain and texture, is of economic value in the manufacture of glass. Such sands are not common in India, but in recent years good sands have been obtained from the crushing of pure quartzose Vindhyan sandstones at several localities in Uttar Pradesh, from Gondwana (Damuda) sandstone of the Rajmahal hills, and from Cretaceous sandstones and Archaean and other pure quartzites of some parts of Tamilnadu and Maharashtra. Sand-deposits of the requisite purity suitable for glass manufacture are found in Hoshiarpur district, Punjab, at Sawai Madhupur in the Jaipur area, Madh in Bikaner, and near Zawar in Udaipur. Good-quality sands suitable for glass-making also occur in the Sabarmati river and at Jabalpur. A pure quartz-grit at Barodhia in the Bundi area, thick deposits of a pure, white, soft, granular quartzite in the Poonch area, and masses of crumbling powdery silica resulting from metasomatic replacement of limestone near Garhi Habibullah, Hazara, are other available supplies. Ordinary white sand is used in India for the manufacture of inferior varieties of glass, while articles of better quality are manufactured out of crushed quartz at Talegaon (Poona), Jabalpur, and at Ambala, Allahabad and Madras. A recent survey[1] of India's resources in silica sands and rocks suitable for the glass industry[1] has produced a useful summary regarding the distribution, localities, extent and purity of the important glass-sand deposits of the country. According to this, several valuable sources of glass sands are Bargarh, south of Allahabad, covering an area of 260 sq. km., where the sand is 96 to 97 per cent pure SiO_2 and only 0.1 to 0.03 per cent

---

[1] Atma Ram and others: *Journ. Sc. Ind. Res.* vol. vii. 4, 1948.

Fe_2O_3; some parts of Rajasthan and the coast of Kerala, where large spreads of white quartz beach-sands cover over 250 sq. km. of the coast. Annual production of glass in India is 425,000 tonnes, value Rs. 45 crores.

Common river-sands are used in mortar-making. Recent calcareous sands, consisting mostly of shells of foraminifers, have consolidated into a kind of coarsely-bedded freestone at some places on the coast of the Saurashtra peninsula—miliolite. (See Ilmenite sand, Magnetite sand, Monazite sand, Gem sand, etc.)[1]

3. LIME, CEMENTS, ETC.

Mortar and Cement—Lime for mortar-making is obtained by burning limestone, for which most kinds of limestones occurring in the various geological systems of India are suitable, but some are especially good for the purpose. Lime, when mixed with water and sand, is called mortar, which, when it loses its water and absorbs carbonic acid gas from air, "sets" or hardens, hence its use as a binding or cementing material. In the plains of India, the only available source of lime is "Kankar", which occurs plentifully as irregular concretions disseminated in the clays. The clay admixture in Kankar is often in sufficient proportion to produce on burning a hydraulic lime. Travertine or calc-tufa, sea-shells, recent coral limestones, etc. are also drawn upon for the kiln, where a suitable source of these exists. When limestone containing argillaceous matter in a certain proportion is burnt, the resulting product is *cement*, in which an altogether different chemical action takes place when mixed with water. The burning of limestone ($CaCO_3$) and clay ($xAl_2O_3.ySiO_2.zH_2O$) together results in the formation of a new chemical compound—silicate and aluminate of lime—which is again acted upon chemically when water is added, hardening it into a dense compact mass. For cement-making, either some suitable clayey limestone is used or more commonly the two ingredients, limestone and clay, are artificially mixed together in proper proportions (Portland Cement). The occurrence of enormous masses of Nummulitic, Vindhyan and older limestones in Sind, Tamilnadu, Madhya Pradesh, Gujarat, Rajasthan, W. Punjab, central India, Assam and other parts, in association with clays and shales, offers favourable conditions for cement manufacture. Natural cement-stones exist in some parts of India, Kankar may be regarded as one of them. Over 20 million tonnes of limestone is consumed in India annually, value 188 million rupees. In 1970 India produced 14 million tonnes of Portland cement.

A high-grade, rapid-hardening cement, rich in aluminous content (*Ciment fondu*) of utility in special structures, can be manufac-

---

[1] *Sands and Crushed Rocks*, A. B. Searle, Oxford Technical Publications, 1923.

tured from aluminous laterites mixed with appropriate quantities of limestone. Pig-iron is a by-product.

4. BUILDING-STONES

Rocks are quarried largely for use as building-stones.[1] Not all rocks, however, are suitable for this purpose, since several indispensable qualities are required in a building-stone which are satisfied by but a few of the rocks from among the geological formations of a country. Rocks that can stand the ravages of time and weather, those that possess the requisite strength, an attractive colour and appearance, and those that can receive dressing—whether ordinary or ornamental—without much cost or labour, are the most valuable. Susceptibility to weather is an important factor, and very costly experiments have been made to judge of the merits of a particular stone in this respect.

With this in view the architects of New Delhi, who required a most extensive range of materials for a variety of purposes, building as well as ornamental, invited the opinion of the Geological Survey of India in regard to the suitability of the various building and ornamental stones quarried in the neighbouring areas of Rajasthan and central India. A special officer of the Survey was deputed to advise on the matter after an examination of the various quarries that are being worked in these provinces.

In northern India, the ready accessibility of brick-making materials in unlimited quantities has rendered the use of stone in private as well as public buildings subordinate. Excellent material, however, exists, and in quantities sufficient for any demand, in a number of the rock-systems of the country, whose resources in rocks like granites, marbles, limestones and sandstones are scarcely utilised to their full extent. An enumeration of even the chief and the more prized varieties of these would form a catalogue too long for our purpose.

Granites—Granite, or what passes by that name, coarsely foliated gneiss, forms very desirable building-stones, very durable and of an ornamental nature. These rocks, by reason of their massive nature and homogeneous grain, are eminently adapted for monumental and architectural work as well as for massive masonries. Their wide range in appearance and colour—white, pink, red, grey, black, etc.—renders the stones highly ornamental and effective for a variety of decorative uses. The charnockites of Madras, the Arcot gneiss, Bangalore gneiss, the porphyries of Seringapatam, and many other varieties of granite obtained from the various districts of the Peninsula are very attractive examples. Its durability is such that the numerous ancient temples and monuments of South India built of granite stand today almost intact after

---

[1] *Stones for Building and Decoration*, G. P. Merrill, 1910.

centuries of wear, and to all appearance are yet good for centuries to come. From their wide prevalence, forming nearly three-fourths of the surface of the Peninsula, the Archaean gneisses form an inexhaustible source of good material for building and oranamental uses[1].

Limestones—Limestones occur in many formations, some of which are entirely composed of them. Not all of them, however, are fit for building purposes, though many of them are burnt for lime. In the Cuddapah, Bijawar, Khondalite and Aravalli groups limestones attain considerable development; some of them are of great beauty and strength. They have been largely drawn upon in the construction of many of the noted monuments of the past in all parts of India. Vindhyan limestones are extensively quarried, as already referred to, in central India, Rajasthan and elsewhere, and form a valued source for lime and cement, as well as for building-stone. The Gondwanas are barren of calcareous rocks, but the small exposures of the Bagh and Trichinopoly Cretaceous include excellent limestones, sometimes even of an ornamental description. The Nummulitic limestones of the extra-Peninsular districts, *viz.* Sind, Hazara, the Salt-Range, Punjab and Assam, are an enormous repository of pure limestone, and when accessible are in very large demand for burning, building, as well as road-making purposes.

Marbles—The marble deposits of India are fairly widespread and of large extent. The principal source of the marbles of India is the crystalline formation of Rajasthan—the Aravalli series. Marble quarries are worked at Mekrana (Jodhpur), Kharwa (Ajmer), Maundla and Bhainslana (Jaipur), Dadikar (Alwar), and some other places, from which marbles of many varieties of colour and grain, including the beautiful chaste white variety of which the Taj Mahal is built, are obtained. It was the accessibility of this store of material of unsurpassed beauty which, no doubt, gave such a stimulus to the Moghul taste for architecture in the seventeenth century.

A saccharoidal dolomitic marble occurs in a large outcrop near Jabalpur, where it is traversed by the Narmada gorge. The famous quarries of Mekrana supply white, grey and pink marbles; a hand-

---

[1] In connection with the building of the Alexandra docks at Bombay, a series of tests on Indian granites was undertaken. These have proved that the granites from South Indian quarries are equal to or better than Aberdeen, Cornish or Norwegian granites in respect of compressive strength, resistance to abrasion, absorption of water, and freedom from voids. The verdict of the various experts consulted was altogether favourable to the use of Indian granites for purposes for which imported granites alone had been considered suitable. (Indian Granites. *Bombay Port Trust Papers*, 1905.)

some pink marble comes from Narmada in the Kishengarh area. Jaisalmer in Rajasthan supplies a yellow shelly marble, while a lovely green and mottled marble of unsurpassable beauty is obtained from Motipura, from an exposure of the Aravalli rocks in the Baroda region. A mottled rose or pink marble is found in the same locality and also in one or two places in the Aravalli series of Rajasthan and of the Narsinghpur district of Madhya Pradesh. The Kharwa quarries of Ajmer produce green and yellow-coloured marbles. Black or dark-coloured marbles come from Mekrana and from the Kishengarh area, though their occurrence is on a more limited scale than the lighter varieties. A dense black marble, capable of taking an exquisite polish, largely employed in the ancient buildings of Delhi, Agra and Kashmir, with highly ornamental effect, is furnished by some quarries in the Jaipur region. Coarse-grained marbles are more suitable for architectural and monumental uses; it is the coarseness of the grain which is the cause of the great durability of marble against meteoric weathering. The fine-grained, purest white marbles are reserved for statuary use, for which no other varieties can be of service.

It is a most regrettable fact, however, that the above-noted deposits of Indian marbles do not find any market to encourage their systematic quarrying. There is no considerable demand for indigenous marbles in India, nor do facilities exist for their export to foreign countries. The deposits, therefore, have to wait the demand of a more thriving and more aesthetic population in the future.

A fine collection of Indian marbles, representing the principal varieties, is to be seen in the Indian Museum, Calcutta.

Serpentine—Serpentine forms large outcrops in the Arakan range of Burma and also in Baluchistan. It occurs as an alteration-product of the basic and ultra-basic intrusions of Cretaceous and Miocene ages. From its softness and liability to weather on exposure it is of no use for outdoor architectural purposes, but serpentines of attractive colour are employed in internal decorations of buildings and the manufacture of vases, statuary, etc. Serpentinous marble (*Verde antique*) is rare in India.

Sandstones. Vindhyan sandstones—The Vindhyan and, to a lesser extent, the Gondwana formations afford sandstones admirably suited for building works. The most pre-eminent among them are the white, cream, buff and pink Upper Vindhyan sandstones, which have been put to an almost inconceivable number of uses. From the rude stone knives and scrapers of Palaeolithic man to the railway telegraph boards and the exquisitely carved monoliths of his present-day successor, these sandstones have supplied for man's service an infinity of uses. It is the most widely quarried stone in India, and being both a freestone as well as a flagstone it can yield, according to the portion selected, both gigantic blocks for pillars from one part and thin, slate-like slabs for paving and roofing from another part. The superb edifices, modern and medieval, of Delhi, Rajasthan and Agra are built of red and white Vindhyan sandstone quarried from a number of sites in the vicinity.

Dr. V. Ball,[1] in writing about Vindhyan sandstones, says: "The difficulty in writing of the uses to which these rocks have been put is not in finding examples, but in selecting from the numerous ancient and modern buildings which crowd the cities of the United Provinces, and the Ganges valley generally, and in which the stone-cutter's art is seen in the highest perfection." Some of the Vindhyan sandstones are so homogeneous and soft that they are capable of receiving a most elaborate carving and filigree work. Centuries of exposure to the weather have tested their durability.

Newer sandstones—Another formation possessing resources in building-stones of good quality is the Upper Gondwana, which has contributed a great store of building-stone to Orissa and Chanda. The famous temples of Puri and the other richly ornamented buildings of these districts are constructed of Upper Gondwana sandstones.

The Mesozoic (Umia) sandstone of Dhrangadhra and the Cretaceous sandstone underlying the Bagh beds of Gujarat (Songir sandstones) furnish Gujarat with a very handsome and durable stone for its important public and private buildings.

Among the Tertiary sandstones, a few possess the qualities requisite in a building-stone, *e.g.* the Murree and Kamlial (Tarki) sandstones; but the younger Siwalik sandstones are too unconsolidated and incoherent to be fit for employment in building work.

Quartzites—Quartzites are too hard to work and have a fracture and grain unsuitable for dressing into blocks.

Laterite—Laterites of South India are put to use in building-works, from the facility with which they are cut into bricks or blocks when freshly quarried and their property of hardening with exposure to air. Its wide distribution from Assam to Kanya Kumari makes laterite a widely used material for road-metal. This stone is not capable of receiving dressing for any architectural or ornamental use.

Slates—Slates for paving and roofing are not of common occurrence in India, except in some mountainous areas, e.g. at Kangra and Pir Panjal in the Himalayas and Rewari in the Aravallis. When the cleavage is finely developed and regular, thus enabling them to be split into thin even plates, the slates are used for roofing; when the cleavage is not so fine, the slates are used for paving.

---

[1] *Economic Geology of India*, vol. liii., 1881.

True cleavage-slates are rare in India; what generally are called slates are either phyllites or compacted shales in which the planes of splitting are not cleavage-planes.

The chief slate-quarries of India are those of Kangra, in the Kangra district; Rewari, in the Gurgaon district; and Kharakpur hills, in the Monghyr district.

Traps—Besides the foregoing examples of the building-stones of India, a few other varieties are also employed as such when readily available and where a sufficient quantity exists. Of these the most important are the basalts of the Deccan, which, from their prevalence over a wide region of Western India, are used by the Railways and Public Works Department for their buildings, bridges, the permanent way, etc. The traps furnish an easily workable and durable stone of great strength, but its dull, subdued colour does not recommend it to popular favour. Of late years some trachytic and other acidic lavas of light buff and cream colours have found great favour in the building of public edifices.

The annual value of the building-stones output in India is 140 million rupees.

5. COAL[1]

Production of coal in India—Coal is the most important of the mineral products raised in India. Within the last fifty years India has become an important coal-producing country, the annual production now nearly supplying her own internal consumption. The yearly output from Indian mines rose in 1969 to 75,000,000 tonnes valued at the present cost of production, Rs. 249,00,00,000.[2] Of post-war outputs, by far the largest part—83.5%—has come from the coal-fields of Bengal, Bihar and Orissa; 9% from the fields of Madhya Pradesh; 3.5% from several fields in Andhra Pradesh and about 2% from three fields in Vindhya Pradesh. This amounts to a total of 98% for the production of coal from the Peninsula. In its geological relations the coal of the Peninsula is entirely restricted to the Damuda series of the Lower Gondwana system. The remainder of the coal raised comes from the Lower Tertiary, Eocene or Oligocene, rocks *viz.* Assam (Makum), Salt-Range (Dandot), Tamilnadu (Neyveli) and Bikaner (Palana). Of these, the Neyveli production is the most important and promising for the future; it averages over $3\frac{1}{2}$% of the total

---

[1] Coalfields of India, *Mem. G.S.I.* vol. xli. pt. 1, 1913 ; C. S. Fox, Jharia Coalfield, *Mem. G.S.I.* vol. lvi., 1930 ; Lower Gondwana Coalfields, vol. lix., 1934; E. R. Gee, Raniganj Coalfield, *Mem. G.S.I.* vol. lxi., 1932; Coal, *Rec G.S.I.* vol. lxxvi. No. 16, 1945 ; Mineral Producton of Indian Union during 1947, *Rec. G.S.I.* vol. 81. pt. 3, 1950.

[2] Mineral Statistics of India, 1970.

ECONOMIC GEOLOGY

Indian produce, while it also approaches Gondwana coal in its quality as a fuel.

The following table shows the relative importance of the various coal-fields of India, with their average annual outputs in recent years in round numbers:[1]

Gondwana Coal.
Bengal, Bihar and Orissa

		Tonnes
1.	Raniganj	2,115,000
2.	Jharia	19,000,000
3.	Giridih	300,000
4.	Bokaro	3,313,000
5.	Karanpura	5,300,000
6.	Talcher	680,000

Madhya Pradesh and Maharashtra

1.	Rewa	2,426,000
2.	Pench Valley	2,000,000
3.	Korea	3,,135,000
4.	Umaria	110,000
5.	Sohagpur and Singrauli	4,440,000
6.	Chanda	804,000
7.	Nagpur	350,000

Andhra

1.	Kothagudam, Singareni etc.	4,132,000
2.	Tandur	6,000,000

Tertiary Coal

1.	Assam (Makum, Nazira, etc.)	498,000
2.	Bikaner (Palana)	60,000
3.	Neyveli (South Arcot)	3,500,000

In the Riasi district of Jammu Province coal, some of anthracite quality, occurs in some widely distributed seams of 30 cm. to 6 m. in thickness in association with Nummulitic strata. The latter occur as inliers in the Murree series (p. 316). The coal-seams are distributed over 60 km. of country in three or four coal-fields. Middlemiss has estimated the quantity available at 100,000,000 tonnes, with mining at ordinary depths. Some of the Riasi semi-anthracites contain 60-82 per cent of fixed carbon.[2]

---

[1] Averages based on Indian Minerals Year Books.

[2] Middlemiss, *Mineral Survey Reports*, Jammu and Kashmir State, Coalfields of Riasi, Jammu, 1930.

In general, the Gondwana coal is a laminated bituminous coal within which dull (durain) and bright (vitrain) layers alternate. Anthracite, *i.e.* coal in which the percentage of carbon is more than 90, and from which volatile compounds are eliminated, is not found in the Gondwana fields. The volatile compounds and ash are, as a rule, present in too large a proportion to allow the carbon percentage to rise above 55 to 60, generally much less than that. The percentage of ash is usually high, 13 to 20, rising to as much as 25 to 33 per cent. Moisture is absent from the coal of the Gondwana fields, but sulphur and phosphorus are present in variable quantities in the coals of the different parts of the Peninsula. Sulphur percentage is generally high in Tertiary coals. In general, the Gondwana coal is good steam or gas coal. Occurrence of coking coal is confined to the Jharia, Giridih and some parts of Karanpura fields.

It is probable that some extent of coal-bearing Gondwana rocks lies hidden underneath the great pile of lavas of the Deccan trap. At several places, chiefly in the Satpuras, the denudation of the latter has exposed coal-bearing Gondwana strata, from which it is reasonable to infer that considerable quantities of the valuable fuel are buried under the formation in this and more westerly parts. Of the coal of younger age, worked from the extra-Peninsula, Assam coal is of a high grade as fuel, while that of the Punjab has a lower percentage of fixed carbon. In the former it rises to as much as 53 per cent, in the latter it never goes beyond 40 per cent. The latter coal, properly a lignite, is more bituminous, friable and pyritous, and contains much moisture. The two last qualities make it liable to disintegration on exposure, and even to spontaneous combustion. With regard to its geological aspects the extra-Peninsular coal is mostly Lower Tertiary. The Salt-Range coal comes from the Ranikot series; and in Assam three horizons occur—one near the bottom and one at the top of the Jaintia series (corresponding to a part of the Kirthar), and a much more important one in the Barail series somewhat above the Eocene-Oligocene boundary. In Burma impure coal occurs at various horizons in the Eocene and Lower Oligocene. The Tertiary coal of Palana (Bikaner) and several other Tertiary areas of Rajasthan and Kutch is properly speaking a lignite (brown coal), though belonging to the Eocene. The lowest thin coal of Assam has been regarded as of Cretaceous age but this now seems improbable and it has been here classed with the Eocene; a few thin seams of brown coal occur in the Jurassic strata of Kutch and possibly some of the coal of the Mianwali district is of this age, although it is more probably of Ranikot age. The newly discovered lignite field of Neyveli in S. Arcot, Tamilnadu, adds materially to the brown coal resources of India. The lignite bed 3-25 m. in thickness has been found to extend over 260 sq. km. and the reserves of brown coal, usable as fuel, have been estimated at 2,000 million tonnes.

Of the total gross resources of coal, 94,000 million tonnes,[1] only one-fourth, that is about 23,600 million tonnes comprise different metallurgical grades of coal, a very small proportion of which, namely 4,600 million tonnes accounts for parent coking coals which are confined only to Jharia coal-field.

Several warning notes have been sounded of late years regarding the small available reserves of good quality coal in India and the approaching exhaustion of coking coal for metallurgical use. Lately revised estimates place the reserves of coal of all types up to a depth of 609 metres in the Indian coal-fields at about 16,796 million tonnes of proved reserves and 130,782 million tonnes of inferred reserves. The student may consult the report of the Government Coal Inquiry Committee of 1946 and G. S. Fox's Memoir on the Lr. Gondwana Coal-fields *Mem. G. S. I.*, Vol. LIX, 1934, and recent publications of the G. S. I.

Proved Reserves of Coking Coal of all Grades

Giridih field	70	Million tonnes
Raniganj field	1,016	,, ,,
Jharia field	3,336	,, ,,
Bokaro and Ramgarh fields	2,163	,, ,,
Karanpura field	427	,, ,,
Panch-Kanhan field	29	,, ,,
Total	7,041	,, ,,

Recent exploration and drilling tests in Bihar and Madhya Pradesh have revealed large additions to previously estimated reserves of 55,000 million tonnes. Experiments have shown that it is possible to upgrade some of this good quality coal to coking coal by washing and blending. To the above-noted reserves of coal must be added the reserves in Tertiary brown coal and lignite, which form a valuable addition to the mineral assets of such outlying States as Assam, Tamilnadu, Rajasthan and Jammu and Kashmir in India and as the provinces of West Punjab, Sind and Baluchistan in Pakistan. Assam Tertiary coal is of high quality and recent estimates of reserves in the coal-fields of Upper Assam and of the Garo and Khasi Hills are over 589 million tonnes. The lignite deposits of Neyveli (South Arcot district, Tamilnadu) are calculated at 2,000 million tonnes; the Rajasthan and Kutch fields are estimated at 190 million tonnes, while the West Punjab and Baluchistan (Pakistan) reserves are put at about 300 million tonnes.

---

[1] Coal of all grades in seams over 0.45 m. thickness and up to 609 metres in depth.

The hitherto coal-less area of Kashmir has been found to contain fuel deposits of considerable size belonging to Pliocene or even newer age. Forty-six million tonnes of moderate-grade lignite are easily recoverable from one area near Baramula. The percentage of combustible matter is generally about 55.[1]

PEAT

The occurrence of peat in India is confined to a few places of high elevation above the sea. True peat is found on the Nilgiri mountains in a few peat-bogs lying in depressions composed of the remains of Bryophyta (mosses). In the delta of the Ganges, there are a few layers of peat composed of forest vegetation and rice plants. In the numerous *Jhils* of this delta peat is in process of formation at the present day and is used as a manure by the people. Peat also occurs in the Kashmir valley in a few patches in the alluvium of the Jhelum and in swampy ground in the higher valleys; it is there composed of the débris of several kinds of aquatic vegetation, grasses, sedges and rushes. Similar deposits of peat are in course of formation in the valley of Nepal. The chief use of peat is as a fuel, after cutting and drying. It is also employed as a manure.

PETROLEUM

Distribution of Oil—The hitherto known petroleum resources of India are on a limited scale. They are confined to the narrow belt of Tertiary strata which constitutes the outer margin of the extra-Peninsular mountains along their whole line of contact with the Peninsular block of the Deccan, from Sind through Baluchistan, N. W. F. Province, the Punjab and Assam, and thence curving southwards along the Arakan chain (on both its sides) to the Bay of Bengal. There are only three areas within this belt which have so far been found to bear petroleum on a commercial scale.[2] These are:

1. *Punjab-Sind-Gujarat Gulf*: The apex of this gulf was at the foot of the Simla Himalayas. From here it widened north-westwards, extending to the Potwar, then curving S.S.W. along the Sind-Baluchistan hill-ranges, and ending at Cambay Gulf. The valley of the lower Indus has gradually supplanted and succeeded this original Tertiary gulf.[3]

---

[1] C. S. Middlemiss, *Rec. G.S.I.* vol. lv. pt. 3, 1924.

[2] E. H. Pascoe, Oil-fields of India, *Mem. G.S.I.* vol. xl., 1911–20 ; H. M. Sale and P. Evans, Geology of the Assam-Arakan Oil Region, *Geol. Mag.* vol. lxxvii., 1940 ; P. Evans and C. A. Sansom, Oil Fields of Burma, *Geol. Mag.* vol. lxxviii., 1941.

[3] E. S. Pinfold, Occurrence of Oil in the Punjab, *Journ. Asiat. Soc. Beng.*, N.S. vol. xiv., 1918.

2. *The Assam Gulf*: This commenced from Digboi and proceeded along the southern side of the Brahmaputra valley, to Sylhet, and along the western flank of Arakan through Eastern Bengal to Akyab. The part south-west of Sylhet is now buried under the alluvium of the Meghna and the delta of the Ganges.

3. *The Burma Gulf*: It extended north-south along the basin of the Chindwin and the lower Irrawaddy, along the eastern flank of Arakan. The latter river valleys are the successors of the Tertiary Burma Gulf.

Burma—The fields are situated in a belt which closely follows the line of the Chindwin and Irrawaddy. In the Yenangyaung field (Magwe district), production is obtained mainly from the Lower Miocene and Upper Oligocene; in the more northern fields of Chauk (Magwe district), Lanywa and Yenangyat (Pakokku district) the production is from the Oligocene.[1] Yenangyaung has yielded about 590 million litres a year, and Chauk about 364 million litres. Lanywa and Yenangyat in the Pakokku district together have produced 91 million litres per year. Further south in the Minbu area small fields yielded about 13 to 18 million litres annually. The Chauk field was formerly known as Singu.

Assam—Surface indications of oil and gas occur at intervals from the Arakan coast north-eastwards through the Chittagong region to the Surma Valley and thence along the north-western side of the Naga Hills almost to the extreme north-east of the Upper Assam Valley. So far through vigorous prospecting four oil-fields have been discovered, Digboi, Nahorkatiya, Hugrijan and Moran in Upper Assam. Deep test wells have been drilled at a large number of other sites but have yielded no sustained production of economic value. The geologists of the Burmah Oil Company and associated companies have carried out much geological mapping and geophysical research in this area.[2]

The Digboi Field. The oil-pools occur along the crest of a sharply folded anticline to the south of the Naga thrust (Fig. 46). The productive oilsands in the Digboi field belong to the Tipam sandstone stage; over 20 separate horizons have proved to be oil-bearing in the field, showing much lateral variation. The Tipam sandstones are about 1,100 m. thick; they are not of marine origin, hence it is improbable that the oil is indigenous to these beds, and the B.O.C. geologists believe that it has migrated from the Barail series, possibly by way of the Naga thrust.

The Nahorkatiya Field. A deep well drilled on a structural sum-

---

[1] G. W. Lepper, *Proc. World Petrol. Cong.*, 1933.

[2] P. Evans, Tertiary Succession in Assam, *Trans. Min. Geol. Inst. Ind.* vol. xxvii., pt. 3, 1932 ; H. M. Sale and P. Evans, Geology of the Assam-Arakan Oil Region, *Geol. Mag.* vol. lxxvii., 1940; E. V. Corps, Digboi Oilfield, *Bull. Amer. Ass. Petr. Geol.* vol. xxxiii., 1949.

mit pointed out by a seismic survey in the Brahmaputra alluvium, about 32 km. from Digboi, has brought to light the existence of the new promising oil-field of Nahorkatiya. Subsequent detailed seismic surveys and the results of extensive drilling in the

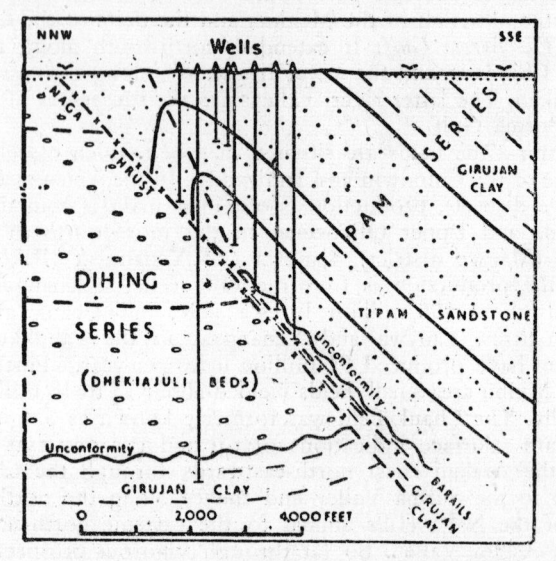

FIG. 46.—Section through the Digboi Oil-field.

area have shown that the structure of the field is much more complex than was at first thought, the Tertiary strata being cut through by many faults of small and large throw (Fig. 47). The base of the Tipam sandstone lies at about 2,800 m. and the oil-sands occur as lenses in the underlying Barails at over 2,950 m. depth. The Barails are nearly 900 m. thick and are underlain by Eocene beds which closely resemble the Kopilis of the North Cachar Hills.

This field may have a potential production of $2\frac{1}{2}$ million tonnes annually,[1] when fully developed.

The Moran Field. Drilling has proved an oil-bearing Barail horizon at a depth of 3,360 m. on a faulted dome near Moran, some 40 km. south-west of Nahorkatiya. Much further drilling will be required to determine the potential of this oil-field for additional one million tonnes.

Gujarat—Geophysical investigations during recent years have proved the existence, underneath the alluvium of North Gujarat

---

[1] W. B. Metre, Petroleum Industry in India, Presidential Address, Geol. Min. Met. Soc. Ind., 1959.

ECONOMIC GEOLOGY

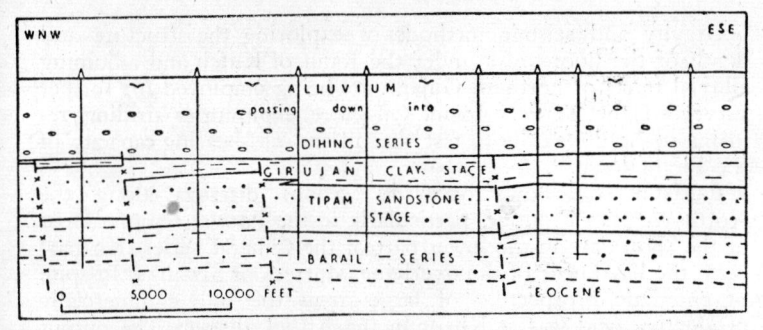

FIG. 47.—Diagrammatic section through the Nahorkatiya Oil-field.
Scale 2 miles to an inch.

and the Rann of Kutch, of a wide basin of post-Nummulitic sediments enclosing productive petroliferous horizons, resting upon a faulted surface of the Traps. This basin, filled with Lower Tertiary sediments, stretches E.-W. from Baroda district across the Gulf of Cambay to beyond Bhavnagar in Saurashtra. Drilling tests have established in Gujarat three oil-fields of which Ankleshwar (annual capacity three million tonnes) is the largest. The north boundary of this Cambay basin probably extends beyond Ahmednagar and the Rann of Kutch a considerable distance towards southern Rajasthan, while its south limit buried under the shallow waters of the Cambay gulf is found to have a structural 'high' of considerable potential off Piram island. The probable reserves of petroleum in this area have been computed at some 60 million tonnes.

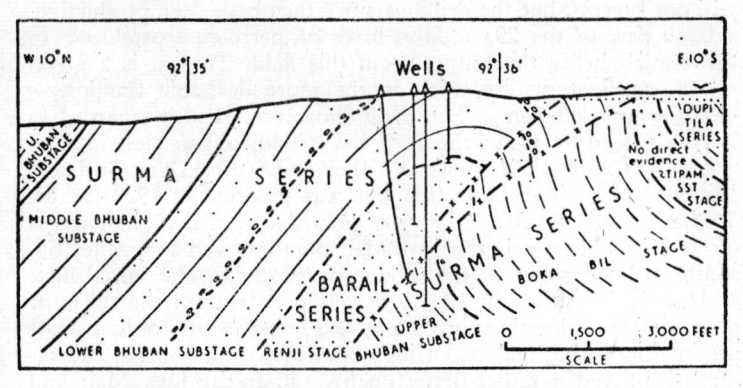

FIG. 48.—Diagrammatic section through the Badarpur Anticline, Assam.

Gravity and seismic methods of exploring the structure and depth of the floor-rocks under the Rann of Kutch and adjoining alluvial tracts of northern Gujarat are being employed for further survey of the Tertiary basin. Off-shore exploratory drilling remains to be done fully to test the oil-and gas-bearing capacity of the Piram dome.

Pakistan—Oil-shows occur in various districts along the North-Western Frontier, particularly in the Potwar region. Most of the shows are in the lower part of the Chharat series (Kirthar) or in the basal beds of the overlying Murrees or Siwaliks. In spite of energetic prospecting of large areas the only commercially productive field was at Khaur in the Attock district. The output reached a maximum of 86 million litres in 1929 but subsequently declined to about 18 million litres. Latterly, successful boring to a depth of 2,165 m. in the adjacent Dhulian dome increased the production again. Although the Khaur production has come from the Murree series, it is believed that the origin is in the underlying Eocene rocks from which the oil has migrated upwards. The deeper drilling has recently proved the occurrence of oil and gas in the Eocene limestones here; in the neighbouring Dhulian area and in the recently discovered fields near Chakwal the production comes from the Eocene.

The Khaur Field. Discovered in 1915, but production was small and fitful till 1922, since when it has fluctuated between 27 million litres to 86 million litres per year. Since 1943, its decline has been steady and the field has approached exhaustion. The numerous borings for new wells in this field have penetrated productive oil-sands at a few levels between 120 m. and 1,770 m. but their distribution and yield were always erratic.

The Dhulian Field. Wells drilled in this field up to 1934 were without success, but the drillings since then have been productive, a large part of the 295 million litres of petroleum produced in the Punjab being the output from this field. The oil is a green oil of excellent quality, rich in the more desirable fractions—gasoline and kerosene. This field, however, is also regarded as having passed its peak; the flow has steadily fallen since 1943.

The Joya Mair Field (Chakwal, Jhelum District). The probable existence of an oil-field in this area was reported in 1929 (D. N. Wadia: *Records, Geological Survey of India, lxi, part iv*). The success of the initial borings in the Joya Mair dome proved a considerable southward extension of the productive region. The initial flow at Joya Mair was about 10,000 barrels per day, but the oil is of poor quality and extremely viscous and consequently only a small production has been taken. Drilling on the adjacent dome of Balkassar has proved a rather better quality oil. In the Joya Mair and Balkassar fields the oil comes from the Sakesar limestone, but the permeability of the reservoir rock is extremely variable, and both at Joya Mair and at Balkassar some of the wells have yielded little or no oil.

As a rule, the Punjab oil wells are distinguished by the absence or scanty presence of natural gas, as compared with Assam and Burma oil wells, where gas is copious and forms a material supplement to the liquid petroleum on condensation. Another remarkable feature of the Punjab oil wells is the extremely high pressures prevailing in the oil-sands and in the overlying beds. Pressures up to 2,722 kg. per 6.5 sq. cm. have been measured in the reservoir beds,[1] and for these pressures no satisfactory explanation has been found as yet. The combined output of the Potwar oil wells is about 66 million gallons at present.

Natural Gas—Natural gas (chiefly marsh gas with some other gaseous hydrocarbons) usually accompanies the petroleum accumulations.[2] The gas may occur in separate sands containing little or no oil, but most of the natural gas of India is found closely associated with the oil, and supplies the propulsive force which carries the oil from the oil-sands into the wells and, if the pressure is sufficient, brings the oil up to the surface. Since gas is essential for the production of the oil and is also valuable as a source of fuel on the oil-fields, care is taken to prevent the waste of gas, which was formerly so common in the oil-fields. In north-east Baluchistan large reserves of natural gas have been discovered in the Laki limestone, the principal accumulation so far known being in the Sui Anticline some 80 km. north-east of Jacobabad, from which several million cubic metres of fuel gas is produced daily for several years.

Potential Sources. There are two or three potential oil-bearing tracts in India to which reference may be made:

1. On the analogy of the great Iran belt of oil deposits at the foot of the Zagros chain, belonging to the same system of upheavals as the Himalaya, and of the Tertiary sedimentary beds stretching from Hazara to Naini Tal and in the Assam foot-hills from the Tista to the Brahmaputra, one can argue the existence of a belt of productive oil-fields in the Himalayan piedmont, chiefly in the Jammu, Kangra, and Nepal foot-hills zone.

2. The Gulf of Cambay and of Kutch on the west coast may be considered another potential area. On either side of the Cambay Gulf are outcrops of Nummulitic strata probably connected underneath the shallow waters of the gulf. Oil and gas in some quantity is encountered in numerous borings in this area. While the occur-

---

[1] E. S. Pinfold, T. G. B. Davies, and W. D. Gill, Development of the Oil-fields in North-West India, *Trans. Nat. Inst. Sci. India*, vol. II. No. 8, 1946.

[2] C. T. Barber, Natural Gas Resources of Burma, *Mem. G.S.I.* vol. lxvi. pt. 1, 1935.

rence of petroliferous beds possessing suitable structures beneath the waters of the Cambay Gulf promise high yields, the visible structures observed along the Saurashtra and Gujarat coast-lines are not of a type which can support oil-fields of large productivity. The ground enclosed in the triangle between Broach, Gogha, and Ahmednagar supports the Ankleshwar field, annual yield nearly 3·8 million tonnes, reserves 60 million tonnes.

3. *Central Rajasthan.* To the north of the Gulf of Cambay and Kutch extend large tracts of southern and central Rajasthan, floored by Mesozoic and Eocene strata. These rocks are covered by desert sands of sub-Recent age and are inaccessible to ordinary geological investigation. Seismic and gravitational surveys are being carried out to map any buried structures suitable for the underground storage of oil.[1]

In the world's total petroleum production annually of nearly 1400 million tons, India's share has been insignificant, average 0.6 per cent of the world's total.

To the above may be added the wide strips of Recent deltaic deposits on the East Coast which may be covering early Tertiary sediment-basins enclosing traps of oil-bearing sands.

6. METALS AND ORES

Neglect of ore-bodies in India—India contains ores of manganese, aluminium, iron, thorium, magnesium and titanium in exportable surplus, and gold, copper, chromium lead and zinc in minor quantities, associated with the crystalline and older rocks of the country. In the majority of cases, however, the ore bodies were up till lately worked not for the extraction of the metals contained in them, but for the purpose of exporting the ores as such in the raw condition, since few smelting or metallurgical plants existed. For this reason the economic value of the ores realised by the Indian miners was barely half the real market-value, because of the heavy cost of transport they had to bear in supplying ores to the European manufacturer at rates current in the latter's country. This serious drawback in the development of the mineral resources of India was pointed out by Sir T. H. Holland as early as in 1908.

Since 1950 there has been a marked growth in the mining and fabrication of metals, both ferrous and non-ferrous in the country; the production of iron and steel, aluminium, ferro-manganese, metal alloys has expanded and the development of resources in copper, lead, zinc, chromium and magnesium is proceeding on systematic lines. Investigation has brought to light new workable

---

[1] A. O. Rankine and P. Evans, Geophysical Prospecting for Oil in India, *Trans. Nat. Inst. Sci. India*, vol. ii. No. 8, 1947.

deposits of copper, titanium, the atomic minerals, beryllium, uranium, thorium and the rare metals, the industrial development of which is being planned. Fair occurrences of nickel, cobalt, cadmium, molybdenum have been located in paragenetic association with some of the above ore-bodies. Of the other industrial metals, there is total absence of tin and mercury and only marginal occurrences of tungsten and silver. Base metals production and the fabrication of metal-based commodities has gone up nearly 300 per cent since the World War. The sulphur content of sulphide ores, which was formerly neglected, is being recovered for elemental sulphur as well as for manufacture of sulphuric acid. Some of the ore-deposits of India, although not of economic value under the conditions prevailing at the present day, are likely to become so at a future day when improved methods of ore-refining and dressing treatment and better industrial conditions of the country may render the extraction of the metals more profitable.[1]

India's annual exports of the various ores and imports of metals have been approximately of the following order in recent years:

Minerals and Metals

Exports:	Quantity		Value (Rupees)
Chromite	111,600	tonnes	2,05,48,000
Iron-ore	15,118,000	,,	87,97,09,000
Magnesite	25,900	,,	80,24,000
Manganese-ore	127,800	,,	11,61,23,000
Mica	20,770	,,	14,49,02,000
Salt	...		
Sillimanite	2,100	,,	13,68,000
Steatite	25,400	,,	71,12,000

Imports:		Value (Rupees)
Ferro-alloys	...	50,51,000
Iron and Steel	...	78,93,27,000
Aluminium	...	1,27,14,000
Copper	...	44,90,88,000
Lead	...	6,29,95,000
Nickel	...	5,06,51,000
Zinc	...	6,95,83,000
Tin	...	7,07,82,000
Sulphur	...	16,50,00,000

Aluminium

Bauxite in laterite—Since the discovery that much of the clayey portion of laterite is not clay (hydrated silicate of aluminium), but

---

[1] Dictionary of Economic Products of India, Coun. Sc. Ind.Res., New Delhi, 1949-59.

the simple hydrate of alumina (bauxite), much attention has been directed to the possibility of working the latter as an ore of aluminium.[1] Bauxite is a widely spread mineral in the laterite cap of the Peninsula, Assam and Burma, but the laterites richest in bauxite are those of Ranchi district, Jabalpur, Balaghat, and Kohlapur, in which the percentage of alumina is 50-58. Other important deposits are those of Saurashtra, Kutch, Bilaspur, Mandla, Sarguja, Mahabaleshwar, Bhopal, Palamau and Salem. The total quantity of ore available is 250 million tonnes roughly, obtainable by simple methods of surface quarrying. Of these 63 million tonnes is high-grade ore.

Extensive deposits of bauxite and aluminium ore, analysing 60 to 80 per cent Al_2O_3, have been discovered in association with the Nummulitics of Jammu and Poonch, where 2 million tonnes of ore are exposed in surface strata. Their mode of occurrence also suggests a lateritic origin, *e.g.* by desilicification of large, subaerially exposed spreads of infra-Nummulitic clay-beds on a series of low, gently inclined dip-slopes. With deposits of the ore so widespread and of such magnitude, so far barely 130,000 tonnes of metallic aluminium per year is being produced in India. Additional production of up to 200,000 tonnes is being planned. The average annual imports of aluminium metal goods into India is about 15,000 tonnes.

Uses—Aluminium has a variety of applications in the modern industries. It is esteemed on account of its low density, its rigidity and malleability. Besides its use for utensils, it has many applications in electricity, metallurgy, aeronautics, etc. It is largely employed in the manufacture of alloys with nickel, copper, zinc and magnesium, which are finding ever widening applications in automobile, aircraft, railway and other engineering construction. The wide range of magnesium-aluminium and other light-metal alloys used in industries has during the last decade transformed the metals position of the world. The mineral bauxite (see page 464) has various industrial uses in the preparation of chemicals, refractories, abrasives, and aluminous cement. The present output of Indian bauxite is 600,000 tonnes and is chiefly consumed in the cement-making industry, refinement of oil and manufacture of metal.

Antimony

Sulphide of antimony, stibnite, is found in deposits of considerable size in Chitral (Pakistan) and at the end of the Shigri glacier in the province of Lahoul, but the lodes are in inaccessible localities.

---

[1] C. S. Fox, Aluminous Laterites, *Mem. G.S.I.* vol. xlix., 1923; *Bauxite and Aluminous Laterite*, London, 1932; Coggin Brown, *Bulletin of I. I. and L.* No. 2, 1921; No. 12, 1921.

The former source was actively worked in the war years, the output being 1,300 tonnes of ore per year. It occurs mixed with galena and blende in the granitoid gneiss in the latter area. Stibnite is also found in Vishakapatnam and in Hazaribagh. But the production of stibnite from these bodies does not appear to be a commercial possibility unless metallic antimony is extracted on the spot. Production of antimony is 680 tonnes from imported ore.

Arsenic

Sulphides of arsenic, orpiment and realgar, form small deposits in Chitral and in Kumaon. The orpiment-mines of the first locality are well known for the beautifully foliated masses of pure orpiment occurring in them, and form the chief indigenous source, but the output has fallen off considerably of late years. The orpiment occurs in calcareous shales and marble in close proximity to a dyke of basic intrusive rock. The chief use of orpiment is as a pigment in lacquer-work; it is also employed in pyrotechnics because of its burning with a dazzling bluish-white light.[1] Arsenopyrite occurs near Darjeeling and in the Bhutna valley, Kashmir.

Beryllium

Beryl is found in the mica-pegmatites of Bihar, Nellore, and Rajasthan. The Bihar mica belt, Jaipur and Ajmer-Merwara contain workable deposits of this mineral from which large crystals, up to 60 cm. in diameter and weighing up to a tonne, are sometimes obtained. The industrial use of beryl lies in the 10–12 per cent of BeO employed in the manufacture of copper-beryllium alloy and in atomic energy reactors as a moderator. The pre-Cambrian rocks of India possess considerable resources in this rare metal of industrial and strategic value, distributed sporadically in pegmatite veins traversing mica-bearing rocks. The annual output in some years is over 1,000 tonnes.

Chromium

Occurrence—Chromite, the principal ore of chromium, occurs as a product of magmatic differentiation in the form of segregation masses and veins in ultra-basic, intrusive rocks, like dunites, peridotites, serpentines, etc. In such form it occurs in Baluchistan (Pakistan), in Karnataka, in several districts of Orissa (chiefly Keonjhar), and in Singhbhum. The Orissa reserves are the largest, computed at over 3.5 million tonnes. Less important deposits have been found in parts of Tamilnadu and in Ratnagiri. The maximum Indian production in recent years is 150,000 tonnes, the bulk of which is exported, only 30,000 tonnes being used locally for manufacturing

---

[1]Coggin Brown, *Bulletin of I. I. and L.* No. 6, 1921.

furnace-bricks, other refractories and chemical products. The Baluchistan deposits are important and are capable of a much larger output. Chromite occurs in the Quetta and Zhob districts in serpentines associated with ultra-basic intrusions of late-Cretaceous age, the annual output being 50,000 tonnes. The Karnataka and Singhbhum deposits produce a relatively small quantity. Some chromite occurs in the "Chalk hills" (magnesite-veins) near Salem, but it is not worked. Large deposits (3.5 million tonnes reserve) of chromite occurring in dunite intrusions forming mountain-masses have been discovered in the Cretaceous volcanics of Burzil and Dras valley of Ladakh, Kashmir.

In all these above occurrences, chromite is a primary ore of magmatic origin.[1]

Uses—Chromite is used in the manufacture of refractory bricks for furnace-linings. Its chief metallurgical use lies in its being the raw material of chromium. An alloy of chromium and iron (ferro-chrome) is used in the making of rustless and stainless steels and armour-plates. A large amount of chromium is used in the manufacture of mordants and pigments, because of the red, yellow and green colours of its salts.

Cobalt and Nickel

Cobalt- and nickel-ores are not among the economic products of India. A sulphide of both these metals is found in the famous copper-mines of Khetri, Jaipur, Rajasthan. The *Sehta* of the Indian jewellers is the sulphide of cobalt, which is used for the making of blue enamel. Cobalt-ore deposits have been reported from Nepal State, and ore averaging 8.7% of cobalt was exported from Nepal during the war; but the mode of occurrence, geology of the deposits and probable reserves yet await investigation. The only notable occurrence of nickel-ores in the Indian region is in the Singhbhum copper-belt and in Nepal, near Kathmandu. Nickeliferous pyrrhotite and chalcopyrite occur in the auriferous quartz-reefs of Kolar, in Kerala, but the occurrences are not of any magnitude. Small deposits of nickeliferous pyrites, containing 1.7 per cent of Ni, have been found in the Purana rocks of Ramsu and Buniyar and in the Carboniferous limestone of Riasi, Kashmir. Import of nickel in India is about 1500 tonnes annually.

Copper

Occurrence—Copper occurs in Singhbhum and Chota Nagpur in Bihar; Nellore and Kistna districts in Andhra ; in Rajasthan—in the Ajmer, Khetri, Alwar regions; and in the outer Himalayas, in

---

[1]*Rec. G.S.I.* vol. lxxvi. 2, 1941.

Sikkim, Kumaon and Nepal. But the only deposits worked with some degree of success are those of the Singhbhum district, Mosaboni mines, which yield about 450,000 tonnes of ore per year, valued at Rs. 287 lakhs. 9,800 tonnes of refined copper are produced from the ore mined at Mosaboni, the most important copper mines in India. In Singhbhum the copper-bearing belt of rocks is persistent for about 130 km. along a zone of overthrust in the Dharwar schists and intrusive granite. The deposits worked at the Mosaboni mine consist of sulphide ore assaying 2.4 per cent of copper. More extensive copper-ore deposits have been recently explored in the Singhbhum Thrust Belt associated with some uranium, nickel and molybdenum. There was a flourishing indigenous copper-industry in India in former years, producing large quantities of copper and bronze from the Rajasthan, Sikkim, and Singhbhum mines, the sites of which are indicated by extensive slag-heaps and refuse "copper-workings". Important copper-mines existed in the Alwar, Ajmer and Khetri regions within historic times. Exploratory tests have revealed nearly 200 million tonnes of ore reserve with an average of 1 per cent copper. Annual consumption of copper in India is about 83,000 tonnes.

The copper ores of Singhbhum and Rajasthan occur as veins or as disseminations in the Dharwar schists and phyllites. In a great number of cases, the ore occurs in a scattered condition. The most common ore is the sulphide, chalcopyrite, which by surface-alteration passes into malachite, azurite, cuprite, etc.[1]

Native copper occurs at some places in South India. In Kashmir isolated masses of pure native copper have been found in the bed of the Zanskar river, but their source is unknown. They occur there as water-worn nodules, weighing up to 22 lbs.

New copper deposits—The copper deposits of Sikkim attracted much attention once. In this area, valuable copper lodes (Cu 3-7%) are proved to exist in association with bismuth and antimony, together with pyrrhotite, blende and galena. Lack of adequate communications over high mountains has been the chief obstacle to successful exploitation of these ore-bodies. In recent years exploration at Khetri, Dariba (Rajasthan), Rakka area in Singhbhum and Guntur district in Andhra have proved reserves of copper-ores (Cu 1%) to the extent of over 250 million tonnes. When brought into production, these deposits are calculated to yield 65,000 tonnes of copper per year. In most of the above deposits, the mode of origin of ore-bodies is similar, viz. they have resulted from the metasomatic replacement of the coun-

---

[1] J. A. Dunn, Mineral Deposits of Eastern Singhbhum, *Mem. G.S.I.* vol. lxix. pt. 1, 1937.

try rock by copper-bearing solutions derived from granitic and other intrusions in the Dharwars.

Gold

Occurrence—Gold occurs in India both as native gold, associated with quartz-veins or reefs, and as alluvial or detrital gold in the sands and gravels of a large number of rivers. The principal sources of the precious metal in India, however, are the quartz-reefs traversing the Dharwar rocks of Kolar district (Karnataka), which are auriferous at a few places.[1] The auriferous lodes of the Kolar gold-fields are contained in the above-mentioned quartz-veins, which run parallel to one another in a north-south direction in a belt of hornblende-schists along shear zones. The gold is associated with pyrite, pyrrhotite and arsenopyrite, the ore being hypothermal in origin. The most productive of these is a single quartz-vein, about 1.2 metres thick, which bears gold in minute particles. Mining operations in this reef have been carried to a depth of 3,050 m., some of the deepest mining shafts in the world, and have disclosed continuance of the same mode of distribution of the ore in the gangue. The gold is obtained by crushing and milling the quartz, allowing the crushed ore mixed with water to run over mercury-plated copper boards. The greater part of the gold is thus dissolved by amalgamation; the small residue that escapes with the slime is extracted by the cyanide process of dissolving gold.

Vein-gold—The annual yield of gold from the Kolar fields once averaged 9,638.8 kg., valued at more than £2,000,000. For the last 25 years it has averaged 5,670 kg., but the production is falling. Besides the difficulties due to increasing depth of workings, the mines are experiencing considerable difficulty from rock-bursts, a problem acutely present in these mines. Next to Kolar, but far below it in productiveness, is the Hutti gold-field of Andhra Pradesh which was also worked from a similar outcrop of Dharwar schists. It produced 595.3 kg. of gold in 1914, but the output since has been fitful for some years. A few quartz-veins traversing a band of chloritic and argillaceous schists, also of Dharwar age, support the Anantpur field of Tamilnadu, whose yield in 1915 approached 680.4 kg. This mine ceased operations after several vicissitudes in 1927. At some other places in the Peninsula, besides those named above, the former existence of gold is revealed by many signs of ancient gold-working in diggings, heaps of crushed quartz, and stone-mortars, which have (as has often happened in India with regard to other metalliferous deposits) guided the attention of the present workers to the existence of gold.

---

[1]Kolar Gold-Field, *Mem. G.S.I.* vol. xxxiii. pts. 1 and 2, 1901.

Alluvial gold—The distribution of alluvial gold in India is much wider. Many of the rivers draining the crystalline and metamorphic tracts in India, Sri Lanka and Burma are reputed to have auriferous sands, but only a few of them contain gold in a sufficient quantity to pay any commercial attempt for its extraction. The only instance of successful exploitation of this kind is the dredging of the Upper Irrawaddy, in search of the gold-bearing gravel in its bed, for some years; but the returns fell off and operations were closed down in 1918. In this way some 141.7 to 170 kg. of gold were won a year. Alluvial gold-washing is carried on in the sands and gravels of many of the rivers of Madhya Pradesh, and in sections of the Indus valley at Ladakh, Baltistan, Gilgit, Attock, etc., but none of them are of any richness comparable to the above instance. The quantity won by the indigent workers is just enough to give them their day's wages with only occasional windfalls.

The present-day production of gold in India has declined to 5286 kg. (190,000 ozs.), valued at Rs. 5.6 crores.

Iron

Occurrence—Iron-ore occurs on a large scale in India, chiefly in the form of the oxides: haematite and magnetite. It prevails especially in the Peninsula, where the crystalline and schistose rocks of the Dharwar and Cuddapah systems enclose at some places ferruginous deposits of an extraordinary magnitude. Among these, massive outcrops of haematite and magnetite of the dimensions of whole hills are not unknown. But the most common mode of occurrence of iron is as laminated haematite, micaceous haematite and haematite-breccia; lateritic haematite also forms large deposits, together with haematite- and magnetite-quartz-schists, the metamorphosed products of original ferruginous sands and clays. The high-grade haematitic ore-bodies of Singhbhum, together with those of Bastar, Keonjhar, Bonai and Mayurbhanj,[1] Karnataka, Goa and Tamilnadu are of Upper Dharwar or newer age, the remarkable concentration of the metal iron in them being ascribed to post-Cuddapah metasomatic action, to original marine chemical precipitation of the oxides, carbonates and other compounds of iron, to volcanic action and other agencies.[2] These ore-bodies, many of them containing 60–65 per cent of iron, are thought to be the largest and richest deposits of iron perhaps in

---

[1] The discovery of important deposits of iron-ore in the Mayurbhanj State is due to P. N. Bose (*Rec. G.S.I.* vol. xxxi. pt. 3, 1904).

[2] H. C. Jones, Iron-ore Deposits of Bihar and Orissa, *Mem. G. S.I.* vol. lxiii. pt. 2, 1934; J. A. Dunn, *Mem. G.S.I.* vol. lxix. pt. 1, 1937; M. S. Krishnan, Iron-ore, Iron and Steel, *Bull. G.S.I.* No. 4, 1954.

the world, surpassing in magnitude the Lake Superior ores. They are now estimated to total 21,000 million tonnes, containing about twelve thousand million tonnes of metallic iron.

The Damuda series of Bengal holds valuable deposits of bedded or precipitated iron-ore in the ironstone shales. Some iron-ore is enclosed in the Upper Gondwana haematitic shales. The Deccan Traps, on weathering, liberate large concentrates of magnetite sands on long stretches of the sea-coast. Iron is a prominent constituent of laterite, and in some varieties the concentration of limonite or haematite becomes so high that the rock can be smelted for iron. In the Himalayas, likewise, there occur large local deposits of this metal in the Purana formations as well as in association with the Eocene coal deposits.

Geographical distribution—A list of localities which contain the most noted deposits of iron ore will be interesting.

In south India the most important deposits consist of those of Salem, Madurai, Karnataka (Bababudan), Goa, Cuddapah and Kurnool, while Singhbhum, Manbhum, Sambalpur, Bastar and Mayurbhanj are the iron-producing districts of Bihar and Orissa. In Bengal proper, the Damuda ironstone shales contain a great store of metallic wealth, which has been profitably worked for a long time, both on account of its intrinsic richness as well as for its nearness to the chief source of fuel. In Assam also iron occurs with coal. In Madhya Pradesh the most remarkable iron deposit is that of the Bastar district where there is a hill, Khadolar, with the biggest concentration of iron-ore in the world. Jabalpur, Drug, Raipur, and Bhilaspur have likewise large aggregates of valuable haematitic ores which have been so far prospected only in part. In Maharashtra, Gujarat and parts of Andhra Pradesh the chief sources of iron are laterite and the magnetite-sands of rivers draining the trap districts, both of which are largely drawn upon by the itinerant *lohars*. Important reserves of high-grade ores of Dharwar age are met with in Goa and Ratnagiri, with low percentage of silica and of phosphorus below the Bessemer limit. In the Himalayas the Kumaon region has been found to possess some deposits. Workable iron-ore is met with in the Riasi district, Jammu hills, in association with the Nummulitic series, which supported a number of local furnaces for the manufacture of munitions of war during the last two centuries.

The deposits that are profitably worked at the present day are the ironstone shales of Burdwan, the high-grade ores of Mayurbhanj, Singhbhum and Manbhum, Bababudan hills (Karnataka) Salem, Goa, Orissa and Madhya Pradesh.

Iron seems to have been worked on an extensive scale in the past, as is evident from the widely scattered slag-heaps which are to be seen in many parts of India. The iron extracted was of high quality and was in much demand in distant parts of the world. The fame of the ancient Indian steel, *Wootz*—a very superior kind of steel exported to Europe, in days before the Christian era, for the manufacture of swords and other weapons—testifies to the metallurgical skill of the early workers.

Annually India imports iron and steel materials (hardware, machinery, railway plant, bars and sheets, etc.) to the value of nearly 150 crores of rupees.

At present India produces 22 million tonnes of iron-ore, the manufactured products from which are: pig-iron 2 million tonnes, steel 9,500,00 tonnes, and ferro-manganese about 100,000 tonnes. About 15 million tonnes of iron-ore is exported, value Rs. 870 million.

Lead and Silver

Very little lead is produced in India at the present time, though ores of lead, chiefly galena, occur at a number of places in the Himalayas, Tamilnadu, Rajasthan and Bihar, enclosed either among the crystalline schists or, as veins and pockets, in the pre-Cambrian and Vindhyan limestones. Lead was formerly produced in India on a large scale. The lead ores of Mewar, Hazaribagh, Manbhum and also some districts of Madhya Pradesh are on a fairly large scale, and they are often argentiferous, yielding a few decagrams of silver per tonne of lead. Large mounds of slag, found in Mewar, Jaipur and in parts of Bihar, indicate that a considerable amount of ingot lead was produced in several parts of India for centuries. The Zawar lead-zinc mines, near Udaipur, have re-opened extensive ancient workings for lead and zinc and have exposed promising ore-bodies. The annual production of lead, however, is yet small from these mines—4,000 tonnes. The annual consumption—about 35,000 tonnes—is met from imports. The Bawdwin Mines, Burma, supplied 15,000 tonnes of lead to India up to 1940 annually.

Zawar Lead-zinc mines—Considerable deposits of lead-zinc ores occur in the metamorphosed limestones belonging to the Aravalli System, near Udaipur in Rajasthan. These are being mined at Zawar, where extensive ancient workings exist. The main orebodies are galena and sphalerite, with pyrite dissemination, in dolomitic limestones along with traces of cadmium and silver sulphides. The mineralised zone is fairly extensive, spread over an area with 30 m. wide lode extending in depth to some hundred metres. Total ore reserves are about 10 million tonnes of mixed lead-zinc ores, the lead content ranging over 2 to 14 per cent, the zinc 2 to 13. The deposits are chiefly of the fissure-filling type showing metasomatised walls with numerous branching veins.

The Zawar mineralisation of the Aravalli limestone is believed to be due to intrusion of a granite body beneath the limestones during post-Delhi orogeny. The lead ores mined at Zawar are locally smelted to 99 per cent purity, silver content ranging from 700 to 850 gm. per tonne.

The Zawar ores belong, geologically, to the class of metasomatic replacements, the original country-rock, an Aravalli limestone, being substituted chemically by the sulphides and carbonates of lead and zinc, by the process of molecular replacement.

Silver—India is the largest consumer of silver in the world, the extent of its average annual imports used to be £10,000,000. But with the exception of the quantity of silver won from the Kolar gold-ores (aggregating some 425 to 708.7 kg.), and lately obtained as a by-product from the smelting of the Zawar lead-ores (about 4,110 kg. per annum), no silver is produced in the country. The production of silver from the rich argentiferous lead-zinc ores of the Bawdwin mines of Burma, in 1929, touched the figure 206.2 tonnes, valued at over a crore of rupees, used to meet part of the demand.

Magnesuim[1]

Mode of occurrence of magnesite—Large deposits of magnesite ($MgCO_3$) occur in the district of Salem as veins associated with other magnesian rocks such as dolomite, serpentines, etc. The magnesite is believed to be an alteration-product of the dunites (peridotite) and other basic magnesian rocks of Salem. When freshly broken it is of a dazzling white colour and hence the magnesite-veins traversing the country have been named the Chalk hills of Salem. The magnesite of Salem is of a high degree of purity (MgO 46.4 per cent), is easily obtained and, when calcined at a high temperature, yields a material of great refractoriness. Other places in India also contain magnesite-veins traversing basic rocks, *viz*. Coorg, Coimbatore, Karnataka, Almora and parts of Eastern Himalayas. The total reserves of this mineral, a principal source of the light metal magnesium, are over 100 million tonnes. Dolomite also occurs in extensive deposits in many parts of the Himalayas and South India, thus adding to the potential reserves of magnesium metal in the country, of use in manufacturing aluminium-magnesium and other light-metal alloys. The industrial uses of magnesite are in the manufacture of caustic magnesia, refractory materials for use in the steel industries and as a source of carbonic acid gas. It is also manufactured into cement (Sorel cement) for artificial stone, tiles, etc. The combined outputs of the Salem and Karnataka magnesite workings reach a total of about 245,000 tonnes, valued at Rs. 4,800,000.

---

[1]Middlemiss, *Rec. G.S.I.* vol. xxix pt. 2, 1896 (Magnesite); *Rec. G.S.I.* vol. xxvi. 7, 1942.

Manganese

Production of manganese in India—With the exception of Russia and Brazil, India is the largest producer of manganese in the world. Within the last forty years, the export of manganese ore has risen from a few thousand tonnes to 1,500,000 tonnes annually. The output has in recent years, however, fluctuated considerably. The major part of this output was exported in the ore condition, but now a significant part of it is treated in the country for the production of the metal, or for its manufacture into ferromanganese, the principal alloy of manganese and iron.

Distribution. Geographical—The chief centres of manganese mining, or rather quarrying (for the method of extraction up till now resorted to is one of open quarrying from the hillsides), are the Balaghat, Bhandara, Chhindwara, Jabalpur and Nagpur districts of Madhya Pradesh, which yield nearly 60 per cent of the total Indian output of high-grade ores. Sandur and Vishakapatnam in Tamilnadu take the next place, then come the Panch Mahal district in Gujarat and the Belgaum district of Karnataka and Singhbhum, Keonjhar and Gangpur in Bihar and Orissa, Chitaldurg and Shimoga districts of Karnataka, and Jhabua in central India.

Geological—Fermor has shown that manganese is distributed, in greater or less proportion, in almost all the geological systems of India, from the Archaean to the Pleistocene, but the formation which may be regarded as the principal carrier of these deposits is the Dharwar. The richly manganiferous facies of this system— the *Gondite* and *Kodurite series*—contain enormous aggregates of manganese ores such as psilomelane and braunite, pyrolusite, hollandite, etc. Of these the first two form nearly 90 per cent of the ore masses. The geological relation of the ore bodies contained in these series and their original constitution have been referred to in the chapter on the Dharwar system (pp. 103-4). Besides the Dharwar system, workable manganese deposits are contained in the laterite-like rock of various parts of the Peninsula, where the ordinary Dharwar rocks have been metasomatically replaced by underground water containing manganese solutions. According to the mode of origin, the two first-named occurrences belong to the *syngenetic* type of ore bodies, *i.e.* those which were formed contemporaneously with the enclosing rock, while the last belong to the *epigenetic* class of ores, *i.e.* those formed by a process of concentration at a later date.

A voluminous memoir on the manganese-ore deposits of India by Sir 'L. L. Fermor, published by the Geological Survey of India,[1] contains valuable information on the mineralogy, economics and the geological relations of the manganese of India.

---

[1] *Mem. G.S.I.* vol. xxxvii., 1909; *Rec. G.S.I.* vol. lxxvi. 9, 1942.

Ores which contain from 40 to 60 per cent of manganese are common and are classed as *manganese ores*. There also exist ores with an admixture of iron of from 10 to 30 per cent: these are designated *ferruginous manganese ores*; while those which have a still greater proportion of iron in them are known as *manganiferous iron ores*.

Uses—Manganese is chiefly used in making steel, and in the manufacture of ferro-manganese and spiegeleisen, both of which are alloys of manganese and iron, and of alloys with other metals. Manganese is employed in several chemical industries as an oxidiser, as in the manufacture of bleaching powder, disinfectants, preparation of gases, etc. Manganese is employed in the preparation of colouring materials for glass, pottery-paints, etc. The pink mineral, rhodonite (silicate of manganese), is sometimes cut for gems on account of its attractive colour and appearance.

Economics—The minimum reserves of richer-grade ores (chemical grade with Mn > 70 per cent, and first grade with Mn > 48 per cent) are not large, computed at only 80 million tonnes. Reserves of lower-grade ores (Mn 40—30 per cent) are on a much larger scale. Beneficiation of the latter by modern ore-dressing methods is raising the Mn content of a large proportion of leaner ore-bodies and increasing their commercial value. Of the present annual production of 1.3 to 1.6 million tonnes, around 700,000 tonnes of ore are utilised by domestic steel and ferro-manganese industries. Seven units are now producing ferro-manganese, their total capacity being 188,000 tonnes. India exports standard-grade, high-carbon ferro-manganese, shipments in 1969 being nearly 100,000 tonnes.

In 1953, the peak year, India exported 1,900,000 tons of Mn-ore valued at over 29.5 crores of rupees.

Strontium

A fairly large deposit of the mineral celestite (96 per cent $SrSO_4$) has been found in the Tiruchirapalli district, estimated to contain a million tonnes, and another in the Mianwali district (West Punjab) of equal purity with estimated reserves of half a million tonnes.

Thorium

The main source of this rare heavy metal, likely substitute for uranium as fission metal in atomic reactors, is monazite. The resources of India in thorium are of considerable magnitude. ThO_2 is a constant ingredient of monazite occurring in the form of beach sands and placer deposits in association with the ilmenite sand spreads (pp. 470-71) of the Kerala coast and those of Tuticurin and Ganjam, etc. on the Tamilnadu coast. The extensive placer posits of monazite lately discovered, lying on the denuded surface of the Archaean plateau of Ranchi-Purulia, in Bihar, are estimated to carry some 200,000 tonnes of ThO_2. The thorium content of Indian monazite ranges from 8 to 10.5 per cent ThO_2; in some unnamed variants of monazite it is as high as 19 per cent and in the rare mineral *cheralite* it is 31 to 33 per cent. Thus India possesses in its monazite deposits thorium resources of high potential value, because of the strategic importance of thorium as a source of atomic energy. The reserves of thorium available in India are estimated at nearly 500,000 tonnes.

The very rare mineral thorianite (ThO_2—70 per cent), found in the crystalline rocks of Sri Lanka in commercial quantity, is found as a rare constituent of some ilmenite sands.

Tin

Tin-ore of Mergui and Tavoy—With the exception of a few isolated occurrences of cassiterite crystals in Palanpur and its occurrence *in situ* in small deposits in the gneissic rocks of Hazaribagh, no commercially workable ore of tin is found in India.

In the neighbouring region of Burma, from which probably India derived its supplies of tin in the past, there occur deposits of tin-ore of workable proportions (the Mergui and Tavoy districts of Lower Burma)[1] which have supplied a large quantity of tin from a remote antiquity. The most important tin ore is cassiterite, occurring in quartz-veins and pegmatites, associated with wolfram in granitic intrusions traversing the Mergui series. But the greater proportion of the tin-ore is obtained, not from the deposits *in situ*, but from the washing of river-gravels (*stream-tin* or *tinstone*) and from dredging the river-beds of the tin-bearing areas, where the ore is collected by a process of natural concentration. India's requirement of tin is 5,000 tonnes annually.

Titanium

Titanium occurs in its two compounds, ilmenite and rutile, the former of which is of wide distribution in the charnockitic and other gneisses of the Peninsula, Bihar and Rajasthan. It occurs on the beaches of the Kerala and Tamilnadu coasts as black heavy sand, along with monazite, zircon and other heavy minerals. Here the volume and degree of concentration of heavy mineral grains are on a scale unknown elsewhere. The largest deposits are on the Kerala beaches between Kanya Kumari and Quilon. Smaller

---

[1] *Rec. G.S.I.* vols. xxxvii. and xxxviii. pts. 1, 1908 and 1909, Annual Reports Coggin Brown and A. M. Heron, Ore Deposits of Tavoy, *Mem. G.S.I.* vol. xliv. pt. 2, 1923.

patches of the sand concentrates occur at Ratnagiri and further north on the Malabar coast and also on the east coast, at Tuticurin, Waltair and Ganjam. A large quantity of ilmenite sand occurs as alluvial placer deposit on parts of the Hazaribagh plateau, Bihar, under the soil cap. The ilmenite (including arizonite) is rich in TiO_2, the average content of titania ranging between 54 to 62 per cent. The total reserves of ilmenite, extractable from the beach and placer sands by magnetic separation is estimated at over 250 million tonnes. Associated with the ilmenite sands are rutile, zircon, monazite, columbite-tantalite, garnet and sillimanite. Monazite forms roughly 1 to 3 per cent of the sand grains. Titanium also occurs as titaniferous magnetite in large masses in Singhbhum and Mayurbhanj and in some quantity in pegmatites, at several localities. Rutile, mainly TiO_2, is obtained to the extent of 1,000 to 2,000 tonnes annually in the magnetic separation of ilmenite from the raw beach sands.

The chief use of ilmenite is in the manufacture of white paints, the opacity and covering power of titanium oxide being very high. It is now increasingly utilised for manufacture of titanium metal, which possesses some remarkable properties and is regarded as the metal of the future, especially in aircraft and chemical engineering industries.

Tungsten

Wolfram—Previous to 1914, Burma contributed nearly a third of the total production of wolfram (the principal ore of tungsten) of the world, but subsequently it increased its output to a much larger extent, heading the list of the world's producers of tungsten, with 3,600 tonnes of ore per annum. The most important and valuable occurrences of wolfram are in the Tavoy district of Lower Burma,[1] where the tungsten-ore is found in the form of the mineral wolframite in a belt of granitic intrusions among a metamorphic series of rocks (*Mergui series*). The tin-ore, cassiterite, mentioned on page 451, is present in the same group of rocks, at places associated with wolframite. Wolfram chiefly occurs in quartz-veins or lodes, associated with minerals like tourmaline, columbite, and molybdenite.

Wolfram is also found in India in Nagpur, Tiruchirapalli and at Degana in Rajasthan, but not in quantities sufficient to support a mining industry in normal times.

Uses of tungsten—Tungsten possesses several valuable properties which give to it its great industrial and military utility. Among these the most important is the property of "self-harden-

---

[1] *Mem. G.S.I.* vol. xliv. pt. 2, 1923 ; *Rec. G.S.I.* vol. xliii, pt. 1, 1913; vol. 1, pt. 2, 1919.

ing", which it imparts to steel when added to the latter. Over 95 per cent of the wolfram mined is absorbed by the steel industry. All high-speed steel cutting-tools have a certain proportion of tungsten in them. Tungsten-steel is largely used in the manufacture of munitions, of armour plates, of the heavy guns, etc., and enables them to stand the heavy charge of modern explosives. Tungsten, by repeated heating, is given the property of great ductility, and hence wires of extreme fineness and great strength, suitable for electric lamps, can be manufactured. In the last war Indian Ordnance factories produced some tungsten-steel, along with other ferro-alloys, for munitions use.

Uranium

Compounds of this highly strategic metal of increasing value as atomic fuel are found in India associated principally with crystalline, igneous and metamorphic rocks. Uranium mineralisation in India is associated with sulphidic copper and oxidised iron rather than with gold, lead, zinc or vanadium ores. No appreciable quantity of uranium has been found in Gondwana sedimentary rocks, though its occurrence in significant quantities has been reported in the phosphatic deposits of Mussoorie. Bituminous and lignitic deposits at several places have been noted to contain low amounts of uranium. The uranium ores of India belong to three categories:

(1) Pegmatitic: Pitchblende and complex niobates, tantalates and titanates of uranium, *e.g.*, samarskite, fergusonite, brannerite, etc.

(2) Uranium compounds impregnating rocks that have been involved in Eparchaean orogenic movements giving rise to large shear and thrust planes, *e.g.*, the Singhbhum Copper Belt in Bihar and the tightly compressed rocks of the Aravalli and parts of central Himalaya. Here the uranium compounds occur in thin disseminations, yielding from about .25 to 1 kg. of uranium to the tonne of rock.

(3) Monazite occurring in the large beach-sand deposits on the East and West coasts of India and in inland placer deposits contains a small fraction, from 0.2 to 0.4 per cent, of uranium oxide. The rare mineral cheralite, a variant of monazite, contains 4 to 6 per cent U_3O_8. Over 15,000 tonnes of uranium is estimated as probable reserve from this source.

Pitchblende (uraninite) occurs in nodular aggregates and patches of basic segregations in pegmatite veins in the Singar Mica Mines of Bihar and also in the mica pegmatites of Nellore and Ajmer. Autunite, torbernite, carnotite, also columbite, samarskite, triplite and allanite are usual associates. In the mica pegmatites of Nellore, masses of samarskite weighing up to 91 kg. have been met with, while pitchblende in nodules up to 16.3 kg. was found in the Gaya pegmatites. The main uranium reserves of the country, however, are not in the pegmatites but they lie in deposits of categories (2) and (3) stated above. Radio-activity of varying intensity is observed in pre-Cambrian of the central Himalayas of Kulu and Kumaon: surveys are carried out to locate uranium concentrations of commercial value.

Vanadium

Vanadium-bearing iron-ore, containing V_2O_5 in quantity varying from 1 to 4 per cent, has lately been discovered in deposits of considerable size, but with a fitful distribution of the vanadium content, in Singhbhum and Mayurbhanj region. Its exact paragenesis and relation with the country-rocks are not yet known, but the ore occurs in association with basic intrusions in Dharwar schists.

The vanadiferous ores are titanium-bearing iron oxides (V— 0.8 to 3 per cent), the deposits being estimated to be 33 million tonnes. Radiometric analyses of vanadium-ore concentrates reveal the presence of minute amounts of uranium.[1]

Zinc

Zinc-ore occurrences are fairly widespread in India associated with the base metal lead. Till lately India's requirement in this metal was met by imports, except for a small output at Zawar (Udaipur region). Promising deposits of lead-zinc have been revealed at Riasi (Kashmir); Almora; Tehri-Garhwal and Bhotang (Sikkim). The main ore-bodies are, sphalerite, calamine and hemimorphite, in association with lead compounds and minor amounts of cadmium and silver. In the new smelter plants, the recovery of cadmium along with sulphur (hitherto lost in the treatment of sulphidic ores in India) for manufacture of sulphuric acid, will be carried out.

Annual consumption of zinc in India is over 80,000 tonnes. Zinc lodes (blende) occur in association with lead-ores in the re-opened Zawar mines in the Mewar region, which produced considerable amounts of metallic zinc a century or two ago. About 18,000 tonnes of zinc-ore are smelted from the ore mined at Zawar with 75 tonnes of cadmium.

A workable deposit of zinc-blende of considerable purity occurring in lenticular veins and lodes has been discovered in the Riasi district of Kashmir in association with a Palaeozoic limestone. The veins sometimes swell to nests of 14.2 cu.m., and some thousand tonnes of float ore occur in the vicinity. Some zinc-ores occur with the antimony deposits of Shigri.

---

[1] *Rec. G.S.I.* vol. lxxvi., 14, 1942.

7. PRECIOUS AND SEMI-PRECIOUS STONES[1]

Diamonds

Panna and Golconda diamonds—In ancient times India had acquired great fame as a source of diamonds, all the celebrated stones of antiquity being the produce of its mines, but the reputation has died out since the discovery of the diamond-mines of Brazil and the Transvaal, and at the present time the production has fallen to a few stones annually of but indifferent value. Even so late as the times of the Emperor Akbar, diamond-mining was a flourishing industry, for the field of Panna alone is stated to have fetched to his Government an annual royalty of 12 lakhs of rupees. The localities noted in history as the great diamond centres were Bundelkhand (for "Panna diamonds"); the districts of Kurnool, Cuddapah, Bellary, etc. in Tamilnadu (containing the "Golconda diamonds"); and some localities in Madhya Pradesh such as Sambalpur, Chanda, etc. The diamondiferous strata in all cases belong to the Vindhyan system of deposits. A certain proportion of diamonds were also obtained from the surface-diggings and alluvial-gravels of the rivers of these districts. Two diamond-bearing horizons occur among the Upper Vindhyan rocks of Madhya Pradesh: one of these (Panna region) is a thin conglomerate-band separating the Kaimur sandstone from the Rewah series, and the other, also a conglomerate, lies between the latter and the Bhander series. The diamonds are not indigenous to the Vindhyan rocks but have been assembled as rolled pebbles, like the other pebbles of these conglomerates, all derived from the older rocks. The original matrix of the gem from which it separated out by crystallisation probably lies in the dykes of basic volcanic rocks associated with the Bijawar series,[2] some of which have been mapped recently. The most famous diamonds of India from the above-noted localities are the "Koh-i-noor", 186 carats; the "Great Mogul", 280 carats; the "Nizam", 277 carats; the "Orloff", 193 carats; the blue "Hope"; the "Pitt", 410 carats. The value of the last-named stone, re-cut to $136\frac{3}{4}$ carats, is estimated at £480,000.

At present stones (of gem quality as well as industrial diamonds) are produced from Panna mines; the annual returns show about 12,000 carats of stones, per year valued at about Rs. 50 lakhs. It does not appear that the Indian diamond deposits are all exhausted. Intensive prospecting and mining by modern methods, in place

---

[1] Goodchild, *Precious Stones* (Constable); L.A.N. Iyer, *Rec. G.S.I.* vol. lxxvi., 6, 1942.

[2] Vredenburg, *Rec. G.S.I.* vol. xxxiii., pt. 4, 1906; K. P. Sinor, *The Panna Diamondfield*, Bombay, 1932.

of the crude and primitive diggings of old, are being employed to revive the alluvial and placer mining in Panna, Kurnool, Bellary and in Madhya Pradesh.

A pipe of ultra-basic rock, resembling the kimberlite of South African diamond fields, has been located in one of the Panna fields. Mining this pipe-rock has given encouraging results, both for gem and industrial diamonds.

Rubies and Sapphires (Corundum)[1]

Burma and Sri Lanka—Crystallised and transparent varieties of corundum, when of a beautiful red colour, form the highly valued jewel ruby, and, when of a light blue tint, the gem sapphire. Rubies of deep carmine-red colour, "the colour of pigeons' blood", and perfect lustre, are often of greater value than diamonds. Rubies are mined in the Mogok district (Ruby Mines district) of Upper Burma, north of Mandalay, which has been a celebrated locality of this gem for a long time. The best rubies of the world came from this district from an area covering some 64 to 78 sq. km., of which Mogok is the centre. The matrix of the ruby is a crystalline limestone—ruby limestone (see pp. 78-79)—associated with and forming an integral part of the surrounding gneisses and schists. The rubies are found *in situ* in the limestone along with a number of other secondary minerals occurring in it. Some stones are also obtained from the hill-wash and alluvial detritus. The output of the Burma ruby-mines amounted, some years ago, to over £95,000 annually, but it has declined of late years.[2] The average annual royalty of Rs. 1,70,000 indicates the state of the industry before the war.

Gem-gravels of Sri Lanka—Sapphires of good water and colour, and to a less extent rubies of indifferent colour, are the more prized stones found in the gem-sands and gravels in the Ratnapura district of Sri Lanka occurring within a hundred feet from the surface, often less, in late-Tertiary and post-Tertiary deposits. These alluvial gem-bearing gravels of Sri Lanka, which have supported precious and semi-precious stone-mining centres for centuries, without apparent exhaustion, must be counted amongst the most prolific gem-fields of the world. The other more common gem-stones found in these beds are topaz, spinels, zircon (hyacinth and jargon), aquamarine, chrysoberyl (alexandrite and cat's eye), tourmalines (rubellite and indicolite), garnets (pyrope and almandine), moonstone, amazon-stone (the gem varieties of felspar), amethyst and rock-crystal.

---

[1] T. H. Holland, Corundum, *G.S.I.* 1898.

[2] One ruby from the Mogok mines, $38\frac{1}{2}$ carats in weight was sold for £20,000 in London in 1875.

The late age of the existing gem-beds of central Sri Lanka is indicated by the presence of fossil ungulate, proboscid and other mammals of the Pliocene and Pleistocene periods, but there is no doubt that the gem-stones enclosed in the alluvial gravels represent the products of weathering of many geological ages. The parent rocks from which they have been derived by erosion and weathering were most probably pegmatites traversing the Archaean gneisses and khondalites of the central Sri Lanka highlands, as, in a few cases, actual occurrence of sapphires, zircons, moonstones and garnets has been observed in the original matrix of the pegmatites.

Sapphires of Kashmir—The Burma ruby locality also yields sapphires occasionally; a sapphire weighing 1,000 carats was found in 1929 and another of 630 carats in 1930 from Mogok, but a larger source of sapphires in India was up till lately Kashmir. The gem was first discovered in Kashmir in 1882; it there occurs as an original constituent of a fine-grained highly felspathic gneiss at Padar in the Kishtwar district of Jammu and Kashmir State, at a high elevation. Transparent crystallised corundum occurs in pegmatite veins cutting actinolite-schist lenticles in Salkhala marble, at an altitude of 4,500 m. Associated minerals in the pegmatite are prehnite, tourmaline, beryl, spodumene and lazulite. Sapphires were also obtained from the talus-débris at the foot of the hillslopes. Stones of perfect lustre and high degree of purity have been obtained from this locality in the earlier years, but the larger and more perfect crystals, of value as gems, appear to have become exhausted since 1908; later discovery, however, by the Mineral Survey of Kashmir has revealed a large quantity of crystallised transparent corundum. The bulk of the output from the mines is confined to what are called "rock-sapphires", valueless for gems and of use as abrasives, watch jewels, etc.[1]

Spinel

Spinel when of sufficient transparency and good colour is used in jewellery; it constitutes the gem ballas-ruby when of rose-red colour and spinel-ruby when of a deeper red. Rubicelli is the name given to an orange-red variety. Spinel-rubies occur in the Burmese area associated with true rubies; also in Sri Lanka, in the well-known gem-sands of Sri Lanka, along with many other semiprecious and ornamental stones.

Jadeite

Jade is a highly-valued ornamental stone on account of its great

---

[1] L. A. N. Iyer, Indian Precious Stones, Calcutta, 1946; Gems and Semi-precious Stones of Ceylon, *Prof. Pap.* 2, *Rec. Dep. Min. Ceylon*, Colombo, 1945.

toughness, colour and the high lustrous polish it takes. A large number of mineral compounds pass under the name of *jade*, but the true mineral, also named *nephrite*, so much sought after, is a comparatively rare substance. Its occurrence is not known in India, but a mineral greatly similar to it in many of its qualities, and known as *jadeite*, is largely quarried in Burma. True jade comes into India from the Karakash valley of South Turkestan.

Sang-e-Yeshm, regarded as jade in the Punjab, is only a variety of serpentine. It differs from the genuine mineral in all its characters, being not so tough, much softer and incapable of receiving the exquisite polish of jade.

Emerald and Aquamarine

Beryl when transparent and of perfect colour and lustre is a highly valued gem. Its colour varies much from colourless to shades of green, blue or even yellow. The much-prized green variety is the *emerald*, while the blue is distinguished as *aquamarine*. Emeralds are rare; the only locality in India is in Mewar, where crystals of exquisite colour and water were discovered in 1953 in a band of biotite-gneiss in hornblende-schist. This locality has yielded about one million rupees' worth of first-quality emeralds. Aquamarines suitable for use as gems are obtained from pegmatite-veins crossing the Archaean gneiss at some places in Bihar and Nellore. Good aquamarines also occur in Coimbatore district and in Mewar, Ajmer and Kishengarh (Rajasthan), from both of which localities stones of considerable value were once obtained. A highly productive locality for aquamarines was discovered in the Kashmir State in the Shigar valley in Skardu, whence crystals of considerable size and purity were recovered. The gem occurs in coarse pegmatite veins traversing biotite-gneiss.

Common beryl occurs in very large crystals, sometimes 30 cm. in length, in the granite-pegmatite of many parts of India, but only rarely do they include some transparent fragments of the required purity.

Chrysoberyl

Chrysoberyl is a stone of different composition from beryl. It is of greenish-white to olive-green colour. A few good stones in the form of platy crystals of tabular habit are obtained from pegmatite-veins in Kishengarh in Rajasthan, which also yield mica and aquamarines. They are found in some felspar-veins in the nepheline-syenites of Coimbatore. Usually they are too much flawed and cracked to be suitable for cutting as gems. Chrysoberyl crystals when possessing a chatoyant lustre are known as "cat's eyes". *Alexandrite* is the deep emerald-green variety found in Sri Lanka it has exquisite colour and pleochroism, showing green by reflected light and deep red by transmitted light.

Garnets

Garnet as a gem-stone—Garnet possesses some of the requisites of a gem-stone—a high refractive index and lustre, a great hardness, a pleasing colour, transparency, etc.—and would be appreciated as such, were it but put on the market in restricted quantities. Garnets are most abundant in the metamorphosed rocks of Rajasthan and Sri Lanka, especially in the mica-schists, and large transparent crystals are frequently found. Quantities of garnets are exported to foreign countries for use in cheap jewellery. The variety used for this purpose is almandine, of crimson to red and violet colours. Crystals of large size, derived from Purana mica-schist, are worked at Jaipur, Delhi and at Kishengarh, where they are cut into various shapes for gems. Those of Kishengarh are considered to be the finest in India, and support a regular industry of about a lakh of rupees yearly.

Zircons

Zircons occur in various parts of India, but nowhere quite flawless or with the degree of transparency required in a gem. *Hyacinth* (the transparent red variety) is found at Kedar Nath on the upper Ganges.

Blue, green, yellow and colourless zircons are common gemstones of Sri Lanka. Zircon has an adamantine lustre and high dispersion power, its refractive index being very high, 1.92–1.98.

Tourmalines

Red and green tourmalines—Pellucid and beautifully coloured varieties of tourmaline, red, green or blue, are worked as gems. The fine red transparent variety *rubellite* is obtained from the ruby-mines district of Burma, where it occurs in decomposed granite veins. The green variety known as *indicolite* occurs in Hazaribagh (Bihar) and in the Padar district of Kashmir, where also some transparent crystals of rubellite are found. The latter tourmalines possess greater transparency, but are much fissured. Gem-tourmalines are also obtained from Sri Lanka from the noted gem-sands or gravels of that island.

Other gem-stones of India

Besides the above-named varieties, other crystallised minerals, when of fine colour and attractive appearance and possessing some of the other qualities of gems, *e.g.* hardness, transparency, etc., are cut for ornamental purposes in different parts of the country.

Among such minerals are the pleochroic mineral iolite or cordierite of Sri Lanka kyanites or cyanites found at Narnaul in the Patiala region; rhodonite (pink manganese silicate) of some localities of Madhya Pradesh; apatite (a sea-green variety) met with in the kodurites of Vishakapatnam. Moonstone and amazon-stone are ornamental varieties of felspar, the former a pearly opalescent orthoclase, met with in Sri Lanka and the latter a green microcline occurring in Kashmir and elsewhere. Turquoise, opaque, of fine blue colour, usually uncut, which is commonly sold in the bazaars of Kashmir and Darjeeling, is a product of Iran or Tibet, occurring in seams or patches in trachytic rocks.

Gem-cutting is a regular industry in places like Delhi, Jaipur and Sri Lanka.

Agates

Various forms of chalcedonic silica, agates, carnelian, blood-stone, onyx, jasper, etc., are known under the general name of *akik* (agate) in India. The principal material of these semiprecious stones is obtained from the amygdaloidal basalts of the Deccan, where various kinds of chalcedonic silica have filled up, by infiltration, the steam-holes or cavities of the lavas. The chief place which supplies raw *akik* is Ratanpur in the Rajpipla region, where rolled pebbles of these amygdules are contained in a Tertiary conglomerate. On mining, the stones are first baked in earthen pots, which process intensifies the colouring of the bands in the agates. The cutting and polishing is done by the lapidaries of Cambay, who fashion out of them (after a most wasteful process of chipping) a number of beautiful but small articles and ornaments. The annual output at Ratanpur is about a hundred tons. Cambay used to be a large market of Indian agates in medieval times for different parts of the world.

Agate wedges, pivots and bearings of scientific instruments are now being cut in India.

Rock-Crystal

Rock-crystal, or crystallised, transparent quartz, is also cut for ornamental objects, such as cheap jewels (*vallum* diamonds), cups, handles, etc. The chief places are Tanjor, Kashmir, Kalabagh, etc., whence crystalline quartz of the requisite purity and transparency is obtained. Flawless, water-clear, untwinned, right-and left-handed quartz-crystals are in demand, because of their piezo-electric property, by the radio and electronic industries.

Amethyst and Rose-Quartz, the purple and pink-coloured varieties of rock-crystal, are cut as ornamental stones and gem-stones. Amethyst occurs in some geodes in the Deccan Trap, filling up lava-cavities near Jabalpur, and in the Bashar region, Punjab. Rose-quartz occurs in Chhindwara and Warangal, Madhya Pradesh.

Amber

Amber is mineral resin, *i.e.* the fossilised gum of extinct coniferous trees. It is extracted by means of pits from some Miocene clay-beds in the Hukawng valley of North Burma. A few quintals are produced annually, from 10 to 100, with an average value around some 150 rupees per quintal. It occurs in round fragments and lumps, transparent or translucent, often crowded with inclusions and with veins of calcite. Amber is employed in medicine, in the arts, for jewellery, etc., and is highly prized when of a transparent or translucent nature.

8. ECONOMIC MINERALS AND MINERAL PRODUCTS

Here we shall consider the remaining economic mineral products, mostly non-metallic minerals of direct utility or of application in the various modern industries and arts. They include salts and saline substances, raw materials for a number of manufactures, and substances of economic value such as abrasives, soil-fertilisers, the rare minerals, etc. With regard to their geological occurrence, some are found as constituents, original or secondary, of the igneous rocks; some as beds or lenticles among the stratified rocks, formed by chemical agencies; while others occur as vein-stones or gangue-materials occurring in association with mineral-veins or lodes or filling up pockets or cavities in the rocks. The more important of these products[1] are:

1. Alkaline Salts.
2. Alum.
3. Asbestos.
4. Barytes.
5. Bauxite.
6. Borax.
7. Corundum.
8. Fluorspar.
9. Graphite.
10. Gypsum.
11. Kyanite and Sillimanite.
12. Limestone.
13. Mica.
14. Mineral Paints.
15. Monazite.
16. Phosphatic Deposits.
17. Pyrite.
18. Rare Minerals.
19. Reh or Kalar.
20. Salt.
21. Saltpetre or Nitre.
22. Steatite.
23. Sulphur.

---

[1]Bulletins and Prof. Paps. on Industrial Minerals in *Rec. G.S.I.* vols. lxxvi. and lxxvii., 1940-45; H. Crookshank, Minerals of Rajputana Pegmatites, *Trans. Min. Geol. Met. Inst. Ind.*, vol. xlii. 2, 1948.

Alkaline Salts

Large amounts of alkaline sodium salts—carbonate, bicarbonate and sulphate—occur as soil efflorescences in many districts of scanty rainfall and low humidity in North India. The principal sources are: (1) the *reh* efflorescences of many parts of Bihar and Uttar Pradesh (p. 473). The estimated potential yield of sodium salts annually available from the top layers of *reh*-infected soils of these parts is 1½ million tonnes, made up of 600,000 tonnes of sodium bicarbonate, 500,000 tonnes of carbonate and 300,000 tonnes of sodium sulphate. (2) The Sambhar, Didwana and Pachbhadra lakes of Rajasthan. The salt-bitterns of these lakes contain notable amounts of Na_2CO_3 and Na_2SO_4 in the upper layers of saline mud. At the bottoms of these lakes millions of tonnes of these two salts are held. (3) The alkaline lakes and depressions of Sind (*Dhands*); these are numerous and the amount of trona available in the larger of these depressions has been estimated by G. de P. Cotter to be about 25,000 tonnes in each. (4) The Lonar lake of Buldana district, Berar, containing alkaline mud at the bottom of the hollow with a few thousand tonnes of these salts. Though the quantity available is large, these salts have not found full industrial use. *Khari*, the crude sodium sulphate recovered from the Rajasthan lake brines and from refining of saltpetre, is employed in various chemical industries (about 20,000 tonnes yearly).[1] The average annual imports of alkali salts from abroad amount in value to Rs. 7 millions. The consumption of soda ash in India in late years has averaged 650,000 tonnes and that of caustic soda 360,000 tonnes. In 1968-69 the production of soda ash was 405,000 tonnes and that of caustic soda 304,000 tonnes.[2]

Alum

Alums are not natural but secondary products manufactured out of pyritous shales or "alum shales".

[*Production*—Pyritous shales when exposed to the air, under heat and moisture, give rise to the oxidation of the pyrites, producing iron sulphate and free sulphuric acid. The latter attacks the alumina of the shales and converts it into aluminium sulphate. On the addition of potash-salts, such as nitre or common wood-ashes, potash-alum is produced, and when common salt or other soda-salts are introduced, soda-alum is produced. In this way several alums are made, depending upon the base added.

The natural weathering of the shales being a very slow process, it is expedited in the artificial production of alum by roasting them. The roasted shale is then lixiviated and concentrated. A mixture of various soda- and potash-salts is then added and the alum allowed to crystallise out.]

---

[1] *Rec. G.S.I.* vol. lxxvii., 1, 1942.
[2] Fourth Five year Plan, p. 300.

The most common alums produced in India are soda and potash alums. There was a flourishing alum industry in the past in Kutch, Rajasthan and parts of the Punjab. But it is no longer remunerative in the face of cheap chemically manufactured alums, and is carried on only at two localities, Kalabagh[1] and Kutch. The principal use of the alum manufactured in India is in the dyeing and tanning industries.

Soluble sulphates of iron and copper—copperas and blue-vitriol —are obtained as by-products in the manufacture of alums from pyritous shales.

Asbestos[2]

Two quite different minerals are included under this name: one a variety of amphibole resembling tremolite and the other a fibrous variety of serpentine (chrysotile). Both possess much the same physical properties that make them valuable as commercial products. Asbestos (both the real mineral and chrysotile) has been discovered at many places in India, but at only a few localities is it of commercial use, *viz*. Pulivendla (Cuddapah), where excellent chrysotile asbestos occurs at the contact of a bed of Cuddapah limestone with a dolerite sill; in the Hassan district of Karnataka; in Rajasthan; and the Saraikela region of Singhbhum. Much of the latter, which is of the actinolite variety, however, does not possess the softness or flexibility of fibre on which its industrial application depends. Asbestos has found a most wonderful variety of uses in the industrial world of to-day, *viz*. in the manufacture of fire-proof cloth, rope, paper, millboard, sheeting, belt, paint, etc., and in the making of fire-proof safes, insulators, lubricants, felts, etc.

Asbestos (amphibole) occurs in pockets or small masses or veins in the gneissic and schistose rocks. The chrysotile variety forms veins in serpentine. The available supplies in India are sufficient to meet any expansion of the indigenous asbestos industry, the Tamilnadu deposits being capable of considerable development for manufacturing asbestos-cement. About 7,000 tonnes are produced from Rajasthan and Karnataka yearly. Imports of raw and manufactured asbestos total over 53 million rupees annually.

Barytes

Barytes occurs at many places in India in the form of veins and as beds in shales, in sufficient quantities, but with few exceptions the deposits were not worked till lately because of the absence of

---

[1]*Rec. G.S.I.* vol. xl. pt. 4, 1910.

[2]Coggin Brown, *Bulletin of I.I. and L.* No. 20, 1922; A.L. Coulson, Asbestos in Madras, *Mem. G.S.I.* vol. lxiv. pt. 2, 1934.

any demand for the mineral. The chief localities for barytes are Cuddapah and Kurnool[1] districts; the Alwar region; Salem; and Sleemanabad (in Jabalpur district). Barytes is used as a pigment for mixing with white lead, as a flux in the smelting of iron and manganese, in paper-manufacture, in pottery-glazes, etc. The whiter and better-quality barytes is used in the local manufacture of paints (lithophone); the coloured variety is used in making heavy drilling mud by the oil companies. The yearly output of recent years is about 50–60,000 tonnes, valued at Rs. 11,00,000.

Bauxite

Besides its use as the principal ore of aluminium, bauxite is mined for various industrial purposes—manufacture of chemicals, abrasives, refractories, cement and in the refining of petroleum. Annual production for these uses is about 800,000 tonnes. Available supplies are large; the better grade of bauxite averages 55% alumina, 28% combined water, 8% titania, 6% ferric oxide, and below 3% silica.

Borax

Borax from Tibet—Borax occurs as a precipitate from the hot springs of the Puga valley, Ladakh, which occur in association with some sulphur deposits. Borax is an ingredient of many of the salt-lakes of Tibet, along with the other salts of sodium. The borax of the Tibetan lakes is obtained either by means of diggings, on the shores of the lakes, or by the evaporation of their waters. The original source of the borax in these lakes is thought to be the hot springs, like those of Puga mentioned above.

Like the nitre, alum and similar trades, the borax trade, which was formerly a large and remunerative one, has seriously declined owing to the discovery of deposits of calcium borate in America, from which the compound is now synthetically prepared. The industry consisted of the importation of partly refined borax, about 9,000 quintals, from Ladakh and Tibet and its exportation to foreign countries. There is a project to refine the large reserves of crude borax in the Puga valley lakes locally and to transport the product by air-craft to the industrial centres in India. Borax is of use in the manufacture of superior grades of glass, artificial gems, soaps, varnishes and in soldering and enamelling.

Corundum and other Abrasives[2]

Occurrence. Distribution—Corundum is an original constituent of a number of igneous rocks of acid or basic composition, whether plutonic or volcanic. It generally occurs in masses, crystals, or

---

[1] A. L. Coulson, *Mem. G.S.I.* vol. lxiv. pt. 1, 1933.
[2] T. H. Holland, Corundum, *G.S.I.*, 1898; *Rec. G.S.I.*, vol. lxxvi-12, 1942.

irregular grains in pegmatites, granites, diorites, basalts, peridotites, etc. The presence of corundum under such conditions is regarded as due to an excess of the base Al_2O_3 in the original magma, over and above its proper proportion to form the usual varieties of aluminous silicates.[1] India possesses large resources in this useful mineral, which are, for the most part, concentrated in Karnataka and Tamilnadu. Other localities are Singhbhum; Rewah (Pipra), where a bed of corundum 732 m. long, 64 m. wide and 9 m. thick is found; the Mogok district (Ruby Mines district) in Upper Burma; Assam (Khasi hills); some parts of Bihar; the Zanskar range in Kashmir, etc. In Burma the famous ruby-limestone contains a notable quantity of corundum as an essential constituent of the rock, some of which has crystallised into the transparent varieties of the mineral, ruby and sapphire. In Tamilnadu there is a large area of corundiferous rocks covering some parts of Tiruchirapalli, Nellore, Salem and Coimbatore. Mostly the corundum occurs *in situ* in the coarse-grained gneisses, in small round grains or in large crystals measuring some inches in size. It also forms a constituent of the elaeolite-syenites of Sivamalai and of the coarse felspar-rock of Coimbatore.

Uses—The chief use of corundum is as an abrasive material because of its great hardness. Emery is an impure variety of corundum, mixed with iron-ores and adulterated with spinel, garnet, etc. The abrading power of emery is much less than that of corundum, while that of corundum again is far below that of the crystallised variety sapphire. As an abrasive, corundum has now many rivals in such artificial products as *carborundum*, *alundum*, etc. Corundum is used in the form of hones, wheels, powder, etc. by the lapidaries for cutting and polishing gems, glass, etc.

The total annual production in India is fitful, averaging 450 tonnes, valued at about Rs. 2,50,000.

Other abrasives. Millstones—While dealing with abrasives, we may also consider here the materials suitable for millstones and grindstones that are raised in India. Massive garnet and garnet-sand occur in many parts of India in sufficient quantity to be used as an abrasive. Flint has a hardness almost equal to garnet and is used in abrasive cloth and paper, and as flint-pebbles in grinding mills. Suitable material for abrasives occurs in most provinces. Quartz-sand and quartzites, of universal occurrence in geological formations, are used in sand-blasting, glass-surfacing and for burnishing. Fused alumina or bauxite is a hard abrasive fit for grinding steel or other metals. A number of varieties of stones

---

[1] In the above instances corundum occurs as an original constituent of the magma, but the mineral also occurs in many cases as a secondary product in the zones of contact-metamorphism around plutonic intrusions.

are quarried for cutting into millstones, though the rocks that are the most suitable for this purpose are hard coarse grits or quartzites. There is a scarcity of such rocks in most parts of the country, and hence the stones commonly resorted to are granites, hard gritty Vindhyan sandstones, and Gondwana grits and sand-stones chiefly of the Barakar stage.[1]

Grindstones—Grindstones, or honestones, are cut from any homogeneous close-grained rocks belonging to one or another of the following varieties: fine sandstones, lydite, novaculite, hornstone, fine-grained lava, slate, etc.

Fluorspar

This mineral is of restricted occurrence in India. Veins of fluorite occur in the rocks of some parts of Peninsula and the Himalayas, in the Vindhyan limestone in Rewah, in granite in Simla Himalayas; but quite substantial deposits have been located in carbonatite in the *Bagh* beds and Deccan Trap. Fluorite (estimated reserves of 10.5 million tonnes) is found in Gujarat and parts of Rajasthan. The chief use of fluorspar is as a flux in the manufacture of steel and aluminium and synthetic cryolite. Annual production : 1,500 to 1,800 tonnes.

Graphite

Occurrence—Graphite occurs in small quantities in the crystalline and metamorphic rocks of various parts of the Peninsula, in pegmatite and other veins, and as lenticular masses in some schists and gneisses. It forms an essential constituent of the rock known as khondalite in Orissa, *i.e.* a quartz-sillimanite-garnet-graphite-schist. But the majority of these deposits are not of workable dimensions. Graphite occurring under such conditions is undoubtedly of igneous origin, *i.e.* a primitive constituent of the magma or more probably a product of interaction of magmatic gases from igneous bodies (charnockites) with Khondalite limestones.[2] Graphite resulting from the metamorphism of carbonaceous strata, and representing the last stage of the mineralisation of vegetable matter, is practically unknown in India, except locally in the highly crushed Gondwana beds of the outer Himalayas. The largest deposits of graphite are in Sri Lanka, which has in the past supplied large quantities of this mineral to the world, its yearly contribution being nearly a third of the world's total annual produce. The graphite here occurs as filling veins in the granulites of the Khondalite series. The structure of the veins is often colum-

---

[1]M. R. Sahni, *Prof. Pap. G.S.I.* vol. lxxvi., 12, 1942.

[2]D. N. Wadia, Age, and Origin of Graphite Deposits of Ceylon, *Prof. Pap., Rec. Dept. Min.*, Ceylon, Colombo, 1943.

nar, the columns lying transversely to the veins. Kerala until lately was another important centre for graphite-mining, supplying annually about 13,000 tonnes of the mineral (valued at Rs. 7,80,000). The graphite industry has practically ceased in Kerala of late years owing to the increasing depths to which mining operations have become necessary.

A few other localities are known among the ancient crystalline rocks, *viz*. a few localities in Orissa, Bihar, Rajasthan and Sikkim, but the quantity produced so far is not large. The present output of graphite is about 3,000 tonnes per annum. After beneficiation it is found to be useful for crucible-making.

Uses—The uses of graphite lie in its refractoriness and in its high heat conductivity. For this reason it is largely employed in the manufacture of crucibles. Synthetic graphite is now largely employed as a moderator in atomic reactors for generating power. Its other uses are for pencil manufacture, as a lubricant, in electrotyping, etc.

Gypsum

Gypsum forms large bedded masses or aggregates occurring in association with rocks of a number of different geological formations. Large deposits of pure gypsum occur in the Salt-Range and Kohat in association with rock-salt deposits of Pakistan and in the Tertiary clays and shales of Rajasthan, Kutch and Tamilnadu, though in less pure state. In Jodhpur, Nagpur and Bikaner, beds of gypsum are found among the silts of old lacustrine deposits and are of considerable economic interest. Millions of tonnes of gypsum, the alteration-product of pyritous limestone of Salkhala age, are laid bare in the mountains of the Uri and Baramula area of Kashmir in a stretch of about 40 km. along the strike. In Spiti, Sirmur, Kumaon and other Himalayan areas, the gypsum occurs in large masses replacing Carboniferous or other limestones. In some cases gypsum occurs as transparent crystals (*selenite*) associated with clays. The handsome massive and granular variety, known as *alabaster*, is used in Europe for statuary, while the silky fibrous variety, known as *satin-spar*, is employed in making small ornamental articles.

The industrial use of gypsum is in the manufacture of synthetic fertilisers, plasters, distempers and in the cement industry. In America it is increasingly used for fire-proofing wallboards as a building material. It is also used as a surface-dressing for lands in agriculture, and as a fertiliser, with considerable benefit to certain crops. Gypsum has begun to be used as a source of sulphur in the manufacture of fertilisers. Available gypsum reserves are: Pakistan, 200 million tonnes; Indian Union, 1,100 million tonnes in accessible localities; large reserves in the Himalayas. Annual production in recent years has been around 1.3 million tonnes.

Kyanite and Sillimanite

These aluminous silicates, owing to their possessing certain valuable properties as refractories at high temperatures, especially in the manufacture of ceramics and glass, have come into prominence of late years. India possesses considerable resources in both these minerals. Kyanite occurs mainly in Singhbhum as kyanite-quartz rock and as massive kyanite-rock in beds of enormous size in the Archaean schists; sillimanite occurs also in the same rock-system in the Rewah area (Pipra village) in beach sands and in Assam. Important workable deposits are found in Kharsawan and other localities in Orissa. Total reserves are computed at over a million tonnes so far. Corundum occurs with these in close relationship, forming a group of highly aluminous schists and gneisses. A high degree of purity, with percentages of aluminium silicate reaching 95 to 97, characterises both these minerals from Rewah, Assam and Singhbhum. Refractories manufactured in India from kyanite are of the order of 80,000 tonnes.[1]

Limestone

Besides their uses as building stones, and as lime and cement raw material, limestones, if of the required purity, have important uses in the chemical, alkaline, sugar and metallurgical industries. Slaked lime is an alkali and is an essential raw material in some chemical industries. Chemically pure limestones, containing over 96 per cent $CaCO_3$ or free from harmful proportions of MgO, silica or iron, can be obtained from Katni, Maihar, Rewah, Bisra, Khasi hills, Jodhpur, Bikaner, Wardha and Chanda in N. India, and several localities in Andhra and Tamilnadu.[2] Pakistan's resources in pure limestone are very large, distributed over Sind, N. W. F. Province, Salt-Range and Hazara.

For use as flux in iron smelting, pure limestone, or in its absence dolomitic limestone, is in demand.

Dolomite—Limestones with more than 10 per cent $MgCO_3$ are called dolomitic; when the percentage rises to 45, they are true dolomites (CaO 30.4 per cent; MgO 21.7 per cent). Both dolomitic limestones and true dolomites are fairly widely distributed in India and in the Himalayas, from which supplies are readily available. Economic uses of dolomite in India are chiefly metallurgical, as refractories (dead-burned dolomite is used in iron, lead and copper smelting furnaces); as blast-furnace flux; as a source of CO_2 gas and magnesium salts; as lime-mortars and other minor uses.

---

[1] J. A. Dunn, *Mem. G.S.I.* vol. lii. pt. 2, 1929; and *Mem. G.S.I.* vol. lxix., 1937.
[2] High-calcium limestones of India, C.S.I.R. Publication, New Delhi, 1957.

Mica

Mica-deposits—India is the largest producer of mica in the world, contributing, of late years, more than 75 per cent of the world's requirements. It appears likely that, despite the threat of synthetic mica, certain grades of Indian mica will remain vital to the world's electrical industries. The exports of mica during the post-war years have fluctuated in quantity from an average of 86,000 quintals (of block and splittings) to about 305,000 quintals, and in 1959 the value realised was over Rs. 120 million.

The mica-deposits of the Indian peninsula are considered to be the finest in the world, because of the large size and perfection of the crystal plates obtainable at several places. This quality of mica is due to the immunity from all disturbances such as crumpling, shearing, etc. of the parent rocks. Crystals more than a metre in diameter are obtained occasionally from the Nellore mines, from coarse pegmatite veins traversing Archaean schists and gneisses, from which valuable flawless sheets of great thinness and transparency are cloven off.

Uses of Mica—Mica (muscovite) finds uses in many industries, and is a valuable article of trade. The chief use is as an insulating material in electrical goods; another is as a substitute for glass in glazing and many other purposes. As a glass substitute, however, only large transparent sheets are suitable. Formerly an enormous amount of *scrap-mica* (small pieces of flakes of mica), the waste of mica-mines and quarries, was considered valueless and was thrown away. A use has now been found for this substance in the making of *micanite*—mica-boards—by cementing small bits of scrap mica under pressure. Micanite is now employed for many purposes in which sheet mica was formerly used. Scrap mica is also ground for making paints, lubricants, etc.

Scientific mining methods and mechanisation of mines with increasing depth of the pegmatites, and rigid control of standards of dressing and grading of finished mica, together with local processing of part at least of split and block mica, are urgently needed reforms.[1] The use of ground mica for firebricks, paints, etc. from scrap and the manufacture of micanite from the waste of mines will make this important mineral industry of India more profitable.

Most of the output of the mines is exported, indigenous industry to absorb any part of the produce, or for the manufacture of micanite being yet in infancy. Although muscovite is a most widely distributed mineral in the crystalline rocks of India, marketable mica is restricted to a few pegmatite-veins only, carrying large

---

[1] C. M. Rajgarhia, *Mining, Processing and Uses of Indian Mica* (McGraw-Hill), 1951.

perfect crystals, free from wrinkling or foreign inclusions. These pegmatite veins cross the Archaean and Dharwar crystalline rocks, granites, gneisses and schists, but they become the carriers of good mica only when they cut through mica-schists. The principal mica-mining centres in India are the Hazaribagh, Gaya and Monghyr districts of Bihar, the Nellore district of Tamilnadu, and Ajmer and Mewar in Rajasthan. Of these Bihar is the largest producer, 74 per cent, Tamilnadu and Rajasthan contributing the rest.[1] The dark-coloured mica, biotite, has no commercial use, but phlogopite—amber-mica—occurring in Tamilnadu and Sri Lanka has industrial uses as a heat and electrical insulator.

Lepidolite, lithia mica, the source of lithium oxide, occurs in pegmatite veins in Rajasthan, Hazaribagh and in lenses 275 to 375 m. in length in the Bastar region of Madhya Pradesh; they contain over 2 per cent of lithium oxide. The mineral is of use in the chemical, glass and porcelain industries and is of use in experiments on atomic fusion.

Mineral Paints

Substances used for mineral paints—A number of rock and mineral substances are employed in the manufacture of paints and colouring materials in Europe and America. Substances which are suitable for this purpose include earthy forms of haematite and limonite (ochres, *geru*); refuse of slate and shale quarries, possessing the proper colour and degree of fineness; graphite; laterite; orpiment; barytes; asbestos; mica; steatite; etc. Many of the above substances are easily available in various parts of India and some are actually utilised for paints and pigments, *viz*. a black slate for making black paints; laterite and *geru* (red or yellow levigated ochre) for red, yellow or brown colouring matters; barytes as a substitute for white lead; orpiment for yellow and red colours in lacquer work.[2] Large quantities of red and yellow ochre in association with graphite-bearing slate occur in the Salkhala system of deposits in the Uri Tehsil of Kashmir State.[3]

Monazite

Monazite is a phosphate of the rare earths, cerium, yttrium, lanthanum, didymium, etc., with a variable percentage of thorium and uranium oxides, which it contains as accessory. It is a portmanteau compound carrying some 15 rare-earth oxides. Monazite

---

[1] Mica Deposits of India, *Mem. G.S.I.* vol. xxiv. pt. 2, 1902; *Bull. of I. I. and L.* No. 15 ; *Rec. G.S.I.* vol. lxxvi., 10, 1942.

[2] Coggin Brown, *Bulletin of I.I. and L.* No. 20, 1922.

[3] C. S. Middlemiss, *Mineral Surv. Rep. J. and K. State* (Graphite and Ochre), Jammu, 1926.

occurs in the ilmenite sand of the beaches, brought there by rivers draining the hinterland and concentrated by wave action on the littoral. It also occurs inland as residual placer deposits on peneplaned surface of some Bihar plateaus. It is found over long stretches of the coast-line both on the Malabar and Coromandel coasts. On the beaches, wave action has concentrated these heavy-mineral sands into rich placers.

The monazite is derived from the pegmatite-veins crossing the charnockites and allied rocks of the high ground. It origin is ascribed to pneumatolytic agencies during the later period of consolidation of the igneous magma. Monazite-bearing pegmatites, carrying associated zircon, rutile, ilmenite and garnet are widely distributed through the charnockite provinces of the Deccan shield—Kerala, Nilgiri, E. Ghats and Hazaribagh. Monazite is also a small accessory constituent of the main granitic and gneissic rocks of the area. The percentage of thoria in monazite varies from 8 to 10.5; about 0.3 per cent of U_3O_8 is associated with it. In the newly discovered mineral *cheralite* (p. 451), found in some pegmatites in Kerala, it ranges from 19 to 33 per cent, with 4 to 5 per cent of U_3O_8. The present annual output of monazite is 1,500 to 2,030 tonnes, mainly employed in the manufacture of cerium compounds, the thorium and uranium being recovered for use as atomic fuel. The total Indian reserves of monazite are now estimated at over 5 million tonnes.

Other industrial uses of monazite are in the incandescent properties of thoria, in the oxide of the rare earths, and in alloys with magnesium metal.

Phosphatic Deposits[1]

Native phosphates, as apatite, or rock-phosphates, as concretions, are highly valued now as artificial fertilisers of manures, either in the raw condition or after treatment with sulphuric acid to convert them into acid or superphosphates. The main occurrences of phosphatic deposits in India on a workable scale are: lime phosphate septarian nodules in clay beds associated with the Cretaceous of Trichinopoly (P_2O_5—15 to 20 per cent), total quantity available 2,030,000 tonnes; a 2-metre thick deposit near Mussoorie, overlying the Krol series limestones, extending for many kilometres and another near Udaipur, Jaisalmer (Rajasthan) with reserves estimated at 20 million tonnes; and massive apatite occurring as a constituent of Dharwar rocks of Dalbhum in Singhbhum, Bihar, where the quantity available is estimated at one million tonnes. A low-grade guano deposit of recent origin is found on some of

---

[1] Sir Edwin Pascoe, India's Resources in Mineral Fertilisers, *Bull. I . I. and L.* No. 42, 1929 ; *Rec. G.S.I.* vol. lxxvi., 4, 1941.

the Laccadive group of coral islands. Preliminary estimate of reserves is some 3 million tonnes yielding 16 per cent P_2O_5, capable of use as a direct fertiliser.

A source of phosphatic material for use as a mineral fertiliser exists in the basic slag formed in the manufacture of steel. Over 50,000 tonnes of this slag (P_2O_5—10 per cent) are being dumped annually at the steel works for want of any present demand.

Superphosphate manure is manufactured in India from rock-phosphate (imported from Morocco and Egypt) and bone-meal to the extent of half a million tonnes per annum. The consumption of chemical fertilisers, however, is steadily rising, and the manufacture of synthetic ammonium sulphate and other artificial nitrogenous and phosphatic fertilizers from the indigenous petro-chemical products is likely to meet the rising demand of Indian agriculture.

Pyrite

Pyrite is a mineral of very wide distribution in many formations, from the oldest crystalline rocks to the youngest sediments, but nowhere is it sufficiently abundant to be of commercial utility in the preparation of sulphur and sulphuric acid, The economic value of pyrite[1] lies in its being a source of sulphur. The occurrences on any considerable scale are those of the pyritous shales deposits lately found in the lower Son valley, Amjor, Bihar, where reserves are computed at 384 million tonnes of pyrites, with average sulphur content of 40 per cent. Others are: Chitaldrug district of Karnataka, and Saladipura in Rajasthan. No attempt was made to develop the elemental sulphur from these deposits because of the cheapness of imported sulphur. With falling supplies from this source, the local manufacture of sulphur and sulphuric acid from the large Amjor deposits is being undertaken, an annual output of 100,000 tonnes being planned. Stores of sulphur exist in connection with metallic sulphides, notably of zinc, copper and lead, which, when worked for the recovery of the metals, will liberate the sulphur as well in large amounts.

Rare Minerals

The rare minerals of India—The pegmatite veins of the crystalline rocks of India contain a few of what are called the rare minerals as their accessory constituents. The rare elements contained in them have found use in modern industries such as electronics, high-grade refractories, the manufacture of special steels, alloys and other products of highly specialised uses in the present-day industries.[2]

---

[1] Fox, *Bulletins of* I. I. *and* L. No. 28, 1922.
[2] Cahen and Wootton, *Mineralogy of the Rarer Metals*, 1912 (C. Griffin).

The most common of these are wolfram, beryl, pitchblende and monazite, a compound of some 15 rare earths, which have been already dealt with; columbite and tantalite (niobates and tantalates of the rare earths), torbernite, clarkeite, hatchettolite, brannerite, annerodite, aeschynite, cheralite, allanite and triplite, which occur in the mica-pegmatites of Hazaribagh, Nellore, and in Kerala and Rajasthan; samarskite, fergusonite and other allied rare minerals, which occur also in these areas; gadolinite (a silicate of the yttrium earths), in a tourmaline-pegmatite associated with cassiterite in Palanpur; and molybdenite, in the crystalline rocks of Chhota Nagpur, Singhbhum, Madurai and in the elaeolite-syenite-pegmatite of Rajasthan and of Kerala. Thorianite has been found in Kerala and Sri Lanka, containing from 60-80 per cent of thoria, uranium (10-30 per cent) and helium. Uraniferous allanite occurs in the pegmatites of Nellore, with sipylite, a niobate of erbium with other rare earths.

Zircon is found, with baddeleyite, as residual grains in ilmenite sands in large amounts (over ten million tonnes), and less commonly with uranium minerals and with triplite in the mica-mines of Gaya and in the nepheline-syenites of Coimbatore. Cyrtolite is a radioactive variety found in some of these localities. Zirconium metal is growing in importance as an alloy metal.

Platinum and iridium occur as rare constituents of the auriferous gravels of some parts of Burma.

Reh or Kalar

The origin of *reh* **salts**—*Reh*, *Usar*, or *Kalar* are the vernacular names of a saline efflorescence composed of a mixture of sodium carbonate, sulphate and chloride, together with varying proportions of calcium and magnesium salts, found on the surface of alluvial soils in the drier districts of the Gangetic plains. At the present day *Reh* is not an economic product, but it is described here because of its negative virtues as such. Some soils are so much impregnated with these salts that they are rendered quite unfit for cultivation. Large tracts of the country, particularly the northern parts of Uttar Pradesh, Punjab and Rajasthan, once fertile and populous, are through its agency thrown out of cultivation and made quite desolate. The cause of this impregnation of the salts in the soil and subsoil is that the rivers draining the mountains carry with them a certain proportion of chemically dissolved matter, besides that held in mechanical suspension, in their waters. The salts so carried are chiefly the carbonates of calcium and magnesium and their sulphates, together with some quantity of sodium chloride, etc. In the plains-track of the rivers, these salts find their way, by percolation, into the subsoil, saturating it up to a certain level. In many parts of the hot alluvial plains, which have got no underground drainage of water, the salts go on accumulating and in course of time become concentrated, forming new combinations by interaction between previously existing salts. Rain water, percolating downwards, dissolves the more soluble of these salts and brings them back to the surface during the summer months by capillary action, where they form a white efflorescent crust. The reclaiming of these barren *kalar* lands into cultivable soils by the removal of these salts would add millions of hectares to the agricultural area of India, and bring back under cultivation what are now altogether sterile uninhabited districts.

The carbonate and sulphate of sodium, the chief constituents of *Reh*, were formerly used as a source of salts of alkalis, and were produced in some quantity for local industry. Their production on an industrial scale for utilisation in the expanding chemical industries of the country is being explored.[1] (See p. 483-484).

Salt

Sources of salt.[2] **Sea-water. Brine-wells**—There are three sources of production of this useful material in India: (1) sea-water, along the coasts of the Peninsula; (2) brine-springs, wells and salt-lakes of some arid tracts, as of Rajasthan and Uttar Pradesh; (3) rock-salt deposits contained in Kutch and Mandi Himachal Pradesh. The average annual production of salt from these sources is rather over eight million tonnes, the whole of which is consumed in the country. The first is the most productive and an everlasting source, which contributes about 75 per cent of the salt consumed in India. The manufacture is carried on at some places along the coasts of Maharashtra, Gujarat and Tamilnadu, the process being mere solar evaporation of the sea-water enclosed in artificial pools or natural lagoons. A solid pan of salt results, which is afterwards refined by recrystallisation. Concentration from brine springs and wells is carried on in various parts of Uttar Pradesh, Bihar, the delta of the Indus, Kutch and in Rajasthan. The principal sources of salt in the last-named province are the salt-lakes of Sambhar in the Jaipur region,[3] Didwana and Phalodi in the Jodhpur region, and Lonkara-Sur in the Bikaner region. The salinity of the lakes in this area of internal drainage has been for long a matter of conjecture, whether it is of local origin, or is due to constant dropping of wind-borne salt as dust from the coast, or from the Rann of Kutch (p. 31). The Sambhar lake is however known to have a large

---

[1] *Rec. G.S.I.* vol. lxxvii., *Prof. Pap.* 1, 1942.
[2] K. H. Vakil, Salt, its sources and supply in India, Bombay, 1945.
[3] H. B. Dunnicliff, *Journ. Ind. Chem. Soc.* vol. viii. 1, and 2, 1945.

reserve of salt in its bottom layers of mud and silt. Up to a depth of 4 m. the reserves are computed at 50 million tonnes at least.

Rock-salt mines—The rock-salt deposits of the Salt-Range mountains and Kohat in Pakistan constitute an immense source of pure crystallised sodium chloride. At Khewra, in the Jhelum district, two beds of rock-salt 170 m. thick are worked; they contain five seams of pure salt totalling 85 m., intercalated with a few earthy or impure layers unfit for direct consumption. The horizontal extension of these beds or lenticles is not known definitely, but it is thought to be some kilometres. Smaller salt-mines are situated at some other places along the Salt-Range. A salt deposit of even greater vertical extent than that worked at Khewra is laid bare by the denudation of an anticline in the Kohat district, north-west of the Salt-Range. Here the salt is taken out by open quarrying in the salt-beds at the centre of the anticline near Bahadur Khel. The thickness of the beds is 300 m. and their lateral extent 13 km. The salt is nearly pure crystallised sodium chloride, with a distinct greyish tint owing to slight bituminous admixture. Salt-beds of considerable size occur in Mandi (Himachal), while some millions of tonnes of pure rock-salt, produced by evaporation of sea-water in enclosed basins, occur embedded in the sands of the Rann of Kutch and in the alluvial tract south-east of Sind.

The average annual amount of rock-salt extracted from the mines in the Salt-Range and Kohat is about 203,000 tonnes. The Mandi salt mines only produce some 6,100 tonnes yearly. The total requirement of salt, domestic and industrial, in India, roughly 10,000,000 tonnes per year, is met by manufactured sea-salt, 5,800,000 tonnes, brines and Rajasthan lake salt and rock-salt, the deficit being met by imports. In 1963 India manufactured 9,750,000 tonnes of salt from the above sources.

Other Salts—The Salt-Range deposits contain, besides sodium chloride, some salts of magnesium and potassium. The latter salts are of importance for their use in agriculture and some industries. Numerous seams of potash-bearing minerals (containing a potassium percentage from 6 to 14 per cent), such as *sylvite*, *kainite*, *langbeinite*, etc., have been found, generally underlying the layers of red earthy salt (*kalar*).[1]

Saltpetre or Nitre (Potassium Nitrate)[2]

India, principally the province of Bihar, used to export this compound in very large amounts before the introduction of arti-

---

[1] *Rec. G.S.I.* vol. xliv. pt. 4, 1914.

[2] Hutchinson, Saltpetre, its Origin and Extraction in India, *Bulletin* 68 (1917), Agricultural Department of India.

ficially manufactured nitrate, and constituted a very important source of supply to Europe and the United States.

Mode of occurrence of nitre—Saltpetre is a natural product formed in the soil of the alluvial districts by natural processes under the peculiar conditions of climate prevailing in those districts. The thickly populated agricultural province of Bihar, with its alternately warm and humid climate, offers the most favourable conditions for the accumulation of this salt in the subsoil. The large quantities of animal and vegetable refuse gathered round the agricultural villages of Bihar are decomposed into ammonia and other nitrogenous substances; these are acted upon by certain kinds of bacteria (*nitrifying* bacteria) in the damp hot weather, with the result that at first nitrous and then nitric acid is produced in the soil. This nitric acid readily acts upon the salts of potassium with which the soil of the villages is impregnated on account of the large quantities of wood and dung ashes constantly being heaped by villagers around their habitations. The nitrate of potassium thus produced is dissolved by rain-water and accumulated in the subsoil, from which the salt re-ascends to the surface by capillary action in the period of desiccation following the rainy weather. Large quantities of nitre are thus left as a saline efflorescence on the surface of the soil along with some other salts, such as chloride of sodium and carbonate of sodium.

Its production—The efflorescence is collected from the soil, lixiviated and evaporated, and the nitre separated by fractional crystallisation. It is then sent to the refineries for further purification. In past years Bihar alone used to produce more than 20,000 tonnes of nitre per year, value Rs. 10 million. The present export of refined nitre from Bihar, Punjab and other parts of India is insignificant.

Uses—The chief use for nitre or saltpetre was in the manufacture of gunpowder and explosives, before the discoveries of modern chemistry brought into use other compounds for these purposes. Nitre is employed in the manufacture of sulphuric acid and as an oxidiser in numerous chemical processes. A subordinate use of nitre in India is as manure for the soil.

Steatite

Mode of origin of steatite—Massive, more or less impure, talc is put to a number of minor uses. From its smooth, uniform texture and soapy feel, it is called soapstone. It is also known as potstone from its being carved into plates, bowls, pots, etc. Steatite is of wide occurrence in India, forming large masses in the Archaean and Dharwar rocks of the Peninsula and Burma; workable deposits occur in Bihar, Jabalpur, Salem, Idar and Jaipur (Rajasthan). The Rajasthan deposits carry the mineral in thick lenticular beds of wide extent in the schists. Some of these beds persist for kilometres. At most of these places steatite is quarried in considerable quantities for commercial purposes. In its geological relations, steatite is often associated with dolomite (as in Jabalpur) and other magnesian rocks, and it is probable that it is derived from these rocks by metamorphic processes resulting in the conversion of the magnesium carbonate into the hydrated silicate. In other cases it is the final product of the alteration of ultra-basic and basic eruptive rocks. At Jabalpur and other places it is carved into bowls, plates and vases; it is also used in soap-making, toilet powder, paints, in pencils, in the paper industry, and as a refractory substance in making jets for gas-burners. The substance has also of late come into use as a special type of refractory, resistant to corrosive slags, and as a paint of high quality for protecting steel. The reserves of good-quality steatite in Jaipur and Jabalpur are believed to be large. The annual production at present is about 152,000 tonnes, with a value of about Rs. 45 lakhs.

Sulphur

Sulphur in small quantities is obtainable as a sublimation product from the crater of Barren Island volcano, and from some of the extinct volcanoes of Western Baluchistan. A reserve of about 254,000 tonnes exists at these localities. Sulphur occurs in the Puga valley of Ladakh, found there as a deposit from its hot springs.

These sources are, however, too insignificant to meet the demand for sulphur in the country which is satisfied largely by imports from foreign countries, amounting to 638,000 tonnes a year. The late discovery of over 355 million tons of pyrite (40 per cent sulphur) in the pre-Cambrian of the Son valley in Bihar, near Amjor, and 115 to 117 million tonnes at Saladipura in Rajasthan has changed the situation and the local production of elemental sulphur and sulphuric acid from this source is projected to supply the major part of the country's requirement in this vital commodity. The quantity of sulphuric acid used annually in India is over 1,500,000 tonnes, which is only a part of the quantity estimated to meet the demands of the growing fertiliser, petroleum-refining, chemical and metallurgical industries.

Sulphuric acid—Sulphur has many important uses, much the most important being the manufacture of sulphuric acid. With regard to this compound we may quote the following valuable statement which was made in 1915 by Sir Thomas Holland, but which is materially true to-day. "Sulphuric acid is a key to most chemical and many metallurgical industries; it is essential for the manufacture of superphosphates, the purification of mineral oils, and the production of ammonium sulphate, various acids and a host of minor products; it is a necessary link in the chain of operations involved in the manufacture of alkalis, with which are bound up the industries of making soap, glass, paper, oils, dyes, and colouring matter; and, as a by-product, it permits the remunerative smelting of ores which it would be impossible otherwise to develop. During the last hundred years the cost of a ton of sulphuric acid in England has been reduced from over £30 to under £2, and it is in consequence of the attendant revolution in Europe of chemical industries, aided by increased facilities for transport, that in India the manufactures of alum, copperas, blue vitriol and alkalis have been all but exterminated; that the export trade in nitre has been reduced instead of developed; that the copper and several other metals are no longer smelted; that the country is robbed every year of over 90,000 tons of phosphate fertilisers, and that it is compelled to pay over 20 millions sterling for products obtained in Europe from minerals identical with those lying idle in India."[1]

The present capacity of the country for the manufacture of sulphuric acid from native sulphur, gypsum and sulphide, lead-, zinc- and copper-ores is about 1,240,000 tonnes per year in about 45 units.

9. SOILS

Soil formation—The soils of all countries are, humanly speaking, the most valuable part of the regolith or surface rocks and constitute in many cases their greatest natural asset. They are broadly speaking, either the altered residue of the underlying rocks, after the soluble constituents have been removed, mingled with some proportion of decomposed organic matter (*residual soil*); or the soil-cap may be due to the deposition of alluvial débris brought down by the rivers from the higher grounds (*drift soil*). The origin and growth of soils, however, is a subject of great complexity involving a long series of changes ending in the production of the clay-factor and other colloids of the soils. The soil of the Peninsula, for the greater part, is of the first description, while the great alluvial mantle of North India, constituting the largest part of the most fertile soil of India, is of the second class. We can easily imagine that in the production of soils of the first kind, besides the usual meteoric agencies, the peculiar monsoonic conditions of India, giving rise to alternating humidity and desiccation, must have had a large share. These *residual* soils of the Peninsula show a great variety both in their texture and in their mineralogical composition, according to the nature of the subjacent rock whose waste has given rise to them. They also exhibit a great deal of variation in depth, consistency, colour, etc. However, the soils of India, so far as their geological peculiarities are concerned, show far less regional varia-

[1] *Rec. G.S.I.* vol. xlvi., 1915, p. 295.

tion than those in other countries, because of the want of variety in the geological formations of India.[1]

Broadly speaking the soils of the Indian Peninsula differ markedly from the soils of European countries, which are largely of post-glacial growth and in which the pedogenic processes have not been in operation long enough to mature them. The latter soils have close affinities with their rocky substratum, both as regards composition and morphology. Podsolisation is a common character of these soils. In both these respects the soils of Peninsular India offer a contrast and, being far older than the Glacial Period of Pleistocene age, have attained full maturity. The effect of these factors is to introduce many changes in the composition, structure and texture, and to modify profoundly the clay-factor of the soils. This is best seen in the two characteristic Indian soils—laterite and black-cotton soil. Podsols, except among some mountain and forest soils of North India, are uncommon in the rest of the country. The alluvial soils of the vast Indo-Gangetic plains likewise differ from Peninsular soils, and from the majority of European soils, in having undergone but little pedogenic evolution since their deposition by river agency so late as in sub-Recent times. They are still largely immature and have not developed any characteristic soil profile, or differentiation into zones.

The soils of South India—Over the large areas of metamorphic rocks the disintegration of the gneisses and schists has yielded a shallow sandy or stony soil, whereas that due to the decomposition of the basalts of the Deccan, in the low-lying parts of the country, is a highly argillaceous, dark loamy soil. This soil contains, besides the ordinary ingredients of arable soils, small quantities of the carbonates of calcium and magnesium, potash, together with traces of phosphates, ingredients which constitute the chief material of plant-food that is absorbed by their roots. The Deccan soil is, therefore, much more fertile as a rule than that yielded from the metamorphic rocks, which is thin and shallow in general (except where it has accumulated in the valley-basins), because of the slowness with which the gneisses and schists weather. The soil in the valleys is good, because the rain moves the decomposed rock-particles and gathers them in the hollows. In these situations of the crystalline tract the soils are rich clay-loams of great productiveness.

Soils of sedimentary rocks—The soils yielded by the weathering of the sedimentary rocks depend upon the composition of the latter, whether they be argillaceous, arenaceous or calcareous, and upon their impurities. Soils capping the Gondwana outcrops are

---

[1] The Geological Foundations of the Soils of India, *Rec. G.S.I.*, vol. lxviii. pt. 4, 1935.

in general poor and infertile, because Gondwana rocks are coarse sandstones and grits with but little cementing material. They are thin sandy soils, capable of supporting tillage only with copious manuring. Argillaceous and impure calcareous rocks yield good arable soils. Reference must here be made to the remarkable black soil, or *regur*, of large areas of the Deccan which has already been described on page 386. The greater parts of Rajasthan, Baluchistan and the Frontier Provinces are devoid of soils, because the conditions requisite for the growth of soils are altogether absent there. The place of soil is taken by another form of regolith, *e.g.* widespread scree and talus-slopes, colluvial gravels, blown sand and loess. In the Himalayan region soil-formation is a comparatively rapid process, the damp evergreen forests playing an important part in the generation and conservation of the soil-cap. The unforested southern slopes of these mountains are generally devoid of soil covers. Likewise deforestation of some tracts of the outer Himalayas has been followed by a stripping of their soil-cover, due to accelerated erosion of the unprotected surface.[1]

Alluvial soils—The alluvial soils of the great plains of North India, as also those of the broad basins of the Peninsular rivers, are of the greatest value agriculturally. They show minor variations in density, colour, texture, porosity, and moisture-content and in the composition of their clay-factor. In spite of minor differences in composition from district to district, in general they are light-coloured loamy soils of a high degree of productiveness, except where it is destroyed by the injurious *reh* salts. The loess caps of the higher parts of the Punjab possess many of the qualities of an excellent soil, but the high porosity tends to lower the underground water-table to inaccessible depths. There are, however, a number of physical and organic factors which determine the characters and peculiarities of soils and their fertility or otherwise; this subject is, however, beyond the scope of this book and cannot be discussed further.[2]

---

[1] Within the past few years attention has been forcibly drawn to the increasing aridity of parts of the Hoshiarpur district of the Punjab, the northward progress of the sands from the southern desert, the deepening of the water-table and the gullying and erosion of tracts that were, three or four generations ago, covered under a fertile soil-cap. These adverse effects are ascribed to the destruction of forests which once clothed the Siwalik foot-hills. Similar effects have been noticed in other sub-montane districts also and serve to impress the important role played by forests in moderating the denudation by rain, in regulating the run off, in conserving the sub-soil water and in binding and protecting the soil-cap from wind and water erosion.

[2] The following books on the study of soils may be consulted: G. W. Robinson, *Soils—Their Origin, Constitution and Classification*, London, 1932; P. Vegeler, *Tropical Soils*, London, 1933; A. N. Puri, *Soil Science*, Simla, 1951.

THE MAIN SOIL GROUPS OF INDIA

On the whole, the principal characters of the main soil groups of India are generally deducible from the nature of their geological foundation. This is well established in the association and genesis of such well-marked groups as black soils, red soils, lateritic soils, no less than in the constantly changing patchy soils of the Himalaya mountains, where changes in the geological nature of the substratum are reflected in the composition, depth and profile of their soil-caps.

Red Soils—This comprehensive term designates the largest soil group of India, comprising several minor types. They cover the Archaean basement of Peninsular India, from Bundelkhand to the extreme south, an area of 2,072,000 sq. km. embracing south Bengal, Orissa, parts of central India and Madhya Pradesh, east Andhra Pradesh and Karnataka and the major part of Tamilnadu.

The parent rocks are acid granites and gneisses, quartzitic and felspathic, with only subordinate rock-types rich in iron and magnesium-bearing minerals. Also the ancient sedimentary sandstones and clays of the Cuddapah and Vindhyan systems have contributed secondarily to the formation of the red soils. The colour of these soils is generally red, often grading into brown, chocolate, yellow, grey, or even black. The red colour is due more to the wide diffusion rather than to a high percentage of the iron content. Many of the so-called "red soils" of South India have no red colour. On the other hand, some red-coloured soils are of quite different constitution, being derived from the surface capping of lateritised rocks. These are generically quite distinct from the true red soils, and are described below. The red soils formed on limestone terrains are also quite different; they form the residue left after dissolution of the bulk of the rock, *i.e.* its clayey and sandy impurities.

A number of subordinate groups and types come under the general designation of red soils. Conditions of free or restricted drainage determine whether salts leached out form a zone of accumulation in the soil profile. Also differences of texture, depth, porosity, humus, and presence or absence of calcareous segregations (*Kankar*) or ferruginous layers (*iron pan*), or of soluble free salts in the soil profile, distinguish these types. For example, there are the light-coloured yellow sandy soils of the dry steppe areas, the dry and wet upland and valley soils, and there are areas of grey semi-desert saline soils (*sierozem*) enclosing patches of non-saline soil.

In general, the red soils derived from acid gneisses are poor in lime, magnesia, phosphates, nitrogen and humus, especially in the drier areas, but are fairly rich in potash. In their chemical composition they are mainly siliceous and aluminous, with free quartz as sand; the percentage of iron is not high; the alkali content is fair, some parts being quite rich in potassium derived from the muscovite and orthoclase of the gneisses. In comparison with the black soils, red soils are as a group deficient in iron oxide, lime, and phosphatic content.

The red soil group is, generally speaking, deficient in its content of soluble exchangeable bases, and their total base-exchange capacity for K, Ca, etc. is generally low.

Black Soils (Regur)—This is another large group of soils of the Deccan, including several distinct sub-varieties and types. It is a general term applied to the large group of black or dark soils common in the Deccan region of Maharashtra, the Malwa and Berar region, the western parts of Madhya Pradesh and Andhra Pradesh Gujarat and Saurashtra, with extensions to central India and Bundelkhand. Isolated patches of *regur*, though less typical, occur also in Tamilnadu. The most characteristic black soils cap the volcanic plateau of the Deccan Traps, forming a mantle of rich residual soil of no great thickness or depth of profile. But the area of black soil is by no means conterminous with the boundary of the Deccan Trap formation, and a wide extent of this soil is found over the surrounding granitic and basic gneissic and other formations, Vindhyan or Cuddapah sandstones and slates. There are also large tracts of transported black soil. The cause of the prevailing dense black colour of these soils over such a wide area is not yet definitely known; partly it may be due to iron or to some minute quantity of a titanium-iron compound, partly to carbon and organic matter. But many black soils contain very little organic substance.

Typical black soil, the familiar *black cotton soil*, is highly argillaceous, with a large clay-factor, 62 per cent or more, without gravel or coarse sand. It is very tenacious of moisture and exceedingly sticky when wet. Owing to considerable contraction on drying, large and deep cracks are formed after the monsoon. The clay-factor of *regur* contains 60 per cent of silica, 25 of alumina and 15 of ferric oxide, on the average. The heavy black soil of the cotton districts of Gujarat, Berar, etc., derived from the basalts, are by reason of their hydrology and climatic conditions very suitable for cotton cultivation. They are characterised by a high proportion of lime and magnesium carbonates (6–8 per cent), iron oxide (9–10 per cent) and fairly constant alumina (10 per cent). Potash is variable (less than 0.5 per cent), and phosphates, nitrogen and humus are low as a rule. Areas of *regur* are credited with high fertility and do not require manuring for long periods, but some upland *regur* grounds are not very productive.

The base-exchange capacity of *regur* is fairly high throughout its profile. These soils are fairly well supplied with replaceable bases and a number of observers in India have studied this aspect of *regur*.

Laterite and Lateritic Soils—These soils do not form such well-defined groups as the two described above. Though laterite itself is something of the nature of a soil-cap, passing down by clear gradation into subjacent rock, the pedogenic processess have stopped short beyond a certain stage and the resulting product, a deeply ferruginous compact clay, porous and vesicular, is not a soil. By further modification and the action of biologic and other soil-forming agencies, laterite is converted into a red-coloured soil with a profile of 30 cm. to 1 m. sometimes more, of disintegrated loam, charged with segregated iron nodules or sometimes an iron pan. Though the laterite cap may be of great thickness, extending to 30 m. or even 60 m. on some plateau tops, laterite soils are usually thin, rarely having a profile of more than 60–90 cm. Regionally the laterite soil group forms a belt of variable width round the Peninsula, in its best development capping the hills and plateaus of the Deccan, central India, Bangladesh and extending through Assam to Burma.

Laterite being largely a product of monsoonic regions, with their alternate dry and moist conditions, leaching action in these soils is complete, with the result that they are denuded of exchangeable bases and other fertilising constituents, giving to the soil a more or less marked acid reaction. The SiO_2: Al_2O_3 ratio is low in mature laterite, in true laterite combined SiO_2 being almost totally absent.

Because of the intensive leaching and the low base-exchange capacity, typical laterite soils are lacking in elements of fertility and are of little value for crop production. But tillage and secondary changes have produced fair soils; parts of the Bombay Deccan have acquired enough potash, nitrogen and phosphates to support good cultivation. The laterite soils developed on the summits of plateaus of Malwa, Malabar, central India, Madhya Pradesh, the Eastern Ghats, Rajmahal hills, Orissa, and parts of Assam and Burma are generally poor agriculturally.

There is an extensive literature on laterite and lateritic soils, both Indian and foreign. But it cannot be said that the subject of the geological origin of laterite, or the agrological nature of the laterite soils, is yet fully known and wide differences of opinion prevail.

Alkaline Soil (Reh Usar, Kalar Soils)

These are salt-impregnated, or alkaline, soils which form an important, albeit a negatively important, group in India. Soils on the drier parts of north Bihar, Uttar Pradesh, the Punjab and Rajasthan tend to saline and alkaline efflorescences. There are yet in these soils many undecomposed rock and mineral fragments, which with weathering liberate sodium, magnesium and calcium salts and sulphurous acid. Such soils are notably impervious and, therefore, have impeded drainage. Large areas, once fertile, have become impregnated with these salts (*reb*, *kalar*), destroying the agricultural value of the ground. The salts are confined to the top layers of the soil, being transferred from below by capillary action. Irrigation by canal water facilitates this transfer, thus a fairly large extent of such salt-charged soils has resulted in the canal irrigated areas of the Punjab and elsewhere within the last two or three decades. Such lands are known as *usar* or *reh* in the north, *kalar* in Sind and *chopan* in Maharashtra and Gujarat. The alkali content is high and there is a large excess of free salts, combined with poverty in nitrogen and organic plant-food material. The salts most common in the *reh* ground are sodium carbonate and sulphate, together with calcium and magnesium compounds. The reclamation of *usar* lands of the Indo-Gangetic plains would add millions of hectares to the cultivable area of North India.

The *nitre-impregnated soils* of Bihar and Punjab are a variant of the above, the difference being that the salt (potassium nitrate) is introduced into the soil-cap from above by the activities of man in densely populated cultivated districts under a warm humid climate.

Mountain and Forest Soils of the Himalayan Region

These soils are found in the depressions within the mountains, in valley-basins and on the less steeply inclined slopes. Generally, it is the north-facing slopes of the Himalayan ranges which support a considerable soil-cap; the south faces of the mountains are too precipitous and exposed to the denuding agencies to be commonly covered with soils. Much the larger extent of these soils is of the wet forest type, under heavy growth of perennial forests of conifers. They are of a heterogeneous nature, varying with parent rocks, climate and local conditions, *e.g.* prevailing wind, rain, snow, ground-configuration, cultivation, etc. They do not form a compact soil-group and, in the inner mountains, support but little agriculture, except in the valleys of the more inhabited parts.

The soil-caps on the broad zone of Tertiary sandstones and clays, constituting the Siwalik foot-hills from Afghanistan to Assam, are primary soils, shallow and immature, containing a large proportion of undecomposed mineral grains; they are sandy or gravelly, porous, devoid of humus and frequently impregnated with lime and soda salts. The higher ranges to the north of the foot-hill belt are clothed in thick forests of pines and rhododendrons; the soils here are various types of mountain-forest soil, podsol soil, mountain-meadow and highland-steppe soil. In the broad lateral valleys of this zone—the *duns*—as in Dehra Dun, Kashmir, Nepal, etc., are alluvial soils of high fertility. In the regions above the limit of forest vegetation, above 4,300 m. the soils are frozen for the greater part of the year; they are thin, clayey and podsolised, with a fairly developed profile, containing a prominent ash-grey horizon due to excessive leaching action of ground waters. The most productive mountain soils are met with in the Middle Himalayas. Some of the wet, deep, upland soils of the Central and Eastern Himalayas, with their high humus content, when cultivated make good tea soils in Kangra, Darjeeling and the Assam ranges.

Soil Erosion—The soil-caps of the middle and outer Himalayan ranges perform a great service in conserving the perennial flow of water in the great rivers descending to the plains at their foot. The role of the sub-Himalayan forests in building these soils and binding them to their parent rocks and then of protecting them from erosion is no less important. The relative ease with which the newly formed soils is eroded from such geological foundation and the necessity of conserving the exiguous soil mantle against the effects of rain, frost and wind are problems which involve serious consequences.

The ravages by floods in many Indian rivers can be moderated, if not checked, by conservation of forests and grass-lands in their upper reaches and in their catchments. These protect the soil cover of the hill-slopes from rapid denudation, and help it in holding back a large part of the rain-water from rushing down the river precipitately and choking its outlets. These rushing floods take away with them millions of tonnes of silt, which is comminuted soil.

Index

Abrasives, 457, 464-465
Abur beds, 252
Aeolian action, in Rajasthan, 31, 372, 373, 474
Aeolian basins, 411
Africa, land connection with India, 33, 164, 167, 169, 213, 254
Agate conglomerate, 291
Agates (*Akik*), 285, 291, 460
Agglomerate slates, Panjal, 208, 209, 210
Ahmednagar sandstone, 271, 274
Ajabgarh series, 116
Akik, 460
Alabaster, 467
Alaknanda flood, 50
Alexandrite, 458
Allah Bund, 44
Allanite, 473
Alluvial basins, 411
    deposits, 26, 335, 381, 382, 383, 386
    gold, 445
Alpine orogeny, 401
Altaid orogeny, 401
Alum, 462
    shales, 462
Aluminium, 440
Alurgite, 104
Alwar series, 96
Amazon-stone, 460
Amb beds, 198, 200
Amber, 295, 461
Amethyst, 111, 460
Ammonites, Cretaceous, 256, 261, 269
    in Productus limestone, 201
    in Spiti shales, 239, 240
    Jurassic, 240, 249
Anaimalai hills, 19
Anantpur gold-field, 444
Anceps beds, 250
Andaman islands, Cretaceous of, 264

    jords of, 44
Angaraland, 169
    flora, 169
Ankaramite, 278
Ankleshwar oilfield, 438
Anorthosite, 77, 85
Antecedent drainage, of the Himalayas, 27
Anthracolithic group, 160, 202
    systems of India, 202
Anthropoid apes, fossil, 348
Antimony, 440
Apophyllite, 281
Aquamarines, 458
Arabian Sea coast, 32, 33, 34, 41
Arakan coast, 34, 37, 413
Arakan Yoma, 235, 264, 320, 333
Aravalli mountains, 4, 18, 94-95, 113, 406
    formation of, 94-95, 113, 406
    former limit of, 95
    glaciers of, 172, 198-199
    relation to Vindhyan formation of, 126, 406
    system, 94-98, 425, 426
Archaean system, 75-88
    crystalline complex of, 73-75
    gneiss of, 75, 76, 86-88, 266, 267, 273, 424
    of Himalayas, 74-75, 84-88
    of Kashmir, 86-88
    See also Dharwar system.
Arcot gneiss, 83, 424
Ariyalur stage, 267-268
Arkose, 117
Arsenic, 441
Arsenopyrite, 441
Artesian wells, 372, 417-418
Aryan era, 193, 195, 203
Asbestos, 463
Assam, 99, 255, 263-264, 272, 318, 323, 332, 420
    coal in, 319, 430, 431, 446

INDEX

corundum in, 465
Cretaceous of, 255, 263-264
earthquakes in, 38, 42
Eocene of, 318-319
fault-structure of, 404
oil in, 319, 433-434
Oligocene of, 323
ranges, 5, 8, 9, 318, 433
Syntaxis, 8, 398
Athgarh sandstones, 190
Athleta beds, 250
Attock, gold-washing at, 445
oil deposits of, 436-437
slates, 128
Auden, J. B., 123, 395, 400
Augite-syenite, 333
Australia, connected with India, 32, 164, 167, 177, 213, 254
Permo-Carboniferous of, 199
Autochthonous zone, of the Himalayas, 393
Autoclastic conglomerates, 108, 140

BABABUDAN hills, iron-ore deposits in, 446
Badasar beds, 252
Bagh beds, 270, 273, 281, 425, 427
conclusions from fauna of, 271
Bagra beds, 188
Bain boulder-bed, 343
Bairenkonda quartzites, 118
Balaghat gneiss, 83
manganese in, 100, 449
Ball, Dr. V., 427
Balmir beds, 252
Baltistan, 13, 17, 208
Báltoro glacier, 22, 23
Baluchistan, 60, 230, 237, 244, 262, 381, 384
chromite of, 258, 442
Cretaceous of, 262
Daman of, 381, 387
igneous action in, 335
Jurassic of, 237, 244

Mesozoic of, 244
oil of, 432
Oligocene of, 329
Triassic of, 230
Banaganapalli beds, 124
Banded jasper, 91, 114
Bangladesh, 416
Banihal, Jurassic of, 243
Bap beds, 199
Barail series, 319, 323
Barakar sandstones, 466
stage, 173, 174
Baripada beds, 295
Barmer (Balmir) sandstone, 274
Baroda, 97, 292
Barren Island, volcano of, 35-36, 389
Barytes, 463
Basalts, 31
of Deccan, 275-277, 285, 386-387, 428, 460
of Rajmahal, 186-187
Basic volcanic series, 117
Basins, lake, 410-412
Bathyliths, 333
Bauxite, 376, 378, 464
Bawdwin, lead-ores of, 447
silver of, 448
zinc, 454
Baxa series, 129
Bayou lakes, 411
Beaches, raised, 30, 43
Baldongrite, 104
Balemnite beds, 240
shales, 262
Bellary diamonds, 456
gneiss, 83
Bengal gneiss, 82
Bengal sea coast, 32-34, 413
Bentonite, 421, 422
Beryl, 111, 441, 458
Bhaber, 371
Bhander series, 123, 126, 455
Bhangar, 370
Bhima series, 123, 124
Bhur land, 371, 381, 384
Biafo glacier, 21

Bihar, earthquake, 40
   saltpetre, 475
Bijaigarh shales, 123, 126
Bijawar series, 115, 117, 455
Bijori stage, 176
Bikaner, coal measures of, 429, 430
   gypsum of, 467
   Jurassic of, 252
   Tertiary of, 294
Bion, H. S., 209
Biotite-gneiss, 87
Black soil, 48, 386, 482
Blaini series, 216–217, 394, 399
Blanford, W. T., 33, 354
Blanfordite, 104
Block-faults, in Kutch, 248
   in the Salt-Range, 133
Bloodstone, 281, 460
Blown sand, 4, 293, 372–373, 384
Blue vitriol, 463, 478
Bokaro, coal-field of, 178, 429
Bombay, inclination of Traps at, 276–277
   submerged forest of, 43
Borax, 464
Bose, P. N., 445
Boulder-beds, Bain, 343
   Blaini, 217
   Hazara, 218
   Rajasthan, 167, 172, 198, 199
   Salt-Range, 167, 198
   Talchir, 172, 177, 198, 210
   Tanakki, 218
Boulder-conglomerates, 342–343, 345, 357
Boulders, in Kashmir, 17
   Potwar, erratic, 385
Boundary Fault, Main, 40, 336–337, 392
Brahmaputra river, 8, 369
Brick-clay, 178, 420
Brine-wells, 473
Broach, Tertiaries of, 291
Budavada beds, 190
Bugti beds, 328–329
Building stones, 83, 111, 124, 127,
192, 251, 285, 377, 380, 383, 384, 424–428
Bundelkhand diamonds, 455
   gneiss, 83, 95
   lakes, 411
Bunter, 225
Burma, 37, 60, 157, 218, 235, 264 303, 320, 323
   amber in, 461
   Cambrian of, 218–219
   Carboniferous of, 219
   corundum in, 465
   Cretaceous of, 264
   crystalline zone of, 74
   Devonian fauna of, 159
   Eocene of, 320
   gold in, 445
   jadeite of, 258, 264, 457–458
   Jurassic of, 252
   laterite of, 377, 380
   lead-ores of, 447
   Miocene of, 332
   oil-fields of, 324, 433
   Oligocene of, 323
   Ordovician of, 158, 161
   Plateau limestone, 160
   ruby mines of, 456, 457, 465
   sapphires of, 457
   Silurian fauna of, 159
   silver in, 448
   Tertiary of, 303
   tin in, 451
   Triassic of, 235
   wolfram in, 452
Burrard, Sir S., 4, 10, 365
Burzil, Cretaceous volcanics of, 259
   hornblende-granite of, 88.
   orbitolina limestone, 260–261

CAINOZOIC era, 222, 287, 295
Calciphyres, 79
Calcite, 279, 281
Caldera, 31, 35
Cambay, agate manufacture of, 460

INDEX 489

Cambrian system, 132–146
  fauna of, 139, 142–143
  of Kashmir, 140–143
  of Salt-Range, 135–138
  of Spiti, 138–139
Cañon, 408
Carbonatites, 281, 466
Carboniferous system, Lower, 154
  Middle, 155
  Upper, 193–220
  earth-movements in, 162
  glacial period of, 167, 172, 198
  of Burma, 218
  of Kashmir, 205–213
  of Salt-Range, 197–202
  of Spiti, 203–205
  origin of, 193–197
Cardita beaumonti beds, 262, 263, 268, 284, 297, 307
Carnatic gneiss, 82
Carnelian (agates), 281, 285, 460
"Cat's eyes", 458
Cave deposits, 386
Cement, 423
Cenomanian age, marine transgression of, 254, 266
Central gneiss, 74, 85, 86
Ceratite beds, 229, 233
Cercopithecus, 347
Ceylon (Sri Lanka), drainage of, 26
  fauna of, 355
  flora of, 355
  gem-sands of, 456, 459
  Gondwana, 191
  graphite of, 466
  Khondalite, 78
  Miocene of, 295
Chail series, 106
Chalcedony amygdales, 279, 281, 291
Chalk hills, chromite in, 442
  magnesite of, 442, 448
Chamba mountains, 16
Chamberlin, Professor T.C., 222, 326
Champaner series, 97
Champion gneiss, 82
Chamtodong lake, 29

Chandarpur series, 118
Chandpur series, 119
Chari series, 249
Charnockite, petrological varieties of, 77, 84, 424
  series, 77, 84–85
  type-rock, 84
Chaugan stage, 188
Chauk oil-field, 324, 433
Chaung Magyi beds, 161
Cheduba Island, mud-volcanoes of, 37
Chenab, 14, 15
Cheralite, 451, 453
Cherra sandstone, 264, 318
Chert, 104, 114
Cheyair series, 115
Chharat Series, 300, 305
Chichali pass, geology of, 246, 263
Chicharia stage, 173
Chidamu beds, 240
Chideru hills, 199, 229
  stage, 198, 201, 221, 229
Chikiala stage, 189
Chikkim limestone, 259
  series, 256
Chilka lake, 412
Chilpi series, 99
Chinji stage, 299, 343
Chirakhan marl, 270
Chitral, Cretaceous of, 69
  Devonian of, 156–157
  orpiment mines of, 441
Chlorophaeite, 279
Chor granite, 106
Christie, Dr., 31, 136
Chromite, 441
Chromium, 441
Chrysoberyl, 111, 458
Classification, principles of geological, 221–224
Clays, 420–422
  brick, 420
  china, 178, 420
  fire, 178, 421
  fuller's earth, 294, 421
  terra-cotta, 178, 421

Closepet gneiss, 82
Coal, economics of, 428–432
　Eocene, 318, 319, 429
　Gondwana, 177, 428–432
　Jurassic, 246
　Oligocene, 429, 430
　Salt-Range, 429
　Tertiary, 294, 429
Coal-fields of India, 166, 177–178, 294, 318, 319, 429
Coastal system, 189, 247
Coasts, of India, 32, 412–413
Cobalt, 442
Coimbatore, Sivamalai series of, 77, 465
Colouring matters, 470
Columbite, 111, 452, 473
Columnar structure, 187, 280
Composition of desert sands, 373
Conglomerates, autoclastic, 108, 140
　Boulder, 188, 189, 341, 342, 345, 357
Conularia, 198
Copper, 110, 442–444
Copperas, 463, 478
Coral islands, 33
　reefs, 43, 267
Coralline limestone, 267, 270
Coromandel coast, 34, 189, 247, 254, 272, 294
　Tertiary of, 294
　Upper Cretaceous of, 272, 294
Correlation of Indian formations, 55, 222
Corundum, 110, 456, 464
Cotter, Dr. G. de P., 178, 304
Cotton soil, 286, 482
Cretaceous system, 254, 274
　end of, 287
　fauna of, 256, 268–269, 270
　geography of, 254
　igneous action in, 260
　infra-Trappean, 272–273
　of Assam, 263–264
　of Burma, 264
　of Burzil and Aster, 259–260
　of extra-Peninsula, 255–264
　of Hazara, 261
　of Kashmir, 259
　of Malla Johar, 257
　of Narmada Valley, 270
　of Peninsula, 266–274
　of Salt-Range, 263
　of Sind and Baluchistan, 262
　of Spiti, 255
　of Trichinopoly (Tiruchirapalli), 266–267
　south-eastern, 266
Cuddalore sandstone, 295
　series, 294, 383
Cuddapah diamonds, 455
　gneiss, 83
Cuddapah system, 113–119
　distribution of, 115
　earth-movements of, 113
　economics of, 119
　Lower, 115
　Upper, 118
Cumbum slates, 118
Cuttack sandstone, 190, 192
Cyrtolite, 473

Dacites, 277
Dagshai series, 300, 331
Daigoan series, 173
Dainelli, Professor G., 53, 358
Daling series, 106
Dalma traps, 103
Daman deposits, 387
Daman slopes, 387
Damodar valley, 174, 175, 178
Damuda coal-field, 177, 428
　flora, 175
　ironstone, 445–446
　series, 174–178
Dandot coal-field, 428
Daonella limestone, 225
Darjeeling series, 107
Deccan lavas, horizontality of, 276–277

INDEX

petrology of, 277
Deccan Peninsula, 404
Deccan plateau, 19
Deccan Trap, 251, 275-286, 404, 482
  age of, 284
  area of, 275
  basalts of, 275-281, 428
  composition of, 277
  fauna of, 282
  fissure-dykes of, 283
  formation of, 275
  inter-Trappean beds of, 281
  laterite of, 377
  thickness of, 276
Dehra Dun, 10, 11, 39
Delhi, stone for, 424
Delhi system, 116
Delta, Ganges, 370, 432
  Indus, 371
Denudation in India, 48
Denwa beds, 188
Deoban limestone, 129
Deola marl, 270
Desert of W. Rajasthan, 48-49, 372
  erosion of, 48-49, 372
  topography of, 372-373
Desiccation, in India, 4, 372
  of Tibetan lakes, 30
De Terra, Dr. H., 358, 361
Devonian system, 148
  of Burma, 159
  of Chitral, 156-157
  of Hazara, 151-152
  of Kashmir, 153
  of Spiti, 147-148
Dharwar system, 89-112
  distribution of, 92
  earth-movements in, 89, 94
  economics of, 110-111
  formation of, 90
  gold-fields of, 444
  homotaxis of, 108
  lithology of, 90
  manganiferous nature of, 100, 103, 449
  plutonic intrusions of, 91

Dhauladhar range, 12
Dhok Pathan zone, 343
Dhosa oolite, 249
Dhurandhar falls, 25
Diamonds, occurrence of, 117, 124, 128, 455
  origin of, 117
  production of, 455
  vallum, 460
Diener, Dr. K., 220, 236
Digboi oil-field, 433
Dihing series, 302, 349
Dinosaurs, fossils, 272
Diorite, 333
Disang series, 318, 319
Dissolution basins, 411
Dogra slates, 128, 140
Dolerite, 187, 277
Dolomites, 138, 200, 224, 234, 468
Dome gneiss, 82
Drainage system, of Himalayas, 26-29, 51
  of Peninsula, 25-26
Dravidian earth-movements, 162
  era, 162
  group, 72
Drew, Frederick, 10
Drowned valleys, 44
Dubey, V. S., 117
Dubrajpur sandstone, 187
Dunes, sand, 372-373, 384
Dunites, 80, 442
Dunn, J. A., 100, 103
Duns, 11, 407
Dupi Tila stage, 302, 349
Durgapur stage, 173
Dust-storms (-mounds), 384-385
Dwarka beds, 292
Dykes, in Archaean gneisses, 83
  Deccan Traps, 251, 283

EARTHENWARE, 420
Earth-movements, Dravidian, 162
Eocene, 287

in Cuddapah age, 113
in Dharwar age, 89
in Tertiary, 287
in Upper Carboniferous, 162
in Vindhyan age, 122, 125
Earthquakes, 37–42
Earthquake zone of India, 37
Eastern coast, 32
Ghats, 19, 33, 114
Economic geology, 414–484
   metals and ores, 438–454
   other minerals, 461–478
   precious stones, 455–461
   soils, 478–484
Elephants, Indian fossil, 338–339, 347–348
Emeralds, 458
Emery, 465
Eocene system, 307–320
   earth-movements in, 287
   of Assam, 318
   of Burma, 320
   of Coromandel coast, 294
   of Gujarat, 291
   of Hazara, 313
   of Kashmir, 314–317
   of Kathiawar (Saurashtra), 292
   of Kirthar, 308
   of Kohat, 312
   of Ladakh, 301
   of Laki, 308, 309
   of Ranikot, 307
   of Salt-Range, 309–310
   of Subathu, 317
Epsomite, 137
Erinpura granite, 116
Erratics, 385
Escarpments, Middle Gondwana, 182
   Satpuras, 182
European geological divisions, 222, 224
Eurydesma beds, 199
Evans, P., 47, 302
Everest, Mt. 6, 9, 240, 241
   limestone series, 224, 241
Exotic blocks of Johar, 258

"Exotic" Trias of Malla Johar, 227
Extra-Peninsula, Cretaceous of, 255–263
   crystalline zone of, 86, 393–394
   Dharwars of, 105
   Gondwanas of, 170
   physiograpic differences from the Peninsula, 1, 2, 3, 197
   Tertiary of, 295–306

FACIES, 57, 254, 290
Fan-talus, 362
Fatehjang zone, 330
Faults, block, 133, 247–248
   Boundary, in the Aravallis, 127
   Main Boundary, in the Himalayas, 40, 336, 392
   reversed, in the Outer Himalayas, 335, 395
   trough, of the Gondwana basins, 166
Fauna, Anthrocolithic, 202
   Australian, 199
   Bagh, 270
   Cambrian, 142
   Ceylon (Sri Lanka), 355
   Cretaceous, 268, 270, 272
   Deccan Trap, 282
   Devonian of Burma, 159
   of Chitral, 156, 157
   Gondwana of Kashmir, 213
   Haimanta, 139
   Jurassic, 242–245, 248–253
   Kirthar, 308
   Kota, 188
   Maleri, 184
   Mesozoic, 227
   Miocene, of Burma, 332
   Muschelkalk, 226
   Nari, 323
   Oligocene and Lower Miocene, 323, 328–329
   Panchet, 182
   Productus limestone, 201–205

INDEX

Ranikot series, 309
  S. E. Cretaceous, 268, 270, 272
  Silurian of Burma, 159
    of Kashmir, 160
  Siwalik, 346–348
  Spiti, 148, 204, 225
  Tertiary, 287
  Triassic, 221, 225, 230, 233–234
  Umia, 191, 250
  Zewan, 215
Fenestella series, 155
Fermor, Sir L. L., 100, 103, 279, 380
Fermoria, 124
Fire-clay, 178, 421
Fissure-dykes, 283
  eruptions, 283
Fjords, Andaman Islands, 44
Flexible sandstone, 91
Flood, Indian rivers, 50
  Indus, 385
Flora, Ceylon (Sri Lanka), 354, 355
  Damuda, 175
  Gondwana, 168, 213
  Jabalpur, 188
  Jurassic, 188, 237
  Kota, 188, 189
  Rajmahal, 187
  Raniganj, 175
  Talchir, 172
  Umia, 191
  Vemavaram, 190
Fluorspar, 466
Flysch, Cretaceous, 256
  Oligocene, 256

Forest, submerged, of Bombay, 43
  of Pondicherry, 43
  of Tinnevelli coast, 43
Fox, Dr. C. S., 178, 378
Fuller's earth, 294, 421
Fusulina limestone, 218, 219, 231

GABBRO, intrusion of, 257, 261, 333
Gadolinite, 473
Gaj series, Sind, 327

Galena, 118, 443, 447
Gangamopteris beds, 211–212
  flora, 213
Ganges delta, 370, 432
  river, 28, 407
  reversal of flow of, 51, 52
Gangotri glacier, 24
Gangpur series, 101, 103, 119
Garhwal nappe, 217, 395, 397, 400
  thrust, 397
Garnets, 459, 465
Garo hills, 264, 318
Garwood, Prof., 29
Gaya, mica-mines of, 470
  pitchblende of, 453
Gee, E. R., 135, 311
Gem-sand, 456, 459
  Ceylon (Sri Lanka) 456
  stones, 455–461
Geodetic Survey of India, 46
Geographical distribution of minerals, 415
  India, 415
  Pakistan, 416
Geography of India, Cretaceous, 254
  early Tertiary, 287–288
Geoid, 45
Geological distribution of minerals, 415
Geological division of India, 1–3
  formation of India, table of, 65–72
  record, imperfections of, 56
Geosyncline, Indo-Gangetic, 365
  meaning of, 195
  Potwar, 329
  Spiti, 138
Gersoppa (Jog) falls, 25, 410
Ghats, Eastern, 19, 33, 43, 114
  Western, 19, 43
Ghosh, P. K., 292
Gigantopteris flora, 169
Gilgit, 13, 14, 27, 50, 445
Giridih coal-field, 429
Girnar hills, 278
Giumal sandstone, 255, 257

494   GEOLOGY OF INDIA

Glacial age, record of, 17, 167, 171, 198, 342–343, 353–363
  boulder-beds, 167, 170, 198, 342–343
  dam, Shyok, 51
  lakes, 50, 411
Glacial deposits of Kashmir, 358
  correlation of, 360
Glaciers, 20–24
  of Himalayas, 20–24, 356
  of Kashmir, 16, 358
Glass-sand, 422
Glauconite, 279, 281
Glennie, Col. E. A., 47, 366
Glossopteris flora, 175
Gneiss, 80–88
  Archaean, 74–76, 80–88
  Arcot, 83, 424
  as a building stone, 424
  Balaghat, 83
  Bellary, 83
  Bengal, 82
  Bundelkhand, 83, 95
  Carnatic, 82
  Central ("Fundamental"), 74, 85, 86
  Himalayan, 74, 85–87
  Hornblende, 75, 87
  Hosur, 83
  Magok, 75, 79, 158, 162
  Peninsular, 82
  Salem, 82
Godavari basin, 188
  region, 184
  river, 284
  valley, 176
Gohna lake, 50, 411
Gokak falls, 25
Golabgarh pass, Lower Gondwanas of, 213
Golapilli sandstone, 189
Golconda diamonds, 117, 455
Gold, 110, 444–445
Golden oolite, 245
Gondite, 78, 104
  series, 100, 104

Gondwana system, 164–192
  climate during, 167
  coal-measures of, 174, 177, 428–430
  distribution of, 170
  fauna of, 184, 188, 190
  flora of, 172, 175, 187, 191
  geotectonic relations of, 165
  homotaxis of life of, 168
  Lower, 172–180
  Middle, 181–186
  of Darjeeling, 240–241
  of Himalayas, 179–180
  of Kashmir, 211–213
  origin of, 166
  sandstones of, 427
  Upper, 186–192
Gondwanaland, 164, 169, 254
Gorges of the Himalayas, 14–15, 27
Grandite, 104
Granite, Himalayan, 9, 258
  hornblende of Burzil, 88
  Jalor and Siwana, 123, 125
  post-Cretaceous, 88
Granites, 77, 125, 257–258, 424
Granophyre, 278
Granulite, 76, 78
Graphite, 78, 466
Graptolites, 158
Gravimetric surveys, 46, 47
Great limestone, 216, 238
Greywackes, 152
Grindstones, 466
Guano, 471
Gujarat, 291, 418, 427
Gwadar stage, 349
Gwalior series, 115, 117
Gymnites beds, 233
Gypsum, 135, 311, 467

HAEMATITE, 101, 446
Haimanta system, 139
Halobia beds, 225
Hanging valleys, 28, 356

INDEX

Harappa, 369
Hayden, Sir H. H., 10, 139, 204, 225
Hazara, 128, 152, 228, 242, 261
  geological map of, Plate XIX
Hazaribagh, lead-ores of, 447
  mica deposits of, 470
  tin of, 451
Hedenstroemia beds, 225
Heim, Arnold, 259
Heliotrope, 281
Hercynian earth-movements, 169
Heron, Dr. A. M., 90, 94, 97, 116, 118
Heulandite, 281
"Hidden Range", 47
High-level laterite, 377
Hill limestone, 300
Himalayan orogeny, 391–402
Himalayas, 5–15, 288, 391
  antecedent drainage of, 26
  Archaean of, 85–87
  Cambrian of, 138–146
  Cretaceous of Northern, 255–286
  crystalline zone of, 86
  Devonian of, 148, 153
  Dharwars of, 106
  diagrammatic section through, 5
  Eocene of, 301
  glaciers of, 16, 20–24
  Ice Age in, 17, 356
  Jammu section of, 316
  Jurassic of, 238–243
  limits of, 7
  meteorological influence of, 7
  Miocene of, 331
  nappe-structure of, 394
  Permo-Carboniferous of, 202–217
  physical features of, 6
  rise of, 288–289
  snow-line and glaciers of, 20–24
  stratigraphical zones of, 9
  structural features of, 391–401
  subaerial erosion of, 53
  syntaxis of, 8, 398, 401
  Tertiary of, 298–302
  thrust-plains of, 392
  Trias of, 224
  valleys of, 14, 24, 27, 408
Himgir beds, 176
Hippurites limestone, 256, 262
  of Iran, 262
Holdich, Sir T. H., 5
Holland, Sir T. H., 31, 84, 108, 129, 222, 380, 438, 477
Hollandite, 104
Homotaxis, 56
Hornblende-gneiss, 76, 87, 195
Horsts, 3, 5, 83
Hosur gneiss, 83
Hsipau series, 252
Human epoch, 387, 389
  implements, 379, 382, 388
Hundes, 228, 288, 301
Hutti, gold-mines of, 444
Huxley, Professor T. H., 56
Hyacinth, 459
Hyperite, 77

ICE Age, records of, 23, 172, 198, 353–358, 383, 385
Idar granite, 116
Igneous action, in Bijawar series, 117
  in Carboniferous of Kashmir (Panjal Trap), 207
  in Cretaceous, 257
  in Damuda coal-measures, 175
  in Deccan Trap, 275–278
  in Dharwar age, 91
  in Malani series, 124–125
  in Oligocene and Miocene, 332
  in Rajmahal series, 186
Ilmenite, 452
Implements, stone, 382, 388
Inconsequent drainage of the Himalayas, 26–27
India, industries, undeveloped, 438–439, 477
Indian Ocean, Expedition, 34
  floor, 34
  Ridge, Central, 34
  UNESCO publications on, 34

Indicolite, 459
Indo-African continent, 164, 168, 254
Indobrahm river, 51
Indo-Gangetic alluvium, 364-375
  plains, 1, 364-367, 418, 473
  trough structure of, 364, 367
Indus river, basins of, 4-5
  delta of, 370
  floods of, 50, 385-386
  gorge at Gilgit, 27, 50
Infra-Krol series, 217
Infra-Trappean beds, 272-273, 285
Infra-Trias of Hazara, 207
Interglacial periods, 357, 360
Inter-Trappean beds, 211, 276, 281, 282
Inter-Trappean flora, 284
Iolite, 460
Iran, 256, 262
  Hippurites limestone of, 262
  turquoise of, 460
Iridium, 473
Iron, distribution of, 445-447
  occurrence and production of, 110, 119, 178, 380, 445-447
Iron-ore series, 101
Iron-ores, origin of, 101, 103
Ironstone shales, 174, 446
Irrawaddy oil-fields, 433
  system, 304, 351
Islands, coral, 33
  volcanic, 35-36, 477
Isostasy, 45-47

JABALPUR falls, 382
  flora, 188
  iron, 445
  marble rocks of, 99, 425
  stage, 188
  steatite, 476
Jade, 457
Jadeite, 258, 264, 457
Jaffna beds, 295
Jaintia hills, 264, 318
Jaintia Series, 318, 319

Jaipur, garnets of, 459
Jaisalmer, clay, 421
  limestone, 252
  marble, 426
Jalore and Siwana granite, 123, 125
Jammu, coal-measures of, 316, 429
  hills, 11, 216, 315, 316
  Kishtwar district of, 457
  sapphires in, 457
  Siwaliks, 344
  Tertiary, 300-302
Jasper, 114, 117, 281, (agate, 460)
  banded, 91, 114
Jaunsar series, 128, 146, 207
Jharia coal-field, 177, 178, 429
Jhelum river, 11, 231, 300, 382
Jhils, 29, 411, 432
Jhiri shales, 123, 127
Jind, flexible sandstone of, 91
Jog falls, *see* Gersoppa
Joya Mair oil-field, 436
Jumna river, 368
Jurassic system, 237-253
  fauna of, 239, 242, 249-250, 252
  marine transgression in, 247-248
  of Baluchistan, 244
  of Hazara, 242
  of Kashmir, 242-243
  of Kutch, 248-251
  of Rajasthan, 251-252
  of Salt-Range, 245-247
  of Spiti, 238-240
Jutogh series, 105
Juvavites beds, 225

KAGHAN, inter-Trappean limestone of, 211
  Permo-Carboniferous, of, 218
Kaimur sandstone, 455
Kaimur series, 123
Kainite, 137, 475
Kalabagh, alum industry of, 463
  coal of, 246
  rock-crystal of, 460

INDEX

salt of, 311
  stage, 198
Kala Chitta hills, 229, 242, 262
Kaladgi series, 115, 118
Kalar, 372, 473, 483–484
Kamlial stage, 343
Kampa system, 240, 256
Kamthi beds, 173, 176
Kanchenjunga, 9
Kangra, earthquake of, 39
  slates, 427
Kankar, 369, 423
Kaolin, 75, 420
Karakoram glaciers, 21–24
  (mustagh) range, 13, 21–23, 88, 239
Karanpura coal-field, 429, 430, 431
Karewas, 359–361, 382
Karez, 387, 418
Karharbari stage, 173, 174
Karikal beds, 295
Kasauli series, 300, 331
Kashmir, 10–18, 86, 106, 151, 205, 231, 242, 259, 300, 314, 344, 358
  aquamarines in, 458
  Archaean of, 86–88
  Cambrian of, 140
  Carboniferous of, 205–213
  coal in, 315
  copper in, 442
  Cretaceous of, 259
  Devonian of, 153
  Eocene of, 314
  geological records of, table of, 71–72
  geotectonic features of, 398
  glaciers of, 16, 398
  Gondwanas of, 176, 211
  Himalayas of, 10–14
  homotaxis of, 71–72
  Ice Age deposits in, 358–362
  Jurassic in, 242–243
  Karewas of, 359–60, 382
  lakes, 15, 359, 411, 412
  Murree series of, 329, 330
  nappe, 394, 399

  Palaeozoic of, 140
  peat in, 432
  Permian of, 213
  physical features of, 10–17
  Pleistocene of, 358–362
  rivers of, 14–15
  sapphires of, 457
  Silurian of, 152
  Siwaliks of, 344
  stratigraphy of, 143–146
  Subathu series of, 314
  Tertiary of, 300–302
  Triassic of, 231–235
  Valley, 407
  Valley of, 231–235
  volcanic action in, 207, 259
Kathiawar (Saurashtra), 191, 278, 292
Katrol series, 249, 250
Katta beds, 198, 200
Kayals, 32
Kaz Nag gneiss, 87
Kerala (Travancore), 32, 295
  graphite of, 466
  ilmenite, 451
  Miocene of, 295
  Monazite of, 450, 471
  Tertiary beds of, 294, 295
Keuper, 225
Khadar, 370–371
Khadolar hill, 446
Khasi hills, 318
Khewra, salt-mines of, 136, 475
Khondalite, 78, 82, 466
Khusak, Cambrian section at, 138
Kioto limestone, 238, 240
Kirana hills, 97, 125
Kirthar series, 296, 308
  fossils of, 308
  of Kutch, 293
  of Sind, 296
Kishengarh, garnets, 459
Kistna series, 115, 119
Kodurite, 78
Kodurite series, 103, 449
Kohat, salt deposits of, 312, 475
Koh-i-Sultan, volcano of, 36

Kojak shales, 322
Kolar gold-field, 93, 444
Kopili alternations stage, 318
Korea coal-field, 429
Kota Stage, 188
Kothari Dun, 11
Krishnan, M. S., 100, 103
Krol belt, 217, 395
   nappe, 395
Krol belt, series, 216-217, 241
   thrust, 400
Kuling system, 204
Kumaon arsenic, 441
   glaciers, 24
   iron, 446
   lakes, 29, 411, 412
Kurnool series, 123, 124
Kutch, alum of, 463
   earthquake of, 37, 44
   golden oolite of, 245
   Gondwanas of, 191
   Jurassic of, 248-251
   Rann of, 5, 31, 44, 374, 389
   Tertiary of, 293
   trap-flows of, 276-280
Kyanite, 468

LACCADIVE Islands, 33
Lachi series, 205, 224, 240
Lacustrine deposits, Karewas, 359
   360, 382, 383
   of Talchir beds, 167
Ladakh, 13, 14
   borax of, 464
   Cretaceous of, 260
   Jurassic of, 242-243
   Salt lakes, 15, 411
   sulphur of, 477
   Tertiary deposits of, 301
Lagoons, 32, 412
Lakes, borax from, 30, 464
   desiccation of, 30
   glacial, 412
   of Kashmir, 15, 359, 411
   of Kumaon, 29, 411, 412

   of Tibet, 29, 30
   salt, 30, 411, 474
   types of, 410-412
Laki series, 298, 300, 308, 317
   salt and gypsum deposits of, 311, 312
Lam, Triassic section at, 233
Lameta series 272-273, 281
   age of, 272
Lamprophyre, 278
Land bridge, Gondwana-Angaraland, 169
Langbeinite, 475
Laterite, 48, 285, 376-380, 427, 446
   origin of, 378
Lateritic deposits, manganese, 104
La Touche, T. H. D., 22, 158, 252
Laumontite, 281
Lead, 110, 447
Lemuria, 165
Lepidolite, 110, 470
Level, recent alterations of, 42-45
Lignite, 43, 295, 430, 431
Limburgite, 277
Lime, 128, 423, 425
Limestones, 308-312, 316, 425
   as building stone, 425
   crystalline, 79, 99, 456
   fusulina, 218, 219, 231
   origin of, 79, 273
   ruby, 79, 158, 456, 465
Lipak series, 149
Lithomarge, 104, 376
Lochambel beds, 240
Loess, 384
Lonar lake, 31, 411, 462
Lower and Upper Vindhyan system, meaning of, 125
Lower Miocene system, 327-333
Low-level laterite, 377
Lydekker, R., 143, 208
Lyell, 58

MACLAREN, J. M., 378
Macrocephalus beds, 250

INDEX

Madhupur jungle, 44
Madhya Pradesh, 78, 79, 99, 188, 281, 282, 285
Magmatic differentiation, 257-258, 277-278
  of Charnockites, 84-85
Magnesian sandstone, 135, 138
Magnesite, 448
Magnetite sand, 285, 445, 446
Mahadek stage, 264
Mahadevan, C., 119
Mahadev series, 183-184
Main Boundary Fault, 40, 336-337, 392
Makrana (Mekrana) marble, 97, 425
Makum, coal-field, 429
Malabar coast, 32, 33, 295, 377
Malabar coast fault, 33
Maldive Islands, 33
Maleri series, 173, 184
Malla Johar, "exotic" blocks of, 258
Mallet, F. R., 35
Manasarovar lake, 16, 29
Manchar series, 296, 349
Mandhali series, 119
Manganese, distribution of, 449-450
  in Dharwars, 100, 103, 449-450
  mode of occurrence of, 103
Manganese-ores, 103-104, 119, 449, 450
Mangli beds, 184
Mantell, Dr., 338
Marble rocks of Jabalpur, 99
Marbles, as building stones, 97, 267, 425-426
  occurrence in the Aravalli series, 97, 426
  varieties of, 425-426
Marine transgression, Cenomanian, 254-264
  deposits of, 247-248
  Jurassic, 247-248
Marsh gas, 177, 437
Martaban system, 75
Mason, Professor K., 21, 23
Mayo salt mines, 136

Mayurbhanj iron-ores, 445, 446
McMohan, C. A., 85
Medlicot, H. B., 145, 336
Meckoceras zone, 225, 233
Megalodon limestone, 225, 240, 242
Mekran coast, 33
  earthquake, 41
  fault, 33-34
  system, 349
Mekrana (Makrana) marble, 97, 425, 426
Mergui series, 451, 452
Mesozoic era, 222, 268
  Alpine type, 258
  in Baluchistan, 244
  in Salt-Range, 246-247
Metals, 438-454
Metre, W. B., 323
Mica deposits, 82, 469
  economics of, 110, 469
  pegmatite, 91, 453, 469, 473
  peridotite, 175
Middlemiss, C. S., 39, 145, 208, 215, 336, 382
Milam glacier, 21
Miliolite, 293, 383, 423
Millstones, 178, 192, 270-271, 465
Minbu oil-field, 324, 433
Mineral paints, 452, 470
  springs, 117, 420
Miocene system, 327-344
  see also Oligocene
Mogok gneiss, 75, 79, 158, 162
Mohenjo Daro, 369
Mohpani coal-field, 176
Molybdenite, 473
Monazite, 450, 470-471
Monchiquite, 277
Monotis shales, 225
Monsoons, geological work, of, 48, 412
Monzonite, 277
Moonstone, 460
Moraines, old terminal, 23, 356, 357
Morar series, 118
Mortar, 423

Motur stage, 173, 176
Mountain range, of India, 5-14, 18-20
Mountain trend-lines, 8, 401
Mount Everest, *see* Everest, Mt.
Mount Everest Pelitic series, 241
Mud volcanoes, 36-37
Multani mattee, 421
Murray Ridge, 33
Murree series, 298, 300, 329-331
Murree thrust-Eplane, 392, 396, 399
Musa Khel, Mesozoic section near, 199, 230
Muschelkalk, 225, 234
Mushketov, F., 125, 402
Mussoorie and Dehra Dun, epicentrum, 39
Mustagh (Karakoram range), 13, 21, 23, 88, 239
Muth quartzite, 148, 151, 153
Mysore Archaeans, 92-94

NAGPUR, manganese of, 100, 449
Nagri zone, 343
Nagathat series, 146
Nahan series, 343
Nahorkatiya oil-field, 433
Nallamalai hills, 20, 118
series, 115, 118
Nammal ravine, Mesozoic, in, 230
Namshim series, 158, 161
Namyau beds, 252
Nanga Parbat, 9, 13, 105
Nanga Parbat, geology of, 105
Napeng beds, 161, 236, 252
Nappes, 394-400
Nappe zones of Himalayas, 394-398
Narmada (Narbada) river, 25, 283, 382
falls of, 382, 410
older alluvium of, 382
Narmada (Narbada) valley
Cretaceous, 270, 271
fissure-dykes, 283

Narcondam Island, volcano of, 36
Nari series, Sind, 296, 322-323
Natural gas, 326, 437
Naungkangyi beds, 157, 161
Negrais series, 264
Nellore, mica of, 82, 470, 473
Neobolus beds, 136, 137
Nepal, 7, 407, 432
Nepheline-syenite, 77, 278, 473
Neyveli, 295, 429, 430, 431
Nickel, 442
Nicobar Islands, 44
Nilgiri gneiss, 81
mountains, 377, 432
Nimar sandstone, 271
Niniyur stage, 267, 269
Niti limestone, 225
Nitre, 48, 475-476
Nodular limestone, 271
Novaculite, 212, 215
Nummulites, 307, 308, 316, 317, 323
Nummulitic limestone, 288, 297-303, 308-318, 423, 425
of Assam, 303, 319
of Burma, 303
of Hazara, 313
of Jammu, 315, 316
of Kutch, 293
of Pir Panjal, 316-318
of Rajasthan, 294
of Salt-Range, 297, 298, 309
of Sind, 309
Nyaungbaw beds, 161

OCHRE, 470
Oil, mineral, 324
mode of occurrence of, 324-326, 432
Oil-fields, of Assam, 319, 433
of Burma, 323-324, 432
of Pakistan, 436-437
Oldham, R. D., 38, 249
Oligocene and Lower Miocene systems, 322-333

fauna of, 323, 327-329
igneous action in, 332
of Assam, 323, 332
of Baluchistan, 322, 329
of Burma, 323
of Kathiawar (Saurashtra), 292
of Outer Himalayas, 331
of Sind, 322-323, 327-329
Olivine, 278
Olivine-norite, 77
Ongole outcrop, 190
Oolite, golden, 245, 249
Ophiceras zone, 225
Orbitolina, limestone, 257, 260, 261
Ordovician system, 147, 151, 158, 161
Ores, neglect of Indian, 438-439, 477
Orogeny, Alpine, 401
Himalayan, 391-402
Orographic lines of N. India, 399
Orpiment, 441, 470
Orthoclase, 75, 460
Orthoclinical type of mountain, 392
Ossiferous gravels, Narmada, 382
Sutlej, 381
Otoceras zone, 204, 225
Oyster banks, 44

PAB sandstone, 262, 263
Pachaimalai hills, 20
Pachmarhi series, 182, 183
Padaukpin limestone, 161
Paikara falls, 25
Paints, mineral, 452, 470
Pakhal series, 119
Pakistan, 59-60
Cambrian of, 135, 139
Carboniferous of, 197-202
coal in, 431
Cretaceous of, 261-263
chromium in, 441
distribution of minerals in, 416
Eocene of, 309-313
gypsum in, 136, 311, 416, 467

Jurassic of, 242, 244, 245-246
minerally productive rock-systems of, 416
Oligocene of, 322, 327, 329
Permian of, 199-202
petroleum in, 436-437
salt in, 136, 310-311, 475
Tertiary of, 296-297
Pakokku district, oil-field, 433
Palaeolithic implements, 379, 382, 388
Palaeozoic era, of Kashmir, 59, 151-156
of Spiti, 147-151
Palagonite, 279
Palana coal-field, 294, 428
Palaghat gap, W. Ghats, 19
Pali beds, 176
Palnad series, 124
Pamir plateau, 6
Panch Mahal, manganese in, 449
Panchet series, 182
Pangkong lake, 29, 411
Panjal, agglomerate slates, 208, 209
inter-Trappean limestones of, 211
range, 12, 59, 207-211, 382
thrust, 392, 396, 399
Traps, 208-210, 211
Panna diamonds, 128, 455
shales, 123, 126
Papaghani series, 115
Par series, 118
Para stage, 242
Parahio river, section along, 149
Parasnath, 407
Parh limestone, 262
Parsora stage, 173, 178, 185
Pascoe, Sir Edwin H., 51, 326
Patcham series, 249
Patli Dun, 11
Pavalur sandstone, 190
Pawagarh, 277
Peaks, Himalayan, 9, 193
Peat, 43, 432
Pegmatite, carrier of rare minerals, 91, 458, 473

mica-, 91, 469, 473
  veins in Bundelkhand gneiss, 83
Pegu system, 304, 332
Pench valley coal-field, 429
Peneplane, 407
Penganga beds, 118
Peninsula, Cretaceous of, 61, 266–270
  Deccan, structure of, 404–406
  distinction from extra-Peninsula, 1–3, 197, 409
  gneiss, 82
  hydrography, peculiarity of, 3
  origin of, 173
  physical features of, 24–26, 61, 162
Peridotite, 175, 257, (dunite, 442), 448
Perim Island, 292
Permian system, 199–205, 213–216
  of Kashmir, 213–216
Permo-Carboniferous system, 219
  of Burma, 218
  of Everest, 205
  of Hazara, 218
  of Jammu hills, 216
  of Kashmir, 205–216
  of Salt-Range, 198–203
  of Simla, 216–217
  of Spiti, 203–205
  of Umaria, 219
Petroleum, distribution of, 432–438
  mode of occurrence of, 325, 432
  nature of, 324
  theories of origin of, 324
Petrological province, Charnockite, 83–84
Phosphates, 471, 479
Phosphatic deposits, 383, 471
Physiography, principles of, illustrated in India, 390–391
Pilgrim, Dr. G. E., 296, 346, 394
Pinfold, G. S., 321, 333
Pinjor zone, 342
Pir Panjal, 12, 59, 382–383
  map of, Plate XV

  physical features of, 11
  recumbent folds in, 392
  section across, 359
  thrust plains in, 392, 393
Pitchblende, 91, 453, 473
Pitchstones, 277
Plains and plateaus, 3–5, 406–407
Plateau basalts of Deccan, 275–276
  limestone, 160, 219
Platinum, 473
Pleistocene system, 353–380
  and later deposits, 381–389
  European, 353
  Glacial Age during, 23–24, 353–363, 385
  in Himalayas, 356–362
  lakes of, 412
  laterite of, 376–379
  of Kashmir, 358–363
Plutonic intrusions, 91, 100, 254, 257, 278, 284, 332
Po series, 149
Pokaran beds, 199
Poonch, 314–315, 440
Popa, volcano of, 36
Porbander stone, 293, 383
Porcelain, 420
Potassium salts, 136, 475
Pottery clays, 420
Potwar (Puthwar) boulders, 385
  geosyncline of, 328, 329–330
  plains, 132
Primates, fossil, 347, 348
Productus fauna, 201, 202
  limestones, 199–202
  series, 130
  shales, 203, 225
Protoretepora beds (limestone), 212
Puga valley, borax of, 464
  sulphur of, 477
Pulicat lake, 412
Pumice, 277
Punjabian stage, 198
Punjab "wedge", 125
Purana group, 130, 131, 248
  Himalayan, 10, 85, 241

INDEX

Purple sandstone stage, 135, 137
Pyrites, 472, 477
Pyroxenite, 77, 261

QUARTZ, haematite-schist, 79
   reefs, auriferous, 444
   -veins in Bundelkhand, 83
Quartzite, Muth, 148-149, 151, 153-154
Quartzites, 81, 113, 116, 118, 427
Quetta, 41, 244, 418, 442
   earthquake, 41
Quilon beds, 295

RAIALO series, 97
Raipur district, iron-ore deposits of, 446
Raised beaches, 30, 43
Rajamundry, beds, 284
   outcrop, 189
Rajasthan (Rajputana), 4, 60, 251, 294, 410
   Aravalli marble of, 97, 425
   Archaean system of, 95
   boulder-beds of, 172, 198-199
   copper of, 442-443
   desert of, 48, 372-374
   Dharwars of, 94-98
   gems of, 458
   glacial period of, 167, 198
   Jurassic of, 251-252
   Lower Vindhyan of, 124-125
   mica of, 470
   salt lakes of, 31, 474
   Tertiary of, 294
Rajmahal flora, 168, 187
   hills, 186, 187, 377, 421
   clay deposits of, 421
   series, 186, 189
   traps, 187, 377
Rajpipla, agates of, 460
   trap-dykes of, 283

Rajpur series, 118
Rakas Tal Lake, 29
Rama Rao, B., 85, 94
Rama Rao, L., 285
Ramri Island, mud-volcanoes of, 37
Raniganj coal-field, 178, 429
   flora, 175
   stage, 175
Ranikot series, 297, 307, 317
Rann of Kutch, 5, 31, 44, 373, 374, 389
Rare minerals, 472
Ratanpur, agates of, 460
Realgar, 441
Recent deposits, 381-389
Recession of the watershed, 28
Red soil group, 481-482
Regur, 48, 386, 482
Reh(or *Kalar*), 372, 461, 473, 483-484
Rehmanni beds, 250
Rejuvenation of the Himalayan rivers, 26-27, 44-45, 365, 408-409
"Relict" mountains, 2, 391
Reshun conglomerate, 257
Rewah series, 126, 455
Rhyolites, Malani, 124
   Pawagarh, 277
Riasi, coal-measures of, 316, 429
   Permo-Carboniferous inliers of, 216, 315
Rift valleys, 3-4, 365
Rifts, 408
Rimo glacier, 24
Ripple-marks, 122, 137
River action in India, 24-25, 28, 49-51
   capture, 28
   changes in, 367
   erosion of, 49, 50
Roches moutonnées, 17, 199
Rock-basins, 29, 412
   -crystal, 111, 281, 460
   -meal, 17
   -salt, 136, 312, 473
Rohtas limestone, 124
Rubellite, 87, 459

Rubies, 111, 158, 456
Rupshu, Cretaceous rocks of, 259
Rutile, 452

SAHNI, B., 190, 284, 311
Sahni, M. R., 235
Sahyadri mountains, 19, 43
Sakoli series, 99
Salajit, 324
Salem gneiss, 82
Saline series, 135, 311
Salisbury, R. D., 222, 326
Salkhala series, 105, 107
Salses (mud-volcanoes) 36–37
Salt, alkaline, 462
   Kohat, 312
   lakes, 30, 411, 474
   magnesium, 137, 475
   Mandi, 474
   -marl, 135, 137, 311
   potassium, 137, 475
   sources of, 473
   wells, 474
   wind-borne, 31, 474
Salt-beds, 312, 372
Saltpetre, 475
Salt-pseudomorph shales, 135, 138
Salt-Range, Cambrian of, 132–138
   Carboniferous and Permian of, 197–202
   coal of, 428
   Cretaceous of, 263
   dislocation mountain, 404
   Eocene of, 309–310
   Gondwanas of, 199–200
   gypsum of, 136, 467
   Jurassic of, 245–247
   lakes of, 411
   loess of, 384.
   Mesozoic of, 246–247
   mountains of, 59, 132–133, 404
   physical and geological features, 132, 246–247, 404
   sections of, 133, 134, 310
   Siwaliks of, 343
   springs of, 420
   Tertiary of, 297–298
   Triassic of, 229–230
Samarskite, 473
Sambhar lake, 31, 474
Sand-dunes, 373, 384
Sands, 422
   gem, 456
   glass, 422
   ilmenite, 451
   magnetite, 285, 446
   monazite, 451, 470
Sandstones as building stones, 426–427
   Gondwana, 427
   Songir, 271, 427
   Vindhyan, 127, 426
Sang-e-Yeshm, 458
Sangla hills, 97
Sapphires, 158, 456
Saraswati river, 52, 368
Satpura hills, Gondwanas of, 173, 188
   physical features of, 19
   trend-line, 406
Sattavadu beds, 190
Saurashtra (Kathiawar), 191, 277–278, 292
Sausar series, 99
Schuchert, Professor Charles, 170, 177
Schwagerina limestone, 219
Sehta, 442
Semri series, 123
Serpentine, 264, 426, 458
Sewell, Col. R. B., 33
Shali limestone, 216
Shams Abari syncline, 142, 153
Shan States, N., Cambrian of, 161
   geological formation of, 157–163
   Jurassic of, 252
   Palaeozoic sections of, 157, 219
   Silurian of, 158–159
   Triassic of 235–236
Shillong quartzites, 264
   series, 99

INDEX

Shyok glacier dam, 51
Siachen glacier, 21, 24
Sikkim, 7, 409
  copper-ores of, 443
  hanging valleys of, 28
Sillimanite, 78, 468
Sillurian system, 147–160
  of Kashmir, 151–153
  of Shan States, 157–159
  of Spiti, 147–148
Silver, 448
Simla Himalayas, Tertiaries of, 299–300
Simla slates, 106, 128, 140
Sind, 44, 60, 262, 296–297, 309, 322–323
  Cretaceous of, 262
  Eocene of, 309
  Oligocene of, 322–323
  Ranikot of, 309
  Siwaliks of, 349
  Tertiary of, 296–297
Singar mica-mines, 453
Singareni coal-field, 177, 429
Singhbhum, asbestos of, 463
  copper-ores of, 443
  uranium of, 453
Singrauli coal-field, 177, 429
Singu (Chauk) oil-field, 324, 433
Sirhan, Mt., geology of, 228
  limestone, 218
Sirmur belt, 331, 393
  series, 331
Sitaparite, 104
Sivamalai series, 77
Sivapithecus, 348
Sivasamudrum falls, 25
Siwalik river, 51, 341
Siwalik system, 335–352, 427
  Boundary Faults of, 336, 338
  composition of, 340–342
  fauna of, 340, 346–348
  homotaxis of, 348
  of Burma, 351
  of Kashmir, 344–346
  of Salt-Range, 342–343
  of Sind, 335, 349

  structure of, 336–338
Siwana granite, 125
Slates, 91, 119, 427
  quarries of, 428
Snow-line, Himalayan, 20
Soan river, 52
Sodium chloride, 31, 137, 372, 473
Sohagpur coal-field, 429
Soil-creep, 53
Soil erosion, 480, 485
Soils, Indian, 478–485
Solfataras, 117
Songir sandstone, 271, 427
Spandite, 104
Speckled sandstone series, 198, 199
Spinels, 111, 158, 457
Spiti, basin, 60, 138–139
  Cambrian of, 138–140
  Carboniferous of, 202–205
  Cretaceous of, 255, 257
  Devonian of, 148
  geological province, 138
  geosycline, 139
  gypsum of, 467
  Jurassic of, 238–240
  Palaeozoic of, 139, 147–151
  shales, 239, 242
  Silurian of, 147
  Triassic of, 224–227
Springs, 417–420
  mineral, 117, 420
  radio-active, 420
Sri Lanka, *see* Ceylon
Sripermatur beds, 190
Stalagmite, 381, 386
Steatite, 476
Stibnite, 440
Stilbite, 281
Stoliezka, F., 86
Stone Age, in India, 382, 388
  implements, 379, 382, 388
Stratigraphy of India, 1, 55–72
Stromatoliths, 118, 122, 130
Strontium, 450
Structure, Assam, 404
  Assam Tertiaries, 350–351

Himalayas, 391–402
Peninsula, 404
Potwar, 329
Salt-Range, 404
Subathu series, 315, 317
Sub-Himalaya zone, 10, 11
Submerged forests, 43–44
Suess, Eduard, 3, 364
Sui gas-field, 437
Suket shales, 123, 126
Sulcacutus beds, 240
Sullavai sandstones, 124
Sulphur, 472, 477–478
Sulphuric acid, 439, 477–478
Surat, Tertiary deposits of, 291
Surma series, 302, 332
Sutlej, ossiferous alluvium of the, 381
Sven Hedin, 16, 256, 257
Syenite, augite-, 333
nepheline-, 77, 278, 473
Sylhet limestone, 318
Sylvite, 137
Synclinorium, 365
of Aravalli range, 95
of Indo-Gangetic plain, 365
Syntaxis of N. W. Himalayas, 8, 398
Syringothyris limestone, 154–155

Tabbowa beds, 191
Tachylite, 277
Tagling stage, 242
Tal series, 241
Talar stage, 349
Talchir boulder-bed, 172, 177, 198, 209
flora, 172
fossils, 172
series, 172–173
Tamil Nadu, Gondwanas of, 189–190
Tanakki boulder-bed, 218
Tanawal series, 128, 206

Tandur coal-field, 429
Tanr, 381, 383
Tantalite, 473
Tapti (Tapi) river, 25, 291, 381, 382
Tapti series, 291
Tatrot zone, 342
Tavoy, tin of, 451
wolfram of, 452
Tectonic lakes, 29, 411
mountains, 2
valleys, 407
Terai, 371
Teri, 381, 484
Terra-cotta clays, 178
Tertiary systems, 287–306
distribution and facies of, 289–290
earth-movements in, 287
fauna of, 291, 295, 307–309, 323
of Burma, 303
of Coromandel coast, 294
of extra-Peninsula, 295–306
of Gujarat, 291
of Himalayas, 298–302
of Kashmir, 300–302
of Kerala (Travancore and Cochin), 294
of Kutch, 293
of Ladakh, 301
of Rajasthan (Rajputana) 294
of Salt-Range, 297–298
of Saurashtra (Kathiawar), 292
of Sind, 296
rise of Himalayas in, 288, 391
Tethys, the, 162, 165, 193, 241, 258, 317
Thar, the, 252, 372
Thomsonite, 281
Thorianite, 473
Thorium, 450, 470
Thrust-planes, 6, 335, 392
Giri, 399
Krol, 395, 399
Murree, 392, 396, 399
Panjal, 392, 396, 399
Tibetan lakes, 29, 30
plateau, 7

zone of the Himalayas, 258–259, 393–394
Tiki stage, 173
Tiles, 420
Tinnevelli coast, 384
  submerged forest, 43
Tipam series, 302, 349
Tipper, G. H., 257
Tirohan Breccia, 123
Tista river, 29
Titanium, 376, 451
Tonkinella, 142
Tourmalines, 459
Trachyte, 278
Transition system, 130
Traps, as building stone, 428
  Bijawar, 117
  Dalma, 103
  Deccan, 251, 275–286, 428, 482
  Gwalior, 117–118
  Panjal, 208–210, 211
  Rajmahal, 187, 377
  Sylhet, 187
Travertine, 423
Triassic system, 221–236
  fauna of, 221, 225–227, 233
  of Baluchistan, 230–231
  of Burma, 235–236
  of Hazara, 228
  of Himalayas, 221, 224–227
  of Kashmir, 231–235
  of Ladakh, 235
  of Salt-Range, 229–230
  of Spiti, 224–225
Trichinopoly (Tiruchirapalli) marble (limestone), 267, 425
  stage, 267
Trilobites, Kashmir, 142
  Salt-Range, 138
Tripetty (Thirupatti) sandstone, 189
Trona, 462
Tropites beds, 225
Tso Lhamo series, 224, 240
Tsomoriri lake 15, 411
Tungsten, 110, 452
Tura stage, 318

UDAS, G. R., 286
Ultra-basic rocks, 80, 257, 442
Umaria coal-field, 429
  marine Permo-Carboniferous, 219
Umia series, 191, 250–251, 427
Unconformity, Purana, 130–131
  Upper Carboniferous, 150, 193
  Upper Palaeozoic, 193
Underclays, 421
Upper Carboniferous, *see* Carboniferous system
Upper Murree, 330
Uranium, 453
Uranium dating, 58
Usar salts, 473, 483–484
Utatur stage, 267

VAIKRITA series, 106, 139
Valleys, drowned, 44
  erosion-, 408
  hanging, 28, 356
  tectonic, 407
  transverse Himalayan, 27, 408
Vallum diamonds, 460
Vanadium, 454
Varve clays, 360, 383
Vemavaram beds, 190
Verde antique, 426
Vihi district, 211, 213, 214
  plan of, Plate XIV
Vindhya mountains, 19, 121, 406
Vindhyan system, 121–122
  composition of, 121–123
  diamonds in, 124, 128, 455
  earth-movements in, 121
  homotaxis of, 129
  limestones of, 123, 124, 125, 425
  Lower, 123–125
  of extra-Peninsula, 128–130
  sandstones of, 123–128, 426, 466
  Upper, 126–128
Virgal stage, 198

Vishakapatnam (Vizagapatam), manganese in Kodurite series of, 104, 449
  stibnite of, 441
Volcanic basins, 411
  islands, 35-36, 477
  phenomena, 31, 35-37, 117, 124, 186, 207-211, 257, 259, 275-277
Volcanoes, 35-36
  mud-, 36-37
Vrendenburg, E. W., 131, 181, 245, 304
Vredenburgite, 104

Western Ghats (Sahyadri mountains), 19, 33, 43
Wetwin slates, 160, 161
Winchite, 104
Wind-blown sands, 4, 31, 372-374, 384
"Windows" in nappes, 392, 400
Wolfram, 452, 473
Woodward, Sir Arthur Smith, 272, 285
Wootz, 447
Wular lake, 15, 29
Wynne, A. B., 206, 375

WAD, 104
Wadia, D. N., 8, 142, 144, 260, 392, 436
Warkalli beds, 295
Water, as an economic product, 417-420
Water falls, 25, Plate VI, 410
Watershed, of the Himalayas, 27
  of the Peninsula, 25
  recession of, 28
Weathering, 2, 82, 89, 286, 390
Wegener's theory, 169
Wells, 417-418
  artesian, 372, 418
  inverted, 419
  tube, 418-419
West, W. D., 112, 217, 394, 395, 400

YAMDOK Cho lake, 29
Yellaconda hills, 114
Yenangyat oil-field, 37, 433
Yenangyaung oil-field, 37, 433
Yenna falls, 25

ZAMIA shales, 250
Zanskar range, 13, 16, 86, 465
Zanskar river, copper in, 443
Zebingyi series, 158, 161
Zemu glacier, 21
Zeolites, 281
Zeewan series, 213-215
Zhob, chromite in, 442
Zinc, 454
Zircon, 111, 459, 473

PLATE XIII. (After Middlemiss. From *Geological Survey of India, Records,* vol. xl. pt. 3.)

PLATE XIV

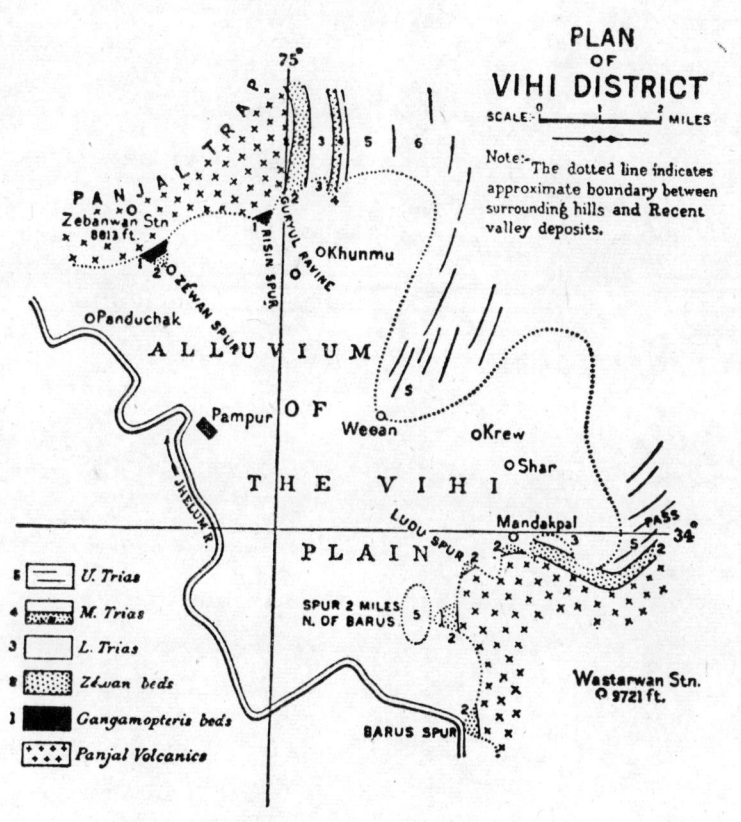

PLATE XV

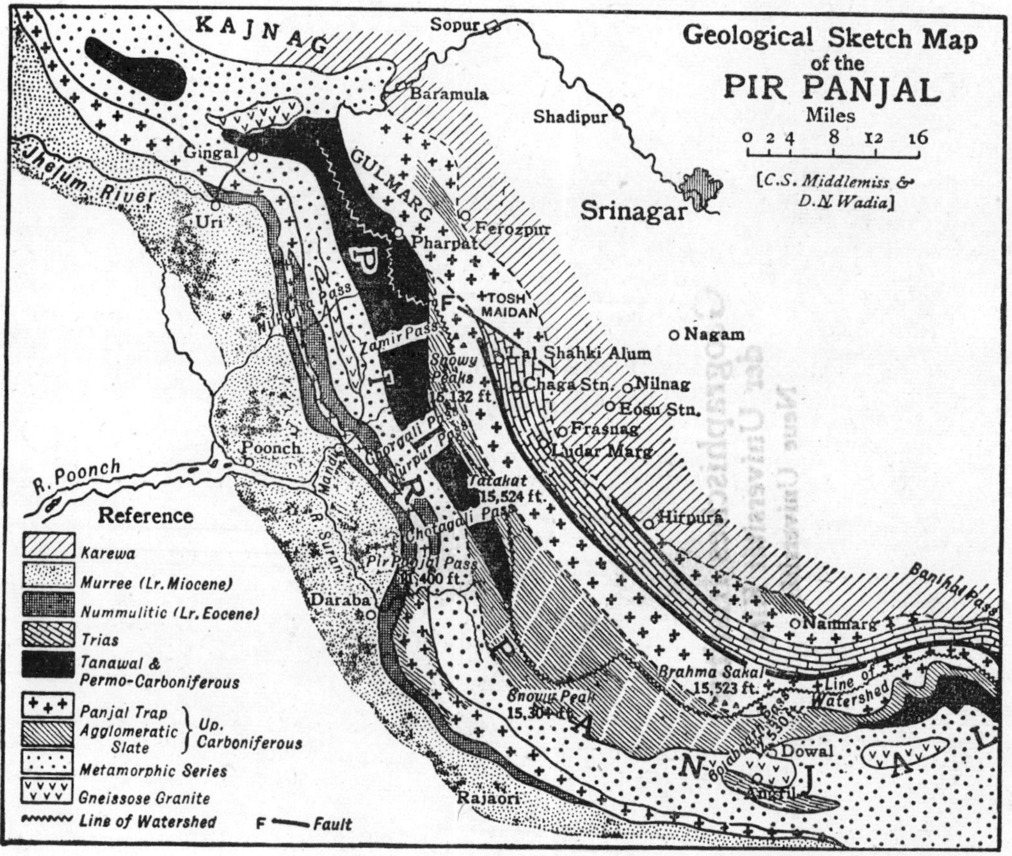
Geological Sketch Map
of the
PIR PANJAL
Miles
0 2 4 8 12 16
[C.S. Middlemiss & D.N. Wadia]
KAJNAG
Sopur
Baramula
Shadipur
Srinagar
Gingal
Jhelum River
Uri
GULMARG
Ferozpur
Pharpat
TOSH MAIDAN
Nagam
Zamir Pass
Lal Shahki Alum
Snowy Peaks 15,132 ft.
Chaga Stn.
Nilnag
Eosu Stn.
Frasnag
Ludar Marg
Poonch
R. Poonch
Tatakat 15,524 ft.
Chotagali Pass
Hirpura
Pir Panjal Pass
Daraba
Banihal Pass
Nanmarg
Brahma Sakal 15,523 ft.
Line of Watershed
Snowy Peak 15,304 ft.
Dowal
Rajaori
Reference
Karewa
Murree (Lr. Miocene)
Nummulitic (Lr. Eocene)
Trias
Tanawal & Permo-Carboniferous
Panjal Trap
Agglomeratic Slate
Up. Carboniferous
Metamorphic Series
Gneissose Granite
Line of Watershed
F Fault

Emery Walker Ltd. sc.

Geographisches Institut
der Universität Kiel
Neue Universität

PLATE XVI

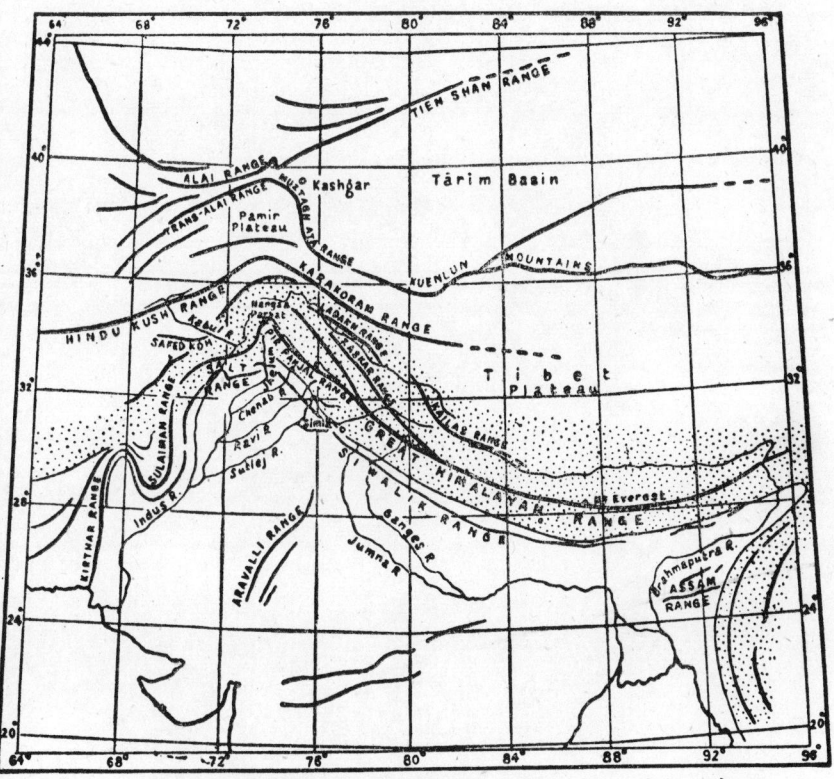

The Himalayan Geosyncline and its relation to adjacent mountain-systems.
(After Burrard & Mushketov).

PLATE XVII

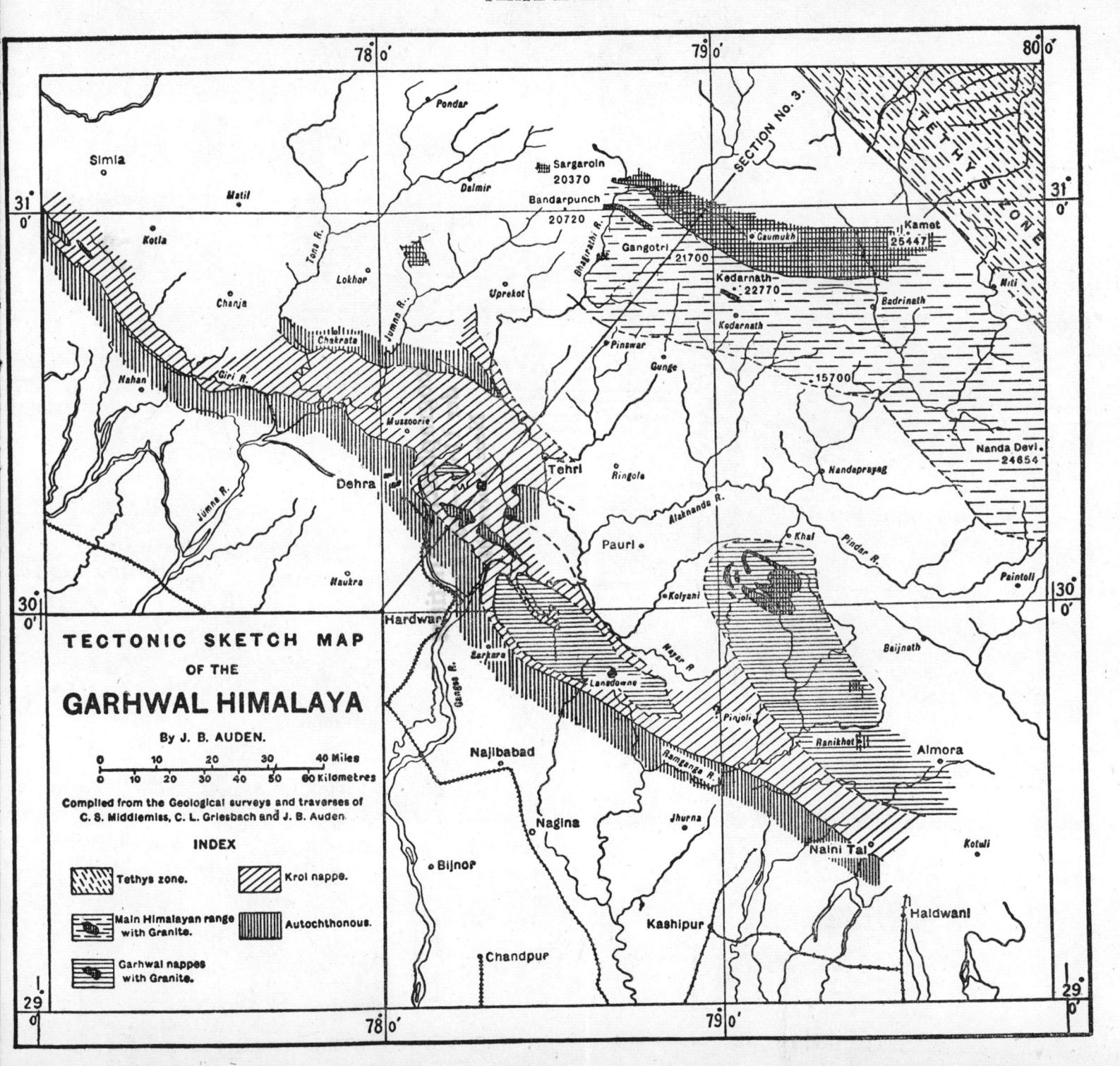

TECTONIC SKETCH MAP OF THE GARHWAL HIMALAYA

By J. B. AUDEN.

Compiled from the Geological surveys and traverses of C. S. Middlemiss, C. L. Griesbach and J. B. Auden.

INDEX
- Tethys zone.
- Main Himalayan range with Granite.
- Garhwal nappes with Granite.
- Krol nappe.
- Autochthonous.

Geographisches Institut
der Universität Kiel
Neue Universität

D N. WADIA GEOLOGY OF INDIA

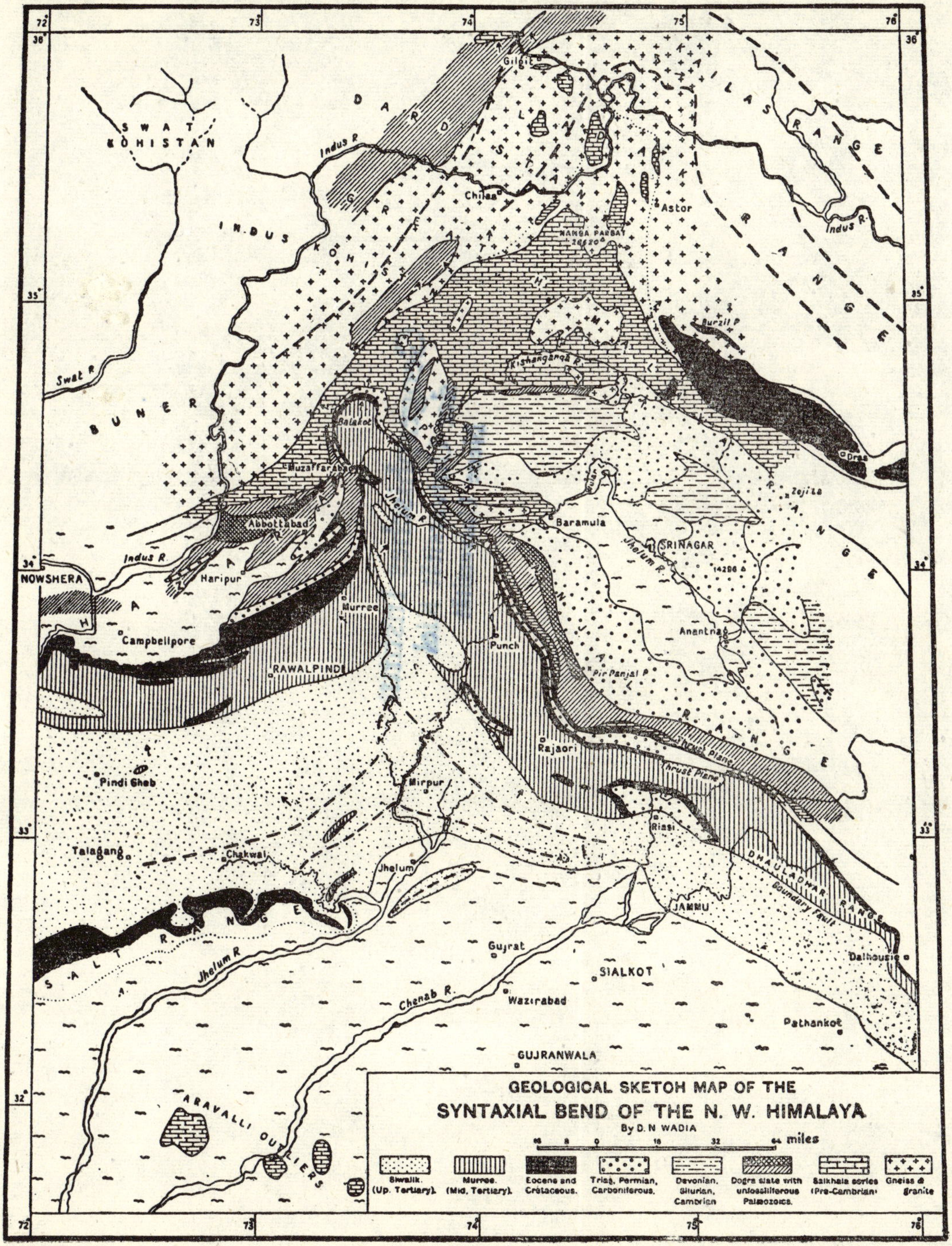

Geographisches Institut
der Universität Kiel
Neue Universität

PLATE XIX

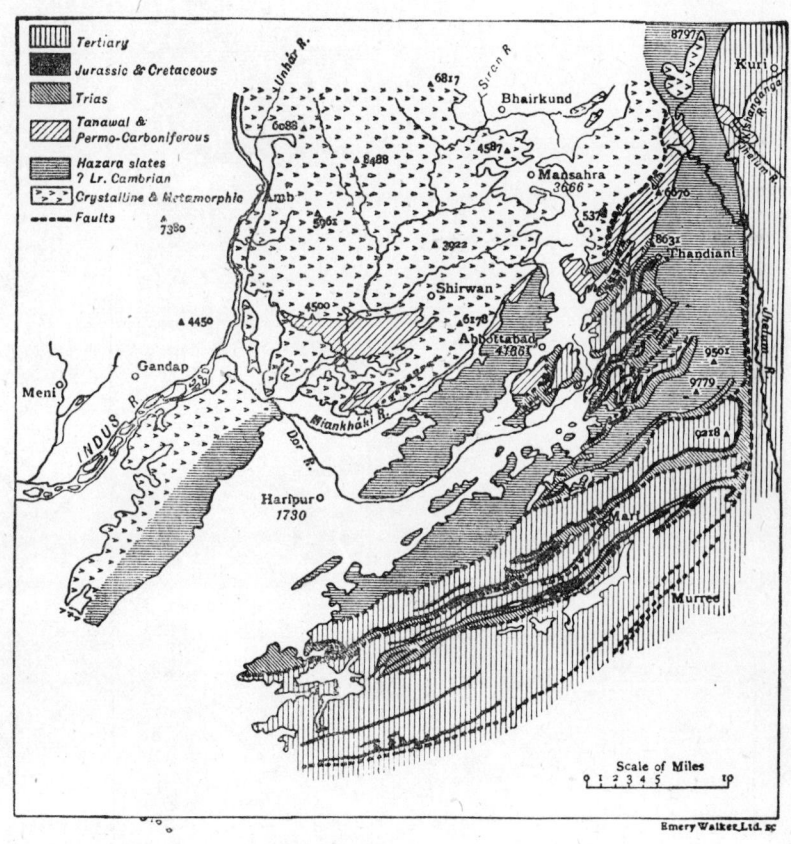

GEOLOGICAL MAP OF HAZARA.

(After C. S. Middlemiss, *Mem. Geol. Survey of India*, **xxvi.**)